# An Ontological and Epistemological Perspective of Fuzzy Set Theory

# AN ONTOLOGICAL AND EPISTEMOLOGICAL PERSPECTIVE OF FUZZY SET THEORY


I. Burhan Türkşen

Director, Knowledge / Intelligence Systems Laboratory

Department of Mechanical and Industrial Engineering
University of Toronto, Toronto, Ontario, Canada



ELSEVIER

Amsterdam – Boston – Heidelberg – London – New York – Oxford
Paris – San Diego – San Francisco – Singapore – Sydney – Tokyo

**ELSEVIER B.V.**
**Radarweg 29**
**P.O. Box 211, 1000 AE Amsterdam**
**The Netherlands**

ELSEVIER Inc.
525 B Street, Suite 1900
San Diego CA 92101-4495
USA

ELSEVIER Ltd.
The Boulevard, Langford Lane
Kidlington, Oxford OX5 1GB
UK

ELSEVIER Ltd.
84 Theobalds Road
London WC1X 8RR
UK



First edition 2006

Library of Congress Cataloging in Publication Data
A catalog record is available from the Library of Congress.

British Library Cataloguing in Publication Data
A catalogue record is available from the British Library.

ISBN-10: 0-444-51891-6
ISBN-13: 978-0-444-51891-0

∞ The paper used in this publication meets the requirements of ANSI/NISO Z39.48-1992 (Permanence of Paper).
Printed in The Netherlands.

This book is dedicated to
*Professor L.A Zadeh* for his vision and seminal ideas,
and to
my son *Altug Noyan Türkşen*,
who I hope will gain a partial insight in to human perception and anticipation via Computing With Words.

# FOREWORD

Professor Türkşen's magnum opus "An Ontological and Epistemological Perspective of Fuzzy Set Theory" or OEP of FST for short, is unlike any other work in the literature of fuzzy set theory and fuzzy logic. Professor Türkşen has long been, and continues to be, a leading contributor to fuzzy set theory, fuzzy logic and their applications. OEP of FST may be viewed as a realization of Professor Türkşen's long-standing endeavor to construct a solid philosophical foundation for fuzzy logic. Between its covers, OEP of FST presents a vast compendium of concepts, ideas and results, extending from the concepts of disjunctive and conjunctive normal forms—Professor Türkşen's specialty—to insightful exposition of Dempster-Shafer theory of plausibility and belief.

In large measure, the content of OEP of FST is shaped by Professor Türkşen's early observation that Principal Disjunctive Normal Form and Principal Conjunctive Normal Form are not equivalent when the variables are labels of fuzzy sets. This observation has led him to development of a theory of interval-valued fuzzy sets of type 2, as well as general fuzzy sets of type 2.

Further extension of his theory led Professor Türkşen to a novel approach to the concepts of plausibility and belief. Another important extension relates to approximate reasoning, cluster analysis, veristic fuzzy sets and Pawlak's rough sets. Ultimately, in OEP of FST, he addresses the basic issue of the nature of truth and provides a novel axiomatic foundation for Computing with Words (CWW).

In his insightful approach to CWW, Professor Türkşen makes an important contribution to a better understanding of its logical underpinnings. Since CWW represents a significant shift in the evolution of fuzzy logic, I should like to add to Professor Türkşen's analysis my own perception of the raison d'etre for CWW and its principal aims.

Computing with words has two levels. In level 1, the objects of computation are words which play the role of granular values of linguistic variables, e.g., small, medium, large, etc. In fuzzy logic, this type of computing with words has been used extensively since the publication of my 1973 paper—a paper in which the concept of a linguistic variable was introduced.

In level 2, the objects of computation are propositions drawn from a natural language, e.g., "It is very unlikely that there will be a significant increase in the price of oil in the near future." Level 2 is much more general than level 1, and should be viewed as the core of CWW.

A key idea in CWW is that of representing the meaning of a proposition drawn from a natural language as what I call a generalized constraint. The concept of a generalized constraint is the centerpiece of CWW.

Why do we need the concept of a generalized constraint? To understand the reason it is necessary to recognize that a natural language is, basically, a system for describing perceptions. What should also be recognized is that perceptions are intrinsically imprecise, reflecting the bounded ability of human sensory organs, and ultimately the brain, to resolve detail and store information.

The intrinsic imprecision of perceptions has an important logical consequence—the intrinsic imprecision of natural languages. This seemingly obvious fact is the compelling reason why classical, Aristotelian logic—a logic which is intolerant of imprecision and partial truth—is not the right logic for dealing with natural languages. What has been largely unrecognized, especially within linguistics and philosophy, is that to deal effectively with natural languages what is needed is fuzzy logic—a logic in which everything is, or is allowed to be, a matter of degree. In essence, this is what CWW attempts to do. More generally, CWW may be viewed as a combination of granular computing and generalized-constraint based semantics of naturalized languages. Although this combination is not discussed explicitly in OEP of FST, Professor Türkşen's work may be viewed as laying the groundwork for it.

What is important to realize is that Professor Türkşen's work abandons bivalence and embraces partiality of truth, along with partiality of possibility, likelihood and most other concepts. There is a profound implication of abandonment of bivalence which is widely unrecognized. More specifically, the implication is that in order to come to grips with the issue of growing complexity, we may have to legitimize the status of conclusions which may have the form of fuzzy theorems, that is, theorems which are partially true. We may be compelled to do this because of what I had called the Principle of Incompatibility—a principle which suggests that high complexity of the kind that we observe in the realm of economic systems, is incompatible with high precision. What this implies is that in the case of systems of high complexity, precise conclusions are likely to have low relevance. Given the deep-seated tradition of quest of precision, this contention may be a bitter pill to swallow.

In sum, Professor Türkşen's work contributes so much and in so many important ways to the development of a better understanding of ontological and epistemological foundations of fuzzy set theory and fuzzy logic. Professor Türkşen and the publisher, Elsevier deserve our thanks and congratulations.

Lotfi A. Zadeh
UC Berkeley
August 30, 2004

# PREFACE

This book is dedicated to the exploration of some of the Ontological and Epistemological issues of Fuzzy sets and Fuzzy logics proposed by L.A. Zadeh. It is essential to state the main difference between Classical and Fuzzy sets and logics at the start. In Fuzzy sets and logics, every word, concept or proposition is allowed to be a matter of degree. In the mathematical treatment, generally, words, propositions and sentences are represented by sets. Sets specify the meaning associated with words, or sentences. Naturally, sets can be represented either as two-valued classical sets or as fuzzy (infinite)-valued sets. Once words are represented by sets, the associated meaning of words gains a *description* in terms of the mapping that is chosen, i.e., crisp or fuzzy. Words represent concepts that are generated and formulated in our "minds" as a result of our perceptions.

Naturally, concepts are combined to represent and communicate more complex concepts. Combinations of concepts are generated by the words and their representation which are combinations of sets with a theory of logic in mathematics. That is, in combining sets, we consider the logical combination of sets to come up with basic expressions that are logically meaningful. But the logical combinations require that sets be verified externally explicitly and crisply. Otherwise, implicitly, it is assumed that the descriptions of sets are already verified crisply. On the other hand, if the verification is fuzzy then there must be a second verification that is crisp.

Both in classical and fuzzy theory, the basic expressions that define the combination of sets and hence words are based on sixteen possible combinations of any two sets that are specified in Table 1. Clearly, there are more complex expressions where fuzzy quantifiers enter into the structure. These are left for future writings. As can be observed, the "AND", "OR", "IMP", etc, are the basic combination operators that form the foundation of the set and logic theories. In terms of mathematical expressions, the basic expressions of each combination generate a Disjunctive Normal Form, DNF, and a Conjunctive Normal Form, CNF, in classical set and logic theory where "AND", "OR" are mapped in a one-to-one correspondence to set symbols "$\cap$", "$\cup$", respectively.

It is well known that these two Normal forms, i.e., DNF and CNF, are equivalent in classical theory. Whereas in fuzzy theory these equivalences break down. We have shown that we obtain Fuzzy Disjunctive Canonical Form, FDCF, and Fuzzy Conjunctive Canonical Form, FCCF, for each combination of any two fuzzy sets where "AND", "OR" do not correspond to "$\cap$", "$\cup$" in a one-to-one mapping. While, *in form only*, FDCF is equivalent to DNF and

FCCF is equivalent to CNF they differ in content. In turn, its is shown that FDCFW FCCF in fuzzy theory. This break down of the equivalence, in Fuzzy set and logic theory, cause the generation of new and additional formulas and "Laws of conservation" in Fuzzy theory. Most fuzzy theory researchers are not aware of this break down of 16 equivalences. Thus, they keep using classical expressions by just fuzzifying them and ignoring the break down of these equivalences. Furthermore it is shown that FDCF$\subseteq$FCCF for certain subclasses of conjugate t-norm, t-conorm and negation based De Morgan Triples. These subclasses of conjugate t-norm, t-conorm and negation De Morgan triples are found to be the strict and nilpotent Archimedean conjugate t-norms, t-conorms and negation based De Morgan triples as well as Max-Min, and standard negation De Morgan Triple, known as the Zadeh Triple. In addition, it is shown that for certain ordinal sums of conjugate t-norm, t-conorm and negation based De Morgan triples, the containment of FDCF and FCCF also holds.

**Table 1**. Sixteen Possible Combinations of any two sets, A and B.

| Number | Meta-Linguistic Expressions |
|---|---|
| 1 | UNIVERSE |
| 2 | EMPTY SET |
| 3 | A OR B |
| 4 | NOT A AND NOT B |
| 5 | NOT A OR NOT B |
| 6 | A AND B |
| 7 | A IMPLIES B |
| 8 | A AND NOT B |
| 9 | A OR NOT B |
| 10 | NOT A AND B |
| 11 | A IF AND ONLY IF B |
| 12 | A EXCLUSIVE OR B |
| 13 | A |
| 14 | NOT A |
| 15 | B |
| 16 | NOT B |

The derivation of the basic expressions in terms of FDCF and FCCF for all sixteen combination of concepts can be obtained by the application of "Canonical Form Derivation" algorithm applied over the Fuzzy Truth Tables. It is also shown that a modified and restricted version of Dempster's multi-valued mapping, identified as T-formalism, generates FDCF and FCCF for the basic forms of the combination of concepts. In these derivations, initially it is assumed that fuzzy sets that represent words are Type 1, i.e., the membership functions are crisp and well defined. But their combination with linguistic "AND", "OR", "IMP", etc., operators generate an Interval-Valued Type 2 fuzzy sets. This is an example of the break down of classical equivalences, i.e., symmetries, in Fuzzy Set and Logic theory. It is suspected that there are other such equivalences and/or similarities that break down in Fuzzy theory. That is, combination of words represented by fuzzy sets and linguistic operators, that do not correspond isomorphically to t-norms and t-conorms, generate second order imprecision and expose the uncertainty associated by the imprecise representation of concepts with fuzzy sets when the membership functions of the fuzzy sets that enter into a combination are represented by Type 1 fuzzy sets.

In classical set and logic theory due to the equivalence of DNF and CNF, there are "Laws of Conservation" and formulas for Belief, Plausibility and Probability which establish the "Principle of Invariance" in classical theory. The introduction of Type 1 fuzzy sets by Zadeh (1965) causes a break down of these laws of conservation and the formulas in "fuzzy set and two-valued logic theory". Hence the "Principle of Invariance" that is at first sight appears to be vanished for fuzzy set and two-valued logic theory in general require further re-structuring.
In such a re-structuring of the ontological and epistemological foundation of fuzzy theory, it will be shown, that at least in "Interval-Valued Type 2" fuzzy set theory there exists new additional formulas and "Laws of Conservation" and therefore the "Principle of Invariance" is re-established in "Interval-Valued Type 2 Fuzzy Set Theory". It is very likely that there are other break down of equivalences and/or symmetries in Fuzzy set theory that cause the generation of additional formulas. Thus it appears, most of us have not yet explored the richness of fuzzy theory as effectively as we should.

Next, we explore the case where membership functions of the fuzzy sets that enter into a combination are represented by Type 2 fuzzy sets in the acquisition experiments of meaning of words. The acquisition of the meaning of the words can be obtained either with measurement experiments or by FCM methods of fuzzy clustering techniques. Thus, the basic expressions obtained by the application of FDCF and FCCF to Type 1 fuzzy sets are next extended over Type 2 fuzzy sets and identified by their upper and lower set definitions. Up to this point, the discussion centers on the assumption that concepts and their representative words and propositions are represented by fuzzy sets and they are *verified and combined* by two-valued logic paradigm. In this theory, only the *descriptions of* words are fuzzified by fuzzy sets but the *verification* is two-valued crisp logic. Thus, in this theory, we do not have general fuzzy logic expression but just fuzzy set descriptive representations with two-valued logical verifications.

It is conjectured that in Full Type 2 fuzzy set representation, the laws of conservation obtained in "Interval-Valued Type 2 theory" breaks down once again. It is further conjectured that the "Principle of Invariance" is re-established in Interval-Valued Type 3 fuzzy set theory. However, these extensions are left for future writings!

Next we consider the case where words and propositions are represented by sets that are verified fuzzily, i.e., with fuzzy logic, and then further combined by two-valued logic paradigm. This entails the formation of fuzzy logic expression over words and propositions. Thus fuzzy logic expressions that are verified by two-valued logic paradigm lead to the generation of basic expressions in the fuzzy logic. It is shown that fuzzy logic expressions also generate FDCF and FCCF expressions as an extension of the FDCF and FCCF formed on fuzzy set representations.

In the next section of the book, it is shown that fuzzy measures such as Belief, Plausibility, Probability, etc., can be formulated on both Type 1 and Interval-Valued Type 2 fuzzy sets. Such integration of fuzzy sets and fuzzy measures enhances and enriches the expressive power fuzzy theory. Next, it is proposed that Belief, Plausibility, Probability, etc., be expressed fuzzily as a matter of degree, i.e., in terms of "low", "moderate", "high", say, Belief, etc.
In the next several chapters of the book, we discuss Type 1, Interval-Valued Type 2 and Full Type 2 reasoning methods for both the product and material implication based reasoning methods and their basic structure which are duplicated due to the breakdown of the equivalence, i.e., FDCFW FCCF.
Finally, we demonstrate bench mark case study applications of Interval-Valued Type 2 fuzzy sets. In addition, we illustrate a theoretical application of Interval-Valued Type 2 fuzzy sets for preference analysis.

At the end, we propose that the axioms of classical set and logic theory be re-stated meta-linguistically to form a meta-linguistic foundation for CWW.
The book is organization as follows:

In the preface, a historical perspective is given from a personal point of view. A brief review of Zadeh's scientific and philosophical contributions are included in Appendixes.
I believe that a "Philosophical Grounding of Fuzzy theories" are long overdue. For this purpose, I have attempted to provide a preliminary personal perspective

on this matter. I am sure this is a naïve approach to this topic for I am not a philosopher but an engineer at heart! However, in this attempt, it is shown that from an ontological and epistemological point of view, the break down of certain equivalences and re-structuring certain formulas and Laws of Conservation require further attention for a sound grounding of fuzzy theory.

In Chapter 1, the notions of description and verification are introduced along with basic notation. In Chapter 2, we discuss computing with words in terms of descriptive and veristic words, the representation of words with numbers and numbers with words, as well as the structure of sentences as atomic and complex sentences.

In Chapter 3, an in-depth treatment of the measurement of membership functions are discussed within the context of several views, the issue of scale strengths and the affect of connectives on membership acquisition.

In Chapter 4, a number of known elicitation method are considered for the membership acquisition from subjects (experts) when a historical data base is not available.

In Chapter 5, fuzzy clustering methods are review for the extraction of membership functions and the fuzzy rules when we have historical data in data bases.

In Chapter 6, we present classes and sub classes of fuzzy sets and logic theories along with the t-norm, t-conorm and corresponding negations.

In Chapter 7, we discuss both the two-valued sets and fuzzy-valued sets and introduce Fuzzy Disjunctive and Conjunctive Canonical Forms, FDCF and FCCF, respectively, in contrast to Normal Forms, DNF and CNF, respectively, of two-valued set theory. In this chapter, we also include a brief discussion of the paradoxes embedded in two valued set and logic theory, i.e., classical theory. It is in this Chapter, we discuss and show the explicit break down of the equivalences of classical set and logic theory.

In Chapter 8, construction of "Fuzzy Truth" tables are introduced. As well, Dempster-Pawlak unification is consider and a T-formalism is introduced as a modification and restriction of Dempsters multi-valued maps in support of the derivation of FDCF and FCCF expressions obtained from the Fuzzy Truth tables. This further validates the break down of the classical equivalences.
In Chapter 9, Archimedean, non-Archimedean t-norms and co-norms are considered, and basic expressions of FDCF and FCCF are re-introduced. The

containment of FDCF in FCCF is shown to exists for continuous Archimedean t-norms and t-conorms and their corresponding negation with an application of additive and multiplicative generating functions.

In Chapter 10, the consequences of fuzzy set descriptions and two valued verification base theory are considered by exposing the novel results such as "Fuzzy Middle", "Fuzzy Laws of Conservation" and "Fuzzy Invariance Properties". That is to say while the classical equivalences break-down in fuzzy theory, new Laws of Conservation and invariances are generated in fuzzy theory. This exposes at least a part of the richness of the fuzzy theory.

In Chapter 11, the relationship between "Compensatory 'AND' " and FDCF and FCCF are explored within the context of Interval-Valued Type 2 Fuzzy theory.

In Chapter 12, Belief, Plausibility, and Probability measures are discussed on the Interval-Valued Type 2 Fuzzy Sets. This result is an assessment of lower and upper Belief Plausibility and Probability measures on Interval-Valued Type 2 Fuzzy Sets. Furthermore, it is proposed that Belief, Plausibility and Probability notions be fuzzified as "low", "moderate" and "high" to be a matter of degree within the basic philosophy of fuzzy theory.

In Chapter 13, Veristic fuzzy sets of Truthoods are explored within Meta-theory of Modal Logic. This provides an alternate approach for the derivation of FDCF and FCCF expressions for the Zadeh Operators as an example. Here, we observe that various transformation of "worlds" lead to rather unexpected results and reveal an alternate view for the realization of FDCF and FCCF expressions. This is yet another example to show again that certain equivalences break down in fuzzy verification studies.

In Chapter 14, we explore various versions of approximate reasoning within the context of GMP, Generalized Modus Ponens. Rule and operation decomposition schemas discussed. As well, FATI and FITA, First Aggregate Then Infer vs First Infer Then Aggregate schemas, respectively, are reviewed. This chapter includes few text book case studies.

In Chapter 15, Interval-Valued Type 2 GMP is discussed with several alternative schemas where the notion of Information Gap that is caused by the break down of classical equivalences is introduced within the scope of Interval-Valued Type 2 GMP and System modeling.

In Chapter 16, a theoretical application is demonstrated for the Interval-Valued Type 2 representation for preference analysis.

In Chapter 17, we propose that the axiomatic foundation of the classical theory be adapted to fuzzy theory by a re-statement of the classical axioms in terms of their meta-linguistic expressions by specifying that the connectives "AND", "OR", "IMP", etc., be linguistic to form a foundation for CWW.

In Chapter 18, our conclusions are stated in a summary form.

The author acknowledges materials extracted and modified from his previously published papers. As well the contents of the book are influenced by several visionary colleagues who have contributed directly or indirectly with discussions and suggestions. I particular, I am indebted to L.A. Zadeh, A. Kandel, G., Resconi for their discussions and suggestions. Furthermore thirty Ph. D's and forty two Master's student have again contributed directly or indirectly through various discussions that were held during my graduate classes. Amongst these, I need to identify several unique individuals, namely, Taner Bilgic, Kudret Demirli, Kemal Kilic, and Ozge Uncu. I believe they in turn have begun and will continue to make worthwhile contributions of their own to Fuzzy Theory. Finally, I sincerely express my appreciation of Dr. E. Hacimenni's and Ms A. Celikyilmaz' dedicated efforts. They have gone through several versions of this book and prepared the final version as required by Elsevier.

I extend a hearty thanks to all those who have contributed in their own personal way.

I.B.Tűrkşen, Toronto, December, 2004

# CONTENTS

# Chapter 0

# FOUNDATION

In his Keynote address at FUZZY-IEEE 2001 Melbourne, Australia, Dec. 2-5, 2001, Lotfi A. Zadeh suggested that researchers should attempt to develop "Expressions" in order to structure a sound foundation for CWW, Computing With Words, based on CWP, Computing With Perceptions.

This book synthesizes my investigation over the last twenty years or so on the structure identification of "Expressions" in combination of concepts as a bases for CWW. The unique perspective of this work can be summarized as follows: *Theories of set and logic are and should be understood as separate and independent disciplines. They interact in the derivation of "Expressions", i.e., new (additional), formulas for the combination of linguistic concepts and their linguistic variables, and linguistic connectives, "AND", "OR", "IMPLIES", etc. They ought to be at least represented with a multi-valued one-to-many mapping. When treated in this perspective, one is able to respond to the two essential questions by the sceptics of the fuzzy theory. These questions are: (1) Since fuzzy theory deals with imprecision, why are the membership functions so precise?, and (2) since fuzzy theory axioms are much fever than two-valued theory axioms, why are fuzzy theorists still applying two-valued (Boolean) formulas in the combination of concepts and in reasoning? It is shown in detail that (1) membership values, and hence membership functions could not be precise and that they must be at least represented by Interval-Valued Type 2 or Full Type 2 fuzziness, and (2) there are Fuzzy Disjunctive and Conjunctive Canonical forms that capture the boundaries of imprecision and uncertainty in reasoning. As a result, knowledge representation and reasoning schemas, i.e., their "expressions", ought to approximately capture both the representation and reasoning and execute computing with imprecision and uncertainty of membership values and functions.*

These two questions and their answers opens the door for a more fundamental question:

**"What are the meta-linguistic expressions of basic axioms for Computing With Words?" That is "do the axioms of classical theory generate new axiomatic expressions once they are re-stated in meta-linguistic terms?"**
Finally, these questions and concerns require that a philosophical grounding of fuzzy theory be explicitly stated in particular for ontological and epistemological grounding of this "Grand Paradigm Shift".

It should be noted that there are two separate intellectual disciplines under the headings of "Set" and "Logic" in most universities. When we integrate the essential elements of these two unique disciplines, we derive formulas with the one-to-many mapping of connectives. This is analogous to Dempster's (1967) multi-valued mapping but a modified and restricted version of it which is called T-formalism. These formulas provide a structure on the ontological and epistemological levels of scientific inquiry for the relations between linguistic concepts and hence form the foundations of knowledge representation, inference schemas and system modeling exercises.

The derivation of the fuzzy set and logic formulas is therefore an essential concern of this work. Another concern of this work is the application of these formulas in approximate reasoning schemas. For this purpose, we first inquire into the implicit and explicit statements of linguistic expressions, which are essential in Computing With Words, to identify "descriptive" and "veristic" frameworks. As well, we question the nature of truth. We treat acquisition and extraction of membership values and functions after showing that there are difficulties in establishing the validity of measurement theory axioms. This leads to the identification of Type 2 membership representation. After this, we investigate the derivation of formulas for five unique set and logic theories by a generalization and extension of "classical truth tables" to "fuzzy truth tables". As well, we demonstrate a unification of Dempster's multi-valued maps, Pawlak's rough sets, and Türkşen's Interval-Valued Type 2 knowledge representation under T-formalism. In turn, we discuss the application of these formulas in approximate reasoning schemas.

Finally, we start a discussion on the meta-linguistic expressions of the axioms that suggests future studies on the foundation of fuzzy theory.

## 0.1. A Personal Perspective

When I decided to study fuzzy sets and logics in 1977, I asked: *"What are the consequences of fuzzifying the Disjunctive and Conjunctive Normal Forms, known as DNF and CNF, respectively, of the two-valued theory?"* To my surprise, I found that DNF and CNF expressions which are known to be equivalent in two-valued theory, are separated when directly fuzzified for all the sixteen possible cases that are generated in the combination of any two concepts A and B. Furthermore, I found that for certain special cases of t-norms, $\Delta$, and t-conorms, $\nabla$, there is a containment relationship between the two normal forms, i.e., $\mathrm{DNF}(\cdot) \subseteq \mathrm{CNF}(\cdot)$ for all the sixteen possible cases that arise in the combination of any two concepts A and B that are fuzzy. These were explained in Fuzzy Sets and Systems, 20 (1986), 191-210, under the title of "Interval-Valued Fuzzy Sets based on Normal Forms". In that paper, I had also observed that $\Delta[\nabla(a,b), \nabla(a, n(b))] \geq a$ and its dual $\nabla[\Delta(a,b), \Delta(a, n(b))] \leq a$

hold true for a, b∈[0,1] for some classes of Δ-norms and ∇-conorms. Next, I found that "Compensatory AND" reported by H.J. Zimmermann and P. Zysno (1980) can be formally explained by fuzzified DNF and CNF of "AND" and/or "OR" combinations of two fuzzy sets A and B. This was published in "Interval-Valued Fuzzy Sets and Compensatory AND", Fuzzy Sets and Systems, 51, (1992), 295-307. Zimmermann and Zysno experiments and my DNF and CNF analysis of "Compensatory AND" demonstrate that "AND", "OR", etc., linguistic connectives are fuzzy (imprecise) connectives and cannot be represented by a one-to-one mapping to t-norms, Δ, and t-conorms, ∇, respectively.

With these results, I became further curious and asked: *"Can the Fuzzy Normal Forms be directly derived from the Fuzzy Truth Tables?"*

Initially, I came across "Fuzzy Logic and Its Applications to Switching Systems", by P.N. Marinos, IEEE-Trans. Compt.C-18, 4 (1969), 343-348, where it was stated that since the fuzzy variables can assume infinitely many values within the interval [0,1], the use of maps or other graphical techniques which play an important role in two-valued set and logics theory are not possible. Motivated by this statement G.W. Schwede and A. Kandel investigated "Fuzzy Maps", IEEE-Trans., Syst. Man Cyber., SMC-7, 9 (1977), 669-674. They proposed fuzzy maps with $4^n$ elements where n is the number of variables. On the other hand, motivated by general concerns of fuzzy theory, A. Kaufmann, in "Theory of Fuzzy Subsets", Academic Press, New York 1975, proposed, under the subheading of "Table of values of a function of fuzzy variables", pp. 198-199, that to study a function of n variables one should consider $n!2^n$ possible cases. In particular, Kaufmann noted that for each fuzzy variable, one needs to examine two cases, i.e., $a \leq b$, $a>b$, $a,b\in[0,1]$.
Based on Kaufmann's proposal and re-interpreting a statement of N. Rescher, in "Many-Valued Logics", McGraw Hill, New York, 1969, pp. 122-123, I then proposed the derivation of Fuzzy Normal Forms directly from an initial structure of fuzzy truth tables in "**Fuzzy Normal Forms**", Fuzzy Sets and Systems, 69 (1995), 319-346.

In the same year, A. Kandel, et.al., published "Knowledge Representation by Conjunctive and Disjunctive Normal Forms Based on n Variable – m dimensional Clauses and Phrases", Fuzzy Sets and Systems 76 (1995), 73-89.
Continuing my work related to "Fuzzy Normal Forms", in 1995, I published "Type I and Interval-Valued Type II Fuzzy Set and Logics" in Advances in Fuzzy Theory and Technology", (ed), P.P. Wang, Duke University (1995), pp. 31-82. In this work, different classes of Fuzzy Normal Forms are identified with respect to the axiomatic foundations of t-norms, Δ, and t-conorms, ∇. As well, I had pointed out that Klein-Four-Group has a particular structure in the light of fuzzy normal forms. Furthermore, an interpretation of the classical

laws of "Excluded Middle and Contradiction" is presented with the results obtained by the application of the fuzzy normal forms. Finally, the semantics of the Interval-Valued Type II fuzzy sets were discussed. Examples of the impact of the fuzzy normal forms on fuzzy inference were presented together with certain applications in cognitive categorization, preference analysis, etc.
In 1995, one of my Ph.D. Candidates, Taner Bilgic, completed his doctoral dissertation where he showed that DNF(.) $\subseteq$ CNF(.) for strict and nilpotent Archimedean t-norms and t-conorms.

In this regard, my expositions of Fuzzy Normal Forms were not satisfactory for some skeptics as well as for my own gut feelings. Hence I was encouraged to clarify my ideas in a series of papers that have been developed in 1996-1997 period and are published under the headings of:

"**Universal Truth Tables and Normal Forms**", by I.B. Türkşen, A. Kandel and Y.Q. Zhang, IEEE-Trans-Fuzzy Systems, 6 (1998) 295-303.

"**Normal Forms of Fuzzy Middle and Fuzzy Contradiction**", by I.B. Türkşen, A. Kandel, and Y.Q. Zhang, IEEE-Trans. SMC, Vol. 29, Part B No. 2 (1999), 237-253.

These last two works on normal forms were developed during my research leave at the University of South Florida in collaboration with Prof. A. Kandel. In these works, I have discovered that there are two new Laws of Conservation that are upheld in Interval-Valued Type 2 fuzzy set theory. During a recent discussion with Prof. G. Resconi, I realized that these laws of conservation are in fact the new formulas that uphold the "Principle of Invariance" in Interval-Valued Type 2 fuzzy set theory which was camouflaged in Type 1 fuzzy set theory.

Afterwards, I discovered an error in these works and as well began to discuss "descriptive" and "veristic" works and their impact on the "combination of concepts". I presented these in "**Theories of set and Logic with Crisp and Fuzzy Information Granules,** J. of Advanced Comp. Intel., 3,9 (1999) 264-273". In this work and thereafter, I began to use Fuzzy Disjunctive and Conjunctive Canonical Forms, FDCF and FCCF, to differentiate them from DNF and CNF of classical Theory.

In 2001, two articles were published that show further progress in the field: (1) "**Computing with Descriptive and Veristic Words: Knowledge Representation and Reasoning**" by I.B. Türkşen, in Computing With Words: (ed) P.P. Wang, Chapter 10, John Wiley and Sons, (2001) 297-328; (2) "**Canonical Forms with Fuzzy truthoods by Meta-Theory based upon**

**Modal Logic**" by a Resconi and I.B. Türkşen, Information Sciences, 131 (2001) 157-194.

More recently, I have investigated a unification of Dempster's "Multi-valued Maps", Pawlak's "Rough sets" and "Interval-Valued Type 2 Fuzzy Sets" in "Upper and Lower Memberships in CWW" – Fuzz-IEEE'2001, Melbourne, Australia, Dec. 2-5 (2001). An extended version of this work is published in I.J. Applied Mathematics and Computer Science (2002) under the heading of "Upper and Lower Set Formulas: A Restriction and Modification of Dempster-Pawlak Formalism" (Türkşen, 2002) 101-111.

In 2002, another one of my Ph.D. Candidates, Kemal Kılıç, completed his doctoral dissertation where he showed that FDCF(.) $\subseteq$ FCCF(.) for some ordinal sums of t-norms and conorms.

In addition, I have been concerned with the "measurement of membership functions" since 1982. (Norwich and Türkşen, 1982). Our continuing investigations on this topic (Türkşen, 1991, Bilgic and Türkşen, 1995) has lead to "Measurement of Membership Functions: Theoretical and Empirical Work", in Handbook of Fuzzy Theory, Dubois and Prade (eds.) (1999). Our studies on the measurement of membership functions gave me further insight in separating descriptive and veristic assignments with "descriptive" and "veristic" words within the perspective of computing with words (Zadeh 1995). This lead to the understanding that there is an interaction between descriptive assignments of elements to a set and the veristic assignments that affirm the descriptive set assignments in the formation of logical constructs.
It is the combined treatment of the descriptive and veristic assignments that give us a better understanding and hence a new approach to the formation of more general "Truth Tables" that help us derive the canonical forms that are essential to formal knowledge representation and approximate reasoning schemas with "Basic Expressions". As indicated above, further support for the derivation of fuzzy canonical forms comes from a unified but restricted and modified re-interpretation of Dempster-Pawlak formalism as upper and lower sets derived from certain information granules.

***All these articles as well as some other conference proceeding papers point out the impact and consequences of set and logic formulas that allow one-to-many representation of linguistic connectives on system modeling and approximate reasoning. Finally, these activities naturally lead to investigations on the philosophical grounding of fuzzy theories on ontological and epistemological level and open further concerns for the axioms of CWW. Furthermore, totality of all these investigations demonstrate the progress of ideas and concepts from one stage to the next and***

***demonstrate some of the modifications, improvements and restructuring of the results over the years of introspection and discussions with colleagues.***

The current work is a compilation of all the previous ideas that have been published. But they have been further modified and improved with more recent considerations and discoveries toward a restructuring of the foundations of fuzzy theory with a perspective on Ontological and Epistemological foundations of the theory with new concerns on the axioms of fuzzy theory that are generated by meta-linguistic expressions. In particular, I have included a discussion on the "Nature of Truth" from logicians point of view and attempted to give a structure to "Computing With Words" (Zadeh, 1995) by a further assessment of "descriptive words" and "veristic words" and their representation by two versus infinite-valued set assignments. The separation of descriptive assignments from the veristic assignments and then the combined treatment of descriptive and veristic assignments give rise to the identification of at least five classes of "set and logic" theories. It is the goal of this work to bring to the surface a comprehension of this simple but fundamental issue in the treatment of set and logic theories with an ontological and epistemological perspective as well as question the axiomatic foundations of fuzzy theory. Finally at the end in Chapter 17, propose a new axiomatic foundation for CWW.
Along the way, in the introduction we have attempted to re-interpret the definiteness versus indefiniteness of concept representation by symbols due to Peirce and Zadeh. At the end, fuzzy canonical forms as expressions for CWW are applied in a general treatment of inference rules within a formal structure. However. before we enter into our work, it is worthwhile to give a brief review of Zadeh's "Scientific and Philosophical Contribution", as well "Fuzzy System Applications in OR/MS/DSS." These are presented in the Appendix.

In turn, we include a brief philosophical grounding of our own ontological and epistemological perspective on Fuzzy theories with a comparison between classical and Fuzzy theories.

It is worthwhile to note that in recent years Walker and Walker (1999-2002) have investigated Disjunctive and Conjunctive Canonical Forms from the perspective of fuzzy theory with an application of algebraic theory constructs and found other interesting properties of these canonical forms.

There is also the book by Mendel (2001) who discuss Type 2 fuzziness. We are in agreement with Mendel that membership acquisition techniques generate Type 2 membership representation. However, Mendel applies only "Sup-Min" composition in the combination of Type 2 fuzzy sets. His and his students work do not treat the fact that in combination of fuzzy sets one must deal will FDCF(.) and FCCF(.) expressions. That is Mendel ignores the break down of classical equivalences as does most fuzzy researchers. In this sense our work is unique and very different from Mendel's approach to Type 2 fuzzy sets. Our emphasis is an Interval-Valued Type 2 fuzzy set theory, which brings to surface the break down of certain equivalences and re-structuring of certain new Laws of Conservations, and new formulas for Belief, Plausibility, Probability, etc. As well we question the axiomatic foundation of fuzzy theory and propose a meta-linguistic foundation for CWW.

## 0.2. A Perspective on The Philosophical Grounding of Fuzzy Theories

As fuzzy theorists and practitioners, we frequently find ourself confronting significant philosophical issues in our work. Indeed, if we are not doing so, we are probably and possibly missing out a lot. While different fuzzy theories and application approaches may be founded upon different set of philosophical presuppositions, all such theories rest upon *some epistemological and ontological* assumptions, whether explicitly acknowledged or not. In this regard, the dictionary definitions of epistemology and ontology are given as follows.

**Epistemology:** The study or theory of the nature and grounds of knowledge, esp., with reference to its limits and validity.

Epistemology lays the ground work for the assessment of consistency and believability of a set of propositions by either a priory or evidentiary basis. Evidentiary basis could be subjective and/or objective.

**Ontology:** 1) A branch of metaphysics concerned with the nature and relations of being. 2) A particular theory about the nature of being or the kinds of existents.

Ontology lays the ground for the structural equivalences and the Laws of Conservation based on assumptions of existences.

A lack of appropriate treatment of the philosophical grounding creates a situation of discord, or at least a level of misunderstanding, between fuzzy theorists and practitioners on the one hand and the crisp theorists and practitioners on the other. In fact, our lack of clearly communicating our philosophical stance has caused and still causes a lack of understanding and/or rejection of fuzzy theory by those who held on to the classical view of the world. Furthermore, it does not help us to explain effectively to managers and decision makers, how fuzzy theory and fuzzy system models could improve their decision-making practices. Some decision-makers are still hesitant to embrace the fuzzy system models, despite the fact that significant and important applications of fuzzy theory are implemented and installed in many electro-mechanical systems, e.g., robotics, camcorders, washing machines, train break-systems, auto-transmission systems, unmanned helicopter control and navigation, etc., etc. But it is important to note that these successful applications are fuzzy control based and hence different and relatively easier compared to decision-making systems as pointed out earlier.

The problem is that more often than not the underlying philosophical assumptions are overlooked or not dealt with in a sufficiently conscious, explicit, detailed, and reflective manner. Frequently, they are left at a vague, imprecise, i.e., unprecisiated, and implicit level, and occasionally they are disavowed outright as in the atheoretical stance. Yet these hidden assumptions continue to exert a highly significant influence upon the ways in which the researchers and practitioners understandings of a particular case study or a system model, will be framed, organized, or subtly structured. In particular, Type 1 fuzzy theory which was very successful in fuzzy control is not capable to capture "uncertainties" embedded in real life decision problems.

At times, it seems as if we in the fuzzy disciplines have been waiting for philosophers or someone else to come along and help us to make our philosophical unconscious more conscious, while we have played the role of a very cooperative participant in the development of the theory and/or its applications. For this reason, I call on all fuzzy theory researchers to assess their philosophical grounding. I am only providing a particular personal view in this book. It is limited to my particular research that shows that there are at least some equivalences that break down some laws of conservation are re-structured and some basic Belief, Plausibility and Probability formulas need to be re-assessed and re-structured.

It is well known that Lotfi A. Zadeh has provided a continuous stream of novel and seminal ideas, from fuzzy sets, to fuzzy logic, to approximate reasoning to syllogistic reasoning, to computing with words and computing with perceptions. In this regard, we are greatly indebted to him for his continuing leadership. But very few of us have taken up some of his suggestions and clearly stated our particular stance in a systematic and constructive manner. I do not mean to state that we have not made significant progress over the last fourty years or so. We have. But we still have to do a lot more.

Many researchers and practitioners have contributed to the theory and its applications in their specific area of concern. At times some have stated their particular assumptions. But a unified view of the theory has not been stated in explicit philosophical content.

Part of the problem has been methodological. Despite the many significant developments, there have been few systematic or comprehensive attempts made to look at the complex and interweaving relationship among the philosophical and scientific issues in question. In this section, we present a methodology with which one might explore the important philosophical bases of our fuzzy theories in a more structured and perhaps a more rigorous manner.

In applications, philosophical positions are taken up more or less simultaneously on several different levels of theoretical inquiry. It is thus important for us to be able to ascertain that our positions on these different levels of inquiry are consistent with one and the other. That is, we must demonstrate that our theories have some overall coherence to them. The method presented in this section is particularly suited to provide such demonstrations in the rest of the book at times implicit and at other times explicit.

### *0.2.1. Underlying Philosophical Bases*

An overview of a systematic approach to reviewing and observing philosophical issues of fundamental importance to fuzzy theory and its practice is presented below. This method involves an analysis of the stated or implied stances taken by any given fuzzy theory on a structured series of essential philosophical questions.

A hierarchy of levels of theoretical inquiry has been developed, and proposed which includes the Ontological, the General Epistemological, Domain Specific Epistemological, and the Application Levels. Each level of this hierarchy poses its own fundamental philosophical questions. Each of these levels and questions in turn provides the philosophical "grounding" of subsequent ones. A given fuzzy theory, and the many of the philosophical pre-suppositions inherent to it, may then be illustrated and classified by exploring the set of propositions adopted by it on this series of crucial questions. The results of these inquiries may then be summarized hierarchically as will be demonstrated.

I have found this systematic framework to be a useful device with which to analyze a number of fuzzy theories, and thus to increase my reflective awareness of them. To put it another way, it is an approach that may help us to explore the largely pre-reflective, unconscious, or pre-conscious philosophical dimensions of our fuzzy theories more consciously. It may more readily allow us both to asses the internal consistency or coherence of our theories and to philosophically compare and contrast them with each other as well.

### *0.2.2. Hierarchy of Levels of Theoretical Inquiry*

A hierarchy of levels of theoretical inquiry and their questions, where the essential question thought pertinent to each level of inquiry being examined and which has been used as the framework for this approach is depicted in Table 0.2.1.

In this table and in this discussion an important issue is the language we use to state our theoretical inquiry. In this sense, once again Prof. Zadeh (2003) has pointed out the research direction to be Computing with Words, CWW, and the need to develop Precisiated Natural Language, PNL.

The bottom level in this hierarchy are foundational to others: positions from Level 1 form the "grounding" or the conditions for the possibilities of positions on Level 2; those of 2 ground 3; etc, The Table is read from bottom level up: from 1 to 7. Thus, on the application level 7, insights and theories are seen to rest upon a series of positions taken on each of the supporting levels 1 through 6. Let us next proceed to examine this hierarchy level by level.

*0.2.3. Ontological Level*

There are two sub-levels in the Ontological Level. They are called Level 1 and Level 2.

*0.2.3.1. Level 1*

At the bottom, theoreticians of any sort must address the most fundamental of philosophical questions: "Is there any such thing as fuzziness independent or partially independent of us?" As well, "Is there fuzzy truth?" or, "Is there any absolute truth?" These questions about the existence of Reality as such are on what I refer to as the Ontological Level of Inquiry. It seems obvious that whether one answers yes or no to these questions will have profound implications for all other levels of the theory. Type of theories and science that we propose and construct in fact depend on whether we answer "yes" or "no" to these questions. In deed, if one answers yes to these questions, it is arguable that classical theories and science has to be re-assessed and must be rendered relevant on a new grounding. As such, this level is considered most fundamental or foundational. It is well known that Classical set and logic theorist's stance is "yes" that there is the absolute Truth and that there is a crisply defined Reality that exists independent of us. Whereas the stance that fuzzy theorist take is that there is no absolute truth and that there is a fuzzily defined Reality beginning with Zadeh's seminal paper (1965), i.e., that the Truth is a matter of degree and that the Reality is dependent on our perceptions (Zadeh, 1999). More generally, everything in our world of perception can only be assessed to a degree. At this level we have to inquire "**what PNL explicates reality?**". What explicates reality more precisely? Which representation of linguistic variables and their linguistic connectives are more realistic? At this junction of our history, there are essentially two alternatives, i.e., crisp or fuzzy representation of linguistic variables and their connectives that explicate reality.

*0.2.3.2. Level 2*

Still within the realm of ontology, a further, higher-level question then arises:

"What is our position or relation to that Reality"? Are we originally separated or apart from it, or is it the very essence of our being *relational* in this respect? Some philosophical and scientific traditions take up stances very different from others on this still quite a fundamental level. If it is relational, then "What is the nature of that relation"?

The classical view is that our *relational* being to Reality is all or none. That is the elements of reality and their belonging to a set is "all or none". As well as the relation of these elements between sets is "all or none". The fuzzy view is that our *relational* being to reality is *a matter of degree*. That is the elements of reality in their belonging to a set as well as in their relation to each other between sets is *a matter of degree*, i.e., there are partial memberships in a set and partial degrees in participating in relation between sets. And the degree of truth of these membership values are also partial. This is compatible and in agreement with the level 1 stance that partial membership, partial participation in relations and partial truth are all perception based and expressed in our use of words and thus computing with words that capture imprecision in set membership and imprecision in combination of concepts via combination of sets but by imprecise linguistic connectives.

Our position is that until we are able to express our knowledge with the appropriate linguistic expressions we will not be able to capture the true nature of our relation to Reality.

At this level, our inquiry is to be stated as:

**"What linguistic expressions capture our positions to realty**?".

**"What PNL expressions capture our positions to reality?"**

**"What are the basic equivalences and the Laws of Conservation that capture our position to reality?"**

*0.2.4. General Epistemological Level*

There are also two sub-levels here. Let us call them Level 3 and 4.

*0.2.4.1. Level 3*

Next level above the Ontological level is the General Epistemological Level of inquiry. The questions of general epistemology ask:

"What is our access to truth or knowledge?
"Where is truth found in our paradigm? How or from what is it constituted?"

These questions are addressed on this level in order to deal with the nature of human knowing in general.

How do we acquire knowledge? Absolutely or partially?

That is once we take a stance on description of concepts and/or verification of their representation, i.e., "Truthood", being absolute or partial, then we have to explicitly state how we obtain it. At this level, we have to ask:

"**What linguistic encoding allows us to access truth or knowledge**?".

*0.2.4.2. Level 4*

Based upon the stances adopted on Level 3 and still within the realm of general epistemology will be questions of General Validity: "Given our General epistemological position on Level 3, "How do we *validate* our knowledge? How do we know it is true? What criteria do we use to assess its truth-value?" Again these questions are asked from the standpoint of the position and limits on human being or human knowledge in general. At this level, we have to ask:

"**What linguistic expressions cause the assessment of truth and knowledge?**".

As well, we have to ask:

(1) What accounts as good, strong, supportive evidence for **Belief**? What is the degree of **Belief**?

This requires that we have to come up with a "good" in terms of a "matter of degree". "Explication" of criteria of evidence or its justification.

(2) What is the connection between a belief being well-supported by good evidence, and the likelihood that it is true? What is the degree of likelihood and its degree of truth?

This inquires into a new definition of "**Ratification**" and "**Verification**" criteria and their assessment to a degree. In particular, we should investigate "**Belief**" related assessments for "**Ratification**" and "**Verification**".

*0.2.5. Domain-Specific Epistemological Level*

Here again there are two-sub-levels: Level 5 and Level 6.

*0.2.5.1. Level 5*

The next major level of inquiry in the hierarchy is the Domain-Specific or Discipline-Specific Epistemological Level. At this level, our questions take the form not of what human beings can know in general; but rather given our general epistemological and ontological position taken below on level 4, the more specific question is:

"What can we know or hope to know or learn, within a discipline or a setting (e.g., what can we know from within an experimental setting in terms of sampled training and testing data sets)?".

Assertions on this level may include domain specific field experiments delimiting our statements, or positions (proposition) that attempt to define the proper limits or "horizons" to a given domain specific field. At this level of inquiring, we ask:

**"What specific expressions of 'Domain-Specific' PNL could and should be used to specify the limits or boundaries?"**.

*0.2.5.2. Level 6*

As was the case with the General Epistemological Level, the Domain-Specific Epistemological Level positions will also provide a basis for domain specific theories of validity, and thus of appropriate methodologies as well. At this level of inquiry, we ask:

**"What ought to be 'Domain-Specific' PNL that defines explicitly 'Ratification' and 'Verification' criteria and to what degree?'**

*0.2.6. Application Level (Level 7)*

Finally, it is only after implicitly or explicitly addressing all previous levels that we come to the Application Level of Inquiry proper. It is at this level of inquiry that our questions about system analysts and designers, managers, decision-makers, doctors, lawyers, engineers, etc. etc., feel, behave, think, and interact, are raised, and our insights, theories, and different emphases are debated in our attempts to better understand them and support them.

Naturally, one might additionally propose various sub-levels within this complex hierarchy suggested in Table 0.2.1. But for the purpose of this section, the key point is that at the Application Level 7, issues should not be tackled in isolation from their philosophical underpinnings.

For as important as the theoretical, pragmatic and methodological controversies and different emphases that are within this level, changes of position that occur at lower levels of the hierarchy may even more profoundly shake the foundation of all that rest upon them. Such changes may be similar to a theoretical earthquake, which necessitates a great deal of subsequent rebuilding. In fact the introduction of the fuzzy set (Zadeh, 1965) has caused such a revolutionary (Grand) paradigm-shift (Khun, 1962) and subsequent rebuilding of fuzzy theory, approximate reasoning and computing with words.

In the following discussion of classical vs Fuzzy theory we will concentrate only on Ontological and Epistemological concern and in particular the break down of equivalences caused by viewing everything in our world as a matter of degree on the ontological level and their consequences on the epistemological level.

### *0.2.7. Classical vs. Fuzzy Theory*

Classical set and logic theory, at times known as Aristotelian theory, in contrast to Fuzzy set and logic theory, at times know as Zadehian theory, will be sketched out and articulated below in terms of the philosophical hierarchy discussed in the previous section. The purpose of this sketch at this juncture is *primarily methodological*; that is, it is presented more as a demonstration of the sort of analysis that may be facilitated by the hierarchical method described above in Section 0.2.2. It is not an exposition that reveals an exhaustive study of classical set and logic theory versus fuzzy set and logic theory. But it is believed that it forms a reasonable grounding.

In particular, we briefly point out some of the basic equivalences known as "**normal forms**" and laws of classical conservation, as well as basic Belief, Plausibility and Probability formulas that exist in classical set and logic theory. In parallel, we then point out how these equivalences break down in fuzzy set and logic theory which in turn lead to the realization of new laws of conservation, as well as Belief, Plausibility and Probability formulas that exist in fuzzy set and logic theory.

*0.2.8. A Hierarchical Sketch of the Classical Set and Logic Theory Model*

Classical theory is a well known *reductionist* theory that has helped us over more than two thousand years with many applications in explanations of physical and natural phenomena and electro-mechanical men made devices. In Table 0.2.2, we now give a sketch of classical theory.

On the Ontological Level, Level 1, of the proposed hierarchy, we must believe firmly in an *observer-independent* and absolute reality. The classical theory subscribes to the **Cartesian "sRo"** paradigm of describing our primary relation to that Reality on level 2. By the sRo paradigm, we refer to that ontological model that posits the subject, s, and the object (or world), o, as initially (at least in priciple) separate from each other but subsequently connected by some sort of relational, R, event.

The prototypical relational event of this sort has been accepted as that of the primary cognitive act of "knowing"; i.e., subject and object come to be related through the subject's coming to *know* the object. The Cartesian split, between the subject and the object as well as between the mind and the body, appear to have been embraced by Classical thinkers throughout most of their theorizing. Such dualistic dichotomous thinking is well recognized and accepted by the defenders of the classical theory. Without reiterating all of their arguments here, Cartesian dualism, or sRo ontological model, inherit much of the terminology of the two-valued set and logic theory with its well known axioms that are exhibited in Table 0.2.3.

Briefly, every element belongs to a concept class, say A, either with full membership or none, i.e., $\mu_A: X \to \{0,1\}$, $\mu_A(x) = a \in \{0,1\}$, $x \in X$, where $\mu_A(x)$ is the membership assignment of an element $x \in X$ to a concept class A in a proposition. This is a *descriptive* assignment.

Furthermore, this descriptive assignment, $D_{\{0,1\}}$, is verified or asserted to be absolutely True, T, or False, F, i.e., $\mu_V: \mu_A \to \{T,F\}$, where $V_{\{T,F\}}$, is the veristic assignment which is the atomic building block of two-valued logic expressions.

A particular consequence is the Law of Excluded Middle, $A \cup c(A) = X$ and Law of Contradiction $A \cap c(A) = \varnothing$.

As a result of this, there exist the law of Conservation, i.e., $a \vee \bar{a} + a \wedge \bar{a} = 1$ and hence the "Principle of Invariance" in classical set and logic theory.

On the Ontological Level, Levels 1 and 2, the acceptance of the descriptive assignment $D_{\{0,1\}}$ and the Veristic assignment, $V_{\{0,1\}}$ provide us the grounding for the formation of Two-valued "Truth Tables" and in turn for the derivation of the combination of concepts for any two crisp sets A and B to be

"A AND B" = "A∩B",

"A OR B"="A∪B",

and

"A IMP B" = A→B = c(A) ∪B, etc.

This means that linguistic connectives "AND", "OR", "IMP", etc., are interpreted in a one-to-one correspondence, isomorphic, to be equal to "∩", "∪", "c(.), ∪", respectively.

There are sixteen basic equivalences for each of the sixteen combination of concepts shown in Table 1. In particular DNF, Disjunctive Normal Forms, and CNF, Conjunctive Normal Forms, of every one of the sixteen combination of concepts, shown in Table 1 in terms of their meta-linguistic expressions, are equivalent to each other. For example, three of these are:

$$(1) \qquad \text{"A AND B"} = \begin{cases} \text{DNF(A AND B)} = A \cap B \\ \text{CNF(A AND B)} = (A \cup B) \cap (c(A) \cup B) \cap (A \cup c(B)) \end{cases}$$

together with the equivalence of DNF and CNF,

i.e., DNF(A AND B) ≡ CNF(A AND B)

$$(2)\ \text{"A OR B"} = \begin{cases} \text{DNF(A OR B)} = (A \cap B) \cup (c(A) \cap B) \cup (A \cap c(B)) \\ \text{CNF(A OR B)} = A \cup B \end{cases}$$

together with equivalences of DNF(A OR B) ≡ CNF(A OR B)

$$(3)\ \text{"A IMP B"} = \begin{cases} \text{DNF(A IMP B)} = (A \cap B) \cup (c(A) \cap B) \cup (c(A) \cap c(B)) \\ \text{CNF(A IMP B)} = c(A) \cup B \end{cases}$$

together with equivalences of DNF(A IMP B) ≡ CNF(A IMP B).

**Table 0.2.1**. Hierarchy Of Levels Of Theoretical Inquiry And Their Questions.

| | |
|---|---|
| APPLICATION LEVEL | 7. How do people, decision-makers, feel, think, behave, and interact? How can we provide them with better decision-making tools?<br>"How can we provide them with a better PNL?" |
| DOMAIN-SPECIFIC EPISTEMOLOGICAL LEVEL | 6. How do we validate knowledge appropriately in this domain specific field? What methodological approaches are appropriate to it?<br>"What ought to be 'Domain specific' PNL for validity of our investigations?"<br>"What expressions of 'Domain-Specific' PNL could and should be applied for<br>"Ratification and Verification" of our results?"<br>5. What can we know or hope to learn within this domain-specific field or discipline? What are the limits or boundaries to it?<br>"What specific expressions of 'Domain-specific' PNL could and should be used to specify the limits or boundaries of our knowledge?" |
| GENERAL EPISTEMOLOGICAL LEVEL | 4. How do we validate our knowledge? How do we know it is true? What criteria do we use to assess its truth-value?<br>"What PNL expressions cause the assessment of truth and knowledge?"<br>3. What is our access to truth and knowledge in general? Where is knowledge and its truth to be found? How or from what are they constituted?<br>"What PNL encoding allows us to assess truth or knowledge?" |
| ONTOLOGICAL LEVEL | 2. What is our position or relation to that Reality (if we do assume that it exists on level 1 below)?<br>"What PNL expressions capture our positions to reality?"<br>1. Is there any reality independent or partially independent of us? Does any absolute truth exist? Does fuzziness exists?<br>"What PNL explicates reality?" "Are they crisp or fuzzy representations of linguistic variables and their linguistic connectives?" "What are the basic equivalences generated by PNL expressions for crisp or fuzzy representations?" |

**Table 0.2.2.** Positions Taken by Classical Set and Logic Theorists on the Hierarchy of Levels of Theoretical Inquiry.

| | |
|---|---|
| APPLICATION LEVEL | vii. Emphasis on mechanistic Super additive systems theory of interactions, relations, equations, etc. |
| DOMAIN-SPECIFIC EPISTEMOLOGICAL LEVEL | vi. Validity and methodology dictated by meta-physical theories.<br>e.g., principle of determinism, symmetry, invariance and randomness.<br>v. Objective facts and truth accessible, but limited only by<br>e.g., subjective distortions (introduction of uncertainty) |
| GENERAL EPISTEMOLOGICAL LEVEL | iv. Correspondence theory of Validity only Objective<br>iii. Objectivist, empiricists, certain |
| ONTOLOGICAL LEVEL | ii. sRo Cartesian dualism<br>i. Realism, crisp meaning representation of linguistic<br>Variables and connectives are defined with two-valued<br>sets and logic theory. Equivalences in "normal forms" together<br>with classical laws of conservation, as well as formulae for Belief, Plausibility, Probability, etc. |

It will show later that these sixteen equivalences break down in Fuzzy set and logic theory, i.e., they no longer hold in fuzzy theory. In turn these equivalences generate certain laws of conservation. Some of these equivalences are:

**Table 0.2.3**. Axioms of Classical & Set & Logic Theory, where A, B are crisp, two valued, sets and c(.) is the complement, X is the universal set and $\phi$ is the empty set.

| | |
|---|---|
| Involution | $c(c(A)) = A$ |
| Commutativity | $A \cup B = B \cup A$ |
| | $A \cap B = B \cap A$ |
| Associativity | $(A \cup B) \cup C = A \cup (B \cup C)$ |
| | $(A \cap B) \cap C = A \cap (B \cap C)$ |
| Distributivity | $A \cup (B \cap C) = (A \cup B) \cap (A \cup C)$ |
| | $A \cap (B \cup C) = (A \cap B) \cup (A \cap C)$ |
| Idempotency | $A \cup A = A$ |
| | $A \cap A = A$ |
| Absorption | $A \cup (A \cap B) = A$ |
| | $A \cap (A \cup B) = A$ |
| Absorption by X and $\phi$ | $A \cup X = X$ |
| | $A \cap \phi = \phi$ |
| Identity | $A \cup \phi = A$ |
| | $A \cap X = A$ |
| Law of contradiction | $A \cap c(A) = \phi$ |
| Law of excluded middle | $A \cup c(A) = X$ |
| De Morgan Laws | $c(A \cap B) = c(A) \cup c(B)$ |
| | $c(A \cup B) = c(A) \cap c(B)$ |

DNF(A OR c(A)) ≡ CNF(A OR c(A)),
DNF(A AND c(A)) ≡ CNF(A AND c(A)),

and the associated law of conservation is:

$$\mu[\text{DNF}(A \text{ OR } c(A)) \equiv \text{CNF}(A \text{ OR } c(A))] \\ + \mu[\text{DNF}(A \text{ AND } c(A)) \equiv \text{CNF}(A \text{ AND } c(A))] = 1$$

Other equivalences such as:
DNF(A AND B) ≡ CNF(A AND B)
DNF(A OR B) ≡ CNF(A OR B)
generate other laws of conservations for T-norms and co-norms, Belief, Plausibility, Probability, etc., as:

T(a,b) = 1 – S(n(a), n(b)),
***Bel***(A) + ***Pl***(c(A)) = 1
***Pl***(A) + ***Bel***(c(A)) = 1
***Pr***(A) +***Pr***(c(A)) = 1

These equivalences, together with the laws of conservation are obtained with the inherent assumption that imprecise and varying meanings of linguistic words and linguistic connectives are precisitated in an absolute manner. All this is done for the sake of developing a "reductionist" theory of knowledge representation and reasoning with it!

On the General Epistemological Level 3, we have the foundation of the objectivism. Its stance is that real truth exists, is potentially accessible, and is to be found on the object side of the Cartesian split. The "subjective" elements, although understandable or symbolically meaningful, and of immense scientific interest, were nevertheless seen as basically unwanted distortions to be removed or eliminated (as much as possible) in order to get at the "objective truth" assumed to be obscured.

On this General Epistemological Level, first at level 3, observation based data are to be obtained from objects with the use of measurement agents such as sensors. Thus representations of objects are developed on measurement based data warehouses. This is done with the assumption of the fact that measurement based models so developed are to stand on the foundation developed on the Ontological level. They are consequently descriptive representations of model concepts that stand on two-valued set theory, i.e., $D_{\{0,1\}}$ which are verified with the two-valued logic theory, $V_{\{0,1\}}$.
Not surprisingly, on level 4 of the proposed hierarchy, dealing with General Theories of Validity, a "correspondence theory of truth" is accepted in most classical scientific investigations. According to this theory, a perception, or an observation, or a judgment is considered valid or true insofar as it may be shown to match with the objective or factual reality of the world around us not withstanding Heisenberg's "Principle of Uncertainty".

Hence on level 4, the "correspondence theory of truth" essentially is based on the two-valued set and truth framework, i.e., Description and Verification, $\{D_{\{0,1\}}, V_{\{0,1\}}\}$, which are accepted on the Ontological level. This means that models developed on the General Epistemological level are to be accepted as true depictions of real system behavior. Furthermore, test data are to be used to validate results obtained from the model build on the level 3. Thus results are assumed to computationally determine as acceptable outcomes for given inputs of the test data.

Therefore on the General Epistemological level, we first have a model expressed in a general form as "A→B" as a descriptive model $D_{\{0,1\}}$ which is verified as $V_{\{0,1\}}$ within the frame work of a classical inference schema such as Modus Ponens which is stated as (A→B)oA = B such that the premise, $\{D_{\{0,1\}}, V_{\{0,1\}}\}$ for "A→B", combined, "o", with the premise $\{D_{\{0,1\}}, V_{\{0,1\}}\}$ for "A" results in a consequence which exactly matches the right hand side of the crisp rule, i.e., B, described and verified as $\{D_{\{0,1\}}\ V_{\{0,1\}}\}$, provided that the observation "A" matches exactly to the left hand side of the rule "A→B". The validation is based on a comparison of the actual output for a given test input data and model output for the same test input data. The error is usually accepted to be a true, $V_{\{0,1\}}$, verification based on a statistical risk which is to be a crisply evaluated assessment dependent on a crisp test of hypothesis.

On the Domain-Specific Epistemological level, the classical system development models that stand on observations which are measurement based data stored in data warehouses for a specific field of inquiry. Measurement data that is suppose to capture the system behavior and its associated concepts that are represented with linguistic variables and linguistic terms are expressed in two-valued descriptive set theory $D_{\{0,1\}}$. Therefore system models so developed are assumed to be universally valid representations of domain specific object world that is captured by measurement based data only! That is, they are verified to be absolutely true, i.e., $V_{\{0,1\}}$, by measurements only.

On level 5 of the hierarchy, dealing with the Domain-Specific Epistemology, for a particular investigation, the scientists appear to find limits to knowledge only on the *subjective side* of the Cartesian split. It is assumed that such subjective distortions may originate in the (system) analyst, but the "untruths" are all seen as coming from the s-or subject- side of the sRo split. The scientists seem to see few or no limits on the object side. That is, the object seems to be just what it is factually; it is not ambiguous, or imprecise in itself, only in our misperceptions of it or in our measurement errors of it. This naturally discards the "Principle of Uncertainty" proposed and verified on the level of atomic physics by Heisenberg.

It is interesting, to note, however, in the case of the Domain-Specific Epistemology endorsed by classical thinking, such "objective facts of the world" (which by definition stand as true, regardless of our will, mood, or perception on them) may include both external states in the world as well as certain "internal" (though nevertheless still "objective"), be abstract, states or realities. Classical science, articulated from the epistemological view-point, appears to have just discovered the internal abstract states, as it were, *out there* in the world. That is Classical science, in this perspective, does not accept that they have been created as concepts and have been expressed linguistically by subjects. Such internal, abstract, concepts are then considered to be universally valid features of the objective world. That is linguistic variables, their terms and linguistic connectives are assume to have crisp well defined meanings.

In terms of Level 6 of the proposed hierarchy, science deals with the question of validity and methodology in the domain specific field of study with reductionism and commitments to specific cause and effect hypothesis and metaphysical theories, e.g., drive a theory and investigate its derivatives. At times, this tended to arbitrarily restrict what was to be considered valid data and not within the particular domain of interest. For example, a good scientist of this persuasion should be actively searching for "derived derivatives" and attempt to find their origin in more primary theoretical precepts in accordance with the classical model.

On the Domain-Specific Epistemological Level, we find various developments of system models with application technologies known as statistical methods, such as multi-variate regression equations, programming methods such as linear and non-linear optimization algorithms or optimal control schemas developed on objective data that are obtained by measurement devises and depend on description and validation frameworks that are given as $\{D_{\{0,1\}}, V_{\{0,1\}}\}$.

As well the validation of the Domain-Specific models are assessed with domain-specific test data that are assumed to be standing on descriptive and verified framework of $\{D_{\{0,1\}}, V_{\{0,1\}}\}$. The validation of the domain specific models are executed with the classical inference schemas such as Modus Ponens as indicated above. This may entail a re-computation of, say, regression, or programming or control models with test data. Results obtained from such models are assumed to be on $\{D_{\{0,1\}}, V_{\{0,1\}}\}$ framework based on some level of statistical risk.

Furthermore, “ratification” or “verification” of such system models are assumed in terms of certain criteria that are developed by classical statistical methods. For example, some of these are known as

- RMSE – Root Mean Square Error
- R2 – How successful the fit is in explaining the variation in the data
- Accuracy of Prediction
- Power of Prediction

which are defined as:

$$RMSE = \sqrt{MSE}, \quad MSE = \frac{SSE}{n}, \quad R^2 = 1 - \frac{SSE}{SST}, \quad SST = \sum_{i=1}^{n} (y_i - \overline{y}_i)^2, \quad SSE = \sum_{i=1}^{n} (y_i - \hat{y}_i)^2$$

MSE: Mean Square Error
SSE: Sum of Square Errors
SST: Total Sum of Squares

Accuracy t (%) = X / P
Power t (%) = X / A

X: Total number of predicted values that are hit correctly in the interval t
P: Total number of predicted values at interval t
A: Total number of actual values at interval t

This approach, however, may be seen by many to be quite problematic in that the theories and methods most properly belonging to the Application Level 7 appear here to be dictating what qualifies as good and valid data on the Domain Specific Epistemological Level 6. Such situation in which the position on lower levels of the hierarch are overtly dictated by higher ones is generally not recommended in this hierarchical approach.

Furthermore, the Application Level 7 theory determines or dictates what to be considered as valid knowledge in the field, the less likely it will be that anything not in conformance with the theory will ever be found, noticed, or admitted as evidence. This in fact was the case for the rejection of fuzzy theory by orthodox, doctrinaire defenders of the classical theory. While such a hegemony of the Application Level 7 theory may appear to increase the coherence by guaranteeing a certain amount of conformity, it will also provide a basis for the systematic neglect of other potentially valid but contradictory observations such as the existence of gray between the black and white dichotomy.

### *0.2.9. A Hierarchical Sketch of Fuzzy Set and Two-valued Logic Theory Model*

Indeed several critical thinkers of the classical theory have argued that it restricts or reduces reality to be "objective"; and that classical theory leads inadvertently to many paradoxes in which the theory becomes over-structured, over-selective, and sometimes even overshadows the real life data, e.g., Russell's paradox, Barber's paradox, Flackross paradox, etc, etc. Such paradoxes ought to be demonstrations for the importance of our being as conscious and aware as we can be of the philosophical presuppositions embodied in our application theories based on classical theory.

This is particularly so with respect to real-life experimental data, and thus the call of fuzzy theorists for more "experience-near" concepts and "expert-insight" approaches for the formation of fuzzy-expert system models. These models are either formed by expert interviews or by fuzzy data mining exercises more generally a combination of both.

The fuzzy theory is a non-reductionist theory that captures the gray information granules between black and white and helps us to cope with the complexity of our modern world in decision-making processes in a manner akin to human decision-making.

In this theory, the paradoxes of two-valued classical theory are explained by admission of the gray information granules between black and white and hence allowing overlaps between classes and categories.

In Table 0.2.4, we next give a sketch of the fuzzy theory in terms of the proposed hierarchy as we interpret them.

On the Ontological level, level 1, of the proposed hierarchy, we must believe firmly in an *observer-dependent and relative* reality that is perception based and communicated and computed with words and their numeric meaning representation with continuous memberships.

On level 2, the classical sRo Cartesian dualism is modified and extended to be "$\mathbf{s} \xleftrightarrow{R} \mathbf{o}$" paradigm that describes a subject's, **s**'s, primary relation to an object's, **o**'s, relation to a subject, **s**, to be a two way relation R. By $\mathbf{s} \xleftrightarrow{R} \mathbf{o}$ paradigm, we refer to an ontological model that posits subject, **s**, and object, **o**, to be inter connected and interacting with relations **R**.

The prototypical relation **R** is interpreted as the primary cognitive act of "knowing" based on perception of subjects articulated in a natural language, i.e., Computing With Words, CWW, followed by objects', **o**'s, meaning representation in terms of continuous membership values that are next processed by relation R that generates a new cognitive interpretation in the act of "knowing" based on meaning representation caused by the inverse relation R.

Thus the Cartesian split between the subject and the object as well as mind and body are unified. Dualistic, dichotomous, thinking is discarded and overlap of categories are accepted. Consequently, "$\mathbf{s} \xleftrightarrow{R} \mathbf{o}$" Ontological fuzzy theory terminology moves beyond the restrictions of two-valued set and logic theory and eliminates most of the axioms exhibited in Table 0.2.3. Instead it rests mainly on the limited set of axioms shown in Table 0.2.5 which are for t-norm and co-norm based structures. There are naturally further restrictions for pseudo t-norm and co-norm based structures which are not stated here. At the end of this book, in Chapter 17, we shall return to this topic of axioms and ask:

**Table 0.2.4.** Position Taken by some of Fuzzy Set and Logic Theorists on the Hierarchy of Levels of Theoretical Inquiry

| | |
|---|---|
| APPLICATION LEVEL | vii. Emphasis on humanistic Decision and Control Systems that contain highly complex non-linear interactions, relations, equations, etc. |
| DOMAIN-SPECIFIC EPISTEMOLOGICAL LEVEL | vi. Validity and methodology dictated by Meta theories of Modal Logics. e.g., principle of non-determinism and overlapping patterns. v. Subjective and objective facts accessible by perceptions and meaning representation of linguistic terms of linguistic variables, linguistic quantifiers and linguistic connectives. Principle of uncertainty |
| GENERAL EPISTEMOLOGICAL LEVEL | iv. Correspondence theory of Validity both objective and subjective. Approximate Reasoning iii. Subjective-objective, experimental and empiricist, e.g., expert and fuzzy data mining based. |
| ONTOLOGICAL LEVEL | ii. $\mathbf{S} \xleftrightarrow{\mathbf{R}} \mathbf{O}$ schema gives credence both the Level subject and the object interaction i. Realism - fuzzy and uncertain Generation of "Fuzzy Canonical Forms" that are not equivalent to each other in contrast to "Classical Normal Forms". Generation of new Laws of Conservation for t-norms, co-norms, Belief, Plausibility, Probability, etc. |

"what are the basic axioms of CWW?", and introduce "Meta-Linguistic Axioms" as a foundation for CWW.

**Table 0.2.5**. Main Stream Axioms of General Fuzzy and Logic Theory

| | |
|---|---|
| Involution | $c(c(A)) = A$ |
| Commutativity | $A \cup B = B \cup A$ |
| | $A \cap B = B \cap A$ |
| Associativity | $(A \cup B) \cup C = A \cup (B \cup C)$ |
| | $(A \cap B) \cap C = A \cap (B \cap C)$ |
| Absorption by X and ∅ | $A \cup X = X$ |
| | $A \cap \phi = \phi$ |
| Identity | $A \cup \phi = A$ |
| | $A \cap X = A$ |
| De Morgan Laws | $c(A \cap B) = c(A) \cup c(B)$ |
| | $c(A \cup B) = c(A) \cap c(B)$ |

Briefly, every element belongs to a concept class, say A, to a partial degree, i.e., $\mu_A$: $X \rightarrow [0,1]$, $\mu_A(x)=a \in [0,1]$, $x \in X$, where $\mu_A(x)$ is the membership assignment of an element $x \in X$ to a concept class A in a proposition. Most of all concepts are definable to be true to a degree.

**Corollary:** Fuzzy set representations of most concept classes can not be reduce to crisp sets; i.e., $\mu_A:X \rightarrow [0,1]$ is not reducible to $\mu_A:X \rightarrow \{0,1\}$.
Furthermore, the descriptive assignment $D_{[0,1]}$ is verified or asserted to be true, T, or false, F, i.e., $\mu_V:\mu_A \rightarrow \{T,F\}$ absolutely in *Descriptive fuzzy set theory*, where $V_{\{T,F\}}$, or $V_{\{0,1\}}$, is the veristic assignment which is the atomic building block of the two-valued logic.

On the other hand, if the descriptive assignment $D_{\{0,1\}}$ or $D_{[0,1]}$ is verified or asserted to be partially true, i.e., $\mu_V:\mu_A \rightarrow [T,F]$ or $[0,1]$ in *Veristic fuzzy set theory,* i.e., fuzzy set theory of truthood, which needs to be further verified or asserted to be absolutely True, T, or False, F, i.e., $\mu_{V'}$: $[\mu_V:\mu_A \rightarrow [0,1]] \rightarrow \{0,1\}$ where $V_{[0,1]}$ is a partial veristic truth assignment but $V'_{\{0,1\}}$ is a secondary absolute veristic assignment which is once again the atomic building block of the two-valued logic!

On the ontological level, levels 1 and 2, the acceptance of the descriptive assignments $D_{[0,1]}$ in *Descriptive fuzzy set theory* and the graded veristic assignment $V_{[0,1]}$ in *Veristic fuzzy set theory* with veristic assignment of $V_{\{0,1\}}$ for *Descriptive fuzzy set theory* and Veristic assignment of $V'_{\{0,1\}}$ for *Veristic fuzzy set theory* provide us the grounding for the formation of fuzzy-valued

"Fuzzy Truth Tables" and in turn the derivation of the combination of concepts for any two fuzzy sets A and B, when they are represented by a Type 1 fuzzy sets, to be

$$\text{"A AND B"} = \begin{cases} \text{FDCF(A AND B)} = A \cap B \\ \text{FCCF(A AND B)} = (A \cup B) \cap (c(A) \cup B) \cap (A \cup c(B)), \end{cases}$$

and

$$\text{"A OR B"} = \begin{cases} \text{FDCF(A OR B)} = (A \cap B) \cup (c(A) \cap B) \cup (A \cap c(B)) \\ \text{FCCF(A AND B)} = A \cup B, \end{cases}$$

and

$$\text{"A IMP B"} = \begin{cases} \text{FDCF(A IMP B)} = (A \cap B) \cup (c(A) \cap B) \cup (c(A) \cap c(B)) \\ \text{FCCF(A IMP B)} = c(A) \cup B, \end{cases}$$

etc., in analogy to the two-valued set and logic theory where FDCF(.)=DNF(.) and FCCF(.)=CNF(.) *in form only.*

Furthermore, as it will be shown, the equivalence, DNF(.)≡CNF(.), breaks down, i.e., we have FDCF(.) FCCF(.) and in particular we get FDCF(.)⊆FCCF(.) for certain classes of t-norms and t-conorms that are strict and nilpotent Archimedean.

For example, particular consequences that we receive are:(1) FDCF(A OR NOT A)⊆FCCF(A OR NOT A) which is the realization of the law of "Fuzzy Middle" as opposed to the Law of Excluded Middle and (2) the Law of "Fuzzy Contradiction, FDCF(A AND NOT A) ⊆FCCF(A AND NOT A) as opposed to the Law of Crisp Contradiction.

As a consequence of these, we obtain new Laws of Conservation as:

$\mu$[FDCF(A AND c(A))] + $\mu$[FCCF(A OR c(A))] = 1;
(well known but re-interpreted for fuzzy sets) and

$\mu$[FDCF(A OR c(A))] + $\mu$[FCCF(A AND c(A))] = 1.
(New and exists for fuzzy sets only)

Hence we once again observe that the "Principle of Invariance" is re-established in Interval-Valued Type 2 fuzzy set theory, but as two distinct Laws of Conservation.

This means that linguistic connectives "AND", "OR", "IMP", etc., are *not* interpreted in a one-to-one correspondence, i.e., non-isomorphic, to be equal to "∩", "∪", "c(.), ∪", etc. That is the imprecise and varying meanings of linguistic connectives are not precisiated in an absolute manner and there is no absolute precisiation of the meaning of words nor is there an absolute precisiation of the meaning of connectives. This provides a framework for the representation of uncertainty in the combination of words and hence in reasoning with them as a foundation for CWW.

The break down of the equivalences in Fuzzy theory, i.e., FDCF(.) FCCF(.), in turn generates new additional formulae for t-norm-conorms, Belief, Plausibility and Probability. That is, we now obtain:

Two T-Norm-Conorms Formulae for fuzzy sets in fuzzy theory:

1) T(a,b) = 1 – S(n(a),n(b)) (Well known but re-interpreted in fuzzy theory)

2) T[T(S(a,b), S(n(a),b)), S(a,n(b))]
= 1 – S[S(T(n(a), n(b)), T(a,n(b))], T(n(a),b)] (New in fuzzy theory)

Two Belief and Plausibility measures over fuzzy sets at a particular $\alpha$-cut:

(1) ***Pl***[FDCF(A AND B)] + [***Bel***[FCCF(c(A) OR c(B)] = 1
***Pl***[(A$\cup$B) + ***Bel***(c(A) $\cap$ c(B)) = 1 (well known but re-interpreted for fuzzy theory)

(2) ***Pl***[FCCF(A AND B)] + ***Bel***[FDCF(c(A) OR c(B))] = 1
***Pl***[(A$\cap$B) $\cup$ (c(A) $\cap$ B) $\cup$ (A $\cap$ c(B))]
+ ***Bel***[(c(A) $\cup$ c(B)) $\cap$ (A $\cup$ c(B)) $\cap$ (c(A) $\cup$ B)] = 1 (New in fuzzy theory)

Two Probability measures over fuzzy sets at a particular $\alpha$-cut:

***Pr***(A AND B) + ***Pr***(c(A) OR c(B) = 1

(1) ***Pr***[FDCF(A AND B] + ***Pr***[FCCF(c(A) OR c(B)] = 1 (well known but re-interpreted for fuzzy theory)
***Pr***[(A$\cup$B) + ***Pr***(c(A) $\cap$ c(B)) = 1
(2) ***Pr***[FCCF(A AND B)] + ***Pr***[FDCF(c(A) OR c(B))] = 1
***Pr***[(A$\cap$B) $\cup$ (c(A) $\cap$ B) $\cup$ (A $\cap$ c(B))]
+ ***Pr***[(c(A) $\cup$ c(B)) $\cap$ (A $\cup$ c(B)) $\cap$ (c(A) $\cup$ B)] = 1 (New in fuzzy theory)

This particular interpretation of measures at all $\alpha$-cut levels of fuzzy sets together with knowledge representation and reasoning formulae form a unique foundation for Type 2 fuzzy set theory in general and in particular for Interval-Valued Type 2 fuzzy set theory generated by the combination of linguistic concepts with linguistic connectives even if the initial meaning representation of words are to be reduced to Type 1 membership representation. More general representations start with Type 2 representation schema and then form Type 2 reasoning schemas to capture both imprecision and uncertainty.

On the General Epistemological Level 3, we have the foundation of an integrated subjectivist-objectivist perspective. Its stance is that real truth is relative and context dependent. It is potentially, partially and approximately accessible and it is to be found on the subject-object integrated interaction. On this level 3, the observation based data are obtained from subject-object interaction with perceptions as well as measurements. That is, data is provide by both humans as well as electro-mechanical sensors.

Thus representations of objects are developed on perceptions of humans and measurements of sensors for the use of human decision makers. This is done with the assumption of the fact that perception-measurement based models are developed to stand on the foundation proposed on the Ontological level, i.e., "$\mathbf{S} \xleftrightarrow{R} \mathbf{O}$" paradigm.

They are consequently descriptive representations of model concepts on fuzzy (infinite)-valued, i.e., $D_{[0,1]}$, sets which are verified with the two-valued logic theory, as $V_{\{0,1\}}$ for the *Descriptive fuzzy set theory*.

On the other hand, descriptive propositions whether they be $D_{\{0,1\}}$ or $D_{[0,1]}$ if they are verified with fuzzy (infinite)-valued truthoods, as $V_{[0,1]}$ and then they are verified with the two-valued logic theory as $V'_{\{0,1\}}$, then we have a *Veristic fuzzy set theory*.

Next on the level 4, the correspondence theory of truth "is basically based on fuzzy valued sets, whether they be for *Descriptive fuzzy sets*, $D_{[0,1]}$, or *Veristic fuzzy sets*, $V_{[0,1]}$, paradigm and two-valued truth (verification) paradigm with either $V_{\{0,1\}}$ for *Descriptive fuzzy set paradigm* or $V'_{\{0,1\}}$ for *Veristic fuzzy set paradigm* which are accepted on the Ontological Level. This means that models developed on the General Epistemological level are to be accepted as true (but approximate) depictions of a real system behaviour. Furthermore, test data are to be used to validate results obtained from the models build on the level 3. Thus results are assumed to computationally but approximately determined as acceptable outcomes for given inputs of the test data.

Therefore on the General Epistemological level, we first have approximate reasoning models expressed in particular as Interval-Valued Type 2 fuzzy set as:

$$\text{A IMP B} = \begin{cases} \text{FDCF(A IMP B)} \\ \text{FCCF(A IMP B)} \end{cases}$$

as a descriptive model, i.e., an Interval-Valued Type 2 rule, a premise. That is $\{\{D_{[0,1]}\ V_{\{0,1\}}\}$ IMP $\{D_{[0,1]}\ V_{\{0,1\}}\}\}=\{D_{[0,1]}\ V_{\{0,1\}}\}$ which is within the framework of a fuzzy inference schema such as Generalized Modus Ponens, GMP, originally proposed by Zadeh as Compositional Rule of Inference, CRI, such that the first premise $\{D_{[0,1]}\ V_{\{0,1\}}\}$ for "A IMP B" combined with a second premise $\{D_{[0,1]}\ V_{\{0,1\}}\}$ for "A′" result in a consequence $\{D_{[0,1]}\ V_{\{0,1\}}\}$ for $B^*$, where the fuzzy similarity of A′ to A together with the t-norm and co-norm, that is chosen, result in $B^*$. The validation is based on a fuzzy comparison of the actual output for a given test input data and model output for the same test input data. The error is usually accepted to be a true, $V_{\{0,1\}}$, verification but based on a risk statistically but fuzzily evaluated assessment dependent on a fuzzy test of hypothesis. It should be noted that all of the proceeding exposition which is made for the *Descriptive fuzzy set paradigm*. A similar exposition is applicable to the *Veristic fuzzy set paradigm* as we have explained earlier!

On the Domain-Specific Epistemological Level, Level 5, we find various developments of system models with applications of technologies known as fuzzy statistical methods, such as fuzzy multi-variate regression equations, fuzzy linear and non-linear optimization algorithms, or fuzzy optimal control schemas developed on subjective-objective data that are obtained by expertise and measurement which are dependent on description and verification frameworks that are given as $\{D_{[0,1]}\ V_{\{0,1\}}\}$.

At this level, we raise the "Principle of Uncertainty" proposed by W.K.Heisenberg at the sub-atomic level to be applicable at the human decision-making level. It is clear that our perception based knowledge that comes either from our senses or from electro-mechanical measurement instruments capture only part of the reality hence at best they are uncertain.

As well, the validation of the Domain-Specific models on Level 6 are assessed with domain-specific test data that are assumed to be standing on a descriptive and verified framework $\{D_{[0,1]}\ V_{\{0,1\}}\}$. The validation of the domain-specific models are executed with fuzzy inference schemas such as Generalized Modus Ponens as indicated above. They may entail re-computations of, say, fuzzy regression, or fuzzy programming, or fuzzy control models with, test data. Results obtained from such models are assumed to be on $\{D_{[0,1]}\ V_{\{0,1\}}\}$ framework for *descriptive fuzzy set models* based on some level of risk and on $\{V_{[0,1]}\ V'_{\{0,1\}}\}$ framework for *veristic fuzzy logic models*.

However, it is to be noted that, at this juncture most fuzzy theory applications are verified and/or validated still with classical statistical measures such as, RMSE, $R^2$, accuracy and power of prediction that is stated in the classical theories "Domain Specific" verification, validation exercises.

As for as I know, at this point in time, there are no fuzzy verification and validation measures, such as, fuzzy RMSE, fuzzy $R^2$, fuzzy accuracy and power of prediction to be found in the literature. These are still open research directions.

On Level 7 of the proposed hierarch, Application Level proper, we are all quite aware of the vast contributions made to fuzzy sets and systems field by Lotfi A. Zadeh as a foundation for numerous applications. The details of his numerous seminal ideas are well beyond the scope of this sketch. A synopsis of his contributions are briefly summarized in my Plenary talk given at IFSA-2001 World Congress (Türkşen, 2001). Let it suffice to point out here that his consistent emphasis on the foundations of fuzzy set and logic theories have provided a grounding for the Application Level 7 includes linguistic variables, their representation with fuzzy sets, i.e., their precisiation with membership functions, and reasoning with imprecise linguistic terms of linguistic variables that are precisiated with membership functions. This emphasis in turn has led to the notions of Computing With Words, CWW, and more recently Computing With Perceptions, CWP.

It is in these respects that many of the familiar revisions and alternatives to classical thinking, suggested by Black, Lukasiewicz, Kleene, etc., were preliminary break away strategies from the classical paradigm. With the grand paradigm shift caused by Zadeh's seminal work and continuous stream of visionary proposals, it is now clear that most of them reflect very different stances adopted at the more fundamental levels of our proposed hierarchy. Those changes, it appears, have sometimes been made only in a more tacit and implicit manner. In my studies, it became obvious that the most radical revisions are likely to be the ones that stem from modification to be made at the low levels of the proposed hierarchy.

It is in this sense that I had investigated measurement theoretic foundations of membership functions and provided experimental existence of Type 2 fuzzy sets in membership acquisition experiments. Also with this conviction I raised the Conjunctive and Disjunctive Normal (Canonical) Form representations in the combination of concepts. In turn, it became to the forefront that linguistic connectives "AND". "OR", "IMP", etc., are also imprecise and that they need to be precisiated in a graded manner and thus "AND" does not correspond to t-norm and "OR" does not correspond to a t-co-norm in a one-to-one isomorphic correspondence.

Therefore, each combination of concepts has its own Conjunctive Normal (Canonical) Form that is not equivalent in general but larger in the set sense and greater in the membership sense or equal to its Disjunctive Normal (Canonical) Form for particular subclasses of t-norms and t-co-norms that are isomorphic to Archimedean norms which are nilpotent or strict, and special cases of (Max-Min) and some ordinal sums. This, on the one hand, increases the uncertainty, but on the other hand, exposes risk associated with the indefiniteness in the combinations of imprecise concepts in knowledge representation and reasoning in an approximate manner. As well, it provides us with additional laws of conservation and new formulae for t-norm-co-norms, Belief, Plausibility and Probability measures. These new formulas expose part of the richness of the fuzzy theory.

Overall, it appears that these philosophical issues on Ontological and Epistemological levels leave Classical thinkers as well as Type 1 fuzzy thinkers with significant dilemmas. Much of the classical work may be seen as an example of a theory that was very heavily committed to sRo model of Cartesian dualism. If one adopts that position on Ontological Level 2, however, one is then left at the General Epistemological Level 3 to choose between two responses to the crucial scientific question: "From where does truth come?" One must ascribe the *origin of truth* or valid knowledge to either the "object side" or the "subject side" of the sRo scenario with the former position being known as "objectivism" an the latter "subjectivism".

With the hindsight over the last 40+ years of debate, one can say that classical thinkers are confronted with a dilemma: because their sympathies are clearly with the "objectivism" side of the epistemological debate, for the electromechanical systems. On the other hand, the subject matter rests more precisely on the "subjectivity" side in all its richness for the human decision-making systems. Thus for the objectivism, it is no wonder that their mechanistic meta-physical explanations demanded revisions. Nor is it surprising that their critical writing have quite a different tone as well and perhaps even a different implicit epistemological leaning.

For all these reasons and more, we have proposed that at the very foundation of the Ontological level, we need to consider the "$\mathbf{s} \overset{\mathbf{R}}{\longleftrightarrow} \mathbf{o}$" paradigm to push to the surface the integrated subjective-objective stand that is investigated by most fuzzy set and logic theory researchers in many different directions.

### *0.2.10. Application of Hierarchical Sketches*

Hierarchical sketches as shown in Tables 0.2.1, 0.2.2, 0.2.4 can summarize an important set of philosophical stances adopted by any given set and logic theory as we have illustrated with the examples given for the Classical Set and Logic theory and Fuzzy Set and Logic theories. Such sketches may be useful for a variety of purposes, some of which are suggested below.

### *0.2.10.1. Explication*

Particularly with set and logic theories that do not discuss their philosophical assumptions directly or explicitly, this structured approach may be helpful in looking for an explication what philosophy seems to be implicit in the texts. Since we have argued all set and logic theories rest upon some philosophical assumptions, it is only consistent for us to argue that if these presuppositions are not dealt with explicitly in our writings, they must still be there implicitly. It then behooves us to bring them to light as much as possible, i.e., to make the philosophical unconscious grounding more conscious.

### *0.2.10.2. Assessing Internal Consistency*

It is only when a multi-leveled overview of a complex theory is available that its internal consistency can be readily evaluated. While a given set of assumptions or hypotheses may appear to be coherently connected with each other at one level of inquiry, they may still be quite incompatible with positions assumed by the same theory on other levels of hierarchy. If that is the case, the internal consistency or coherence of the model as a whole will appear as lacking. When this is the case, it seems that trouble can not be far behind!

For example, if we were to assert a subjectivist position on the General Epistemological level 3 of the hierarchy, we should not then go on to adopt the principles of classical set and logic theory on the Application Level 7, for a theory that places all its epistemological emphasis on the subjective or the internal mental model of meaning representation is not compatible with an application model that focuses only on external realizations of reductionist approach that dichotomizes our world arbitrarily. An even stronger incoherence would be found in the case of a theory, i.e., fuzzy theory, that took a radical relativistic stance on Ontological Level 1, i.e., one that denies the existence of any reality independent of one self, and combine this with a correspondence theory of truth as its major criterion for validity on the General Epistemological Level 4. That is one that judges the truth of something by comparing it with an objective independent reality, i.e., classical theory, while the lack of internal consistency may be obvious in those examples, in other cases it may not be. By developing a hierarchical sketch of a given set and logic theory, as was done above, we may provide ourselves with means of a theory analysis that readily lends itself to the detection of more subtle defects in the overall coherence of the model as a whole. As was indicated, just relying on Type 1 fuzzy theory, theorists and application oriented researchers over look that there are never formulas in Interval-Valued Type 2 fuzzy theory. Hence by omitting internal uncertainties of the theory, researchers are likely to produce results with subtle but serious defects.

### *0.2.10.3. Assessing the Level of Inquiry for an Argument*

One's assessment of how convincing a given argument is dependent on the level of inquiry. In some writings authors themselves are not clear about this. This has a critical importance, in evaluating the accuracy or suitability of theories when one proposes certain degree of fuzziness in their propositions. If one takes up a "fuzzy" stance on the ontological Level 1, which in an effort argues against any concept of objectivity at all, one would take up very extreme position that would be unlikely to convince many researchers. On the other hand, one might choose not to state, in principle, that any Reality exists independent of us on the Ontological Level 1, but still assert fuzziness on the General Epistemological Level 3. In this stance, one might maintain, for example, that one can never "objectively know", or have "untainted access" to any pristine, "absolute truth" of that Reality. This is a far more defensible position than the previous one. Still in a less disagreeable argument, fuzziness is still adopted; but it would be limited only to the Domain-Specific Epistemological Level 5. Such an argument might take the form of an assertion that within the specific field of, for instance, the consumer behavior in financial domains, one can never be entirely objective about the truth, or validity of what is argued for in that domain. This might be seen as merely an acknowledgement of the uncertainty bounds of what one can definitely know from a certain situation or perspective. Such a position is clearly less controversial than the first stance discussed above.

Such arguments would be further supported if we could develop fuzzy RMSE, fuzzy $R^2$, fuzzy accuracy and power of prediction measures.

Each of the three arguments may be accurately described as a fuzzy theory argument of one sort or another, but to fully assess them, we must ask the crucial question, "At what level of inquiry does one propose that their stance is properly applied?"

Unfortunately, the answer is not always readily available in any explicit writing. It is for this reason that the clarifying studies are advocated by the hierarchical sketch method outline in this section and for the rest of us who are concerned about these matters.

In the realm of current fuzzy theories, I feel that such clarifications are particularly important when considering such theories as the various versions of structure identification in fuzzy system modeling.

*0.2.10.4. Philosophical Comparison among Different Fuzzy Theories*

The standard approach taken by the proposed hierarchical sketch method may be particularly useful for comparing the philosophical bases of different fuzzy theories. This helps to structure our observations about the fuzzy theories and lends itself readily for making comparisons amongst them. Diverse fuzzy theories may appear to differ primarily at one level of the hierarchy or at times at many levels of inquiry.

Most common comparison of fuzzy theories will, of course, focus primarily on their differences of emphasis and approach at the Application Level 7 of Fuzzy Logic. But I believe, one needs to explicitly state the fundamental assumptions made on other significant levels as well. This may provide fuzzy researchers with a greater understanding of the core differences involved. Some examples of these divergences of opinion amongst fuzzy theories at a variety of levels on our proposed hierarchy are stated briefly next. Many or most of the major fuzzy theories appear to adopt a realist position on the Ontological Level 1. This group accepts the seminal paper of L. A. Zadeh and relaxes the LEM and LC of the classical set and logic theory and generally applies the connectives ($\vee$, $\wedge$, -), i.e., (Max, Min, St. negation) De Morgan Triple. On the Ontological Level 2, it seems that most "traditional" fuzzy theorists have at least implicitly accepted the non-Cartesian $\mathbf{s} \xleftrightarrow{R} \mathbf{o}$ model, as descried above. For example, in many current fuzzy rule bases, the experts determine the membership functions and the rules subjectively based on their experience and expert knowledge or fuzzy clustering methods are used to elicit membership function and rules with an added review by experts.

On the General Epistemological Levels 3 and 4, many of the philosophical disagreements become more obvious while it is known that the classical set and logic theory model takes up an objectivist stance at this level, it is clear that fuzzy theories do not in general accept this position. For example, subject-object relation theory, $\mathbf{s} \xleftrightarrow{R} \mathbf{o}$, with the emphasis on "internal representation of mind" of an expert and various projections and introjections of the expert in terms of membership functions and selection of t-norms and t-conorms for the formation of fuzzy rules, certainly based on quite a subjectivist stance. A third option, a "co-constitutional" epistemology, in which the truth is seen to emerge in the "interaction between" the so-called subject and object model is also practiced by many fuzzy researchers. Particularly for those who are involved in clustering techniques for the elicitation of membership functions and fuzzy rules fall into this category. Furthermore, they allow the experts to review the results of the clustering techniques and modify them with their experience based mental models.

In level 4 of the hierarchy, we find theories that adopt objectivist epistemology on Level 3 and then usually embrace a "correspondence theory of truth" on the Level 4. On the other hand, those that are less objectivistic will tend to rely on "coherence" or "pragmatic" theories for their general validation criteria.

We have also pointed out our stance that differ from most of the fuzzy researches, On the General Epistemological level, we have proposed our own unique representation of the combination of concepts that generate Interval-Valued Type 2 schemas from the combination of Type 1 schemas by exposing “uncertainty” generated by the combination of Type 1 fuzzy sets with linguistic connectives. Whereas most of the current fuzzy researchers who are application and/or theory oriented take up the stance that combination of two Type 1 fuzzy concepts still generate a Type 1 fuzzy concept where they assume that linguistic “AND”, “OR” correspond isomorphically to a t-norm or co-norm, respectively. Our approach entails a unique approximate reasoning schema of Generalized Modus Ponens and in general a unique Type 2 fuzzy theory which expose associated uncertainties. It should be noted that all these result are the unique Interval-valued interpretation of meta-linguistic concepts based on the stance taken on the Ontological Level.

On the Domain-Specific Epistemological Level 5, One asks several question such as: "What are the limits of what one can possibly acquire (or extract), represent and reason with within the known Domain-specific fuzzy theory?" We may find different answers offered by a variety of theories we have at our disposal. An engineer with classical perspective might argue that all we can really know is what we can empirically observe or measure. Therefore, this is the sort of data that is most important to us and it should be taken as hard evidence. In contrasts, an engineer with a fuzzy theory perspective usually takes the stand that data assessed by perception of the knowledge domain expert or aided by fuzzy cluster analysis exposes degrees of data belonging to fuzzy patterns (clusters). Thus our knowledge of system behavior are generally enhanced by extracting hidden rules of behavior that are graded and capture hidden uncertainties.

Thus, the case of fuzzy theory emphasizes interaction of data elements in fuzzy clusters and subjective expert assessment of validity at the Domain -Specific Epistemological Level 5. This allows "cognitive insights" of experts to help us get a better approximation to a system behavior.

In Level 6, the appropriate sets of validation criteria and methodologies that are endorsed by a particular theory will tend to be tightly tied to with the assumptions made on the preceding Level 5.

At this level, one needs to question, for example, whether t-norms and t-conorms properly represent "AND", "OR", etc., linguistic connectives in a one-to-one correspondents. Alternately, as I have done in most of my work, one argues that t-norms and t-conorms do not effectively represent the linguistic "AND", "OR", etc., connectives. This case also brought to surface by Zimmermann and Zysno (1980) experiments and many other linguistic studies that demonstrate the graded meanings of "AND" ness, "OR" ness, etc., in human natural language usage. Therefore we have to come up with alternate representations to specify what is meant by "AND", "OR", etc. In this regard, I have proposed that FDCF, Fuzzy Disjunctive Canonical Forms, and FCCF, Fuzzy Conjunctive Canonical Forms might be used as a starting point of departure in the right direction.

However, it is unfortunate that in most "Domain Specific" validation exercises, most of us are still using RMSE, $R^2$, accuracy and power of predictions measures that are classical theory based. I encourage and challenge future researcher to develop fuzzy versions of these validation measures.
Finally, having dealt with each of the preceding levels, we then reach the concerns of the Application Level 7, itself. It is here that we find ourselves on our most familiar ground. It is on this level that the well-known differences of emphasis and techniques among the various fuzzy theories are traditionally debated.

At the Application Level 7, we assess the comparison of: (1) the purely subjective, expert based specifications of fuzzy set membership functions and rules vs. (2) the purely objective, fuzzy clustering based specification of membership functions and rules vs. (3) the interactive models that take into account both the subjective and the objective models. And finally at this level, we also confront the fact that membership acquisition gives us basically a Type 2 membership, instead of a Type 1 membership which should not be avoided in application domains.

Our general perspective is that such comparison and discussions will generate deeper and enriched quality insights within the broader context of our unified "philosophical paradigm" development upon which each of them is to rest.

### *0.2.11. Theory Development*

One last possible usage of the hieratical method is the employment of this framework for either: (i) further development of existent fuzzy theories or (ii) to build new ones, i.e., CWW and CWP. It is suggested that by proceeding in a step-wise manner up the levels of the proposed hierarchy one might be able to develop a fuzzy theory that is particularly strong in terms of its clarity and explicitness about the assumptions we make in our work. The philosophical directness and openness of such an approach might also provide a fuzzy theory developed in this manner with a high degree of internal consistency and coherence for CWW and CWP. In this regard, in Chapter 17, we propose the start of a new discussion on the axiomatic foundations of fuzzy theory within the context of CWW and CWP.

### *0.2.12. Future Work and Conclusions*

I find this hieratical approach to be quite useful in exploring the philosophical grounding of a number of major contemporary fuzzy theories. Some of the theories that I have proposed and investigated with this hierarchical approach are: (i) classical set and logic theory where set description D and set verification V are specified by classical two valued set and logic theories, i.e., $\{D_{\{0,1\}}, V_{\{0,1\}}\}$ set; (ii) set description is fuzzy but verification is crisp, i.e., $\{D_{[0,1]}, V_{\{0,1\}}\}$. The $\{D_{[0,1]}, V_{\{0,1\}}\}$ theory mostly structured around the assumption that two Type 1 fuzzy set combination entail a new Type 1 fuzzy set. This is the common practice in most of the current fuzzy set applications. I have proposed that the combination of two Type 1 fuzzy sets entail at least an internal-valued Type 2 fuzzy set whose upper and lower bounds are specified by FCCF and FDCF, respectively. This approach opens the potential of fuzzy theory by exposing the uncertainty entailed in fuzzy reasoning.

We all know that in the acquisition of memberships, we are confronted with Type 2 fuzzy sets represented either simply as Interval-Valued Type 2 fuzzy sets or full Type 2 fuzzy sets that exposes the distribution of membership values between the upper and lower membership values by second order fuzzy sets. Naturally, there is much work required to expose all the assumptions that are made in the development and analysis of Type 2 fuzzy sets.

Next, we ought to investigate fully, the philosophical grounding of the theories where the descriptions are given either crisply or fuzzily but the verification is made fuzzily which in turn require yet another verification crisply, i.e., $\{\{\{D_{\{0,1\}}$ or $D_{[0,1]}\}\ V_{[0,1]}\}\ V'_{\{0,1\}}\}$. Such theories are concerned with fuzzy truthoods of descriptions which are inturn verified crisply (Resconi and Türkşen, 2001). Again much future work is needed to clarify the grounding of these theories.

Furthermore, much work is needed in the domains where we combine fuzziness with Probability, Possibility, Belief and Plausibility measures. It is generally known that investigations in these domains are limited and applications are rather scarce.

*0.2.12.1. Conclusions*

In order to expose the philosophical grounding of fuzzy theory we proposed a hierarchical level of theoretical inquiry. We have shown that such a hierarchy exposes our stances with respect to crisp and fuzzy set and logic theories. It is hoped that the proposed hierarchy of our philosophical grounding will help both the researchers and the practitioners of fuzzy sets and logics to determine their particular stance on basic issues of concern.

**Appendix 1. Scientific and Philosophical Contribution of L.A. ZADEH**

L. A. Zadeh's initial contributions expressed in his seminal papers, e.g., "Fuzzy Sets" (1965), "Probability Measures of Fuzzy Events" (1968), "Outline of a New Approach to the Analysis of Complex Systems..." (1973), "Fuzzy Sets as a Basis for a Theory of Possibility" (1978), "Theory of Approximate Reasoning" (1979), etc., had a direct impact on mathematics, science and engineering and caused the development of fuzzy-neural system modeling and creation of novel solutions for electro-mechanical systems.

However, I believe, his main thesis has been that, in humanistic systems, human reasoning and decision making is not just "measurement" based, although it is an important component, but it is rather linguistic and perception based. The concepts and notions embedded in his celebrated papers, e.g., "Concept of a Linguistic Variable..." (1975-1976), "The Role of Fuzzy Logic..." (1983), "Syllogistic Reasoning..." (1985), "Computing with Words" (1996-2001), suggested novel approaches to complex humanistic systems for their potential analysis and solutions in humanistic terms in manners akin to "Human-like Reasoning".

In order to understand the deeper meanings of Fuzzy Theory, let us delve very briefly into the etymological origins of the word fuzzy.

R. Hodge (2001) states that " 'Fuzzy Logic' was born out of Zadeh's acute sense of different logic(s) inherent in human languages. ... his concern for the strengths as well as weaknesses of natural languages in scientific thought...". Zadeh uses `fuzzy` "to apply to categories of language or thought, not to the nature of (physical, mechanistic) phenomena". His use of "fuzzy" is an "example of his genius with language" with a background in " Indo-European languages", i.e., Russian and Iranian, and Turki languages, i.e., Azeri-Turkish.

Hodge (2001) further states that 'fuzzy' comes from a word "fusus" that refers to fire and water and their effects, to energies particles as well as liquids: to a

world with unstable outlines, a world in flux. The family of English words that descend from it reflects the range of meanings of "fusus". They include 'infuse', 'con-fuse' and 'transfuse' from 'melted or joined' with 'de-fuse' part of the same branch ...(from) remove; diffuse' from 'spread out, extended' and 'profuse' and 'effuse' from 'pour out in abundance'. All these words are formed by the addition of a prefix to 'fusus', to limit or constrain the fuzzy range of meanings of fusus to a more specific (but still somewhat fuzzy) meaning.
In general, initially, "Zadeh used 'fuzzy' to apply to categories of language and thought, [but] not to the nature of phenomena. However, recently, he has developed a typology of edges (boundaries or borders in phenomena, not in categories) in terms of categories that describe them" (Hodge, 2001).

This typology of edges, in terms of categories, is introduced in "Toward a Theory of Information Granulation..." (1997). For example, Zadeh provides description of how humans perceive and identify in categorical terms such natural phenomena as nose, cheek, etc., of a human face with fuzzy boundaries.

Edges, which are distinct, can be described with fuzzy membership functions. This is the basis for Type 1 fuzzy theory. But the edges that can be described with fuzzy membership categories can be represented with Type 2 or Type 3, etc., fuzzy theories. That is "crisp" edges can be represented with Type 1 membership functions, where as "fuzzy" edges can be represented with Type 2 or higher levels of fuzziness.

These Type 2 or higher levels of fuzzy membership functions represent membership of membership values, i.e., imprecision of imprecise membership functions and contain uncertainty associated with membership values for varying shades of meaning in words. In real life human communication words have imprecise meanings, sometime known as vagueness, ambiguity, etc., even in a given context such as human decision processes and descriptions of natural phenomena.

In particular, Type 2, and higher levels of fuzzy theory expose risks associated with managerial decision making in OR and MS studies. (Türkşen, 1986, 2001).

It should be noted that there are two prior aspects to Zadeh's contributions.
First of all, let us recall that Zadeh came, I believe, to realize that there are limitations placed on scientific though process by the classical paradigms while he was working on his publications that appeared prior to 1965, e.g., "Linear Systems Theory-The State Space Approach"(1963) "frequency analysis"(1950), "Wiener's theory of prediction"(1950), "sample-data systems"(1952), etc.

Secondly, his concern therefore with human decision-making processes in scientific thought brought him in contact and initially in conflict with the defenders of Probability theory and later with OR and MS. (I remember OR Conferences where Professor Zadeh was an invited guest speaker addressing large audiences.) (A further personal note: I was introduced to Professor Zadeh in summer 1970 at NATO Conference on OR Education that was held in Istanbul, Turkey, by R. Machol, then the President of IFSA, Operation Research Society of America.)

Let me next attempt to articulate Zadeh's scientific and philosophical contributions in a bit more detail.

In his seminal paper "Fuzzy Sets" (1965), Zadeh introduced the notion of "...a continuum of grades of membership" along with the "complement", "containment", "union" and "intersection" operations with "Max-Min" and with "Algebraic sum and Product", as well as "convex combination of " fuzzy sets, "Fuzzy sets induced by mappings", and "separation of convex fuzzy sets".

Implicit within these introductory concepts are the relaxation of the "Law of Excluded Middle", LEM, and its dual the "Law of Contradiction", LC. This naturally is unacceptable in classical set theory. But, it is known throughout human struggle, one needs to break away from traditional ways of thinking for the discovery and development of novel theories. L.A. Zadeh, in fact, broke away from the essential axiom of the classical theory in 1965.

In "Probability Measures of Fuzzy Events", he introduces the notion of a fuzzy event with examples such as "It is a *warm* day", "X is *approximately* equal to 5", "in twenty tosses of a coin there are *several* more heads than tails". These expressions "are fuzzy because of the imprecision of the meaning of the underlined words" (1968). Moreover, he generalizes the mathematical expressions of mean, variance and entropy in probability theory to "the mean, variance, ...(and) entropy of a fuzzy event...".

In "Decision-Making in a Fuzzy Environment", Bellman and Zadeh (1970), they introduce "...a decision process in which the goals and/or constraints, but not necessarily the system under control, are fuzzy in nature".

Furthermore, they illustrate a new framework "...by examples involving multistage decision processes...".

In "Similarity Relations and Fuzzy Orderings", Zadeh (1971) discuss the *similarity relation* to be "a fuzzy relation which is reflexive, symmetric and transitive" together with "*fuzzy linear ordering* ... fuzzy preordering, (and) fuzzy weak ordering...".

Until the early 70's, Zadeh and his follower's essentially developed the foundations of fuzzy mathematics of fuzzy sets. This may be considered the first stage of fuzzy theory which had no practical applications.

In "Outline off a New Approach to Analysis of Complex Systems and Decision Process" (1973), he introduces the concepts of "A linguistic variable..." and "the compositional rule of Inference...". He then stresses that this new "...approach provides an approximate and yet effective means of describing the behaviour of systems which are too complex or too ill-defined to admit of precise mathematical analysis. Its main applications lie in economics, management science, artificial intelligence, psychology, linguistics, information retrieval, medicine, biology and other fields in which the dominant role is played by the animate rather than inanimate behavior of system constituents". In this paper, we are also introduced to the notion of "Computation of the Meaning of Values of a Linguistic Variables".

"The Outline of a New Approach..." is a landmark paper. It is on the bases of this paper, Mamdani and Assilian (1975) developed first practical laboratory version of an applied fuzzy system model and its use in industrial fuzzy control. This gave rise to the wide spread "fuzzy control" application in electro-mechanical systems.

In "A Fuzzy-Algorithmic Approach to the Definition of Complex or Imprecise Concepts" (1976), we read "The high standards of precision which prevail in mathematics, physics, chemistry, engineering and other 'hard' sciences stand in sharp contrast to the imprecision which pervades much of sociology, psychology, political science, history, philosophy, linguistics, anthropology, literature, art and related fields". In this paper, we also find the definitions of "fuzzy truth", as well as S and Π membership functions.

As well, in this paper, we find an exposition on the relation between classification and attribution questions, their analytical representations together with a graphical interpretation that demonstrates "cylindrical extension" and "projection" which shows what we recently come to recognize as "projection anomaly" in fuzzy clustering techniques in fuzzy system modeling. (Uncu, Türkşen, 2001).

In "Fuzzy Sets as a Basis for a Theory of Possibility" (1978), he states that "...when our main concern is with the meaning of information-rather than with its measure (in Weiner and Shannon sense of the statistical theory of Communication) the proper framework for information analysis is possibilistic rather than probabilistic in nature...". In this paper, we are introduced to "the concept of a possibility distribution", "possibility measure", truth qualification", "probability qualification", etc.

In his celebrated papers, "The Concept of a Linguistic Variable and Its Application to Approximate Reasoning – I, II, III, (1975-1976), we read, "By a *linguistic variable* we mean a variable whose values are words or sentences in a natural or artificial language."

It is stated that "Given our veneration for what is precise, rigorous and quantitative, and our disdain for what is fuzzy, unrigorous and qualitative, it is not surprising that the advent of digital computers... have proved highly effective in dealing with *mechanistic* systems, that is, with inanimate systems whose behavior is governed by the laws of mechanics, physics, chemistry, and electromagnetism. Unfortunately, the same cannot be said about *humanistic* systems,...".

Professor Zadeh indicates that "...the ineffectiveness of computers in dealing with humanistic systems is a manifestation of what might be called the *principle of incompatibility* – a principle which asserts that high precision is incompatible with high complexity.

In his paper on "the Concept of a Linguistic Variable...", term sets are specified for *Age*, Appearance, Truth, and Probability, etc.

As well we find the notions of *interaction* and *non-interaction* of fuzzy sets, linguistic variables and their linguistic terms, the *extension principle*, *type n*, n = 2, 3,... fuzzy sets, *linguistic truth variables* and *fuzzy logic*, *Truth Tables* and *Linguistic Approximation*, *Linguistic Probabilities* and *their computations*, *composition rule of inference* which was later identified as GMP, etc.

In "A Theory of Approximate reasoning" (1979) rules of inference and approximate reasoning are further discussed in detail as projection principle, entailment principle, semantic equivalence, etc.

In "The Role of Fuzzy Logic in the Management of Uncertainty in Expert Systems", (1983) which is dedicated to Prof. Eli Sanchez, it is stressed that "Management of uncertainty is an intrinsically important issue in the design of expert systems because much of the information in the knowledge base of a typical expert system is imprecise, incomplete or not totally reliable".

In this article, it is shown that there are effects of fuzziness in facts and rules. Types of proposition are discussed with inference in fuzzy logic. Translation rules are revisited. We are also introduced to inference from quantified propositions", e.g., $Q_1$A's are B's, and $Q_2$ (A and B)'s are C's $\geq Q_1 \otimes Q_2$) A's are C's, etc.

Next, we find "Syllogistic Reasoning in Fuzzy Logic and Its Application to Usuality and Reasoning with dispositions" (1985). In this paper, Zadeh views fuzzy logic "...as a generalization of multivalued logic in that it provides a wider range of tools for dealing with uncertainty and imprecision in knowledge representation, inference, and decision analysis. Such topics as "intersection/product syllogism", dispositional Modus Ponens". In this paper, they are treated together with "fuzzy quantifiers", "compositionality", "robustness", and "usuality". In this context, we observe the use of Σ count (.) in many examples. Fuzzy syllogisms and reasoning with dispositions are amplified with "MPR(Major Premise Reversibility) chain Syllogism", "Antecedent Conjunction Syllogism", "Consequent Conjunction Syllogism", etc.

These and other related issues are treated further in "A Computational Approach to Fuzzy Quantifiers in Natural Languages"(1983). Unfortunately, works in this area are very few, e.g., Narazaki and Türkşen. (1994)

Next, Zadeh proposes "A Theory of Commonsense Knowledge"(1984). He states, "The conventional knowledge representation techniques based on the use predicate calculus and related methods are not well-suited for the representation of commonsense knowledge because the predicates in propositions which represent commonsense knowledge do not, in general, have crisp denotations. For example, the proposition *Most Frenchmen are not tall* cannot be represented as a well-formed formula in predicate calculus because the sets which constitute the denotations of the predicate *tall* and the quantifier *most* in their respective universes of discourse are fuzzy rather than crisp. "Meaning representation" is further discussed with "Test-Score Semantic", "Composition of Elastic Constraints", together with "Rules pertaining to modification... composition... quantification" as well as "representation of dispositions" with "inference from dispositions" and the applications of "sigma count" and relative sigma count."

In a series of papers, "Fuzzy Logic=Computing with Words" (1996) "Toward a theory of fuzzy information granulation and its centrality in human reasoning and fuzzy logic" (1997), and "From Computing with Numbers to Computing with Words –from manipulation of measurements to manipulation of perceptions" (2001), Zadeh writes "...the main contribution of fuzzy logic is a methodology for computing with words. No other methodology serves this purpose". He goes on to state "in its traditional sense, computing involves ... manipulation of numbers and symbols. By contrast, humans employ mostly words in computing and reasoning, arriving at conclusions expressed as words from premises expressed in a natural language or having the form of mental perception".

He, then, traces the origins of this development stating "The concept of CWW is rooted in several papers starting with ["Outline of a New Approach..."(1973)] in which the concept of a linguistic variable and granulation were introduced. The concepts of fuzzy constraint and fuzzy constraint propagation were introduced in ["Calculus of Fuzzy Restrictions"(1975)], and developed more fully in ["A Theory of Approximate Reasoning"(1979)], etc.

In these works, there are schemas that show how one gets started with the notion of granulation and first arrive at information and action granules and then apply divide and conquer principle. Next, one identifies crisp and fuzzy information granules, CIG, FIG. The examples of CIG are given as "time → years → months → weeks → days → ..." and FIG as "age → very young + young + middle-aged + old + very old".

As well, examples of mental and physical granulation are demonstrated as – mental granulation: "body → head + neck + left arm + chest + right arm + ..." and physical granulation: "speech, walking, eating."

The generalized constraint that was introduced in previous papers are re-stated as "X isr R, where isr (pronounced ezar) is a variable copula which defines the way in which R constrains X."

The role of R in relation to X in defined by the value of the discrete variable r where r could take on values:

" e: equal (abbreviated to =)
d: disjunctive (possibilistic)
(abbreviated to blank)
c: conjunctive
p: probabilistic
λ: probabilistic value
u: usuality
rs: random set
rsf: random fuzzy set
fg: fuzzy graph
ps: rough set (Pawlak set)"

Next, we are exposed to fuzzy constraint propagation and the rules of inference in fuzzy logic under the headings of: Conjunctive Rule 1, Conjunctive Rule 2, Disjunctive Rule 1, Disjunctive Rule 2, Conjunctive Rule, Projective Rule, Subjective Rule, as well as Derived Rules under the headings of: Compositional Rules, Extension Principle (Mapping Rule), Inverse Mapping Rule, Generalized Modus Ponens, Generalized Extension Principle, Syllogistic

Rule, Constraint Modification Rule, etc. Once again we are treated with the example of "Balls in a Box".

"A box constrains ten balls of various sizes of which several are large and a few are small. What is the probability that a ball drawn at random is neither large nor small?"

"To be able to answer this question, it is necessary to be able to define the meanings of *large*, *small*, *several large balls*, *few small balls*, and *neither large nor small*. This is a problem in semantics, which falls outside the probabilistic theory, neurocomputing and other methodologies.

Zadeh, thus concludes "In our quest for ... Machine intelligence (high MIQ), we are developing a better understanding of the fundamental importance of the remarkable human capacity to perform a wide variety of physical and mental tasks without any measurements and any computations. Underlying this remarkable capability is the brain's crucial ability to manipulate perceptions – perceptions of distance, size, weight, force, color, numbers, likelihood, truth and other characteristic of physical and mental objects. A basic difference between perceptions and measurements is that, in general, measurements are crisp whereas perceptions are fuzzy."..."Humans employ words to describe perceptions ... (in this regard) ... manipulation of perceptions is reduced to computing with words... In coming years, computing with words and perceptions is likely to emerge as an important direction in Science and Technology."

More recently, Zadeh began to suggest the construction of "Protoforms" for CWW (Fuz-IEEE'2001).

## Appendix 2. Fuzzy Systems Applications in Operations Research, Management Science and Decision Support Systems

In order to understand, the development of the "Fuzzy Systems Applications in Operations Research, OR, Management Science, MS, and Decision Support Systems, DSS", it is essential to comprehend the message that Professor Zadeh has been attempting to communicate to us. Naturally what follows is my personal perspective based on my re-reading his papers and listening and re-assessing his many lectures in numerous conferences. I believe his main thesis has been that, in humanistic systems, human reasoning and decision making is not just "measurement" based, as we are taught through out our academic education, rather "perception" based. Furthermore, our knowledge and decisions are communicated between humans with words of natural languages which strongly suggest "Computing with Words", CWW. It is, I believe, this thesis, that resided in the deep recesses of his genius and commenced with

"Fuzzy Sets" in 1965 and came to surface toward the beginning of this Millennium in "Toward a Perception-Based Theory... " (Zadeh, 2000).

Briefly, it can be said that Professor Zadeh initially thought that fuzzy theory would impact human decision making processes and therefore what appeared to be a starting point for this would be OR and MS. Because OR and MS began to introduce mathematical models and solutions to humanistic systems at the end of WWII.

When we trace the successes of fuzzy logic, we find that initially it started to impact OR and MS investigations but later the successes in fuzzy control realized in Japan and later in Europe and North America have shadowed its impact on Decision Sciences. Thus the fuzzy logic applications in management decision support area were fewer in comparison to many successes in fuzzy control.

It should be noted that there are distinctions between fuzzy control and OR, MS, and DSS models.

In fuzzy control:

(a) usually input/output variables are known, such as mass, acceleration, torque, etc.
(b) there may even be a model such as PID, etc., be it solvable or unsolvable but well stated in classical approaches.
(c) Usually, the initial conditions and the target is well known.
Whereas in OR, MS and DSS:
(a) usually input/output variables are not known there may be hundreds or thousand of variables where some are nominal or ordinal as well as scalar; and
(b) there may not even be any model at all other than some initial guesses.
(c) Usually initial conditional and targets are not well defined or defined very vaguely.

In this sense, fuzzy control problems are, relatively speaking, an easier class of problems as compared to OR, MS, and DSS problems.
Consider for example fuzzy control of rigid-link robot arm. Its dynamics can be expressed as:

$$M(\dot{q})q + N(q,\dot{q}) + \tau_d = \tau \tag{0.1}$$

where

$$N(q,\dot{q}) \equiv C(q,\dot{q})\dot{q} + F(\dot{q}) + G(q) \tag{0.2}$$

is the vector of the nonlinear terms; in (1), M(q) is the inertia matrix and $\tau_d(t)$ represents disturbances. In (0.2), $C(q,\dot{q})$ is the coriolis centripetal matrix, $F(\dot{q})$ the friction terms, and G(q) the gravity vector; the control input $\tau(t)$ is the vector of joint torques.

The main objective in motion control of manipulators is to make the robot follow a *prescribed desired trajectory* $q_d(t)$. Define the tracking error e(t) and the filtered tracking error r(t) as:

$$e = q_d - q \qquad (0.3)$$

$$r = \dot{e} + \Lambda e \qquad (0.4)$$

with $\Lambda>0$ a positive matrix.

On the other hand in fuzzy OR, MS, DSS, variables, dynamics of the system and its trajectory are not well known, but ill-defined. the goal or objective is given in general; e.g., client or customer preference, cost minimization, revenue maximization, etc., critical and significant variable must be selected from among an ill-defined set of variables, system behavior model must be discovered, forecast of system response be made for future client and customer requests.

Consider for example fuzzy OR, MS, and DSS for a “fuzzy system modeling” with supervised and reinforcement learning for bandwidth allocation and QOS routing to provide service level agreements, SLA's, between end users and access providers, between ISP's, as well between 2 layers of IT networks, i.e., between the IP layer and optical layer. In these cases, there are generally two main issues:

(i) Service provisioning, i.e., how one configures the network. One must determine decision support models to reserve the necessary resources, set the required parameters on the devices, etc., based on the SLA signed between a user and a provider.

(ii) Service assurance, i.e., how does one monitor the service provided by the network and eventually detect SLA violations, or even anticipate such problems. In this case one is to observe and identify data collection from the network resources, i.e., determine the relevant metrics, and also processing algorithms that will help diagnose troubles with the service. This implies a direct link between a service and the network measures which are not very obvious.

OR and MS studies started with the application of mathematical models to strategic and tactical military operations related to military decision making processes, during WWII. Later, OR and MS studies became an essential component of the curriculums in Departments of Industrial Engineering, Operations Research, OR, and Management Sciences, MS, and Systems Management Engendering at the Universities starting in the '50s.

There are good many operational issues that are investigated in OR and MS with the purpose of discovering and disseminating knowledge about planned, coordinated and controlled activities of people, machines, materials, money, energy and information. Some of these include planning, allocation, and distribution of resources; analyzing, scheduling and controlling these activities; analysis and control of the quality of goods. Some typical examples are planning production of goods, spare parts, inventory control, scheduling processes, tasks and orders, materials, information, energy, and capital; analysis, planning and controlling waiting lines, forecasting and predicting demand, marketing and assessment of consumer and/or client preferences and demand patterns, etc. These operational and logistic issues need to be dealt with in every activity of every day human life by ecutive and managerial decision-making in manufacturing, process and production industries, in healthcare systems, in government agencies, in financial institution, etc. Some of the essential methodologies that are utilized in OR and MS are: Linear and non-linear programming, probability, statistics, and stochastic process, mathematics, i.e., set and logic theories, algebra and calculus, etc.

About ten years after his seminal paper, Professor Zadeh gave further insights with his interpretation of probability theory when he introduced the possibility theory in "Fuzzy Sets as a Basis for a Theory of Possibility"(1978). In that thesis, he advocates that "...when our main concern is with the meaning of information – rather than with its measure (in Wiener and Shannon sense of the statistical theory of communication)- the proper framework for information analysis is possibilistic rather than probabilistic in nature...".

It can be said that the first impact of fuzzy theory was demonstrated in "Decision-making in a Fuzzy Environment" by R.E. Bellman and L.A. Zadeh(1970) where it was stated that "By decision-making in a fuzzy environment is meant a decision process in which the goals and/or the constraints, but not necessarily the system under control are fuzzy in nature". Further, it is stated that "The use of these concepts is illustrated by examples involving multi-stage decision processes in which the system under control is either deterministic or stochastic".

Professor Zadeh next published his very important work "Outline of a new Approach to the Analysis of Complex Systems and Decision Processes"(1973).

He described this "approach...(to be)...a substantive departure from the conventional quantitative techniques of system analysis"...(which) "has three main distinguishing features: 1) Use of so called 'linguistic' variables in place of or in addition to numerical variables; 2) characterization of simple relations between variables by fuzzy conditional statements; and 3) characterization of complex relations by fuzzy algorithms."

In 1975, Professor Zadeh, in his celebrated paper, "Concept of a Linguistic Variable...", states: "One of the fundamental tenets of modern science is that a phenomenon can not be claimed to be well understood until it can be characterized in quantitative terms". He further states "Unquestionably...(this has) proved to be highly effective in dealing with <u>mechanistic</u> systems, that is, with inanimate systems whose behavior is governed by laws of mechanics, physics, chemistry and electromagnetism. Unfortunately, the same cannot be said about <u>humanistic</u> systems, which-so far at least-have proved to be rather impervious to mathematical analysis and computer simulation"..."It may be argued, as we have done in...(previous writings), that the ineffectiveness of computers in dealing with humanistic systems is a manifestation of what might be called the <u>principle of incompatibility</u>- a principle which asserts that high precision is incompatible with high complexity"..."In retreating from precision in the face of overpowering complexity, it is natural to explore the use of what might be called <u>linguistic</u> variables, that is, variables whose values are not numbers but words or sentences in a natural or artificial language"..."What is...important, ..., is that by use of so-called <u>extension</u> <u>principle</u>, much of the existing mathematical apparatus of systems analysis can be adapted to the manipulation of linguistic variables. In this way, we may be able to develop an approximate calculus of linguistic variables which could be of use in wide variety of practical applications". This may be considered the essential message for the beginning of fuzzy systems applications in OR and MS. It is at about this juncture we begin to see works on "optimization of fuzzy Systems", "fuzzy mathematical programming "(Zimmermann, 1978), etc. The years between 1965 and 1975 may be considered incubation years where essential basic works on fuzzy mathematics, cognitive and decision process were being developed by Kaufmann (1975), Zadeh, Fu, Tanaka, Shimura(1975), Neogita and Ralescu, (1975).

Next, Zadeh published "A Theory of Approximate Reasoning" (1979) where he states that "Informally, by approximate or, equivalently, <u>fuzzy reasoning</u>, we mean the process or processes by which a possibly imprecise conclusion is deduced from a collection of imprecise premises. Such reasoning, is, for the most part, qualitative rather than quantitative in nature, and almost all of it falls outside of the domain of applicability of classical logic."

In 1983, Zadeh published "The Role of Fuzzy Logic in the Management of Uncertainty in Expert Systems" where he states "...the conventional approaches to the management of uncertainty in expert systems are intrinsically inadequate because they fail to come to grips with the fact that much of the uncertainty in such systems is possibilistic rather than probabilistic in nature. As an alternative, it is suggested that a fuzzy-logic-based computational framework be employed to deal with both possibilistic and probabilistic uncertainty within a single conceptual System".

During this period, we begin to see substantial works on "decision making and expert systems" and their applications to OR and MS, e.g., Zimmermann, 1987; Negoita, 1981, 1983. There were naturally many additional works on fuzzy mathematics and fuzzy optimization, as well, measurement of membership functions came forward at this period, e.g., D.Dubois and H.Prade, 1980; Negoita and Stefanescu, 1982; Gupta and Sanchez, 1982; Norwich and Türkşen, 1981, 1982; Kaufmann and Gupta, 1985.

In these pioneering works we observe investigations on: membership functions, fuzzy relations, fuzzy logic and inference, classification and similarity measures, expert systems, medical diagnosis, psychological measurements and human behaviour, fuzzy clustering algorithms, individual and group decision-making in fuzzy environments, fuzzy mathematical programming, multi-criteria decision-making, and decision support systems.

In more recent years, we find there are quite a few fuzzy theory applications in OR and MS. Examples of these are "Quality Control and Maintenance", "Ecological Modeling and Data Analysis" "Fuzzy Logic and Possibility Theory in Biomedical Engineering", "...Computer Aided Medical Decision Systems", "Strategic Planning", "Decision and Planning in R&D", "Production Planning and Scheduling", "Fuzzy Sets Methodologies in Actuarial Sciences", "Fuzzy Sets in Human Factors and Ergonomics", "...Software Methodology and Design Tools"(Zimmermann, 1999). As well, we find: "Retrieving Information", " Decision-Making", "Designing and Optimization" (Dubois, Prade and Yager, 1997), "Scheduling Under Fuzziness" (Slowinski, Hapke, 2000), "Optimization and Decision" (FSS, Vol. 119, No 1, 2001).

Thus, we observe that there is a resurgence of fuzzy theory applications in OR and MS. However, we need to work on further developments of fuzzy theory in particular on Type 2 fuzzy knowledge representation and reasoning. This is more acutely needed in the development of humanistic decision making domains which Professor Zadeh have been urging us to direct our attention over the last thirty five years or so.

Over the years: it should be noted that the skeptics of fuzzy theory have been asking two questions:

(i) If fuzziness is to deal with imprecision why are the membership functions so precise?

(ii) Why do fuzzy theory use the same formulas of the two-valued theory for "AND", "OR", "IMP", etc.?

Type 2 theory responds positively to these questions: (i) Type 2 membership exposes uncertainty in the acquisition of membership functions and as well as a representation of perception with information granulation (Burillo and Bustince, 1995, 1996; Bilgic and Türkşen, 1997, 2000; Norwich and Türkşen, 1981, 1982, 1984; Karnik and Mendel, 1998, 1999, 2000; Liang and Mendel, 2000); (ii) Type 2 reasoning brings to surface the increase in uncertainty in the combination of imprecise concepts identified by words by the application of Fuzzy Disjunctive Canonical Forms, FDCF and Fuzzy Conjunctive Canonical Forms, FCCF.(Türkşen, 1999, 2001; Gerhrk, Walker and Walker, 2000; Resconi and Türkşen, 2001). Thus in Type 2 theory we can represent uncertainty more effectively and expose risks associated with decision making and hence provide a more effective tool for managerial decision making in OR and MS.

Furthermore, we respond to the skeptics who question "why the same formulas of the two-valued theory are directly used in fuzzy theory" by exposing the new formulas and new "Laws of Conservation" generated by Interval-Valued Type 2 Theory.

## References


1. Bellman, R., L.A. Zadeh (1970), "Decision-Making in a Fuzzy Environment", *Management Science*, 17:4, 141-164.
2. Bilgic, T., I.B., Türkşen (1995), "Measurement-theoretic Frameworks for Fuzzy Set Theory". In T.P. Martin and A.L. Ralescu, editors, Fuzzy Logic in Artificial Intelligence: Towards Intelligent Systems, Volume 1188 of Lecture Notes in Artificial Intelligence, 252-265, Springer, 1997. Selected papers from IJCAI'95 Workshop Montreal, Canada, August.
3. Hodge, R. (2001), "Key Terms in Fuzzy Logic Deep Roots and New Understanding", *University of Western Sydney*, Australia (Submitted).
4. Norwich, A.M. I.B. Türkşen (1982), "Stochastic Fuzziness", in: M.M. Gupta and E.E. Sanches (eds.), *Fuzzy Information and Decision Processes*. North Holland, Amsterdam, 13-22.
5. Norwich, A.M., I.B.Türkşen (1982a), "The Construction of Membership Functions", in: R.R. Yager (ed.), *Fuzzy Sets and Possibility Theory*. Pergamon Press, New York, 61-67.

6. Norwich, A.M., I.B. Türkşen (1982b), "The Fundamental Measurement of Fuzziness", in: R.R. Yager (ed.), *Fuzzy Sets and Possibility Theory*. Pergamon Press, New York, 49-60.
7. Resconi, G., I.B. Türkşen (2001), "Canonical Forms of Fuzzy Truthoods by Meta-Theory Based Upon Modal Logic", *Information Sciences*, 131, 157-194.
8. Türkşen, I.B. (1999), "Theories of Set and Logic with Crisp or Fuzzy Information Granules", *J.of Advanced Computational Intelligence*, 3,4, 264-273.
9. Türkşen, I.B., A. Kandel, Y-Q. Zhang (1999), "Normal Forms of Fuzzy Middle and Fuzzy Contradiction", *IEEE-SMC, 29-2, Part B, Cybernetics*, 237-253.
10. Türkşen, I.B., A. Kandel, Y-Q. Zhang (1998), "Universal Truth Tables and Normal Forms", *IEEE-Fuzzy Systems*, 6-2, 295-303.
11. Türkşen, I.B. (1992), "Interval Valued Fuzzy Sets and 'Compensatory AND' ", *Fuzzy Sets and Systems*, 51, 295-307.
12. Wang, P.P. (ed.) (2001), Computing With Words, Wiley, New York (to appear).
13. Zadeh, L.A. (2001), "From Computing with Numbers to Computing with Words - From Manipulation of Measurements to Manipulation of Perceptions", in: P.P. Wang(ed.) *Computing With Words*, Wiley Series on Intelligent Systems, Wiley and Sons, New York, 35-68.
14. Zadeh, L.A. (2000), Computing With Perceptions, Keynote Address, *IEEE- Fuzzy Theory Conference*, San Antonio, May 7-10.
15. Zadeh, L.A. (1999), "From Computing with Numbers to Computing with Words—From Manipulation of Measurements to Manipulation of Perceptions", *IEEE-Trans on Curciuts and Systems*, 45, 105-119.
16. Zadeh, L.A. (1995), Keynote address, *IFSA Congress*, Sao Paulo, Brazil.
17. Dempster A.P (1967), "Upper and Lower Probabilities Induced by a Multivalued Mapping", In: *Annals of Mathematical Statistics*, 38, 325-339.
18. Türkşen I.B. (1986), "Interval-Valued Fuzzy Sets based on Normal Forms", *Fuzzy Sets and Systems,* 191-210.
19. Zimmermann, H.J., P. Zysno (1980), "Latent Connectives in Human Decision Making", *Fuzzy Sets and Systems*, 4, 37-51.
20. Türkşen I.B. (2001), "Computing with Descriptive and Veristic Words: Knowledge Representation and Reasoning", *in: Computing With Words*, P.P. Wang(ed.), Chapter 10, Wiley, New York, 297-328.
21. Türkşen I.B. (2002), "Interval-valued Type 2 Fuzzy Sets, Multi-valued Maps and Rough Sets", (in) A Grmela and N.E.Mastorakis (eds.), Advances in Intelligent Systems, *Fuzzy Sets*, Evolutionary Computation, WSEAS, 142-146.
22. Türkşen I.B. (2002), "Upper and Lower Set Formulas: A Modification of Dempster-Pawlak Formalism", *Special Issue of the International Journal of Applied Mathematics and Computer Science,* (submitted).
23. Zadeh, L.A. (1979), "A Theory of Approximate Reasoning", *in J. Hayes, D. Michie, and L.I. Mikulich (eds) Machine Intelligence*, Halstead Press, New York, Vol. 9, 149-194.
24. Zadeh, L.A. (1978), "Fuzzy Sets as a Basis for a Theory of Possibility", *Fuzzy Sets and Systems*, 3-28.
25. Zadeh, L.A. (1973), "Outline of a New Approach to the Analysis of Complex Systems and Decision Processes", *IEEE Trans. Systems Man Cybernet*, 3, 28-44.
26. Zadeh, L.A. (1971), "Similarity Relations and Fuzzy Ordering", *Information Sciences*, 3, 177-200.
27. Zadeh, L.A. (1968), "Probability Measures of Fuzzy Events", *J.Math. Analysis and Appl.*, 10, 421-427.

28. Zadeh, L.A. (1965), "Fuzzy Sets", *Information and Control Systems*, Vol.8, 338-353.
29. Zadeh, L.A., C.A. Desoer (1963), "Linear System Theory-The State Space Approach", *Mc Graw-Hill*, New York.

# Chapter 1

# INTRODUCTION

The ontological and epistemological foundations of fuzzy sets and fuzzy logics are reassessed from the perspective of equivalences of the classical theory which breaks down in fuzzy theory. In turn, it is shown that there are additional new formulae and laws of conservation which demonstrate in part the richness of fuzzy theory beyond the classical theory. In particular, combinations of linguistic variables with linguistic connectives generate part of this richness within the framework of CWW and CWP proposed by L.A. Zadeh.

Another unique perspective of this work is to emphasize the fact that theories of sets and logics are similar but separate even though they are used together in deriving the formulas of combined concepts within the scope of computing with words and information granules for the construction of "expressions". First, it should be noted that there are two separate disciplines in most universities under the headings of "set theories and their courses" which are generally included in a mathematics department and "logic theories and their courses" which are generally included in a philosophy department.

Formalization of logic theories has its origins in Aristolalian works and the formalization of set theory have its origins in Cantor's work. With Zadeh's (1965) seminal work, fuzzy set theory was introduced into the set theoretic studies. Later Zadeh (1973, 1975) introduced fuzzy logic as a foundation for approximate reasoning. Naturally Zadeh's contributions demonstrate the convergence and integration of many earlier investigations and contributions by many world renowned scientists such as Max Black and J. Lukasiewicz, etc.

It is found that in most of the current works in fuzzy theory, fuzzy set membership values and fuzzy logic values are assumed to be the same at times implicitly (in most works). At other times, this assumption is made more explicit. This causes a great deal of confusion in understanding and in restructuring the foundations of fuzzy theory for the construction of expression in CWW. In this regard, Zadeh (1997, Prague IFSA Congress) has pointed out that there are four aspects to fuzziness: (1) fuzzy sets, (2) fuzzy logics, (3) fuzzy relations, and (4) fuzzy semantics. Each of these aspects need to be studied and investigated on their own as well as in relation to others. More recently, Zadeh pointed out the necessity to construct "Precisiated Natural Language, PNL, expressions" for the implementation of CWW. In this work, we present some of the foundational concepts related to the ontological and epistemological aspects of fuzzy theories in general, and in particular to set and logic aspects, and their impact on the construction of "PNL expressions" for CWW. At the end, we propose a re-structuring the basic axioms of set and logic theory by expressing them linguistically in order to provide a foundation for CWW.

It appears that there has been an ongoing discussion amongst the mathematicians and logicians who attempt to distinguish the notions of set and logic theories and their essential normative foundations. For example, there is a response to Quine's (1970) statement that "*second-order logic (with standard semantics) is set theory in sheep's clothing*". In "Metaphysical Myths, Mathematical Practice", Azzouni (1994) points out that "*second order logic with standard semantics is hardly set theory, if by that phrase is meant* _*first-order set theory*_. *They simply don't have the same models*." The approach in this work is from a more practical point of view. That is the extraction and acquisition of membership values and functions of descriptive assignments from experts and/or from data-mining and knowledge discovery of input-output data vectors in order to capture concept definitions, i.e., semantic representation of concepts, and then their verification by independent observers. These are two separate but related assignments of membership values, i.e., descriptive and veristic assignments which address the epistemological foundation of fuzzy theories. They are used in the combination of concepts for the derivation of PNL expressions in order to represent our knowledge both qualitatively and quantitatively. Furthermore, such combination of concepts must allow one-to-many representation of linguistic connectives in order to capture our knowledge properly and to expose the non-linearity and the second order imprecision and its associated uncertainty that can be represented in fuzzy theories.

## 1.1. Description and Verity

In an attempt to identify the assignment of membership functions within the framework of "Computing With Words", (Zadeh, IFSA 1997 and IFSA 1995 Conferences) it needs to be pointed out that either there is an assignment of words to numbers i.e., membership values and hence functions, or there is an assignment of numbers, i.e., memberships values and hence functions, to words. But in either case after one defines or describes the characteristics of an object (entity or event) by these two types of assignments that have been indicated above, one has to affirm or deny, i.e., verify, in the sense of epistemology, the given definition or description by a second assignment, which is generally from words to numbers. The natural language expression that affirms or denies, i.e., verifies, a given natural language expression – a given sentence – forms the basis of logic a assignment asserting the given sentence's degree of validity or

reliability or correctness or one's confidence in that sentence, in the sense of epistemology. In this regard, it should be recalled that logicians start every sentence with the expression "it is the case that ..." or "it is not the case that ...".
Whereas the natural language expression that defines or describes a phenomena, an occurrence, and object, or an abstract concept, in attempting to identify the linguistic term that captures the characteristic of that object, phenomena or occurrence, or abstract concept, to a degree, forms the basis of its descriptive set assignment. Thus there are potentially many basic theories that one could structure. We attempt to formulate a framework for several of these structures as a generalization of two-valued set and two-valued logic based classical framework and identify a "myopic" structure that is in most current works of fuzzy theory.

In natural languages, there are a number of word categories such as "nouns", "adjectives", "adverbs", "verbs", etc.

Nouns are labels that identify general classes of entities and/or their attributes be they concrete or abstract, such as "age", "height", "inventory", "profit", "people", or "number", "zero", "chairs", "building", etc., as well "truth", "belief", "probability", "possibility", "plausibility", "necessity".
Adjectives are modifiers that describe subclasses of entities and attributes that are identified by nouns, such as "old people", "tall buildings", "low inventories", "high profits", "high probability", "low possibility", etc., as well "probable", "true", "false", believable", "possible", "necessary", etc., "positive number", "small positive number", "positive zero", etc.

Adverbs are modifiers of adjectives that qualify subclasses of adjectives, such as "very tall buildings", "somewhat old people", "very low inventories", "very high profits", "very high probability", "very high possibility", etc., as well "highly probable", "very true", "very false", "somewhat true", "highly believable", "very necessary", "definitely possible", "more or less plausible", etc.
It should be recalled that adjectives and adverbs indicate the implicite fact that the nouns and their definitions have to be assessed as a matter of degree. They never imply an absolute meaning. Thus the "meaning representation" becomes an essential concern within the ontological and epistemological foundations of the fuzzy theories.

Verbs that are characteristically the grammatical center of predicates and express an act, occurrence, or mode of being, etc.

In fuzzy set theory, the concern is to identify membership assignments to subclasses of entities described by adjectives or subclasses that are qualified by adverbs in order to precisiate the imprecision caused by these modifiers within the context of PNL. Recall that, in fuzzy theory, nouns are called "linguistic variables" and the adjectives on their own or adjectives with adverbs make up the set of "linguistic terms" of a linguistic variable. Beyond this classification of words, we need to identify yet another classification of the linguistic terms:

(i) First there are expressions that contain linguistic terms that describe an entity, object, occurrence, etc., in a proposition which is formed with a subject (a noun), a verb (predicate) and at least an adjective or an adjective and an adverb, etc. For example:

"John is tall",
"Inventory is very high", etc.

(ii) Secondly, there are expressions that contain linguistic terms which are used to verify a given proposition that describes an attribute of a noun. For example:

"John is tall is true"; or "it is the case that 'John is tall'";
"'Inventory is low' is false"; or "it is not the case that 'inventory is low'", etc.

(iii) Thirdly there are linguistic terms that signify and describe our belief or confidence within the concern of epistemology in a proposition or a given piece of information. For example:
"'John is tall' is highly believable";
"It is highly believable that 'John is tall'";
"'Inventory is low' is rather unbelievable"
"It is rather unbelievable that the inventory is low";
"'John is tall' is somewhat probable";
"It is somewhat probable that 'John is tall'";
"It is highly probable that 'John is tall'";
"It is quite possible that 'John is tall'";
"It is quite plausible that 'John is tall'";
"Probability is very high that 'inventory is low'".

Naturally, there are more complex propositions with a build in hierarchy. For example:
"'John is tall' is highly believable" is true;
"'Inventory is low' is rather unbelievable" is true;
"It is somewhat probable that 'John is tall' is true;
"It is quite possible that 'John is tall' is false;
"Probability is very high that 'inventory is low' is false, etc.

(iv) We need to further note that sentences at times are subject to partial truth qualifications which in turn form truth qualifying sentences. Such sentences would also have to be verified for us to be able to use them in combination of truth qualifying expressions. For example:

"'John is tall' is very true" is true;

It should be noted that "John is tall" is a description of John; but "'John is tall' is very true" is an assessment of the sub-sentence "John is tall" in terms of fuzzy truth qualification. Finally we have:

"'John is tall' is very true" is true.

Such a statement is to be understood as:

"'John is tall' is very true" is verified by "is true".

This hierarchy of descriptive and logic verifications are shown in Diagram 1.1. Thus , we treat descriptive words in sentences that help us identify membership values in determination of fuzzy sets and veristic words that help us to form logical combination of fuzzy sets denoted by descriptive words in a hierarchy of protoform construct.

## 1.2. Nature of Truth

The nature of truth is one of central philosophical concerns in studies of logic. There appears to be three possibilities of handling the concept of truth:

State an explicit definition, or
State an implicit definition, or
Show that no explicit or implicit definition can be stated.

The most basic fact about the concept of truth appears to be that the statement which is formed by attaching the truth predicate to a sentence is true if and only if the sentence itself is true. Thus the essential question is "how does one verify" that "the sentence itself is true?" In the assignment of numbers, i.e., membership values, to words, we can state that:
"$x \in X$ isr A, $\mu_A(x) \in [0,1]$ is true", if and only if,

the sentence "$x \in X$ isr A, $\mu_A(x) \in [0,1]$" is itself true. (1.1)

i.e., the sentence "$x \in X$ isr A, $\mu_A(x) \in [0,1]$" is verifiable by an independent agent, where "isr" is to be interpreted as "belongs to", "is member of", "is compatible with", "is similar to", etc. (Zadeh, 1995).
Thus, how does an independent agent verify that "$x \in X$ isr A, $\mu_A(x) \in [0,1]$" is itself true?

In our work, it is assumed that either:

(1) there are experts who can verify that the sentence is itself true, or

(2) there are necessary and sufficient data vectors obtained from database files and/or sensor readings and an algorithm that constructs membership functions, i.e., $\mu_A(x) \in [0,1]$, by processing the data vectors, and thus via a constructive scheme, an analyst is able to verify that "x∈X isr A, $\mu_A(x) \in [0,1]$" is true. The expression (1.1) is known as T-sentence.

Philosophers have argued about whether or not T-schema is universally valid. Some have argued (Dummett, 1959, 1978; Davidson, 1967, 1969, 1984, 1990; Hintikka, 1975) that some of T-sentences are *not* universally true.

In turn Tarski (1932, 1944, 1956) stated that one can't express the concept of truth, i.e., the verification concept, in the same language that states the sentence of the descriptive concept . Thus there is no expressible absolute concept of truth applicable to all sentences of all languages.

Tarski's attempt is to demonstrate that any natural language does necessarily contain truth qualifying sentences but the "language can't contain its truth predicate".

Consider a natural language that contains truth qualifying sentences s∈S, such that it contains its truth predicate as follows:

$$T(s) \leftrightarrow s \qquad (1.2)$$

where s∈S, s a sentence in the set of sentences S and T is the truth predicate in a natural language L and s∈L, i.e., s is a sentence of L. In such cases, we are confronted with a natural contradiction. The reason is that one can also assume that L contains a 'core' sentence s for which it holds that

$$T(s) \leftrightarrow \neg s \qquad (1.3)$$

where $\neg s$ is the complement of s. Then naturally, we get $s \leftrightarrow \neg s$, i.e., crisp contradiction.

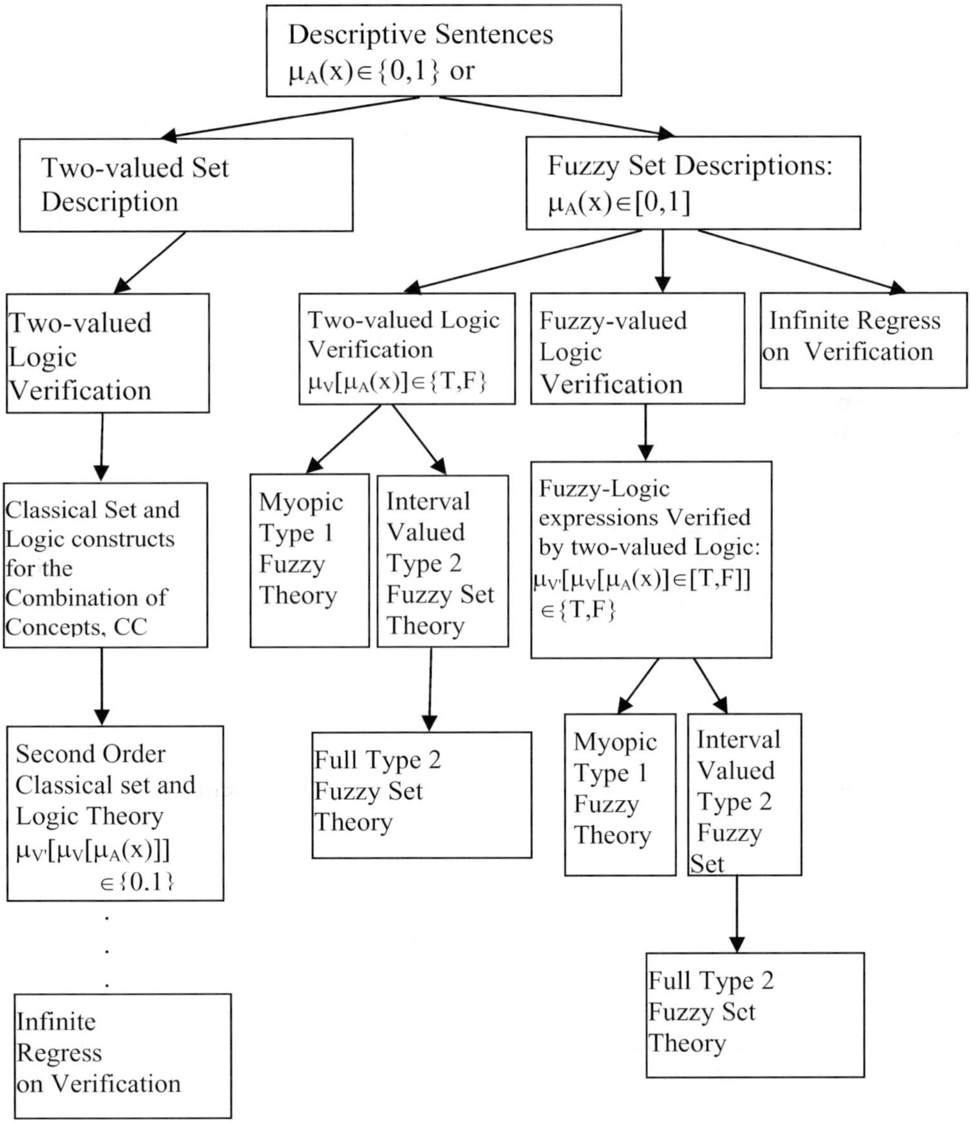


**Diagram 1.1.** Description and Verification Hierarchy

Examples of natural language sentences with the property (1.2) have been discussed since antiquity. The two examples of such a sentence are:

(i) "All Cretans are Liars" (1.4)

as uttered by the Cretan Epimenides; and

(ii) "I am a Liar" (1.5)

as can be uttered by any one of us.

In case (1.4), if we assume that all utterances of Cretans except possibly the utterance of Epimenides are false, then if Epimenides' utterance is true, then (1.4) is false, and vice versa.

Similarly, if (1.5) is true, then I am a Liar and hence my utterance cannot be true, and if it is false, them I am <u>not</u> a Liar and my utterance has to be true.

Thus, Tarski argued that same holds for any <u>reasonably rich</u> formal language. After further investigations, Tarski introduced the concept of <u>satisfaction</u> as a binary relation between linguistic and extra-linguistic items. In addition, Tarski stated a finitely based recursive definition of satisfaction, and then showed how to define truth in terms of satisfaction. To be true is, according to Tarski, to be satisfied by everything which may be considered.

That is for a sentence to be true, it must be verified (satisfied) to be true by everyone who utters the sentence.Current literature provides various perspectives on the nature of truth, e.g., , such as truth as correspondence, truth as (ideal) justifiability, pragmatic theories, T-sentences as the only clue, etc.(Peregrin, 1997)

As indicated above, our approach is in agreement with Tarski's perspective. That is, we assume that either there are observers who feel satisfied by their knowledge and experience that a given sentence is true or false and hence can make a statement to verify the truthood of the sentence or there are necessary and sufficient data vectors that provide an analyst with a data processing algorithm and with necessary and sufficient knowledge in order for the analyst to be satisfied to verify whether or not a given sentence is true or false. Thus, we assume that, for every linguistic statement, which an analyst makes, it can be verified in the sense of Tarski's "satisfaction" to be true or false.
It is clear that the nature of truth as discussed by most logicians and philosophers is two-valued truth. Thus, we have $\mu_{V(S)}(t) \in \{0,1\}$, that is the verification V of sentence S has a dichotomous membership value. This lead us to a discussion of "fuzzy truthood" next.

### *1.2.1. Fuzzy Truthood*

Fuzzy truthood is a "degree of truthood", i.e., $\mu_V(s) \in [F,T]$, $s \in S$, beyond the binary truth $\mu_V(s) \in \{F,T\}(\{0,1\})$ in the sense of Zadeh (1975). This in turn is an extension of Tarski's "satisfaction" to be true or false to a "degree of satisfaction" by an expert, an agent, that the given sentence S is "true to a degree" or "false to a degree". This we explore further in Chapter 13 under the heading of "Veristic Fuzzy Sets of Truthoods".

It is to be noted that the descriptive sentence S may be assigned a two-valued or fuzzy valued membership function. It is also the case that the veristic statement that is fuzzy $\mu_V(s) \in [F,T]$, $s \in S$, may in turn needs to be verified with an expert to be true or false in the original sense of Tarski, i.e., $\mu_{V'(V)}(s'_V) \in \{0,1\}(\{F,T\})$, and so on..., where $s'_V \in S$, i.e., $s'_V$ is a fuzzy truth qualifying sentence in the set sentences of L, the natural language.

## 1.3. Definiteness vs. Indefiniteness

Charles S. Peirce's (1958) logic of vagueness characterize all varieties of indeterminacy and determinacy that affect either the breath (reference, denotation, extension) or depth (sense, connotation, intension) of symbols. Indefiniteness and definiteness (in breath) and precision (in depth) are part of the logic of vagueness. (Brock, 1979; Sheriff, 1989; Hoopes, 1991). Peirce's discussion presupposes that every symbol is capable of determining an interpretant symbol and that symbols are at least potentially general.

In general a symbol S is indeterminate iff (∃P)~(S is P or S is ~P). Thus, Locke's famous idea of the triangle is stated as:

"It is not the case that a triangle in general is scalene or that is it not scalene"
Peirce indicates, such examples must be understood in terms of universal and affirmative predication. Hence Peirce suggests that we substitute universal "subjects", i.e., "every triangle" or "any triangle", for S.
As a result, the sentence defining indeterminacy is satisfied:

For it is false that "any (every) triangle is scalene and also false that any (every) triangle is not scalene".

The symbol S is determinate in at least one respect iff
(∃P) (S is P or S is ~P)

Thus for example the term "man" is determinate with respect to the term "animal" because it is true
"Any (every) man is an animal"

and false that

"No man is an animal" (every man is non animal)
In contrast, consider, for example, the fact that "man" turns out to be indeterminate with respect to "sagacity", i.e., "man has a degree of sagacity".The depth of symbols, i.e., a degree assignment, or belonging, to a qualitative state or property, primarily covers the predicates and/or implicates of symbols. Thus we get the following results:

The subject "man" determines the predicate "animal".

But for Peirce, this means that "humanity" entails "animality" or, if you will, that a man is necessarily an animal. It is also implied that the class or collection of men is one of the "logical divisions" or subsets of the class of animals and this in turn means that "man" falls within and partially determines the breadth of "animal".

While this is true to some extend within the physical characteristics of "man" and "animal", there is an alternate use when one states that "this man is an animal" in which case one is referring to his behavior. In such uses however, we are concerned with the degree to which "this man" behaves like "animal". Hence, we need to clarify or precisiate the degree to which each man behaves like an animal.
The breadth of a symbol is whatever a symbol applies to or whatever implies that symbol. Thus in the proposition in question, the subject determines the predicate's breadth and the predicate determines the subject's depth.

In the indeterminate case, the term "man" does not determine the predicate "sagacious", a man is not necessarily either sagacious or necessarily "non-sagacious"; humanity neither implies sagacity nor no sagacity.

The class of men could be assigned into "logically divisible" classes of the sagacious and the non-sagacious men with different degrees.

Probably the most important points to retain from Peirce and combine it with Zadeh's fuzzy sets are that:

In the determinate case, the subject implies the predicate as part of its logical or essential depth; and at times, this needs to be interpreted as a matter of degree.
In the indeterminate case, the subject neither implies the predicate nor its negation. However, the predicate is usually a possible "further determination" of the subject with a membership assignment and, if it is added, increases the informed depth of the subject.

It should be noted in the light of two-valued versus infinite (fuzzy)-valued set interpretation:

In the determinate case, the subject, set symbol, S, represents the predicate A, i.e., its individual elements, $s \in S$, with the same properties, as part of its content, i.e., the membership $\mu_A(s)=a$, $a \in \{0,1\}$ for every $s \in S$ with $\mu_S(s)=a$ and that this fact is affirmed to be true. That is an individual s belongs to the class of "men", S, and the same individual s belongs to the class of "animal", A, and its membership in both S and A is "1" if it is taken in strict physical sense. But as pointed out at times meaning representation of the predicate A may require that there be a degree precisiation to be specified. In the example given above "this man behaves like an animal say, to degree a".

Whereas, in the indeterminate case, the set symbol, S, represents the predicate A to a certain degree such that while $\mu_S(s)=1$, $\mu_A(s)=a\in[0,1]$ and increases our knowledge of the individual element $s\in S$ and provides us with an "informed depth", say, a=0.6, of a particular element $s\in S$. That is an individual s belongs to the class of "men", S, one hundred percent, but the same individual belongs to the class of "sagacious men" to the degree 0.6.

Furthermore the "truth" verification is the affirmation of the membership degree $\mu_A(s)=a$, $a\in[0,1]$ as a value between "0" and "1", i.e., $\mu_V[\mu_A(s)\in[0,1]]$, depending on our knowledge of the particular element's fitness to the over all class property of the set symbolized, say by A≡"sagacity".

Peirce goes on to raise and answer some interesting general questions about the occupants (elements) of what may be called the universality of symbols.
The first equation:

"Are there any entirely indeterminate general symbols; such that
(S)( ∃P)~(S is P OR S is ~P)"

The second question:

"Are there any entirely determinate symbols", such that (S)( ∃P)~(S is P OR S is ~P)"

Peirce answers both of these questions in the negative and embraces a third, mediating, position according to which every symbol is, at lease potentially, determinate in some respects, but, of course, not in the same respect.
Peirce's thesis may be represented by the conjunction of
(S) (∃P)(S is P OR S is ~P) AND (S) (∃P)~(S is P OR S is ~P)

There are a number of reasons why Peirce rejects the notion of an absolutely indeterminate or determinate general symbol. Peirce's most forceful remark on the subject is: "[No symbol can be] …absolutely universal, since nothing could be truly asserted [about] such a symbol, i.e., everything is a matter of degree. Hence it would be quite meaningless. The principle reason such a symbol would be meaningless is quite clear. Such a symbol would violate Peirce's requirement that every symbol be capable of determining an interpretant symbol. i.e., by being capable of implying something.
What is being ruled out as unintelligible here is notion of an absolutely simple, unanalyzable, indefinable, and hence inexplicable general symbol which is the

stock in trade of classical (Cartesian) philosophy. In classical approach, we are supposed to "just understand" the meaning of such terms without being able to convey them to others. Others must "just understand" them without further explanations. Such terms are called self-defining in order to show their affinity to another intuitionist's delight: the self-evident (self-justifying!) judgment.(Recall our exposition on the classical set and logic theory in Preface 0.2, i.e., a Perspective on the Philosophical Grounding of Fuzzy Theories.)

Peirce rejects this, in fact, his semiotic seems to be constructed so as to deprive intuitionists of their vocabulary. The positive edge of the assertion that there are no completely indeterminate or determinate signs is that all symbols are, in some way or another, analyzable, definable, and explicable to others. This view can now be implemented and expressed with the degree assignment to information granules (Zadeh, 1997) as follows:

(S) $(\exists P)$(S is P OR S is ~P) $(\mu_P(S) \in [0,1])$ AND (S) $(\exists P)$~(S is P OR S is ~P) $(1 - \mu_P(S))$

Let's next consider Peirce's attack on the notion of an absolutely determinate symbol. He calls absolutely determinate symbols "logical atoms", "incapable of logical division". Because of their indivisibility, such symbols must be absolutely singular in breadth of reference. Definite descriptions and proper names would appear to be candidates for the status of "logical atoms". But, Peirce cannot allow that any symbol actually have this status without giving up the thesis that all symbols are, at least potentially, general, that is, indeterminate, or capable of logical division into information granules of Zadeh (1996). {Again recall our stance on fuzzy set and logic theory in 0. Foundations).

## 1.4. Syntax of a Formal Language, a PNL

In the light of the discussion presented in subsections 1.2 and 1.3, it is proposed that we consider a formal language a PNL defined by the following seven items of concern. The details of these are further discussed in the chapters indicated in parenthesis.

1. Assignment of symbols, i.e., set labels, to propositions e.g., A,B,C,... and predicates, e.g., a,b,c...(see chapters 1, 2, 3, 4 and 5)
2. Choice of crisp connectives, i.e., t-norms and co-norms, (Diagram 1.2) (see chapter 6)
3. Choice of "descriptive" assignments, e.g., {0,1}, [0,1] to an element $x \in X$ for a linguistic value A, a set symbol, of a linguistic variable $A$, i.e., $\mu_A(x)=a$, $a \in \{0,1\}$ OR $a \in [0,1]$. (Table 1.1) (see chapters 1, 2, 3, 4, 5, 6 and 7)

4. Choice of "veristic" (truth value) assignments, e.g., {T,F}, [T,F] or alternately {0,1}, [0,1], respectively, to the descriptive assignments specified in 3 above. (Table 1.1) (see chapters 7 and 8)
5. Choice of "measures" crisp or fuzzy, such as "probability", "possibility", "belief", and their fuzzy versions, etc., that add a further refinement to the description given in 3 above. (see chapter 12)
6. Construction of truth tables and derivation of formulae of the normal (canonical) forms to explicitate the imprecise meaning of linguistic connectives, "AND", "OR", "IMP", etc. (see chapter 8)
7. Choice of axioms that dictates operations of crisp connectives, i.e., t-norms and conorms, on "descriptive", "veristic", probabilistic, possibilistic, etc., assignments. (Table 1.2, 1.3)
8. Choice of inference rules with normal (canonical) forms, i.e., DNF≡CNF, v.s., FDCFW FCCF, or in particular FDCF⊆FCCF. (see chapters 8 and 9)
9. A treatment of the combined effect of Fuzziness and Belief is shown in Chapter 12.
10.Choice of "Approximate Reasoning Methods (see Chapters 14, 15).
11. A theoretical application of Interval-valued Type 2 representation is given in Chapter 16.
12. In Chapter 17, we propose "Meta-Linguistic Axioms" as a foundation with CWW and CWP.
13. Finally in Chapter 18, we state our conclusion and further research directions with open problems in our Epilogue.

Given a formal language, i.e., a PNL, then propositions that are inferred with the axioms and inference rules called theorems. Semantics (meaning representation) are imbedded in (2)-(7) and the preservation of (2)-(7) and (8). In this general framework, theorems hold to a degree but not absolutely, e.g., Law of Fuzzy Middle and Law of Fuzzy Contraction. (see Chapter 10).

That is, in fuzzy set and logic theory theorems are not satisfied in the absolute sense, i.e., they are not tautologies. But they are satisfied as a matter of degree in given contexts.
**Thus there are various examples that show the break down of certain classical equivalences which leads to the generation of new formulas that hold to a matter of degree in fuzzy set and logic theories.**

### 1.5. Basic Notations – Type 1 Theory

A fuzzy set A over a universe of discourse X, which itself is a classical set, is defined by its membership function: $\mu_A$: X → [0,1]. Fuzzy set membership functions are thus generalized characteristic functions.
In classical theory, we have characteristic functions that define a crisp set in an analogous manner: $\chi_A$: X→ {0,1}.

It is important to differentiate right at the very beginning that, $\mu_A \in [0,1]$ for the fuzzy sets and $\chi_A(x) \in \{0,1\}$ for the crisp sets, define a *description* of a set in terms of membership or characteristic functions. These descriptive values are not *truth degrees* at all!

In addition, we have to have an external source that verifies that these membership and characteristic functions are true in the sense of our discussion on the "Nature of Truth" a la (Tarski, 1932, 1944, and Zadeh, 1965, 1975).
That is, we have
$\mu_V[\mu_A(x) \in [0,1]] \in \{T,F\}$or
$\mu_V[\chi_A(x) \in \{0,1\}] \in \{T,F\}$.
In addition, truth verification of fuzzy truthoods;
$\mu_{V'}[\mu_V[.]]$ OR $\mu_V[.] \in [0,1] \in \{T,F\}$.

These two different membership assignments, i.e., one for description of a set and the other for the fuzzy verification of the membership functions together with a final two-valued verification and their values are confused in some current literature. At times, we read for example "the membership degrees of fuzzy sets are considered as generalized truth values, i.e., as *truth degrees* of a suitable many-valued logic L."(Gottwald, 1993).

**In our basic theory this is not acceptable and totally wrong!**

For example, we may define a possible continuous membership function for the fuzzy set of real numbers close to "zero" as:
$\mu_A(x) = (1 + 5x^2)^{-1}$

On the other hand, a discrete membership function for a fuzzy set of say, young people, 'Y', over a crisp set of people S = $(S_1, S_2, \ldots\}$may be defined as:
Young (S) = $1/S_1 + .8/S_2 + .7/S_3 + \ldots, S_i \in S$
which may also be written as

Young (S) = $(1,S_1) \cup (.8,S_2) \cup (.7,S_3) \cup \ldots$
Thus in general, for the case in which a fuzzy set A is defined on a universal set X that is finite and countable, we write:

$$A = \sum_1^n \mu_i/x_i \quad \text{or } A = \bigcup_{i=1}^{n} (\mu_i, x_i)$$

$X = \{x_1, x_2, \ldots, x_n\}$

Similarly, when X is a continuous variable, a fuzzy set A is often written in the form:

$A = \int_X \mu_A(x)/x$ *or* $A = \int_X (\mu_A(x), x)$

An α-cut of a fuzzy set A is a crisp set $A_\alpha = \{x \in X \mid \mu_A(x) \geq \alpha.\}$

For example, the fuzzy set of "young" people can be written in terms of its α-cuts as follows:

Young = $(1)\{S_1\} + (.8)\{S_1,S_2\} + (.7)\{S_1, S_2, S_3\} + \ldots$

or alternately

Young = $(1)\{S_1\} \cup (.8)\{S_1,S_2\} \cup (.7)\{S_1, S_2, S_3\} \cup \ldots$

The set of all levels $\alpha \in [0,1]$ that represent distinct α-cuts of a given fuzzy set A is called a *level set* of A. Formally, we write $\Lambda_A = \{ \alpha \mid \mu_A(x) = \alpha$ for some $x \in X\}$.

It is important to note that α-cuts and α-level sets do not transform a fuzzy set into a crisp set. They are in fact weighted crisp set representation of a fuzzy set. It is important once again to emphasize that whenever a definition or a description of a fuzzy set is given, it is implicitly assumed that the definition or the description is verified that it is "true" even though it is not specified to be "true" in an explicit manner. Hence our emphasis on *descriptive* and *veristic* assignments!

Furthermore, for a given fuzzy set say $\mu_A(x) = (1+5x^2)^{-1}$ or

Young (S) = $1/S_1 + .8/S_2 + .7/S_3 + \ldots, S_i \in S$

May be verified fuzzily as

$\mu_V[\mu_A(x) \in [0,1]]$ or

$\mu_V[\mu_{Young}(S)] \in [0,1]$.

In such a case, we need an external verification V′ that $\mu_V(.)$ is true, i.e.,

$\mu_{V'}[\mu_V(.)] \in [T,F]$, etc.

### 1.6. Basic Notations – Type 2 Theory

In Type 1 theory, it is assumed that a membership function is well known and crisp. However, it is known that a membership function is not well known and is not crisp. Both the measurement of membership experiments and fuzzy clustering algorithms have clearly demonstrates that a membership function is imprecise and contains uncertainty.

In Type 2 theory, one captures the imprecision of a membership function by representing it either as Interval-Valued Type 2 fuzzy set or as Full Type 2 fuzzy set. For this purpose, we next introduce the basic notation of Type 2 fuzzy sets.

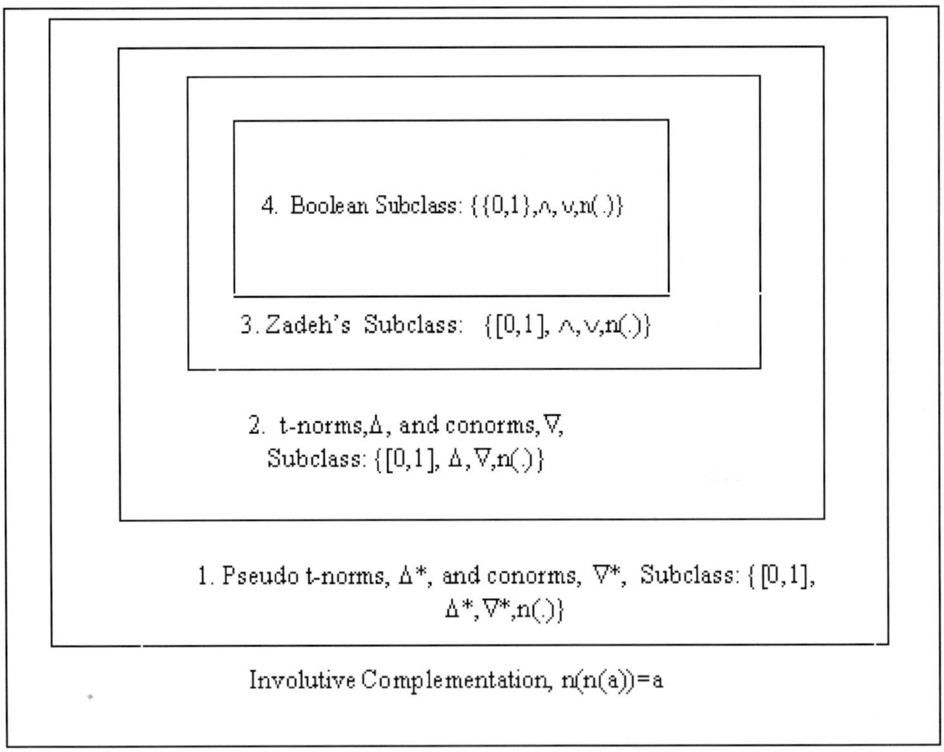


**Diagram 1.2.** Monotonic Logic Systems: Representation and Reasoning with De Morgan Triples

**Definition:** A Type 2 fuzzy set A, is represented by a Type 2 membership function $\mu_A(x,u)$ where $x \in X$ and

$$u \in \{ \underline{\mu}_A(x,u), \quad \overline{\mu}_A(x,u)\} \subseteq [0,1]$$

**Table 1.1.** Representation of definiteness, e.g., absolutism, or precision or certainties, indefiniteness, e.g., tolerance, or imprecision, or uncertainty with respect to concepts, e.g., description, verity, probability, possibility, belief, plausibility, etc.

| CONCEPTS | SYMBOLS | DEFINITENESS | INDEFINITENESS |
|---|---|---|---|
| DESCRIPTION | D | {0,1} | [0,1] |
| VERITY | V | {0,1} | [0,1] |
| PROBABILITY | Pr | {0,1} | [0,1] |
| POSSIBILITY | Po | {0,1} | [0,1] |
| BELIEF | Bel | {0,1} | [0,1] |
| PLAUSIBILTY | Pa | {0,1} | [0,1] |
| . | . | . | . |
| . | . | . | . |

$A = \{\mu_A(x,u), (x,u) | x \in X, u \in \{ \bar{\mu}_A(x), \underline{\mu}_A(x)\}\} \subseteq [0,1]$

such that $0 \le \mu_A(x,u) \le 1$.

In continuous domains A is expressed as:

$$A = \int_{x \in X} \int_{u \in [\underline{\mu}_A(x), \bar{\mu}_A(x)] \subseteq [0,1]} \mu_A(x,u)/(x,u)$$

In discrete domain A is expressed as:

$$A = \sum_{x \in X} \sum_{u \in [\underline{\mu}_A(x), \bar{\mu}_A(x)] \subseteq [0,1]} \mu_A(x,u)/(x,u)$$

**Definition:** An alternate form of Type 2 fuzzy set representation is given as Interval-Valued Type 2 fuzzy sets.

$A = \{\mu(\mu_A(x))/(\mu_A(x),x) | x \in X, \mu(\mu_A(x)) \subseteq [0,1]\}$

Thus alternate representation of A is as:

$$A = \int_{x \in X} \mu(\mu_A(x)) / (\mu_A(x), x) = \int_{x \in X} \Big[ \int_{\mu(\mu_A(x) \in [\underline{\mu}_A(x), \overline{\mu}_A(x)] \subseteq [0,1]} \mu(\mu_A(x) / (\mu_A(x)) \Big] / x$$

In particular, interval valued Type 2 fuzzy set A is expressed as:

$$A = \int_{x \in X} \Big[ \int_{\mu(\mu_A(x) \in [\underline{\mu}_A(x), \overline{\mu}_A(x)] \subseteq [0,1]} 1 / (\mu_A(x)) \Big] / x$$

**Table 1.2.** Axioms of Two-valued Set and Two-Valued Logic Theory

| | |
|---|---|
| Involution: | c(c(A)=A |
| Commutativity: | A∪B = B∪A |
| | A∩B = B∩A |
| Associativity: | (A∪B) ∪C = A∪ (B∪C) |
| | (A∩B) ∩C = A∩ (B∩C) |
| Distributivity: | A∩ (B∪C) = (A∩B) ∪ (A∩C) |
| | A∪ (B∩C) = (A∪B) ∩ (A∪C) |
| Idempotence: | A∪A = A |
| | A∩A = A |
| Absorption: | A ∪ (A∩B) = A |
| | A ∩ (A∪B) = A |
| Absorption of Complement: | A ∪ (c(A) ∩B) = A∪B |
| | A ∩ (c(A) ∪ B) = A∩B |
| Absorption by I and ∅: | A ∪ I =I |
| | A ∩ ∅ =∅ |
| Identity: | A ∪ ∅ = A |
| | A ∩ I = A |
| Law of contradiction: | A ∩ c(A) = ∅ |
| Law of excluded middle: | A ∪ c(A) = I |
| De Morgan's Laws: | c(A∪B) = c(A) ∩ c(B) |
| | c(A∩B) = c(A) ∪ c(B) |

**Table 1.3.** Axioms of t-norms and t-conorms

| | |
|---|---|
| Involution: | $c(c(A)) = A$ |
| Commutativity: | $A \cup B = B \cup A$ |
| | $A \cap B = B \cap A$ |
| Associativity: | $(A \cup B) \cup C = A \cup (B \cup C)$ |
| | $(A \cap B) \cap C = A \cap (B \cap C)$ |
| Identity: | $A \cup \varnothing = A$, $A \cup I = I$ |
| | $A \cap I = A$, $A \cap \varnothing = \varnothing$ |
| De Morgan's Laws: | $c(A \cup B) = c(A) \cap c(B)$ |
| | $c(A \cap B) = c(A) \cup c(B)$ |

## 1.7. Epistemological Concerns

The basic notations specified above in subsections 1.5 and 1.6 essentially address part of the ontological concerns of fuzzy theories. The epistemological concerns generally address issues of:

(i) Foundationalism, (ii) Coherentism, (iii) Reliabilism, (iv) Critical rationalism. Foundationalism admits many and various variations. We re-state classical claims to generalize their scope with fuzziness in several expressions where (′) and (″) indicate alternate expressions:

(i): Some basic ***beliefs*** are justified to some fuzzy degree independently of the support of other beliefs and they are non-empirical in character.

(ii): Some basic ***beliefs*** are justified to somc fuzzy degree not by the support of other beliefs, but by a subjects experience, i.e., they are empirical.

(i′): Some basic beliefs are justified to a some fuzzy degree, not by the support of other beliefs, but because of causal or law-like connection between a subjects' belief and the state of affairs which makes it appear true, i.e., expert knowledge which are considered crisp in the classical perspective but which are intrinsically fuzzy under our $s \xleftrightarrow{R} \acute{e}$ structure proposed at the Ontological level.

(ii′): Some basic beliefs are justified to some fuzzy degree, not by the support of other beliefs; but by virtue of its content, its intrinsically self-justifying

character. Again they may be assumed to be crisp or fuzzy depending on the agenda we work on.
These statements requires the determination of critical, affective variables and the belief that they are independent. Clearly the determination of these approximately independent critical and affective variables require crisp or fuzzy statistical criteria.

(i″): Some basic beliefs are justified to some fuzzy degree interacting with other beliefs.

(ii″): Some justified beliefs are derived and are justified wholly to some fuzzy degree via direct or indirect support of basic beliefs that are inherently fuzzy.

In Chapter 12, we will treat our Epistemological concerns of fuzzy belief from a perspective of foundationalism.

However, in Chapter 17, we return to our Ontological concerns and propose a re-statement of the axioms of fuzzy theory to form a new foundation for CWW.

## References

1. Davidson, D. (1967), "Truth and Meaning", Synthese 17, reprinted in Davidson (1984), 17-36.
2. Davidson, D. (1969), "True to the Facts", Journal of Philosophy 66, reprinted in Davidson (1984), 37-54.
3. Davidson, D. (1984), Inquiries into Truth and Interpretation, Clarendon Press, Oxford.
4. Davidson, D. (1990), "The Structure and Contents of Truth", Journal of Philosophy 87, 279-328.
5. Dummett, D. (1959), "Truth", Proceedings of the Aristotelian Society (New Series) 59, reprinted in Dummett (1978), 1-24.
6. Dummett, M. (1978), Truth and other Enigmas, Duckworth, London.
7. Hintikka, J. (1975), "A Countereexample to Tarski-Type Truth-Definition as Applied to Natural Languages", Philosophia 5, 207-212.
8. Hoopes, J. (ed). (1991), Peirce on Signs: Writings on Semiotics, Chapel Hill and London, *University of North Carolina Press*.
9. Tarski, A. (1932), "Der Wahrheitsbegriff in den Sprachen der deduktiven Disziplinen", Akademischer Anzeiger der Akademie der Wissenschaften in Wien 69, translated as "The Concept of Truth in Formalized Languages" in Tarski (1956), 152-278.
10. Tarski, A. (1944), "The Semantic Conception of Truth", Philosophy and Phenomenological Research 4, 341-375.
11. Tarski, A. (1956), Logic, Semantics, Metamathematics, Clarendon Press, Oxford.
12. Zadeh, L.A. (2002), "Toward a Perception-Based Theory of Probabilistic Reasoning with Imprecise Probabilities", J.of Statistical Planning and Inference, Elsevier Science, V.105, pp.233-264.
13. Zadeh, L.A. (1975), "Fuzzy Logic and Approximate Reasoning", *Synthese*, 30, 407-428.

14. Zadeh, L.A. (1975), "The Concept of a linguistic variable and its application to approximate reasoning", Parts I, II, *Information Sciences*, 8, 199-249, 301-357, 43-80.
15. Zadeh, L.A. (1975), "Calculus of Fuzzy Restrictions", *Fuzzy Sets and Their Applications to Cognitive and Decision Progresses*, L.A. Zadeh, K.S Fu, K. Tanaka and M. Shimura (eds), Academic Press, New York, 1-39.
16. Zadeh, L.A. (1973), "Outline of a New Approach to the Analysis of Complex Systems and Decision Processes", *IEEE Trans. Systems Man Cybernet*, 3, 28-44.
17. Zadeh, L.A. (1965), "Fuzzy Sets", *Information and Control Systems*, Vol.8, 338-353.

# Chapter 2

# COMPUTING WITH WORDS

"Computing with words", CWW, is a phrase coined by L.A. Zadeh (1995) during the IFSA'95 Conference. In our proposed construction of concepts for CWW, we have interpreted CWW with two different approaches. These approaches are the result of our application oriented research on fuzzy set theories that were needed to form a sound foundation for two different approaches in industrial applications.

At the end of this book, in Chapter 17, we propose a new axiomatic foundation for CWW.

In the first of our approaches, we have developed fuzzy expert systems with interviews of "experts", and assigned fuzzy membership functions to explicate the meaning of words, linguistic terms of linguistic variables, that the experts have used during the knowledge acquisition phase of our investigations. This has led to our concern and research on the "measurement of membership functions" (Norwich and Türkşen, 1982, 1984; Türkşen, 1991; Bilgic and Türkşen, 1995, 1999). In these studies, it was assumed, that the membership values assigned by experts were implicitly verified.

In the second of our approaches, we have developed fuzzy expert systems by fuzzy data-mining and knowledge discovery, in particular, fuzzy cluster analysis methods applied to system input-output data; and assigned linguistic terms, words, of linguistic variables of our natural language. That is, numbers that were assigned and thus defined membership functions were the result of input-output data analysis via fuzzy clustering methods. This has lead to our concern and research on the fuzzy cluster analyses methods (Emami, Türkşen and Goldenberg, 1998, 1999; Kılıç, Türkşen, 2001; Uncu, Türkşen, 2001). In the case of this second approach, again implicitly, we had assumed and took it for granted that the algorithms we used were verifying the membership assignments that we had obtained.

In this regards, in this section, we will first review: (i) assignment of numbers, i.e., membership function, to words, i.e., linguistic terms (values) of a linguistic variable, and then (ii) assignment of words, i.e., linguistic terms (values) of a linguistic variable, to numbers, i.e., membership functions. In this context later, we will first review the acquisition of membership functions from a "measurement theoretic" perspective which is concerned with the scale strength of the membership values. Measurement theory addresses the representation and the uniqueness properties of numbers assigned to words, the linguistic terms (values) of linguistic variables, extracted from experts during knowledge acquisition phase. Secondly, we will review the acquisition of membership functions that are obtained from "input-output" data-mining and discovery via fuzzy clustering methods. In this second approach, briefly, the scale strength of the membership values depend on the mathematical expressions used during the number crunching phase of the analyses before such membership functions are assigned to words, i.e., linguistic terms (values) of linguistic variables of our natural languages.

## 2.1. Words to Numbers

Assignment of numbers to words of a natural language may be either with two-valued, dichotomous, or infinite-valued, fuzzy, numbers depending on whether we want to or need to represent them with graded or without graded membership in a set representation. That is whether we need to use a reductionist approach for gross generalization or a non-reductionist approach that reflect the granular nature of information. The choice in Peirce-Zadeh sence is to expose the degree of indeterminacy, imprecision, and hence non-reductionist stance. It is very well known that there are many different categories of words. In this discourse, we consider only the two categories: (i) those that describe an attribute of an object with a linguistic term (value) of a linguistic variable; such as: "TALL", "SHORT", "VERY TALL", "VERY SHORT", " LOW", "HIGH", "EXCELLENT", "POOR", etc.; and (ii) those that verify, i.e., state the logical affirmation or denial of a linguistic expression specified in (i) above; such as: "TRUE", "VERY TRUE", "FALSE", "VERY FALSE", etc.

### *2.1.1. Descriptive Words*

In the first category, a sentence describes an attribute of an object where linguistic expressions are stated as:

"this person is TALL", (2.1)

"that person is HEAVY", etc. (2.2)

where "TALL" describes the "height" attribute of a person and "HEAVY" describes the "weight" attribute of a person. In these examples, "TALL", "HEAVY" are linguistic terms (values) of linguistic variables, "height", "weight", respectively. It is clear that "*semantic representations*" of descriptors, i.e., words such as "tall", "heavy", can be represented either as two-valued, dichotomous, sets or infinite-valued, fuzzy, sets. The choice depends on one's purpose. These two types of representations are developed at times from numbers to words and at other times from words to numbers and they may be represented by crisp or fuzzy sets.

The classical view is that it is better to give a representation in two-valued sets which reduce the complexity and simplifies the operations. Naturally, the view of the fuzzy community is that it is better to give infinite-valued representations simply because there are natural granules of information and hence fuzzy representations fit the reality much more naturally.

### *2.1.2. Veristic Words*

In the second category, a sentence of the first category is taken as the "subject" and a new sentence is formed in order to verify the sentence taken as the object. Veristic sentences are stated as:

"'this person is TALL' is 'TRUE" (2.3)

"'this person is HEAVY' is 'FALSE'" (2.4)

Again "TALL" and "HEAVY" are linguistic descriptors that need to be given a "semantic representation" as explained above; but in these second category sentences, we employ another class of words, i.e., "TRUE", "FALSE" that "verify", "falsify", respectively, in order to affirm or deny the descriptive expressions, i.e., sentences that are stated in the first category. It should be noted that in "logic" books expressions (2.3) and (2.4) are stated slightly differently as alternate expressions such as:
"it is the case that 'this person is TALL'" (2.3′)

"it is not the case that 'this person is HEAVY'" (2.4′)

It is also to be noted that we could verify the descriptive expressions, i.e., sentences, to varying degrees of truthoods such as:

"'this person is TALL' is 'VERY TRUE'", (2.5)

"'this person is HEAVY' is 'SOMEWHAT FALSE'". (2.6)

Furthermore, there may be a hierarchical chain of verifications. That is the statements (2.5) and (2.6) may be subject to further verification such as:

"'this person is TALL' is 'VERY TRUE' is 'TRUE'", (2.5′)
"'this person is HEAVY' is 'SOMEWHAT FLASE' is 'TRUE'". (2.6′)

Again it is clear that "semantic representations" of verity assignments, i.e., imprecise words of affirmation and denial can be given assignments again either as two-valued or infinite-valued for the same reasons stated before.

In summary, therefore it is to be noted that both the descriptive sentences that contain words such as "TALL, VERY TALL, SOMEWHAT TALL, LOW, VERY LOW, etc., …" and the veristic sentences that contain words such as: "TRUE, VERY TRUE, SOMEWHAT TRUE, etc., …" may be represented in the numeric domain by numbers either in crisp, two-valued, dichotomous, manner or in fuzzy, infinite-valued, non-dichotomous, way. Thus both categories of words can be semantically represented by sets whether two-valued, i.e., crisp, or infinite valued, i.e., fuzzy. Hence there are potentially four possible theories of descriptive and veristic sentence combinations.

## 2.2. Descriptive and Veristic Assignments

If we let $D$ stand for descriptor sentences and $V$ stand for veristic sentences that assert the truth of a d$\in$D, then we can identify the four possible theories as:

(1) $\{D_{\{0,1\}}, V_{\{0,1\}}\}$ that identifies the structure of two-valued descriptive sentences that are verified by two-valued veristic sentences. In this case, in expressions (2.3) and (2.4), TALL and HEAVY as well as TRUE and FALSE are interpreted in crisp sense.

(2) $\{D_{[0,1]}, V_{\{0,1\}}\}$ that identify the structure of infinite, fuzzy-valued, descriptive sentences that are verified by two-valued veristic sentences. In this case, in expressions (2.3) and (2.4), TALL and HEAVY are interpreted in fuzzy sense whereas TRUE and FALSE are interpreted in crisp sense.

Similarly;

(3) $\{\{D_{\{0,1\}}, V_{[0,1]}\}, V'_{\{0,1\}}\}$ and

(4) $\{\{D_{[0,1]}, V_{[0,1]}\}, V'_{\{0,1\}}\}$ identify the other two theories.

It should be noted that the structure of (2.5′) and (2.6′) are identified as the further verification of $\{D_{\{0,1\}}, V_{[0,1]}\}$ as $\{\{D_{\{0,1\}}, V_{[0,1]}\}, V'_{\{0,1\}}\}$ and as the further verification of $\{D_{[0,1]}, V_{[0,1]}\}$ as $\{D_{[0,1]}, V_{[0,1]}\}, V'_{\{0,1\}}\}$.

That is "TALL" and "HEAVY" are assigned either two-valued set membership or fuzzy valued set membership in expressions (2.5′) and (2.6′) while "VERY

TRUE" and "SOMEWHAT FALSE" are assigned infinite-valued fuzzy set membership which are further verified as "TRUE".

Let us next discuss each of these assignments, i.e., descriptive and veristic type sentences, a bit more in depth.

*2.2.1. Descriptive Set Assignments*

The numbers to descriptive words are assigned either by an expert(s) who observe(s) or a data-mining and knowledge discovery system (algorithms) that analyse(s) an attribute of an object, say a person's height, and then with experimentally acquired expertise and knowledge about the particular attribute of an object, i.e., the person's height, either the expert(s) or knowledge discovery system (algorithms) provide(s) a linguistic expression, i.e., a sentence, to communicate their knowledge about that attribute of the object, i.e., the person's height. Such a linguistic expression may be represented crisply or fuzzily in the numerical domain.

Consider a descriptive sentence,

"This person is heavy" (2.7)

which is expressed symbolically as:

$p \in P$ is $H \in \mathit{H}$ (2.7′)

such that p is the person in the set of people, P, and H is the linguistic term that stands for "heavy" in the set of linguistic terms of linguistic variable "people's weight", *H*={heavy, very heavy, light, very light, …}, when we have no measurement of the person on the X axis. This provides a semantic representation for the linguistic value "heavy". Alternately, the statement may be qualified with more specific information on the base axis X with a crisp expression such as:

(a) "This person is HEAVY, with the weight of 200kgr"

which is expressed symbolically as

"$p \in P$ is $H \in \mathit{H}$, with physical measurement x=200kgr" or more compactly as:
$x \in X$ is $H \in \mathit{H}$, x(p)=200kgr,

such that for a particular person in a group, or collection of people, $p \in P$, $x \in X$, we have the base axis value when we have measurement information of the

person on the base axis X. As well, it may be a crisply graded imprecise expression such as:
(b) "This person is tall with membership grade, $\mu_H(p) = 0.7$" –without a membership value associated with a physical base axis measurement $x \in X$ being available, which is expressed symbolically as:

"$p \in P$ is $H \in \mathit{H}$ with $\mu_H(p) = 0.7$", such that $\mu_H(p) \in [0,1]$,

when we have no physical measurement of the person on the X axis.

We could also have a statement such as:

(c) "This person is tall, with a physical measurement of x=200kgr and with the membership grade, $\mu_H(x)=0.7$", which is expressed symbolically as

"$p \in P$ is $H \in \mathit{H}$ with $\mu_H(x(p)=200kgr)=0.7$",

if we have the measurement of the person on the X axis as well as his/her membership value in a fuzzy set H.

It is to be noted, for further clarity, that "This person is heavy" (2.7) assigns the person to the class of "heavy" people leaving open the meaning of the class of "heavy" people to be further defined, interpreted, either as a crisp, two-valued, or fuzzy, infinite-valued set.

Whereas the statements (a), (b), (c) identify the person in the class of "heavy" people with additional more specific information to give it a bit more clarity in its meaning; i.e., in case (a) the specification is on the base axis X; in case (b) the specification is in the membership axis $\mu_H$ without any further information on the base axis X; in case (c) the specification is in both the membership axis $\mu_H$ and the base axis X.

Furthermore case (a) can only be associated within the context of the two-valued set interpretation. Whereas the statements (b) and (c) can be interpreted within the context of infinite-valued, fuzzy, set interpretations.

In this manner, we separate and distinguish the object which is a real world entity, i.e., the person and its attribute or property in a linguistic expression of a natural language. Next we transform the linguistic representation of the person and its attribute to a meta-linguistic expression and then transform it to an abstract and symbolic expression, and then we impose an interpretation which actually forms the basis of its "semantic", i.e., meaning representation, as a result of the analysts interpretation in terms of two or infinite valued set representation. (Türkşen, 1995a). Thus in this scheme, we touch upon Charles

Sanders Peirce's semiotic (1931-1958) structure of tricotomy, (Sheriff, 1989; Hoopes, 1991) i.e., an object (person), its representation in terms of a linguistic expression "Person is tall" and its interpretation by the analyst which leads to a set representation either crisp or fuzzy to convey the meaning (semantic) representation of the vague word in the numerical domain. Hence, a word "tall" is assigned a set of numbers depending on whether the analyst is interested in providing semantic (meaning) representation of the word in dicatomous or graded, granular and fuzzy manner. (Zadeh, 1965 – 2002)

### *2.2.2. Veristic Set Assignments*

In the case of veristic sentences, the assignment of numbers are for the verification, affirmation, and/or denial of a linguistic descriptive expression, i.e., a descriptor sentence that represent the knowledge of an expert or algorithmic system about the attribute qualification of the object under observation. In turn, linguistic-symbolic representation, together with its meaning representation in the set domain, would help us to retrieve required information either on the person's weight or his/her membership value in the set via reverse or referential access. As indicated Section 1.2, the study of logic is concerned with the assignment of truth to a given sentence. In this manner, we separate a descriptive sentence given by a group of expert(s) from the veristic sentence that takes the descriptive sentence as the object by another group of expert(s), who specify the confidence, or the certainty of the second group of experts on descriptive sentence stated by the first group of experts. In the case of fuzzy cluster analysis techniques, the unsupervised or supervised learning, i.e., algorithmic systems, generates the membership value assignments that describes the patterns of a variable. In turn they are implicitly verified by the algorithm that is implemented with our trust in the algorithmic systems.

The veristic "set assignments" that define the definiteness or indefiniteness asserted in a linguistic expression could be two-valued which specify that the given expression is "absolutely" "true" or "false". Such a verification by its very nature indicates total confidence or total lack of confidence and hence a dogmatic, absolute, statement or stance on the linguistic expression given in the first instance. Or, the veristic set assignment could be an infinite-valued, graded expression, which signify that the given expression is "true" or "false" to a degree. Such a verification by its very nature indicates a partial confidence or degree of definiteness or indefiniteness placed on the linguistic expression given in the first instance. Thus in the case of veristic assignments, we separate and distinguish between the veristic property of linguistic sentences and their representation with attributes of verity such as "TRUE", "FALSE", "VERY TRUE", etc. and their interpretation with crisp or fuzzy sets that define their semantic, meaning, representation. It should be noted that implicit verifications in current system modeling exercises, by their very nature, are two-valued

whether they are obtained from experts or from an algorithm unless repeated experiments are carried out to elicit degrees of verification.
Let us again consider the linguistic expression:

$p \in P$ is $H \in \mathcal{H}$.

and its three variants as discussed in Section 2.21 but in their compact versions as:

a) $x \in X$ is $H \in \mathcal{H}$ with $x(p)=200kgr$;
b) $p \in P$ is $H \in \mathcal{H}$, with $\mu_H(p)=0.7$, and
c) $p \in P$ is $H \in \mathcal{H}$, with $\mu_H(x(p)=200kgr)=0.7$.

Let $y \in Y$ be any one of these three expressions, i.e., y is a descriptive sentence. Then in terms of verification, we have two alternatives:

(i) verification is two-valued, or
(ii) verification is infinite, fuzzy, valued, and verification of (ii) is two-valued.

Thus, when we see or hear a sentence "$y \in Y$ is true" we can interprete this in two alternative ways depending on the verification information given to us as follows:

(i′) "$y \in Y$ is true, $t=1 \in \{0,1\}$" (see chapters 8, 9, 10, 11, 12)
(ii′) "$y \in Y$ is true, $t=0.8 \in [0,1]$" (see Chapter 13)

In the first case, the verification, i.e., truthood, is restricted to two valued representation whereas in the second case the verification, i.e., truthood, is allowed to be graded in the infinite valued representation.

In turn, (ii′) can be further verified as follows:

(iii′) "$y \in Y$ is true, $t=0.8 \in [0,1]$" is $t'=1 \in \{0,1\}$. (see chapter 13)

Thus, depending on our choice or the availability of information given to us either by experts or sensors, or by algorithms, we have to realize that the veristic set assignment associated with a descriptive set assignment could be either two-valued or infinite, fuzzy, valued. In turn, the fuzzy verification can be further verified in a hierarchical manner.

## 2.3. Structure of Sentences

The structure of a sentence could either be simple, called atomic, or complex. An atomic sentence describes an attribute of an object. A complex sentence generally could have a variety of structures. We state two unique types of complex structures in this section.

### *2.3.1. Atomic Sentences*

An atomic sentence describes an attribute of an object as stated earlier. Thus at times, they are called atomic descriptive sentences. The structure of an atomic sentence is given as follows:

s::= "x$\in$X isr A, $\mu_A(x)\in[0,1]$", i.e., decription is fuzzy, $D_{[0,1]}$. (2.8)

Such an atomic sentence may be verified either by experts or data processing algorithms, i.e., there are experts who are satisfied by their own independent experiences and knowledge that a given atomic sentence, i.e.,

s::= "x$\in$X isr A, $\mu_A(x)\in[0,1]$"

is "true" or "false" and therefore they can assign a truth value to the sentence such that either "s is T" or "s is F" ; or,

there are necessary and sufficient data vectors and a data processing algorithms that analyse the data such that, based on a suitable criteria, a truth value can be assigned to a representation which can be summarized as

s::= "x$\in$X isr A, $\mu_A(x)\in[0,1]$" and,

either "s is T" or "s is F", i.e., verification is two-valued, $V_{\{0,1\}}$.
In these representations, the atomic, descriptor, sentence "s" is verified in the two-valued logic taking on the truth predicates "T or F" depending on either "experts" or "algorithms" are able to do the verification. It is to be noted that we have stated "s" in general, with $\mu_A(x)\in[0,1]$. Since there is the special subcase with the unique boundary value, i.e., $\mu_A(x)\in\{0,1\}$, we are able to construct two possible frameworks for a combined theory of descriptive and veristic sets. More explicitly, we have:

s::= "x$\in$X isr A, $\mu_A(x)\in\{0,1\}$ together with "s is T" or "s is F",

which is the framework for the classical, two-valued, set description and two-valued set verification, i.e., $\{D_{\{0,1\}},V_{\{0,1\}}\}$ or s::= "x∈X isr A, $\mu_A(x) \in [0,1]$" together with "s is T" or "s is F", which is the framework for the fuzzy, infinite-valued set description and two-valued set verification, i.e., $\{D_{[0,1]}, V_{\{0,1\}}\}$.

*2.3.2. Complex Sentences*

An atomic sentence as specified in (2.8) can be made complex in a hierarchical manner by assigning to the atomic sentence a fuzzy truth qualifier and then verifying that the given complex sentence is true or false. Thus we have:

$$s'_V ::= \text{"}s \in S \text{ isr } V,\ \mu_V(s) \in [0,1] \text{ such that} \tag{2.9}$$

$$s''_T ::= \text{"}s'_V \in S \text{ is T" or } s''_F ::= \text{"}s'_V \in S \text{ is F"},$$

where s∈S is given as stated in (2.8); and the first level of verification is stated with $s'_V \in S$ and the next level of verification is stated with $s''_{(\cdot)} \in S$, where (.) = {T,F}, for the truthood of the complex sentence $s'_V \in S$. In a recursive manner such a process may be repeated to generate more and more complex sentences.

Naturally, other types of complex statements can be generated with respect to probability measures, belief measures, etc.

For example, given statement (2.8), we can state

$$s'_P ::= \text{"}s \in S \text{ isr } \Pi,\ \mu_\Pi(s) \in [0,1] \tag{2.10}$$

such that s::= "x∈X isr A, $\mu_A(x)$".

In this case *Π* is a probabilistic linguistic term such as "likely", "very likely", etc., and A is either a crisp or a fuzzy set, depending whether $\mu_A : X \to \{0,1\}$ or $\mu_A : X \to [0,1]$, such as "absolutely greater than 180" or "tall", respectively. Thus the statement (2.10) would specify fuzzy probabilities over crisp or fuzzy sets. In turn, statement (2.10) must be verified by an (independent) judge or experiment to be true or false. Therefore, we get:

$$s''_T ::= \text{"}s'_P \in S \text{ is T" or } s''_F ::= \text{"}s'_P \in S \text{ is F"}$$

In a similar manner, belief measure expression demonstrates a complex sentence structure as follows:

$$s'_B ::= \text{"}s \in S \text{ isr B},\ \mu_B(s) \in [0,1], \tag{2.11}$$

such that s::= "x∈X isr A, $\mu_A(x)$, and $s''_T$ ::= "$s'_B \in S$ is T or $s''_F$ ::= "$s'_B \in S$ is F"

Note that the statement (2.11) would specify "weak belief", "moderate belief", "strong belief", etc.

Again, A is either a crisp or a fuzzy set depending on whether $\mu_A:X \to \{0,1\}$ or $\mu_A:X \to [0,1]$, etc.
It should further be noted that we need to first obtain probabilities or belief over crisp sets and then determine fuzzy probabilities and fuzzy beliefs, etc. But for these we need to develop α-cut expression etc. These complex sentence structure analyses are subject to further investigations in later chapters of this work.(see chapter 12 for Belief, etc., measures over α-cuts of Type 1 and interval-valued Type 2 fuzzy set.).

## References


1. Bilgic, T., I.B. Türkşen (2000), "Measurement of Membership Functions: Theoretical and Empericial Work", in: *Handbook of Fuzzy Theory*, D. Dubois and H. Prade (eds.).
2. Bilgic, T., I.B. Türkşen (1997), "Elicitation of Membership Functions: How Far Can Theory Take us?", *Proceedings of Fuzzy-IEEE '97*, July 1-5, 1997, Barcelona, Spain, Vol. III, 1321-1325.
3. Bilgic, T. (1995), Measurement-Theoretic Frameworks for Fuzzy Set Theory with Applications to Preference Modelling, PhD thesis, University of Toronto, Dept. of Industrial Engineering Toronto, Ontario M5S 1A4 Canada.(supervisor, I.B. Türkşen)
4. Bilgic, T., I.B., Türkşen (1995), "Measurement-theoretic Frameworks for Fuzzy Set Theory". In T.P. Martin and A.L. Ralescu, editors, Fuzzy Logic in Artificial Intelligence: Towards Intelligent Systems, Volume 1188 of Lecture Notes in Artificial Intelligence, 252-265, Springer, 1997. Selected papers from IJCAI'95 Workshop Montreal, Canada, August.
5. Bilgic T., I.B. Türkşen (1995), "Measurement-theoretic Justification of Fuzzy Set Connectives", *Fuzzy Sets and Systems*, 289-308.
6. Emami, M.R., I.B. Türkşen and A.A. Goldenberg (1991) "A Unified Parameterized Formation on Reasoning Process in Fuzzy Modeling and Control", *Fuzzy Sets and Systems*.
7. Emami, M.R., I.B. Türkşen and A.A. Goldenberg (1996) "An Improved Fuzzy Modeling Algorithm Part I: Inference Mechanism", *Proceedings of NAFIPS, 96*, Berkeley, CA, 289-293.
8. Emami, M.R., I.B. Türkşen and A.A. Goldenberg (1996) "An improved fuzzy modeling, Part II: Systems Identification, Proceedings of NAFIPS 96, Berkeley, CA. 294-298.
9. Emami, M.R., I.B. Türkşen, A.A. Goldenberg (1997), "Fuzzy Clustering and Fuzzy Modeling", IEEE-Trans on Fuzzy Systems, (to appear).

# Chapter 3

# MEASUREMENT OF MEMBERSHIP

In this section, assignment and/or acquisition of membership values and hence membership functions are reviewed in order to show that there is an implicit verification of these assignments, i.e., truth qualifications associated with such assignments. Ever science Zadeh (1965) introduced the fuzzy sets one of the main controversies has been on the meaning, i.e., semantics, and measurement of membership functions. In particular, a lack of a consensus on the meaning of membership functions has created confusion. In order to diffuse this cloud of confusion and form a foundation for a sound theory of fuzzy sets and its linkage with truth assignments, a rigorous semantics together with practical elicitation methods for membership functions are necessary. This section aims to review the various interpretations and, theoretical models. The next Chapter discusses the elicitation methods of membership function in order to further emphasize the semantics of the fuzzy theory and its computational realization by way of the acquisition of membership values and explicitly stating the implicit truth assignments.

We first start with the formal (i.e., mathematical) definition of a membership function. A fuzzy set, say A, has a membership function $\mu_A$, defined as a function from a well defined universe (the base axes set), X, into the unit interval as: $\mu_A : X \rightarrow [0,1]$ (see Figure 3.1). (At times, we may be interested in $\mu_A : P \rightarrow [0,1]$ defined over a well-defined population or objects P as we have done in the previous section. But, we will use $\mu_A : X \rightarrow [0,1]$ throughout.)

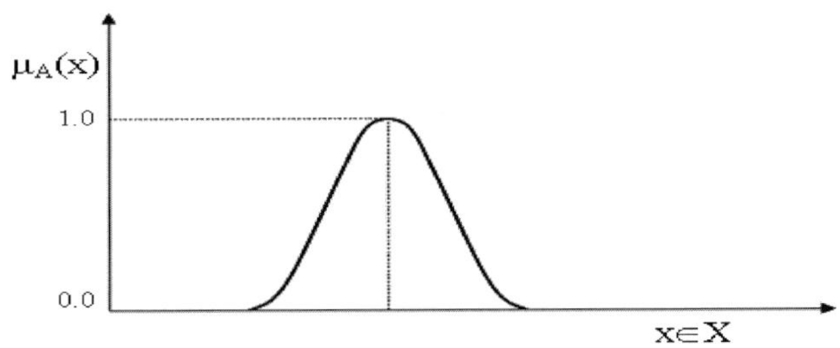


**Figure 3.1.** Membership function, $\mu_A(x)$ Š [0,1], such that A represents a linguistic term, say, medium.

For example, the imprecise predicate "John (x) is big, B", with $\mu_B$(x) is represented by a number in the unit interval, $\mu_B$(x)∈[0,1]. We need to right away point out that there is an implicit statement which generally is not stated. That is, in fact, we need to state explicitly " 'John (x) is big, B', with $\mu_B$(x)= 0.8 is true, T". There are several possible answers to the question "What does it mean to say $\mu_B$(x)= 0.8 is true, T?"

**Likelihood view:** 80% of a given population declares that " 'John is big' is true".
**Random set view:** 80% of a given population describes " 'big' as 'an interval on the base axis X containing John's height' is true".
**Similarity view:** John's size is away from the prototypical object which is "'big' to 'the degree 0.2 (a normalized distance)' is true".
**Utility view:** '0.8 is the utility of asserting' that " John is 'big' is true".
**Measurement view:** When compared to others, "John is bigger than some and 'this fact can be encoded as 0.8 on some scale' is true".

Next, these different interpretations of the concept of the grade of membership are summarized, emphasizing their differences as well as similarities, and the consequences of subscribing to a particular interpretation. Each of the interpretations mentioned above are critically discussed. Furthermore a measurement-theoretic interpretation is given where we emphasize that there is the implicit verification which we make explicit with a truth assignment, which is omitted most of the time in the current literature.

### 3.1. Interpretations of Grade of Membership

When someone is introduced to the fuzzy set theory the concept of graded membership sounds fairly intuitive since this is just an extension of a well known concept: "membership" in a set to "graded memberships". However, the second step requires a fundamental change in the mental structures of our residual memory. Within the classical analysis, the membership assignment is $\mu_A : X \to \{0,1\}$ and the implicit verification, i.e., truth qualification, is also given but usually omitted, i.e., $\mu_V$ : [$\mu_A$ : $X \to \{0,1\}] \to \{0,1\}$. This fact is clearly expressed and stated in the logic books as indicated in the previous section. In order to clarify the change in our mental constructs, we note that in fuzzy theory, we have $\mu_V : [\mu_A : X \to [0,1]] \to \{0,1\}$. Naturally, there is the hierarchical extension of this mental structure where the first level truth verification may be graded and in turn both the graded membership and the graded truth verification may be subject to a crisp verification as $\mu_{V'}:[\mu_V:[\mu_A:X\to[0,1]]\to[0,1]]\to\{0,1\}$. For emphasis, we should be reminded that the graded membership $\mu_A$ is a descriptive membership assignment whereas the graded membership $\mu_V$ is a fuzzy truthood, i.e., a veristic membership assignment, and the dichotomous membership $\mu_{V'}$ is a crisp veristic membership assignment.

Unfortunately, in current literature, at times, membership value (descriptive) assignment and its truth verification are assumed to be the same. This simplifying assumption overlooks the essentially separate human cognitive processes of description and verification. In general, they are orthogonal to each other.

Recall our brief exposition on descriptive and veristic human cognitive processes discussed in the previous chapter, let us return to the general question of membership:

How can a graded membership be measured?

This question has been considered, in the context of many-valued logics, by many people from different disciplines. Although more than two thousand years ago, Aristotle commented on an "indeterminate truth value", the interest in formal aspects of many-valued logics has started in early 1900's (Lukasiewicz, 1920; McCall & Ajdukiewicz, 1967). But the meaning of multiple truth values has not been explained to satisfaction. For some, this is sufficient to discard many-valued logics all together (Kneale, 1962; French, 1984). On the other hand, the intellectual curiosity never lets go of the subject (Scott, 1976). Again it should be noted that these discussions were directed toward graded truth but not the (descriptive) graded membership.

It was Zadeh who introduced the (descriptive) graded membership in his seminal paper "Fuzzy Sets" (1965) and later extended it to veristic membership in 1975.

In the sense of Zadeh, anyone who is to use fuzzy sets must answer the following questions:

(i) What does graded (descriptive and/or veristic) membership mean?
(ii) How are these memberships measured?
(iii) What operations are meaningful to perform on a combined treatment of descriptive and veristic membership values that represent various relations between linguistic concepts?

To answer the first question, one has to subscribe to a certain view of fuzziness. We first summarize some of these views and then present a measurement-theoretic development in which these three questions can be answered.
It is to be noted that both the “likelihood” and the “random set” views of the membership functions stated above implicitly assume that there are more than

one evaluators or experiments are repeated. Therefore, if one thinks of membership functions as "meaning representation", they come close to the claim that "meaning is essentially objective". On the other hand, during the initial phases of the development of fuzzy sets, it has been widely accepted that membership functions are subjective (Zadeh 1965, 1975). The “similarity” and “utility” views of the membership functions differ from the others in their espousing a subjective interpretation. The measurement view is the connection of subjective and objective views in the sense that the problem can be defined in both ways depending on the observer(s) who is (are) making the assessments. The assessments can be results of subjective evaluations or results of objective analysis or a combination of both. Again all these assessments have an implicit verification associated with them which we are making explicit in this writing in order to clarify the set and the logic connection which is required in the derivation of the formulas of the protoforms for the combination of concepts to be discussed later in detail. It is important to emphasize that all these membership assessments are context dependent.

*3.1.1. The likelihood view*

Particularly, Hisdal (1985, 1988) advocates a likelihood view of the concept of a grade of membership. Hisdal's work is a typical example of creating a descriptive (interpretive is the word she uses) theory of fuzzy sets.

Hisdal's TEE (Threshold, Error, assumption of Equivalence) model for membership functions considers several sources of fuzziness:

Errors in measurement (which are essentially statistical in nature).
Incomplete information.
Interpersonal contradictions.

Furthermore, Hisdal argues that, for a single subject equipped with perfect information (about, say, John's size) there cannot be any fuzziness! In this case John would be called “big” if his size is greater than a well-defined or arbitrarily accepted threshold that forms the foundation of dichotomous two-valued assignments.

Hisdal considers three ways of eliciting the membership function:

(i) Label experiment: “John is big”, where big $\in$ {very big, big, small}. A careful observation shows that on such experiments imprecise description is stated explicitly by the subject with an implicit two-valued verification.
(ii) Yes-No experiment: “Is John big?” In these experiments, the analyst asks an explicit but imprecise descriptive question and the subject gives an explicit two-valued verification.

(iii) Grade of membership experiment: "What is the degree of belonging of John to the set of big people?" In these experiments, the analyst asks for a specific membership value that assigns "John" to an imprecisely described set of "big" people. But when the subject gives a specific membership value, there is an implicit two-valued verification, that, what s/he states is true to the best her/his knowledge. That is, it appears, Hisdal's postulate confuses the distinction, we are making between the descriptive and veristic assignments. Essentially, Hisdal's argument goes as follows: when asked to what degree a person is "big", the subject who has an imprecise measurement of the person's weight (say he knows that it is within 100± 10 kg) constructs an error curve to quantify the possible error around 100 kg. Then, s/he would argue as: I verify that "in 75% of the cases the subject's height lie in my quantization interval for 'big' ". Therefore

$$\mu_V\,[\,\mu_{\text{big}}(100\text{kg})] = \mu_v[\text{P}(\text{big}|\text{x} = 100\text{kg}) = 0.75\,] = 1(\text{T}),$$

where P(big | x=100 kg) denotes the (conditional) probability, 0.75, that the label "big" is assigned and verified (attested) to be true $\mu_V[\cdot]=1$ for the subject in question given that the subject's weight is 100 kg.

In this and other interpretations, it should be noted that $\mu_V[\cdot]=1(\text{T})$ is our addition to emphasize that a given descriptive membership value assignment is explicitly verified, V, to be true, T.

Hisdal postulates that a subject's grade of membership assignment is the modification of her/his (crisp) answers to the Label or Yes-No experiments where the modification is made by her/his estimate of the error curve.

These arguments essentially apply to the first type of fuzziness, errors in measurement.

Hisdal also goes on to explore the connectives which are valid under this interpretation. She argues that min and max are not always appropriate as conjunction and disjunction, respectively. This should come as no surprise, since in this probabilistic setting we know that joint probabilities are not always simply recovered from the constituent probabilities aside from the fact that there is the overriding simplification that lumps the descriptive and veristic concerns. Furthermore, it is our view that probabilities are in general orthogonal to set descriptions. That is probability theory and (fuzzy) set theory have different complementary agendas for the explanation of real-life phenomena.

*3.1.2. Random Set View*

The definition given for a membership function in Figure 3.1 is called the vertical representation (Dubois &Prade, 1989). There is at least one other representation of a membership function. It is the, so-called, horizontal representation where a fuzzy set F on a referential set X is represented in terms of its "α-level-cuts" (Figure 3.2):

$$\{A_\alpha | \alpha \in (0,1]\},$$

where $A_\alpha = \{x | \mu_A(x) \geq \alpha\}$. Zadeh (1971) defines the membership function as:

$$\mu_A(x) = \sup\{\alpha \in (0,1) | x \in A_\alpha\}$$

In this representation, the membership function is viewed horizontally as a nested family of α level-cuts. Note that each level-set, $A_\alpha$, is a set in the classical sense.

Alternatively, the membership function can also be represented as an integral:

$$\mu_A(x) = \int_0^1 \mu_{A_\alpha}(x)\, d\alpha.$$

where $\mu_{A_\alpha}(x) = 1$ if $x \in A_\alpha$ and 0 otherwise (Dubois & Prade, 1989). Of course, one must make continuity and measurability assumptions so that the integral is well-defined. In that sense, it is a specialized view of the membership function. However, in this view the membership function can be viewed as a uniformly distributed random set, consisting of the Lebesgue measure on [0,1] and the mapping $\mu_{A_\alpha} : X \to \{0,1\}$.

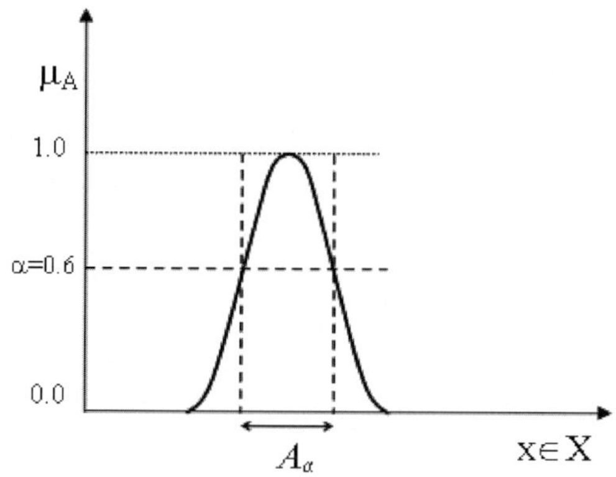


**Figure 3.2.** The horizontal representation of membership function, $\mu_A(x) \geq \alpha$ and $\mu_{A_\alpha} = 1$ for $x \in A_\alpha$ and 0 otherwise.

Therefore, for $\mu_A(x)$, the random set view is stated as: say, 60% of the population defined and verified that a particular interval on the base set, X, is an interval containing x for a given size, e.g., x=John's size, on the basis of a linguistic term A, e.g., bigness, whereas 40% define intervals which exclude x to be A. It should be noted that 0.6 is a value on the membership axis. Note once again, we add an explicit verification to each definition!, i.e., $\mu_{A_\alpha}(x)=1$ is verified, that it is asserted to be true i.e., $\mu_V[\mu_{A_\alpha}(x)=1]=1$. To make our purpose further clear, we note that one could have the following concern: "if the population do not attest to its being true then what is the value of their definition and assignment of such membership values?" That is "on what basis are we to make use of the given membership information"?

### *3.1.3. Similarity View*

The similarity view of the membership functions naturally arise in prototype theory where membership is a notion of being similar to a representative of the category (Lakoff, 1987). It is also in this sense that we are investigating the expressions for the combination of concepts in CWW. In a sense a membership function measures the degree of similarity of an object, subject to the set in question. It is also in this sense that we ought to take the prototype membership functions. That is they measure the degree of similarity of prototype to everyday linguistic expressions of our natural languages. It is assumed that there exists a perfect example of the set (or the category or the natural language expression) that belongs to the set to the full degree. Others belong to the set to a degree measured by their relative distance, or similarity, to the perfect example, i.e., the prototype.

An early theoretical and experimental discussion of this method is given by Zimmermann and Zysno (1985) in which a mathematical model of a membership function is defined on the real line where memberships are assumed to be a function of the distance between an object and an ideal object. In our sense the membership functions are assumed to be functions of distance between our definitions, meaning representation, and ideal meanings of natural languages expressions.

Zwick, Carlstein and Budescu (1987) study different measures of similarity among fuzzy concepts. They consider similarity measures defined both on metric spaces and in a set theoretic manner. They consider 19 such measures and try to experimentally verify the performance of these measures. They report that all measures successfully yield similar and dissimilar concepts in a crude manner. However, when the aim is to distinguish between degrees of similarities, metric measures perform better. From a measurement-theoretic point of view, this is expected. In the crude categorization one is interested only about similar objects not with how similar they are. This leads to two-valued set assignment separating similar and non-similar objects in a crisp manner. This can be done on an ordinal scale. However, when the aim is to find out how similar the concepts are then one needs to move to interval and ratio scales on which metric measures are meaningful and carry more information.

Ruspini (1991) also considers similarity semantics for fuzzy set theory. He builds the notion of similarity on a certain modal logic (Chellas, 1980) in which the accessibility relation is a similarity relation in a metric space. This method provides a formalism of fuzzy set theory based on the possible world semantics of modal logics (Klir, et. al. 1994; Resconi, et. al. 1992, 1993, 1994, 1999). However, it needs to be pointed out that modal logic related investigations provide us verity set memberships in fuzzy sets but not descriptive fuzzy set assignments (Resconi and Türkşen, 2001).

Clearly similarity measure that describes how similar an object is to a prototypical element of a set is one aspect of our knowledge which is the descriptive part and its explicit assertion and/or verification in terms of modal logic accessibility relation is the other aspect of our knowledge which is the verity assessment of the descriptive situation of an object in a category. However intuitive, the similarity view of the membership functions demands too much from the cognitive abilities of human beings. Regardless of the type of fuzziness (subjective, objective, personal, interpersonal) one who subscribes to similarity view assumes that perceived fuzzy concepts can be evaluated on a continuous metric space. The measurement-theoretic discussion that is provided in Section 3.2 poses some questions and doubts as to metric space assumptions (particularly the Archimedean condition).

In the light of demands on the cognitive abilities of humans and concerns about metric space assumptions, our thesis on the expressions for the combination of (imprecise) concepts with (imprecise) connectives, i.e., linguistic connectives "AND", "OR", etc., takes on a particular stance towards CWW and computing with perceptions being approximate with respect to unknown absolute meaning of concept combinations.

### *3.1.4. Utility Theory View*

Giles (1988) offers a decision-theoretic interpretation for membership functions. He argues that a set is equivalent to a property and claims that a sound meaning to the membership function can be given by considering it together with the problem of fuzzy reasoning. Hence, he is following the path of logic to come up with a semantic theory for the concept of a graded truth value. Again we note the confusion that juxtaposes membership values and truth values.

He defines a fuzzy sentence as "a sentence to which we attribute a degree of belief" which is a function of possible worlds (or possible states of mind, nature, etc.). Giles considers sentences that are asserted rather than merely uttered. For example, when one asserts that "John is big" Giles assumes that there exists a pay-off function related to this assertion. In our view, assertion is nothing more than a truth verification. In order to further clarify one's assertion one may want to specify a pay-off function which is a further qualification. This pay-off function offers more if the statement is closer to truth (e.g., you are more credible among your peers if your statement is closer to the truth!). Thus, Giles attempts to explicitly bring to the surface the associated truth verification, that we have brought to the surface explicitly. This assumption provides the utility theory approach for the assessment of the membership function of truth for a given sentence. Giles (1988, p. 304) claims that "Utilities of assertions carry the meanings of assertions" which in fact is not the case. Utility of assertions brings to the surface nothing more than the associated truth degree. Of course, the utility approach yields a membership function for the truth verification on an interval scale (unique up to positive linear transformations) (Fishburn 1970).

When viewed in this way, the connectives for the theory are truth functional in a particular theory! As we shall discuss later, it is by distinguishing and separating membership value assignments and truth assignments, we will be able to show how the classical connectives can simply be applied to give the combination of graded truth, verity, concepts in a truth functional way. In this manner, as it will be shown later, we are able to drive the expressions as disjunctive and conjunctive canonical forms where the mapping of linguistic connectives, such as "AND", "OR" are not one-to-one, isomorphic to t-norms and t-conorms, respectively.

### 3.2. Measurement Theory View

The three problems posed earlier can be re-assessed by using the techniques of measurement theory (Krantz, Luce, Suppes and Tversky 1971, Roberts 1979, Narens 1986, Suppes, Krantz, Luce and Tversky, 1989, Luce, Krantz, Suppes and Tversky 1990). In such a theory one can discuss the representation, the uniqueness, and the meaningfulness of such representations.

Here the measurement theory is applied to formally answer the general questions that was posed at the beginning of this section. The aim is to bring to the surface the axioms that are implicit in subscribing to a particular interpretation, i.e., semantics of membership functions and to discuss them as to their testability and/or intuitiveness.

It should be clearly understood that the basic tenet of measurement theory (as in any axiomatization of a theory) is to lay out assumptions as clearly as possible so that their suitability can be tested and/or they can be accepted on normative grounds.
At this juncture, it is important to differentiate between two related but different measurement problems: (i) membership measurement and (ii) combination of memberships with connectives. We argue that, although the first problem received much attention in the literature the second one is closely related to the question of "which connectives to use in the combination of imprecise, fuzzy, linguistic terms that are represented by membership functions?". Furthermore, we have to explore the relationships between linguistic connectives, "AND", "OR", etc., that are imprecise and do not map directly in a one-to-one correspondence to precise connectives, t-norms, $\Delta$, and t-conorms, $\nabla$, respectively. This, in turn, causes a break down of certain equivalences that exist in classical theory with many consequences. With this background, these problems can be combined to arrive at an axiomatization for fuzzy set theory. For this purpose, with definitions, representation and uniqueness results are stated for ordered algebraic structures to expose issues related to crisp connectives in this chapter. The discussion on imprecise connectives is treated later. The results obtained here provide a technical background to facilitate the discussion of the measurement-theoretic frameworks for fuzzy set theory in the subsequent sections for our proposed framework of combination of concepts and the derivation of their formulas.

### *3.2.1. Membership Measurement*

There are two important (but different) measurement problems in fuzzy set theory. The first kind deals with measuring the degree of membership of several subjects or objects in a single fuzzy set. This problem has been studied in (Yager, 1979; Norwich and Türkşen, 1982b; Norwich and Türkşen, 1984; Türkşen, 1991; Bollmann-Sdorra, Wong & Yao, 1993; Bilgic, 1995) among others.

For this purpose, it is assumed that there are a single fuzzy set A and an infinite number of agents x in X, etc. The question is "To what degree an agent from X belongs to a fuzzy set A?" The first measurement problem is stated as: "an agent belongs more to A than another agent", where A is a linguistic term, usually a noun qualified by an adjective or an adjective and an adverb. The important thing to notice is the fact that this representation compares several agents or several experimental observations over a single fuzzy set, A. The resulting representation measures the degree to which each agent belongs to the fuzzy set A.

To capture this graded membership concept, consider a binary relation, $\underset{\sim A}{\succ}$ on X with the following interpretation:
"'x $\underset{\sim A}{\succ}$ y $\Leftrightarrow$ x belongs to A at least as much as y belong to A' is true (or false)."
Or equivalently, "'x $\underset{\sim A}{\succ}$ y $\Leftrightarrow$ x is at least as much as A as y is A' is true (or false)." Examples of such sentences are:

"'Mary is at least as intelligent as John' is true (or false)."
"'The new generation is less political than the old one' is true (or false)."
"'This task is more important than the other' is true (or false)."
Recall that logicians would state these very same statements as:
"It is (or not) the case that 'Mary is at least as intelligent as John'."
"It is (or not) the case that 'the new generation is less political than the old one'."
"It is (or not) the case that 'this task is more important than the other'."

Norwich & Türkşen (1982b, 1984) consider the relation $\underset{\sim A}{\succ}$, and discuss its possible representations. The structure is assumed to be bounded (i.e., there exists elements $x^-$ and $x^+$ in X such that for all other $x \in X$, $x^- \underset{\sim A}{\succ} x \underset{\sim A}{\succ} x^+$. This amounts to assuming that there are elements $x^+, x^- \in X$ for which the fuzzy assignment of the term A is completely true or completely false, respectively. If A is "big", for example, there may be some elements of A for which "absolutely big" and "absolutely not big" are completely unambiguous and this can be verified for the purpose at hand. However, if $x \in X$ is "beautiful" or "smart", etc., it is very hard to find elements $x \in X$ who are "absolutely beautiful" or "absolutely not beautiful". If the boundary conditions are accepted, then it is clear that an ordinal scale representation exists for (X, $\underset{\sim A}{\succ}$) once the transitivity and connectedness of $\underset{\sim A}{\succ}$ are accepted.

Norwich & Türkşen (1982b) also consider the difference measurement problem. Discussing the axioms of an algebraic difference structure (X x X, $\underset{\sim A}{\succ}$) for membership in a fuzzy set, they argue that membership can be measured on an interval scale provided the axioms of algebraic difference structure can be verified. The extra information required in this case is to be able to compare pairs of agents (e.g., "John is bigger (smaller) than James" and "James is bigger (smaller) than Jack" implies that "John is bigger(smaller) than Jack"). The interval scale uniqueness for this representation guarantees that averaging is a meaningful operation as well as comparison but the notion of an origin is

arbitrary (i.e., the bounds do invariably map to the same numbers in the numerical domain).
Hence, a function, $\mu_A$:X→[0,1] exists, i.e., verified, such that:

"$x \underset{\sim A}{\succ} y \Leftrightarrow \mu_A(x) \geq \mu_A(y)$ is true, T (or false, F)."

The uniqueness is either ordinal or interval depending on the axioms that are accepted and/or verified to be true where $\mu_A(x)$ measures the degree with which "x" belongs to A. In the case of an interval scale, the following representation must also hold:

$(x,y) \underset{\sim A}{\succ} (z,w) \Leftrightarrow \mu_A(x) - \mu_A(y) \geq \mu_A(z) - \mu_A(w)$ is true, T (or false, F).

In Norwich & Türkşen (1984), it is argued that ratio scales are not likely to arise in fuzzy set theory. They suggest that there is the difficulty with the concatenation of psychological attributes and hence no operational meaning can be assigned to the concatenation operator (addition) of extensive measurement which results in a ratio scale. However, Norwich and Türkşen (1984) also raise the possibility of defining "relative ratio scale" assignments for certain context dependent cases. Norwich & Türkşen(1984) also show that one obtains a scatter of membership values in these experiments. This will be further discussed later in the acquisition of Type 2 membership representation. In the next section, another representation theorem is discussed in which ratio scale formally arises.

### *3.2.2. Ordered Algebraic Structures*

In this section basic definitions; and representation and uniqueness results are discussed for algebraic structures called ordered semi-groups. These results form the basis of further discussions on measurement-theoretic semantics to be further discussed later. In this regard, the results of Fuchs (1963) and Schweizer & Sklar (1983) are translated in terms of ordered algebraic structures as is customary in the measurement theory literature.

**Definition 1.** The algebraic structure $<A, \oplus>$ where A is a nonempty set and $\oplus$ is a (crisp) binary operation on A, is called a semi-group if and only if $\oplus$ is associative (i.e., for all a,b,c,∈A, $(a\oplus b)\oplus c=a\oplus(b\oplus c)$. If there exists e∈A such that for all a∈A, $e\oplus a=a\oplus e=a$ the structure $<A, \oplus,e>$, is called a semi-group with identity e or a monoid. Finally, $<A, \oplus,e>$ is a group if and only if it is a semi-group with identity e and any element of A has an inverse, for all a∈A, there exists b∈A such that $a\oplus b=b\oplus a=e$.

When the algebraic structure is also endowed with an ordering, $\underset{\sim}{\succ}$ one then obtains ordered algebraic structures.

**Definition 2.** Let A be a non-empty set, with a binary relation "$\succsim$" on A and $\oplus$ a binary operation on A. $\langle A, \succsim, \oplus \rangle$ is an ordered algebraic structure if and only if the following axioms are satisfied: (weak ordering) the relation $\succsim$ is connected and transitive (monotoncity) for all a,b,c,d,$\in$A, a $\succsim$ c and b $\succsim$ d imply a$\oplus$b $\succsim$ c$\oplus$d. The asymmetric part ($\succsim$) and the symmetric part ($\sim$) of any relation $\succsim_A$ are defined as usual: a $\succsim_A$ b if and only if a $\succsim_A$ b and not b $\succsim$ a and a$\sim$b if and only if a $\succsim$ b and b $\succsim$ a.

Adding more properties to an ordered algebraic structure results in specializations of the concept. In this section, only ordered semi-groups are considered (where the concatenation is associative). These are summarized in the following definition.

**Definition 3.** Let **A**=$\langle A, \succsim, \oplus \rangle$ be an ordered algebraic structure such that $\langle A, \oplus \rangle$ is a semi-group. Then **A** is called an ordered semi-group. Furthermore, it is said to be: .

Weakly Associative (WA): iff a$\oplus$(b$\oplus$c)$\sim$(a$\oplus$b)$\oplus$c
Solvable (Sv): iff whenever there exists c$\in$A such that a $\succsim$ b$\oplus$c.

Strongly Monotonic (SM): iff whenever a $\succsim$ b then a$\oplus$c $\succsim$ b$\oplus$c then c$\oplus$a $\succsim$ c$\oplus$b.

Homogeneous (H): iff whenever a $\succsim$ b iff a$\oplus$c $\succsim$ b$\oplus$c iff c$\oplus$a $\succsim$ c$\oplus$b.

Idempotent (Ip): iff for all a$\in$A, a$\oplus$a$\sim$a.

Bounded (B): iff there exist unique elements u and e in A such that for all a$\in$A, u $\succsim_A$ e, u $\succsim$ a and a $\sim$ e.

Archimedean (Ar): iff for any a,b$\in$A there exists a positive integer m such that $a^m \succsim_A b$ where $a^m$ is recursively defined as $a^{(1)}=a$, $a^{(m)}=a\oplus a^{(m-1)}$.

Continuous (C): iff $\oplus$ is continuous as a function of two variables, using the order topology on its range and the relative product topology on its domain.

By a representation of an ordered algebraic structure, we mean a real valued function that maps the ordered algebraic structure, $\langle A, \succsim, \oplus \rangle$ to a numerical

structure, $<X,\geq,S>$, where X is a subset of R, with the natural ordering of real numbers and $S:X \times X \rightarrow X$ is a function. Since we focus on ordered semi-groups, in the resulting representation, S is necessarily associative.
The boundary condition, asserts the existence of a minimal and maximal element in set A. Hence, given the weak ordering and the boundaries, one can replace the set A by the familiar interval notation [e,u].
The following lemma demonstrates some of the consequences of axioms imposed on a bounded ordered semi-group (Schweizer and Sklar, 1983).

**Lemma 1.** Let $<A, \succsim, \oplus>$ be a bounded ordered semi-group with bounds e and u. Then A also satisfies the following conditions for all $a,b \in A$:

$a\oplus b \succsim \sup(a,b)$,

$u\oplus a \sim a\oplus u \sim u$

$a\oplus a \succsim a$.

Schweizer & Sklar (1983, Section 5.3) define a function on a closed real interval endowed with the natural ordering. With this background, representation theorems with varying uniqueness characteristics can be stated (Bilgic and Türkşen, 1996).

### 3.3. Membership and Connectives

One of the concerns is the relationship between the measurement of membership and the selection of crisp connectives, i.e., t-norms, $\Delta$, and t-conorms, $\nabla$.

The justification of using different operators as connectives for fuzzy sets is considered by many authors. Bellman and Giertz (1973) give the first axiomatic justification for using min and max operators followed by Fung and Fu (1975). Yager (1979) also considers the same problem and proposes axioms from which min and max follow as the only solutions. Furthermore, the introduction of triangular norms, conorms and parameterized families of functions as possible candidates of connectives generated a large body of research (Yager, 1980; Dubois and Prade, 1980, 1982; Weber, 1983; Alsina, 1985).

Dubois and Prade (1989) mention the measurement-theoretic justification of the connectives and this problem was discussed further as a conjoint measurement problem by Türkşen (1991). Bollmann-Sdorra et. al. (1993) discusses the same problem and clearly distinguishes the membership measurement and selection of (crisp) connectives. Bollmann-Sdorra, et. al., justify the use of min and max operators as intersection and union, respectively.

Bilgic (1995) and Bilgic and Türkşen (1994) consider the same problem. They start with a very weak algebraic structure for which only an ordinal scale representation exists. Then, they continue to add axioms on this structure on an "as-needed" basis and show how ratio and absolute scale representations are obtained. At each addition of a new axiom they discuss the testability and plausibility of the new axiom for fuzzy set theory.

### *3.3.1. Scale Strength*

Membership measurement can at best be performed on an interval scale (Norwich and Türkşen, 1984; Türkşen, 1991). The formal existences of scales that are stronger than interval pose the problem of identifying a natural origin. However, since the ranking of properties of an individual is a highly subjective act of the observer, there cannot be universally accepted bounds on the measurement scale (Norwich and Türkşen 1984). This makes all the scales resulting from the measurement, relative to the observer and amounts to concluding that the formally attainable absolute or ratio scales are not likely to arise in the framework of fuzzy set theory (Bilgic and Türkşen 1994). Amongst the axioms imposed on the fuzzy set structure, the boundaries and the Archimedean axiom are the most problematic.

The boundary axiom asserts the existence of (unique) maximal and minimal fuzzy terms and requires that there should be a fuzzy term which is attributed more than any other fuzzy term to the subject. Although this sounds reasonable for some cases, it is unlikely that it will hold for all fuzzy terms.

The Archimedean axiom (Definition 3) means that if Mary is smarter than she is angry, there should be a finite amount of anger which, when attributed to Mary, will make her at least as angry as she is smart. Admittedly this might hold for some cases. However consider the statement "alcohol intake is more hazardous to liver than it is enjoyable". Archimedean axiom states that a finite amount of joy would make alcohol intake more enjoyable than it is hazardous. Is this quantity really finite?

It is important to notice that all the representations that are discussed in this section are "two-way" representations in the sense that the conditions to be imposed on the qualitative structure should be both necessary and sufficient for a representation on the numerical domain. If "one-way" representations are considered, then many of the problematic structural axioms like Archimedean and continuity that are necessary for the representation in the real line can be omitted. However, this approach would not result in the highly fruitful concept of meaningfulness.

If this is the approach taken then one can use t-conorms for all representations since they satisfy all the conditions of the theorems and more. However, such usage must be made with care. The numbers resulting from using t-norms and/or conorms, i.e., the conjunctions and/or disjunctions can only be compared to each other in the ordinal sense and no other arithmetic operation (however tempting) can be performed on them unless one is ready to accept or be willing to verify empirically that the Archimedean and continuity axioms are truly held. If everyone cannot agree on the bounds of the system, the ratio and absolute scale representations that arise may be considered to be relative to the observer. On the other hand, without the Archimedean axiom, there does not seem to be a representation that has a uniqueness stronger than ordinal scale.

In fact, all these questions are partially answered in experiments of membership function. In these experiments, we do not obtain "unique" responses from subjects for the same object. This causes the generation of Type 2 membership representation and reasoning due to the fact that there are multi-valued mappings both in representation and reasoning. These are further reasons to consider expressions as approximation to absolute (ideal) concept combinations.

### *3.3.2. Two or More Subjects*

Another of our concerns is the measurement of membership functions for two or more subjects or objects. In combining such membership functions, it is almost impossible to treat separately the measurement of membership functions and the selection of crisp connectives to be used in the combination of membership functions.

The scales resulting from membership measurement and property ranking problems do not necessarily measure the same entity. The degree with which a subject or object belongs to a certain fuzzy set may not be equal to the degree that fuzzy set is associated with him or her.

One way to pursue this is to combine the membership measurement and property ranking problems by introducing a new structure where the resulting measurement scale necessarily measures the membership degree in a fuzzy set. This path is followed by Bollmann-Sdorra, et al., (1993) Bilgic, (1994) and Bilgic and Türkşen (1995).

Bollman-Sdorra et al. (1993) consider an ordered algebraic structure which justifies the use of min as conjunction and max as disjunction. Their representation is ordinal. Bilgic and Türkşen (1995) introduce two ways this combination can be made. In the first one, the two different problems are simply cast into a bounded semi-group structure. The consequences of this model are analyzed. It is argued that since accepting the Archimedean axiom can be very hard for some fuzzy terms, the ratio scale representations are not likely to arise.

In the literature of the ordered algebraic structures, there are very strong (absolute scale) representations for similar structures that one can use to represent a fuzzy set structure.

However, representations stronger than ordinal scales require unacceptable structural axioms like the strong monotonicity (SM) and the Archimedean (Ar) conditions.

If one gives up the Archimedean axiom and endows the structure with other axioms (particularly the Idempotency axiom, Figure 3.3), then some weak representations can be obtained. These are ordinal scale representations and particularly the function max as originally suggested by Zadeh can be recovered as the unique disjunction satisfying some reasonable axioms.

However, ordinal scale representations exist at the cost of accepting that the equivalence of two fuzzy propositions is transitive. This fact is known to yield a class of paradoxes usually called the heap's or sorties paradox. The equivalence of fuzzy terms cannot be transitive. In order to introduce the intransitivity of equivalence in fuzzy set theory, Bilgic and Türkşen (1995) propose another model in which the weak ordering axiom is replaced by interval ordering axiom which results in a threshold representation. Unfortunately, this representation has some peculiar uniqueness characteristics.

The measurement models show that measurement of membership functions in fuzzy set theory is formally possible. However, the acceptability of each formal model needs to be critically analyzed. Alternatively, one needs to go to Type 2 representation by relaxing formal requirements. By moving to Type 2 representation, one captures uncertainty associated with the measurement of membership beyond imprecision.

On the other hand, once the Archimedean axiom is omitted it is not possible to come up with representations stronger than ordinal scale. This suggests that if one is not ready to accept the Archimedean axiom the only meaningful operation that can be performed on the measurement scale is the comparison (i.e., the results are on an ordinal scale). Any other arithmetic operation is simply meaningless. *This suggests that triangular norms and conorms can be used to model crisp connectives in fuzzy set theory but their results should not be attached any cardinal significance.* Thus they provide an approximate representation and thus approximate reasoning must be implemented.

Archimedean axiom is necessary when a representation into real numbers is sought. If a representation into field extensions of the real number system is considered, then Archimedean condition is not necessary anymore (Narens 1986). In this sense, Type 2 representation is realized as a field extension of real numbers. However, this approach has consequences of a philosophical nature as to why degrees of truth are not real numbers and why do they require non-standard analysis (Robinson 1966). This is still an open question. This is also a ground for the break down of certain equivalences that exist in classical theory.

If one accepts the Archimedean axiom and continuity, the results about the additive generators of Archimedean triangular norms and conorms simply state that using these in the unit interval and using addition on the extended reals amount to the same thing. Therefore, using Archimedean triangular norms or conorms as conjunction or disjunction of fuzzy sets is simply a matter of preference or convenience. For disjunction, one can equivalently use addition on extended reals. Therefore, the justification of continuous, Archimedean triangular norms and conorms require some unnatural structural axioms and in the end they amount to ordinary, additive extensive measurement. In order to move away from additivity one either has to give up the Archimedean axiom which as we have seen does not lead to strong representations (but nevertheless recovers max) or give up some other structural axiom. Luce, et. al. (1990) show that if one accepts the Archimedean axiom then associativity (in presence of positivity) amounts to additivity. Hence, the first candidate to give up seems to be the associativity. The attempts of Fodor (1993) and Fodor and Keresztfalvi's (1995) to generalize triangular norms by dropping associativity require a measurement-theoretic discussion. As for giving up the continuity, the main thrust behind the concept of a fuzzy set is continuous gradation. Hence moving away from continuity is highly counter-intuitive. Why and where discontinuities should occur is an ill-defined problem.

### *3.3.3. Conjoint Measurement*

One particular difficulty with the notion of fuzzy terms is that they can be composed of two or more simple terms (e.g., the term 'comfort' may constitute two parts 'humidity' and 'temperature'). Türkşen (1991) applies conjoint measurement techniques to this problem with the aim of showing that the scale for the higher order term (comfort) is constructed from the individual scales (humidity and temperature) using triangular norms and conorms. Under the assumption of decomposability, Türkşen (1991) shows that all the triangular norms and conorms preserve the monotonicity of the building scales.

Türkşen (1991) considered the problem of justifying triangular norms, conorms, and parameterized families of them as valid under conjoint measurement. In so doing, he considers only "one-way" representations. However, one requires further investigation to find both necessary and sufficient conditions that the main operations such as triangular norms, conorms and the parameterized families must abide by their acceptability in fuzzy set theory .

### *3.3.4. Linguistic Concepts and Linguistic Connectives*

It is to be noted from Section 3.3.2 forward to this section, the discussion has addressed how membership values are combined with t-norms and t-conorms with the aim of justifying the suitability of a given t-norm and/or a t-conorm. A higher level of concern is the combination of linguistic concepts with the linguistic connectives such as "AND", "OR", "IMPLIES", etc. As it was pointed out in the preface, it is known that linguistic connectives "AND", "OR" do not correspond in a one-to-one mapping to any one of the t-norms and t-conorms, respectively. (Zimmermann and Zysno, 1980, 1985; Türkşen,1992). The separation of the "descriptive" membership assignments and the "veristic" membership assignments together with the experimental (Zimmermann and Zysno, 1980, 1985) and formal results (Türkşen, 1992) obtained on linguistic connectives "AND", "OR", etc., lay the necessary grounds for the investigation of one-to-many mappings for "AND", "OR" connectives with the application of t-norms and t-conorms within the derived fuzzy canonical forms to be discussed later in this work. Once again, it will be shown that the equivalence of normal forms in classical theory break down in fuzzy theory with many consequences. These naturally lead to the generation of Disjunctive and Conjunctive Canonical forms, which are not equal to each other and hence, causes the generation of Interval-Valued Type 2 fuzzy set realization in representation and reasoning. These will be discussed in more detail further in this book. Before we get to that treatment it is worthwhile to review next the membership elicitation methods.

## References


1. Alsina C. (1985), "On a Family of Connectives for Fuzzy Sets, *Fuzzy Sets and Systems* 16, 231-235.
2. Bellman, R., M. Giertz (1973), "On Analytic Formalism of the Theory of Fuzzy Sets", *Information Sciences*, 5, 149-156.
3. Bilgic, T. (1995), Measurement-Theoretic Frameworks for Fuzzy Set Theory with Applications to Preference Modeling, PhD thesis, University of Toronto, Dept. of Industrial Engineering Toronto, Ontario M5S 1A4 Canada.(supervisor, I.B. Türkşen)
4. Bollmann-Sdorra, P., S.K.M. Wong and Y.Y. Yao (1993), "A Measurement-Theoretic Axiomatization of Fuzzy Sets", *Fuzzy Sets and Systems,* 60(3), 295-307.
5. Chellas, B.F. (1980), Modal logic: An Introduction, *Cambridge University Press*, Cambridge, UK.
6. Dubois, D., H. Prade (1980), Fuzzy Sets and Systems: Theory and Applications, *Academic Press*, New York,.
7. Dubois, D., H. Prade (1989), "Fuzzy Sets, Probability and Measurement", *European Journal of Operational Research*, 40, 135-154.

8. Dubois, D., H. Prade (1982), "A class of Fuzzy Measures Based on Triangular Norms: A General Framework for the Combination of Information", *International Journal of General Systems*, 8, 43-61.
9. Fishburn, P.C. (1970), Utility theory for decision making, Operations Research Society of America, *Publications in Operations Research*, No. 18, Wiley, New York.
10. Fodor, J.C. (1993), "A New Look at Fuzzy Connectives", *Fuzzy Sets and Systems*, 57: 141-148.
11. Fodor, J.C., T. Keresztfalvi (1995), "Nonstandard Conjunctions and Implications in Fuzzy Logic", *International Journal of Approximate Reasoning*, 12: 69-84.
12. Fuchs, L. (1963), Partially Ordered Algebraic Systems, *Pergamon Press*, London.
13. Fung, L.W. & Fu, S.K. (1975), An axiomatic approach to rational decision making in a fuzzy environment, in H.J. Zimmermann, L.A. Zadeh & B.R. Gaines (eds), *Fuzzy Sets and Their Application to Cognitive and Decision Processes*, Academic Press, New York, pp. 227-256.
14. Giles, R. (1988), The concept of grade of membership, Fuzzy Sets and Systems 25: 297-323.
15. Hisdal, E. (1988), "Are Grades of Membership Probabilities?", Fuzzy Sets and Systems, 25, 325-348.
16. Hisdal, E. (1985), "Reconciliation of the Yes-No Versus Grade of Membership Dualism in Human Thinking", in M.M. Gupta, A. Kandel, W. Bandler & J.B. Kiszka (eds), Approximate Reasoning in Expert Systems, North-Holland, Amsterdam, pp. 33-46. Also in Dubois, D., H. Prade, and R. Yager eds., Readings in Fuzzy Sets for Intelligent Systems, Morgan Kaufmann, pp. 854-860, 1993.
17. Klir, G.J. (1994), "Multi-valued Logics Versus Modal Logics: Alternative Frameworks for Uncertainty Modeling", in P.P. Wang (ed.), *Advances in Fuzzy Theory and Technology*, Duke University, Durham, North Carolina USA, Vol. 2, 3-48.
18. Krantz, D.H., R.D. Luce, P. Suppes and A. Tversky (1971), Foundations of Measurement, Vol. 1, *Academic Press*, San Diego.
19. Lakoff, G. (1987), Women, Fire, and Dangerous Things: What Categories Reveal About the Mind, *The University of Chicago Press*, Chicago.
20. Luce, R.D., Krantz, H.D., Suppes, P. & Tversky, A. (1990), Foundations of measurement, Vol. 3, Academic Press, San Diego.
21. Lukasiewicz, J. (1920), O Logice Trojwartosciowej (On three valued logic) in, N. Rescher, Many-Valued Logic (1969), *McGraw Hill*, New York.
22. McCall, S., K. Ajdukiewicz (eds) (1967), Polish logic, 1920-1939, Oxford, Clarendon P. papers by Ajdukiewicz [and others]; with, an introduction by Tadeusz Kotarbinski, edited by Storrs McCall, translated by B. Gruchman [and others].
23. Narens, L. (1986), Abstract Measurement Theory, *MIT Press*, Cambridge, Mass.
24. Norwich, A.M., I.B. Türkşen (1984), "A Model for the Measurement of Membership and the Consequences of its Empirical Implementation", *Fuzzy Sets and Systems*, 12, 1-25.
25. Norwich, A.M., I.B. Türkşen (1982b), "The Fundamental Measurement of Fuzziness", in: R.R. Yager (ed.), *Fuzzy Sets and Possibility Theory*. Pergamon Press, New York, 49-60.
26. Resconi, G., G.J. Klir and U. St. Clair (1992), "Hierarchical Uncertainty Metatheory Based Upon Modal Logic", Int. J. of General Systems, Vol. 21, 23-50.
27. Resconi, G., I.B. Türkşen (2001), "Canonical Forms of Fuzzy Truthoods by Meta-Theory Based Upon Modal Logic", *Information Sciences*, 131, 157-194.
28. Roberts, F. (1979), Measurement Theory, Addison Wesley Pub. Co.

29. Ruspini, E.H. (1991), "On the Semantics of Fuzzy Logic", *International Journal of Approximate Reasoning,* 5(1), 45-88.
30. Schweizer, B., A. Sklar (1983), *Probabilistic Metric Spaces,* North Holland, Amsterdam.
31. Suppes, P., DII. Krantz, R.D. Luce and A. Tversky (1989), Foundations of Measurement, Vol. 2, *Academic Press*, San Diego.
32. Türkşen, I.B. (1991), "Measurement of Membership Functions and Their Acquisitions", *Fuzzy Sets and Systems,* 40, 5-38.
33. Türkşen, I.B. (1991), "Measurement of Membership Functions and Their Assessment", *Fuzzy Sets and Systems*, 40, 5-38.
34. Weber, S. (1983), "A General Concept of Fuzzy Connectives, Negations and Implications Based on t-norms and t-conorms, *Fuzzy Sets and Systems* 11, 115-134.
35. Yager, R.R. (1980), "On a General Class of Fuzzy Connectives", Fuzzy Sets and Systems, 4, 235-242.
36. Yager, R.R. (1979), "A Measurement-informational Discussion of Fuzzy Union and Intersection", *International Journal of Man-Machine Studies,* 11, 189-200.
37. Zadeh, L.A. (1975), "The Concept of a linguistic variable and its application to approximate reasoning", Parts I, II, *Information Sciences*, 8, 199-249, 301-357, 43-80.
38. Zadeh, L.A. (1971), "Similarity Relations and Fuzzy Ordering", *Information Sciences*, 3, 177-200.
39. Zadeh, L.A. (1965), "Fuzzy Sets", *Information and Control Systems*, Vol.8, 338-353.
40. Zimmermann, H.J., P. Zysno (1985), "Quantifying Vagueness in Decision Models", *European Journal of Operational Research*, 22, 148-154.
41. Zwick, R., E. Carlstein, D.V. Budescu (1987), "Measures of Similarity Among Fuzzy Concepts: A Comparative Analysis", *International Journal of Approximate Reasoning*, 1(2), 221-242.

# Chapter 4

# ELICITATION METHODS

Concerning membership functions, two main trends are detected in the experimental research. One branch tries to experimentally validate the postulates of fuzzy set theory. The other branch adopts a certain interpretation of the concept of the 'grade of membership' and then seeks ways to elicit them. Mainly, there has been six ways used in experiments with the aim of constructing membership functions (Norwich and Türkşen 1982b, Chameau and Santamarina 1987a, Türkşen 1991):

polling: In this method, the analyst asks the experimental subjects to verify explicitly the assignment of an object to a fuzzy set. In response to “Do you agree that John is tall?” with an (Yes/No) answer.

direct rating (point estimation): In this method, subjects are asked to classify color according to its darkness, classify John according to his tallness. In general, the question is: "How A is x?" In this method, the analyst asks the experimental subject to assign a membership value to x in A which is assumed to be verified implicitly since the subjects assignment of the stated membership value ought to be “true” to the best of his knowledge.

reverse rating: In this method, subjects are asked to identify the person who is tall to the degree 0.6? In general, one is asked to identify 'x' who (or what) is A to the degree $\mu_A$(x). In this method, the analyst asks the experimental subject to identify (hence verify implicitly) the object that belongs to the fuzzy set for a specific membership value.

interval estimation (set valued statistics): In this method, subjects are asked to give an interval in which they think an object lies in, or give an interval in which they think the height of John lies in. In these methods, the analyst asks the experimental subject to give an interval to which the objects property falls with the implicit assumption of verification.

membership function exemplification: In this method, subjects are asked "what is the degree of belonging of X in the fuzzy set A" “What is the degree of belonging of John in the set of tall people?” In general, "To what degree 'x' is

A?" In this method, the analyst asks the experimental subjects to determine the degree of an objects belonging to a fuzzy set with implicit verification.

pair wise comparison: In this method, subjects are asked which object x or y is more A (and by how much?) In this method, the experimental subject is to compare two objects and state the comparative membership with an implicit verification.

It should be pointed out once more in the responses to these six experiments, there is an implicit assertion that the given response is verified to be true by the respondent.

## 4.1. Polling Methods

In polling one subscribes to the point of view that fuzziness arise from interpersonal disagreements. The question "Do you agree that 'x' is A?" is asked to different individuals (a number of individuals) for a set of objects x. The answers are polled for each x and an average is taken for each x to construct the membership function. We come back to this approach in meta-theory based on "modal" logic and show that in fact, we obtain interval valued Type 2 fuzzy truthoods (Resconi and Türkşen, 2001).

It should be pointed out that yes/no responses are equivalent to true/false verifications. Thus they are in fact nothing more than logic verification assignments. Thus to take their averages as descriptive membership values clearly points out the confusion in the minds of the researchers who assume descriptive membership values and veristic membership values, i.e., truthood degrees, are the same. These average values can only determine degree of truth or verification, i.e., fuzzy truthood values, associated with the responses.

Hersh and Carmazza (1976) used this approach in their experiment 1. They presented their subjects a phrase such as "small", "very small", "large", "very large", etc., and then showed 12 squares in random order. The subjects responded by 'yes' or 'no' depending on whether or not they think that the phrase applies to the square shown. The subjects were also shown the 12 squares at the beginning of the experiment in ascending order so that every subject was operating within approximately the same context. Although the experimental results justify $(1-\mu(x))$ for the connective NOT, the 'squaring the membership function' was not justified for VERY. In our view, $(1-\mu(x))$, for example, should really be a truth qualification value for truth negation, and thus it should be interpreted as $(1-\mu_T(x,A))$. This means that 'x' belongs to A is false to the degree $\mu_F(x,A)=1-\mu_T(x,A)$ where A stand for, say, "small", but T and F stand for the fuzzy terms "true" and "false" respectively.

Once again, the information obtained from these experiments are not descriptive membership values for terms such as “small”, “large”, “very large”, etc., but they are veristic membership values for truth qualification by individuals who agree or do not agree with the matching of the squares with the linguistic values such as “small”, “large”, etc.

## 4.2. Direct Rating Methods

Direct rating seems to be the most straightforward way to come up with a membership function. This approach subscribes to the point of view that fuzziness arise from individual subjective vagueness. The subject is required to classify 'x' with respect to “A” giving membership value $\mu_A(x)\in[0,1]$ over and over again at different times. The experiment has to be carefully designed so that it will be harder for the subject to remember past answers. Hersh and Carmazza (1976) use this approach in their experiment 2. They report similar results for negation, disjunction, etc. However, it should be noted that in this experiment since descriptive membership values are given explicitly and they are assumed to be verified implicitly, they have obtained descriptive set assignments with the absolute verity qualification of “true”. Recognizing the responses of one particular subject, in this experiment, Hersh and Carmazza have also differentiated between the “descriptive” and the “veristic” interpretations of membership assignments. It appears one particular subject became aware of the distinction we are making, i.e., differentiation between “descriptive” membership and “veristic” membership grades.

One can also use direct rating to obtain membership function Türkşen (1988, 1991). In that case the subject is asked "How A is “x”?" and the subject marks a point on the interval between “0” and “1” on a sliding scale and the experimenter has a perfect reading of the subjects evaluation on the membership axis, scale, for “A”, i.e., $\mu_A(x)\in[0,1]$ is known to the experimenter but not to the subject. The same descriptive question is asked to the same subject over and over again, at different days and times. The membership function is constructed using the assumption of probabilistic errors and by estimating a few key parameters as is usually the case for this type of construction. It is to be noted that there is a scatter of $\mu_A(x)$ given at different times. Thus, the descriptive membership should be captured at least by an interval-valued Type 2 representation. But generally, in most current application, this descriptive membership values are captured in Type 1 representation, Figure 4.2. Our experimental interval-valued Type 2 results are shown in Figures 4.3 and 4.4. Chameau and Santamarina (1987a) also discuss this method (which they call membership exemplification, however, the term is reserved for another method). They used several subjects and aggregated their answers as opposed to asking a single subject the same question over and over again as was done in Norwich and Türkşen (1984) experiments. Norwich and Türkşen (1984) (Figure

4.2) and Chameau and Santamarina (1987a) report that this method results in membership functions with a wider spread (more fuzzy) when compared to polling and pairwise comparison. Once again, there is a clue here. These results indicate that "polling" experiments measure something other than the descriptive membership values. That is they measure verity membership values. The spread of membership function are clearly an indication that one ought to use a Type 2 representation (Figures 4.3 and 4.4) as opposed to reducing the spread to a membership function by curve fitting (see Figures 4.1 and 4.2).

On the other hand, it is clear that in direct measurement experiments, the subject gives two responses one is descriptive response that specify "how A is x, i.e., $\mu_A(x)\in[0,1]$ and the other is the logic response that verify the given descriptive response be it implicitly, i.e., the given membership value $\mu_A(x)$ is "true". Thus, we infact get $\mu_V[\mu_A(x)]=1$, $\mu_V[.]\in\{0,1\}$, i.e., {F,T}.

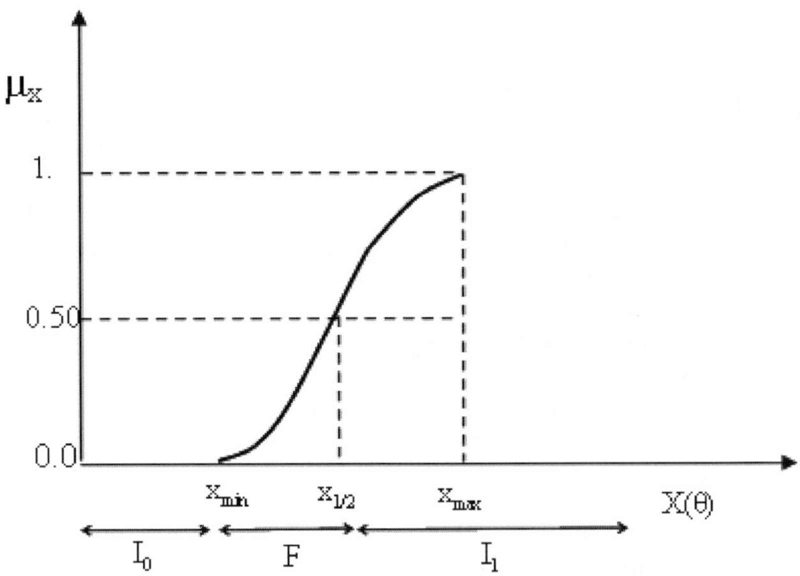


**Figure 4.2**. Experimental results of direct rating for "TALL MAN".

Zimmermann and Zysno (1980) consider the measurement of membership functions and the justification of connectives within an empirical setting based on measurement theory. They argue that since the numerical membership scale is bounded, the scale has to be the absolute scale. They mention the biases that one can have in direct elicitation methods, particularly the "end effect" which is a common problem in all bounded scales (including probability). They opt for an indirect elicitation method in the spirit of measurement theory and admit that only interval scales can be constructed with indirect methods. Then they suggest a combination of a direct rating technique and a Thurstonian scaling method.

This way they create two scales and then try to identify the relationship of the two scales. They also report that neither the minimum nor the product operators are adequate representations of conjunction. This leads to their definition of the “compensatory and” which is later investigated by Türkşen (1992) within the context of Disjunctive and Conjunctive Normal Forms which will be discussed later after the derivation of fuzzy canonical forms. It is also stated that direct rating methods require evaluations to be measured at most on the interval scales (Türkşen 1988, 1991).

## 4.3. Reverse Rating

In this method, a subject is given a membership degree and then asked to identify, i.e., verify, the objects for which the given degree corresponds to the fuzzy term in question (Türkşen 1988, 1991). This method can be used for each individual by repeating the same question for the same membership function as well as for a group of individuals.

Again we note in this reverse rating experiments, the subject(s) verify, identify the object that correspond to a given membership degree. Implicit naturally in such an identification response is the logic assertion of truth.

Once the subject's (or subjects') responses are recorded the conditional distributions can be taken to be normally distributed and the unknown parameters (mean and variance) can be estimated as usual. This method also requires evaluations to be made on at least interval scale. In this method, we observe a spread on the base axis values (see Figure 4.5) This is a one-to-many map. It should be noted that we observe a spread in membership domain and hence Type 2 – membership representation should be selected as a result of "Reverse Rating" experiments (Figure 4.5) in a manner similar to in the ones, we obtained in "Direct Rating" experiments. (Figures 4.2, 4.3 and 4.4)

Chameau and Santamarina (1987a) consider reverse rating as a valuable tool to verify the membership function obtained by using another approach rather than an acquisition method.

## 4.4. Interval Estimation

Interval estimation subscribes to the random set-view (or, one-to-many map)of the membership function (Section 3.1.2). The subject is asked to give an interval that describes the “A-ness of 'x'”. Let $I_i$ be the set-valued observation (the interval) and $m_i$ the frequency with which $I_i$ is observed. Then, $R=(I_i, m_i)$ defines a random set (Dubois and Prade 1989, 1991).

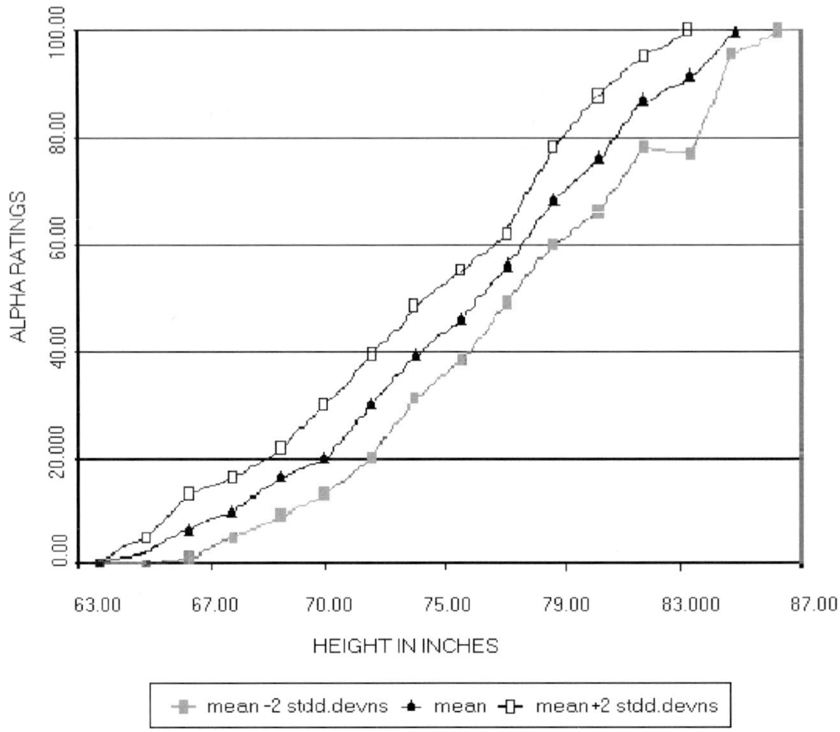


**Figure 4.3. Direct** Rating for "Tall" Man (subject #3).

It is to be noted that $I_i$ is a crisp interval and the subjects assertion that $I_i$ describes the "A-ness of 'x'" is just a logic response verifying that the objects "A-ness" falls in the given interval. This is affirming that the objects "A-ness" falls into a given interval is "true".

Chameau and Santamarina (1987a) find this approach of elicitation particularly advantageous over polling where the answer mode of verification is necessarily crisp (Yes/No), **i.e., verifications are given in two-valued logic paradigm**. Interval estimation is a relatively simple way of acquiring the membership function of "fuzzy truthood" and it results in membership functions that are "less fuzzy" (the spread is narrower) when compared to direct rating and polling.

Interval estimation subscribes to the uncertainty view of membership functions as opposed to the vagueness view, and in that sense, it brings the issues of uncertainty modelling using fuzzy set theory, random sets, possibility measures and their relations to probability theory (Dubois and Prade 1993).

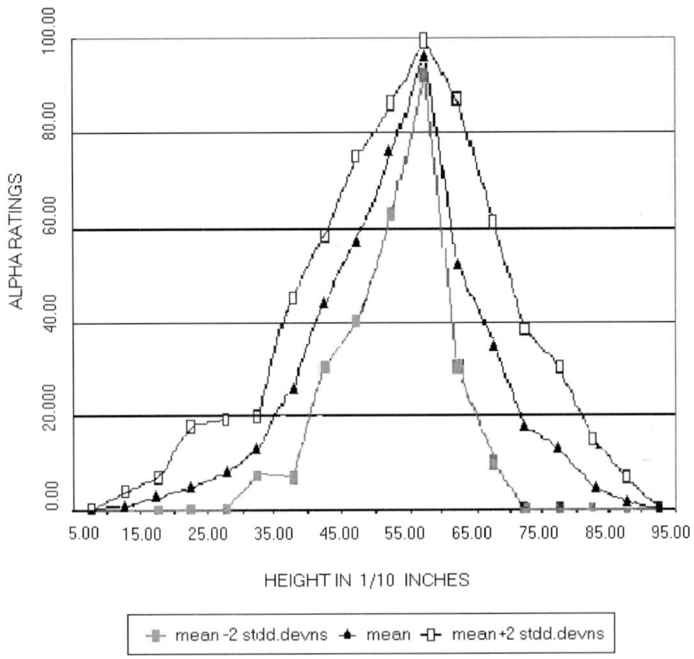


**Figure 4.4**. Direct Rating for "Pleasing" House (subject #1).

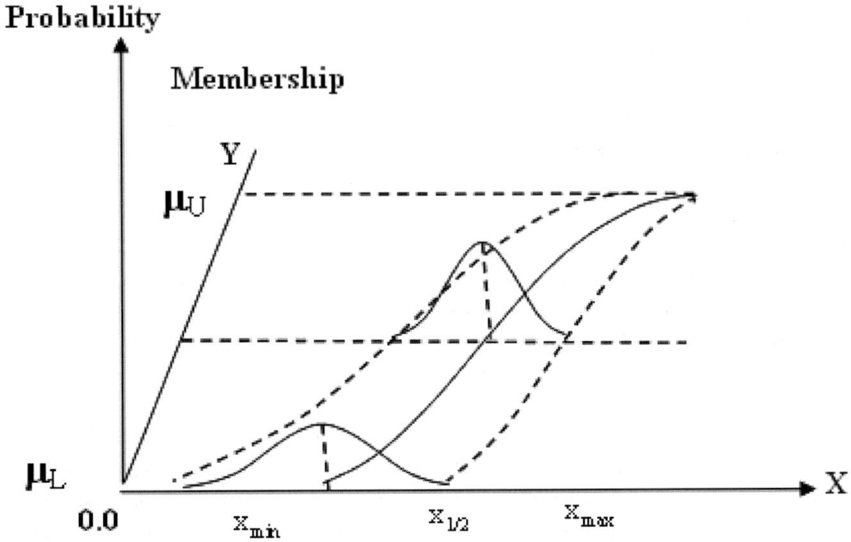


**Figure 4.5.** Experimental distributions over the referential set in "Reverse Rating".

Zwick (1987) also considers the random set interpretation and uses the law of comparative judgments to assess the membership functions, in our view, of the fuzzy truthhood.

## 4.5. Membership Exemplification

In terms of membership function exemplification, Hersh and Carmazza (1976) performed a test for the direct elicitation of the membership functions. Hersh and Carmazza ordered twelve squares in ascending order and indicated each square with an ordinal number. They asked the subjects to write the number(s) which is **(are)** appropriate for “large”, “very large”, “small” etc. Again there is the implicit verification of certain numbers that are assigned to “large”, etc. are “true”. The results are at variance with polling; most likely because there is no repetition in this elicitation method to normalize the effects of error or "noise". As well, the results are in fact elicitation of descriptive membership values and hence the variance to polling results.

However, the use of computer graphics to give an example membership function to be modified by the subject greatly enhanced this procedure as is usually witnessed in commercial applications of "fuzzy expert system shells". Again, subject’s identification of the membership function as “large”, “very large”, etc., is a logic response attesting that the selected linguistic word is applicable and hence it is the “true” membership function to the object’s property that is observed. But the membership function so obtained is a descriptive membership assessment.

## 4.6. Pair wise Comparison

Oden (1979) discusses the use of fuzzy set theory in psycholinguistic theories. He considers comparisons of the form: “which is a better example of a bird: an eagle or a pelican?”, and after the answer to this question (say, an eagle is chosen)**, i.e., verified**, that a particular label, say, “pelican” is “true”; then a second question is asked: “How much more of a bird is an eagle than a pelican?”. **It should be noted that the second question is to extract the descriptive membership assignment.**

Chameau and Santamarina (1987a) also use the same pair wise comparison technique. Following Saaty (1974), they require the subjects to provide pair wise comparisons and the strength of preference. This yields a non-symmetric full matrix of relative weights. The membership function is found by taking the components of the eigenvector corresponding to the maximum eigen value. The values are also normalized. Chameau and Santamarina (1987a) find the requirement that evaluations be on a ratio scale to be unnatural.

However, they espouse a "comparison-based point estimation" which determines the position of a set of stimuli on the reference axis by pair wise comparison and the membership is calculated by aggregating the values provided by several subjects. Although the subjects of Chameau and Santamarina (1987a) experiments ranked this method almost as good as the interval estimation method (which was ranked as the best method), this method also needs the unfortunate assumption of a ratio scale. Furthermore, pairwise comparisons require a relatively simple domain in many comparison experiments.

## 4.7. General Remarks on Subjective Methods

In general, Chameau and Santamarina (1987a, 1987b) report good agreement between direct rating, interval estimation and membership exemplification with the comment that, in most of the cases, fuzzy sets obtained by exemplification method are wider (fuzzier) than the ones obtained by other methods. The main difficulty with the point-estimation method is the contradiction between "fuzziness of the perception" and the crispness of the response mode. Again, this strongly suggest the need for Type 2 representation. This difficulty is overcome by the interval estimation method which in turn needs a minimum number of assessors or assessments. However, Chameau and Santamarina (1987a) report that as low as five assessments are sufficient. Exemplification yields membership functions without further processing which is an advantage. The way they carry out the pair wise comparison method assumes a ratio scale for the measurements which is hardly justified.

The assessors that took part in the experiments of Chameau and Santamarina (subjectively) rated the interval estimation method as the best in terms of expected consistency and expected quality. The age of the assessors affected their response, particularly in the "old-not old" categories.

In all the elicitation methods, Chameau and Santamarina obtain the membership functions based on averaging or aggregation of the responses from several assessors. In that sense they do not subscribe to the individualistic interpretation of fuzziness. Chameau and Santamarina justify this approach by assuming that the fuzziness is a "property of the phenomenon" rather than a "property attributed by the observers", which may be disputed.

One important issue in constructing membership functions is the context. It has always been emphasized that fuzzy set theory and particularly the membership functions are context dependent. Hersh, Carmazza and Brownell (1979) discuss the effects of changing context in determining the membership functions, a problem which eluded linguists for a long time (Kamp 1975). They report that the frequency of occurrence of the elements does not effect the location and form of the membership function (i.e., if one asks "Is John tall?" (is true or false) over and over again the resulting membership is consistently of the same form). On the other hand, the number of unique elements significantly affects the form of the membership function (i.e., the answer to the question "Is John tall?" (is true or false) varies when there is only one more person to consider versus when there is more than one more person to consider. The context changes!). This result tends to suggest that the membership function is not only a function of the object from the universe of discourse but the discourse as well as the persons involved in the discourse. That is, if X is the universe of discourse and $x \in X$, then the membership function for the fuzzy subset A of X is actually of the form:

$$\mu_A(x) = f_A(x, X, p, P)$$

where $p \in P$ is the person in a group of people involved in the discourse.

**This observation is the foundation of our proposal for $s \overset{R}{\longleftrightarrow} o$ interaction in the ontological foundations of fuzzy theory which changes sRo relation of the classical theory.**

Dombi (1990) takes up the problem of determining the membership function. He focuses on providing a theoretical basis for membership construction which is relevant to the problem at hand, described with only a few parameters that are meaningful. He avoids discussing problems of context dependency, measurement and uncertainty.

Chen and Otto (1995) consider constructing continuous membership functions from a given set of discrete points (evaluations). The rationale is that: one can only answer finite amount of questions from which a continuous membership function has to be constructed. They propose that the membership function can be constructed on an interval scale. When one requires membership functions to be continuous and convex, curve fitting methods might yield membership functions that are outside the unit interval and non-convex. In order to obtain continuous membership functions that are invariably bounded, convex and continuous, Chen and Otto propose a constrained interpolation method. This method, which minimally considers the semantic issues, is a powerful way of obtaining a continuous membership function from discrete data, a problem frequently encountered in practice.

As it was emphasized in Section 3, it is worthwhile to emphasize again in this section that in all the elicitation methods, the response of the subjects contains either an implicit or an explicit truth, verification, qualification expression as well as descriptive set membership assignment. In polling experiments the response is explicitly "truth" expression in terms of yes/no response. Thus, as it was indicated before, the average of yes/no answers can only provide a graded "truth" value and nothing more. Such polling exercises therefore do not provide us a schema to elicit descriptive membership functions. On the other hand, direct rating; for example, provide a schema to elicit descriptive membership functions. The responses of subjects *implicitly specify* that the response they give is "true" in general. Otherwise, the response makes no sense, if we do not accept the fact that the given response is a true response of the subject based on his or her knowledge and/or expertise, why should we even bother to use the information provided by such a response.

In summary, therefore, one specifies:

(i) a descriptive set assignment, and
(ii) a veristic truth qualification,

whenever one determines a membership function associated with a fuzzy term and in the cases of polling and interval estimation the veristic truth qualification given us fuzzy membership values of "truth" term, but does not give us descriptive membership values. They just provide the identification of an applicable label. Whereas in direct rating experiments and the acquisition of preference strengths, we obtain descriptive membership assignments of the given fuzzy term with an implicit verification that the descriptive membership is "truth".

As well, we observe that, there are usually a spread of points that are summarized, or approximated, by a Type 1 membership function in reductionist approach. The spread of responses ought to be captured by a Type 2 membership representation (see Figures 4.3, 4.4 and 4.5).

## References


1. Chameau, J.L., J.C. Santamarina (1987a), "Membership Functions Part I: Comparing Method of Measurement", *International Journal of Approximate Reasoning,* 1, 287-301.
2. Chameau, J.L., J.C. Santamarina (1987b), "Membership Functions Part II: Trends in Fuzziness and Implications", *International Journal of Approximate Reasoning*, 1, 303-317.
3. Chen, J.E., K.N. Otta (1995), "Constructing Membership Functions Using Interpolation and Measurement Theory", *Fuzzy Sets and Systems*, 73, 313-327.
4. Dombi, J. (1990), "Membership function as an evaluation", *Fuzzy Sets and Systems,* 35, 1-22.
5. Dubois, D., H. Prade (1993), "Fuzzy Sets and Probability: Misunderstandings, Bridges and Gaps", *Second IEEE International Conference on Fuzzy Systems*, IEEE, San Francisco, California March 28-April 1, 1059-1068.
6. Dubois, D., H. Prade (1991), "Random Sets and Fuzzy Interval Analysis", *Fuzzy Sets and Systems,* 42, 87-101.

7. Dubois, D., H. Prade (1989), "Fuzzy Sets, Probability and Measurement", *European Journal of Operational Research*, 40, 135-154.
8. Hersh, H., A. Carmazza and H.H. Brownell (1979), "Effects of Context on Fuzzy Membership Functions", in M.M. Gupta, R.M. Ragade & R.R. Yager (eds), *Advances in Fuzzy Set Theory*, North-Holland, Amsterdam, 389-408.
9. Norwich, A.M., I.B. Türkşen (1984), "A Model for the Measurement of Membership and the Consequences of its Empirical Implementation", *Fuzzy Sets and Systems*, 12, 1-25.
10. Norwich, A.M., I.B. Türkşen (1982b), "The Fundamental Measurement of Fuzziness", in: R.R. Yager (ed.), *Fuzzy Sets and Possibility Theory*. Pergamon Press, New York, 49-60.
11. Oden, G.C. (1979), Fuzzy propositional approach to psyholinguistic problems: an application of fuzzy set theory in cognitive science, in M.M. Gupta, R.K. Ragade & R.R. Yager (eds), Advances in Fuzzy Set Theory and Applications, North-Holland, Amsterdam, 409-420.
12. Resconi, G., I.B. Türkşen (2001), "Canonical Forms of Fuzzy Truthoods by Meta-Theory Based Upon Modal Logic", *Information Sciences*, 131, 157-194.
13. Saaty, T.L. (1974), "Measuring the Fuzziness of Sets", *Journal of Cybernetics*, 4, 53-61.
14. Türkşen, I.B. (1992), "Interval Valued Fuzzy Sets and 'Compensatory AND' ", *Fuzzy Sets and Systems*, 51, 295-307.
15. Türkşen, I.B. (1991), "Measurement of Membership Functions and Their Acquisitions", *Fuzzy Sets and Systems,* 40, 5-38.
16. Türkşen, I.B. (1991), "Measurement of Membership Functions and Their Assessment", *Fuzzy Sets and Systems*, 40, 5-38.
17. Türkşen, I.B. (1988), "Stochastic Fuzzy Sets: A Survey, Combining Fuzzy Imprecision with Probabilistic Uncertainty in Decision Making", *Springer-Verlag*, New York, 168-183.
18. Zimmermann, H.J., P. Zysno (1980), "Latent Connectives in Human Decision Making", *Fuzzy Sets and Systems*, 4, 37-51.
19. Zwick, R. (1987), "A Note on Random Sets and the Thurstonian Scaling Methods", *Fuzzy Sets and Systems*, 21, 351-356.

# Chapter 5

# FUZZY CLUSTERING METHOD

A particularly important concern in practice is to construct membership functions from a given set of data via unsupervised or supervised learning approaches. In general, membership functions may be constructed from available data when adequate amount of data is already collected in a database or a data warehouse. Naturally, the concern is extended beyond the acquisition or extraction of membership functions to the development of fuzzy system models, i.e., formation of fuzzy rule bases.

These are data-mining and knowledge discovery experiments with fuzzy clustering techniques. In these experiments, fuzzy clustering algorithms that extract and identify fuzzy clusters are the verifiers, i.e., “truth” qualifiers. After the clusters are identified, various curve fitting techniques are applied to determine membership function over the scatter points of clusters. Thus, the curve fitting technique are the secondary verifiers attesting that these membership functions are "true" as best as we can determine with statistical methods. It is clear that both the clustering and curve fitting algorithms are "approximate" verifies. As a result, we acquire an approximate knowledge of a given system. Hence, we have an automatic verification scheme that attests to the truth qualification of membership functions and fuzzy rules so constructed under the assumptions of a given algorithm. Again we note the separation between the assignment of linguistic descriptive words to numbers and their verification, i.e., attesting to their “truth”, via a computer algorithm that helps us construct the membership functions. It should be noted that in these exercises, generally, “truth” is two-valued.

In fuzzy clustering techniques, the data on the base axes of the system under consideration are assumed to be on the absolute scale in order to carry out addition, subtraction and multiplication operations. Furthermore, the membership functions are also assumed to be on the absolute scale in order to carry out exponentials in clustering criteria. There is no research to verify that these assumptions are valid and sound. However, there are a good number of engineering examples that attests to the goodness of these assumptions on empirical grounds.

When database contains nominal and ordinal data they must be treated in a special way. One way is to compute approximate probabilities of certain categories with the logistic curve analysis.

It is to be noted that the elicitation of membership functions via fuzzy data-mining and fuzzy cluster analysis requires that we assign linguistic values to the numbers. Hence this is the assignment of words to numbers.

There are several approaches (Nakanishi, Türkşen and Sugeno, 1993; Emami, Türkşen, and Goldenberg, 1998; Türkşen, 1999; Kılıç and Türkşen, 2001; Uncu and Türkşen, 2001; Türkşen and Bodur, 2001; Türkşen, 2001) for the development of fuzzy system models, i.e., the structure identification of fuzzy rules, via knowledge discovery with the application of fuzzy clustering techniques, known as FCM-Fuzzy C-Means (Bezdek and Haris, 1979; Sugeno and Yasukawa, 1993; Kirshnapuram and Keller, 1993; Gustafson and Kessel, 1979; Younis, 1996; Duda, Hart and Stork, 2001).

There are essentially two basic approaches:

(i) Output clustering and (ii) input-output clustering.

There are also two alternative approaches in output clustering

(i)(a) projection into each input-variable.

(i)(b) projection into the total input-variable space as a whole.

(i)(a) First determine the clusters of the output-variable (performance variable) data and then project the membership values of the output clusters into each of input-variable spaces separetely to determine scatter points of the input clusters. Secondly apply curve fitting techniques over the scatter points to determine the membership functions of the input and output varables for each cluster.

In Figure 5.1. only the output variable is clustered and the membership values of each cluster are projected into each of the input variable spaces.
The number of clusters identified with fuzzy clustering algorithm in an unsupervised learning determines the number of fuzzy rules that displays the behavior of the system (be it approximately).

(i)(b) First determined the clusters of the output variable data but then project the membership values of the output clusters over the input-variable-vector-space as a whole. Secondly apply curve fitting techniques over the scatter points of the output variable and over the scatter points of the input-variable-vector space as a whole.

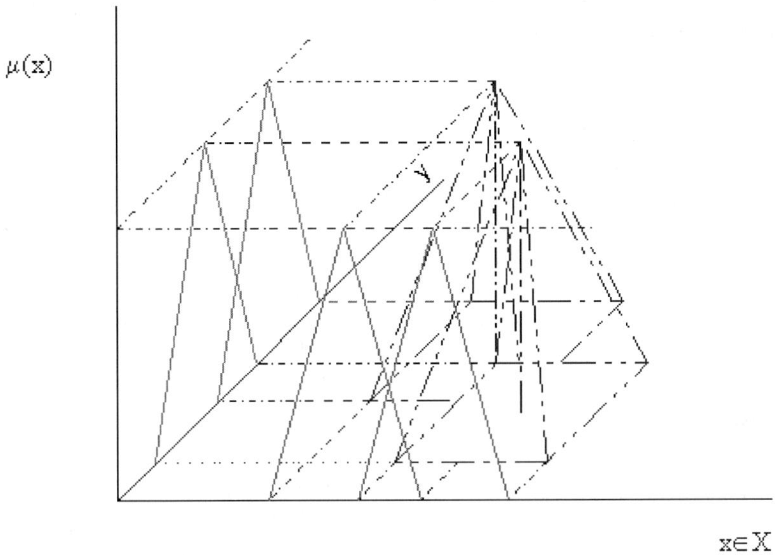


**Figure 5.1.** Output clusters induce input cluster via projection.

Again the number of clusters determined for the output variable identifies the number of fuzzy rules.

As well there are two alternatives in input-output clustering:

(ii)(a) projection into each of the input-output variable spaces and

(ii)(b) do not project but just keep the centers only for each variable.

(ii)(a) First determine the clusters of input-output space as a whole, then project the membership values of the clusters of the input-output space into each of the input variables spaces as well as the output variable's space.

Secondly apply curve fitting techniques over the scatter points of each input and output variable spaces.

The number of clusters determined over the whole input-output spaces identifies the number of fuzzy rules.

(ii)(b) First determine the clusters of input-output spaces as a whole. In this case, there is no need to curve fit over the scatter points to determine membership values. Keep only the cluster centers for each input-output variable. Form a rule base that identifies only the cluster centers for each cluster and hence for each fuzzy rule.

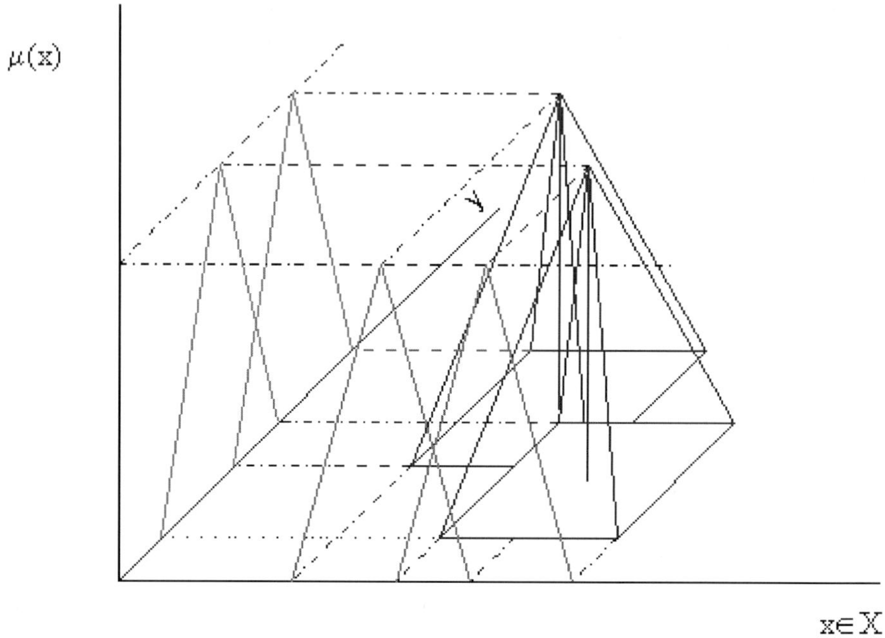


**Figure 5.2.** Input-output variable space is clustered as a whole and membership values are projected into each of the input and output variable space.

Clearly, these four approaches as stated above form a base to help us construct Type 1 knowledge discovery and the associated Type 1 fuzzy rule base representation.

In order to get to Type 2 fuzzy representation, in curve fitting, one needs to take into account the variability of the scatter points and form the appropriate confidence intervals around the membership functions that are obtained by the curve fitting techniques.

An alternate approach is to determine Type 2 scatter points by a suitable search over the number of clusters, c, and order of fuzziness, m. This could therefore, be applied either with just the output clustering approaches, (i)(a) and (i)(b), or with input-output clustering as a whole (ii)(a) and (ii)(b).

## 5.1. Fuzzy Clustering Techniques

The idea of fuzzy clustering is to divide the output data into fuzzy clusters which overlap with each other. Therefore, the assignment of each sampled data vector to each cluster is defined by a membership grade in [0,1]. In formal terms, clustering unlabeled data $X=\{x_1, x_2, \ldots x_n\} \subset R^f$, is the assignment of c number of cluster labels, $\gamma_1, \gamma_2, \ldots \gamma_c$, i.e., words, to the data elements in X where n is the number of samples of data vectors in a training data set and $R^f$ is the feature space. c-Clusters of X are sets with (c,n) membership values $\{u_{ik}\}$ that can be conveniently arranged as a (cxn) matrix $U=[u_{ik}]$.

The problem of fuzzy clustering is to find the optimum membership matrix U for the fuzzy clustering in X in an unsupervised learning experiment. The most widely used objective function is the weighted within-groups sum of squared errors function J which is define as the following constrained optimization problem [Bezdek, 1981; Kandel, 1982]:

$$\min\left\{ J(U,V;X) = \sum_{k=1}^{n}\sum_{i=1}^{c}(u_{ik})^m (\| x_k - v_i \|)_A \right\}, \qquad (5.1)$$

where $0 \le u_{ik} \le 1$,

$\forall$i,k; $0 < \sum_{k=1}^{n} u_{ik} < n, \ \forall i;$

$\sum_{i=1}^{c} u_{ik} = 1, \ \forall k;$

V = $\{v_1, v_2, \ldots v_c\}$ is the vector of (unknown) cluster centers of the cluster labels, $\gamma_1, \gamma_2, \ldots \gamma_c$. $\|.\|$ specifies a norm and A is a positive definite matrix which specifies the shape of the clusters. The usual selection is the Euclidean norm together with A=I the identity matrix. Consequently spherical clusters are identified. There are a variety of distance measures an analyst can experiment with as shown in Figure 5.3; and Table 5.1. If the distance measure is considered to be the Mahalonobis norm then the matrix A is the inverse of the covariance matrix, C, then the algorithm generates models with elliptic clusters [Gustafson and Kessel, 1979; Kaufman, et. Al. 1990; Younis, 1996]. It should be noted that fuzzy clustering falls into a particular similarity view of membership interpretations as discussed in Chapter 3. That is it considers only distance based similarity.

**Table 5.1.** Examples of Distance Functions.

| Name | Function |
|---|---|
| Minkowski | $d_p(a, b) = \left[\sum_{i=1}^{s} \lvert a_i - b_i \rvert^p\right]^{1/p}, p > 0$ |
| Euclidean | $d_2(a, b) = \left[\sum_{i=1}^{s} (a_i - b_i)^2\right]^{1/2}$ |
| Hamming | $d_1(a, b) = \sum_{i=1}^{s} \lvert a_i - b_i \rvert$ |
| Maximum | $d_\infty(a, b) = \max_{i=1 \ldots s} \lvert a_i - b_i \rvert$ |
| Mahalanobis | $D_A(a, b) = (a - b)^T A(a - b)$, where A is a positive definite matrix |

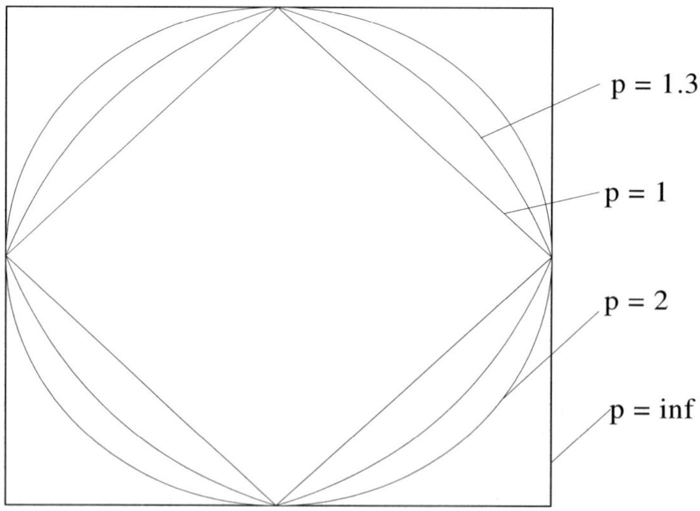


**Figure 5.3.** Graphical realization of different distance formulas for Minkowski measure defined in Table 5.1 above.

### *5.1.1. FCM Algorithm*

The well known FCM algorithm has the following steps.

1. Choose (c , m, iter, ε)
2. Find initial cluster centers .
3. ITERATE

for t=1 to iter

**CALCULATE** $u_{ik,t} = \left[ \sum_{j=1}^{c} \left( \frac{\left\| x_k - v_{i,t-1} \right\|_A}{\left\| x_k - v_{j,t-1} \right\|_A} \right)^{1/m-1} \right]^{-1}$

**CALCULATE** $v_{i,t} = \frac{\sum_{k=1}^{N} (u_{ik,t})^m x_k}{\sum_{k=1}^{N} (u_{ik,t})^m}$

**If error** = $\left\| v_t - v_{t-1} \right\| \leq \varepsilon$ **, THEN stop**

Next t

### *5.1.2. Statistical Background*

When the data is statistically correlated, the distance metric that depends on the distribution of the data is the Mahalonobis distance defined as:

$$dM=(xk-vi )TC-1(xk-vi) \tag{5.2}$$

where C is the covariance matrix of a data $v_i$, i=1,…c, are cluster centre values, and x={$x_k$}, k=1,…,n, represents n-samples of a data vector. Thus, the Mahalonobis distance should be applied when the statistical covariance properties of the data are to be considered.

Figure 5.4 depicts two possible cluster identification of two-clusters by applying the Euclidean and Mahalonobis distance measures. It is clear from the Figure 5.4 that applying the Euclidean distance results in undesirable clustering of the two patterns. On the other hand, applying the Mahalonobis distance allows the natural clustering of the two patterns.

This shows that the modeling capabilities of the Mahalonobis distance-based clustering is much greater than that of the Euclidean distance-based algorithms. This fact needs to be taken into consideration especially as the dimensionality increases wherein the possibility of having hyper-spherically shaped clusters is virtually zero.

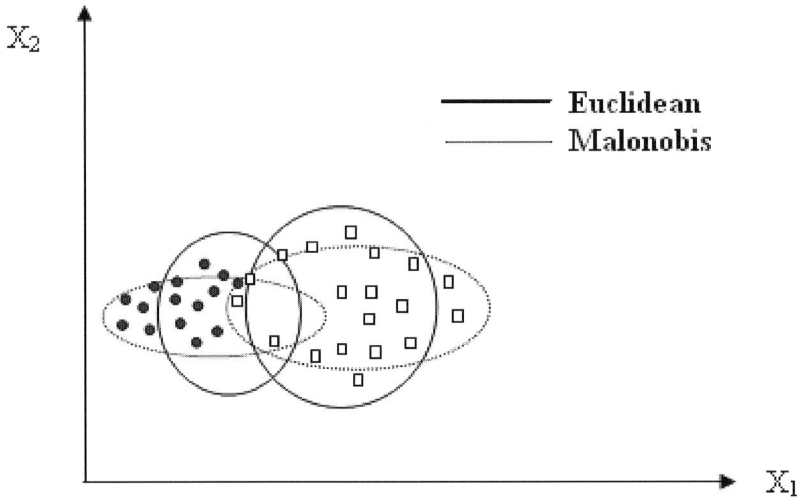


**Figure 5.4.** The results of clustering is either circular or elliptical depending on whether Euclidean of Mahalonobis distance metric is applied.

It is clear from Figure 5.4 that if we could describe the shape of each cluster by its centroid and the sample covariance matrix of the data points assigned to that cluster, then the most useful similarity measure would be the Mahalonobis norm metric.

Therefore the selection of a criterion of fitness is a very important issue in the design of a clustering algorithm. In contrast to the two-dimensional representation given in Figure 5.4, one cannot visually demonstrate how good the resulting clusters are in higher dimensions. The criteria of fitness, or performance indices, are generally specified as a function of the memberships whose minima define a "good" clustering as specified in expression (5.1). The algorithm then becomes a numerical procedure for finding memberships which minimize the objective function.

Suppose D( $x_k$ , $v_i$ ) is the variation between datum $x_k$ and the closest cluster center $v_i$. This can be thought of as the performance index of any data vector

$x_k$ in relation to a cluster centre, $v_i$, the prototype. The performance of the system can be assessed by an average variation, E{D(X,V)}, between the input and the final cluster prototype.

If the process is stationary and ergodic, then the mathematical expectation is, with probability one, equal to the long term sample average

$$E\{D(X,V)\} = \lim_{n \to \infty} \frac{1}{n} \sum_{k=1}^{n} D(x_k, v_i) \tag{5.3}$$

Equation (5.3) implies that if the input statistics do not change with time and if the time average equals the ensemble average then the resulting sample average variation should be nearly the expected value. Therefore the same cluster prototypes used for future data should yield approximately the same average. However, in general, it is not always easy to justify the assumptions of stationarity and egodicity. Hence, again we have to come up with an approximation schema.
The simplest and most widely used measure of variation is the squared error variability measure,

$$D(k,i) = \sum_{i=1}^{c} (x_k - v_i)^2 \tag{5.4}$$

Substituting (5.3) in (5.2) give the mean squared error (MSE) criterion function, Dunn (1974) developed the first fuzzy extension to the least mean square error approach to clustering and this was generalized by Bezdek [1981] to a family of algorithms. Let n be the total number of samples, $v_i$ be the mean of those samples for the cluster label $\gamma_i$. The generalized mean squared error is then defined by:

$$J = \frac{1}{n} \sum_{i=1}^{c} \sum_{k=1}^{n} (u_{ik})^m \| x_k - v_i \|_A \tag{5.5}$$

Where $u_{ik}$ is the membership of the $k^{th}$ data sample vector in the $i^{th}$ cluster, m is a weighting exponent, degree or level or order of fuzziness, which is strictly greater than one, and A is any positive definite matrix (Bezdek, 1981).

It is noted that m=1 causes the identification of crisp clusters. Many algorithms were designed based on minimizing J. In these algorithms, J is monotonically non-increasing function as the member of iterations increases.

Modifying the criterion function to allow the use of a generalized distance as a distance metric is possible by defining Linde et. al. (1980) and Bezdek (1981) generalized A to a positive definite matrix. They showed that if the covariance matrix for all the data is used, the Mahalonobis distance is the basis for the partitioning.

Clustering using the global covariance matrix results in a monotonically decreasing objective function. However, all the clusters may not have the same distribution as that of the whole data set. Hence, using the global covariance matrix for each cluster may not give the best grouping of the data.

Other common criterion functions are the average squared distances between samples in a cluster domain and the average squared distances between samples in different cluster domains. [Julius, et. al. 1974].

*5.1.3. Variations of FCM*

In current literature there are many variations of the Fuzzy C-Means (FCM) algorithm. FCM clustering algorithm suffers from four major shortcomings which may be stated as follows:

The number of clusters (c) is not known *a priori*.
There is no theoretical base for the determination of the weighting exponent (m).
An initial guess of the location of the cluster centers, $V_o$ is not known *a priori*.
There is no known method of selection to determine the norm matrix A suitable for a model.

Conditions specified in FCM algorithm lead to the determination of local extrema of J. But different choices of these four initial conditions might lead to different local extrema. Therefore, the knowledge of a proper initial conditions could improve the FCM algorithm.

For example, one could first introduce a generalization of scatter criteria for the purpose of expressing the compactness and separation of the clusters [Bezdek, et. al. 1980; Kandel, 1982; Sugeno, et. al. 1993; Emami, et, al, 1998; Türk□en, 1999] by analyzing both $S_W$ and $S_B$,

$$S_W = \sum_{i=1}^{c} \sum_{k=1}^{n} (u_{ik})^m (x_k - v_i)(x_k - v_i)^T \quad , \tag{5.6}$$

where $S_W$ is the "within" fuzzy scatter criterion ; and $S_B$ is the 'between' fuzzy cluster scatter criterion:

$$S_B = \sum_{i=1}^{c}\left(\sum_{k=1}^{n}(u_{ik})^m\right)(v_i - \bar{v})(v_i - \bar{v})^T$$

and where both the fuzzy total mean vector, $\bar{v}$, and each cluster centre, $v_i$, are weighted with the membership values, of $u_{ik}$, and which are defined as:

$$\bar{v} = \frac{1}{\sum_{i=1}^{c}\sum_{k=1}^{n}(u_{ik})^m}\sum_{i=1}^{c}\sum_{k=1}^{n}(u_{ik})^m x_k, \quad v_i = \frac{1}{\sum_{k=1}^{n}(u_{ik})^m}\sum_{k=1}^{n}(u_{ik})^m x_k \tag{5.7}$$

It should be noted that in Sugeno and Yasukawa Model [1993] $\bar{v}$ is not a weighted mean of data. Furthermore, in current literature, there are many scatter criteria analyses.

Secondly, "m", which is known as the "weighting exponents", or the order of fuzziness, varies in the range of $(1, \infty)$ and indicates the "level or degree of fuzziness". It is also known that larger values of "m" create large overlaps among them.

Furthermore, it is known that the trace of $S_T = S_W + S_B$ decreases monotonically from a constant value K to zero, as "m" varies from one to infinity:

$$K = TrS_T\left(\sum_{k=1}^{n}\left[\left(x_k - \frac{1}{n}\sum_{k=1}^{n}x_k\right)\left(x_k - \frac{1}{n}\sum_{k=1}^{n}x_k\right)^T\right]\right) \tag{5.8}$$

where tr = trace and K depends only on the data set .

A suitable value of "m" is generally determined experimentally for each case study data. In particular, it is found experimentally that the value of m depends very much on the characteristics of a given input/output data as shown in Figures 5.5. For example, we have found that (i) Region I in Figure 5.5 was a good region to explore and determine m for "spare parts data" which we had 50 data vectors as the training set; (ii) whereas Region II was good for "financial data, production scheduling data and adverse drug reaction data" and "serum lithium concentration data" for which we had around 80 data vector as the training data set; and (iii) finally, Region III was good for "alcohol dependency data" for which we had 23 data vectors. [Türkşen, 1999]

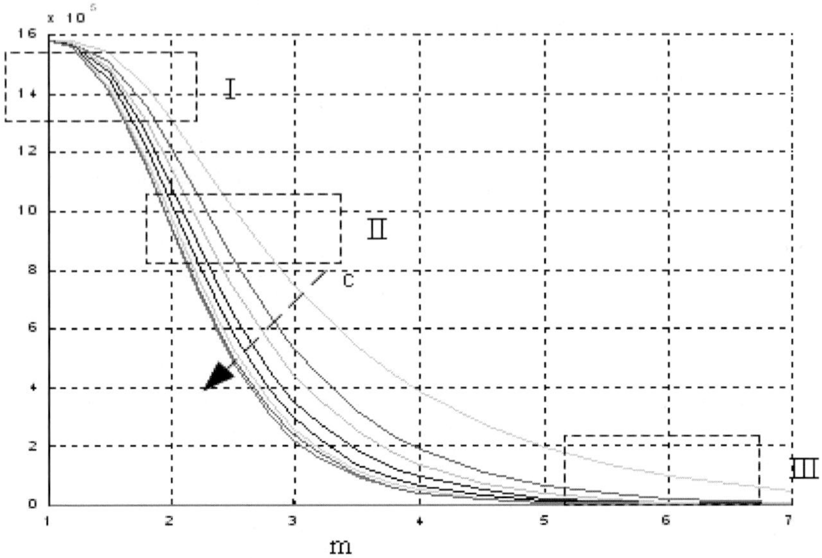


**Figure 5.5.** Reliable solutions Regions are identified as Region I, II, and III with experimental data, where c is the number of output clusters which identify the number of fuzzy rules.

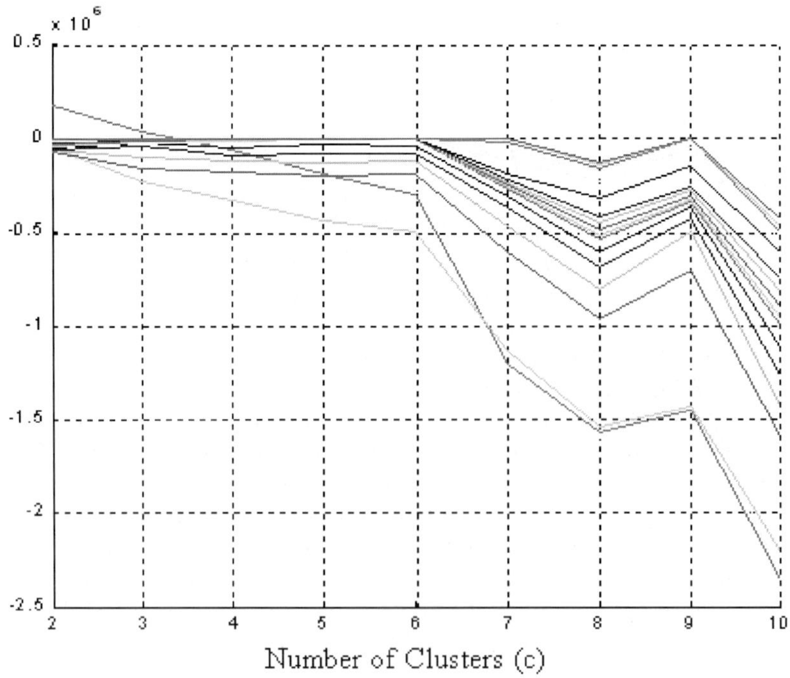


**Figure 5.6.** CVI, Cluster Validity Index vs the number of clusters, c

**Table 5.2.** Some Cluster Validity Indexes

| | |
|---|---|
| Bezdek's partition coefficient; | $v_{pc}(u) = \sum_{k=1}^{n}\sum_{i=1}^{c} u_{ik}^2 / n$ |
| Bezdek's partition entropy; | $v_{pe}(u) = \left\{ \sum_{k=1}^{n}\sum_{i=1}^{c} u_{ik}^2 \log_a(u_{ik}^2) \right\} / n$ |
| Xie-Beni index; | $v_{XB}(u) = \sum_{i=1}^{c}\sum_{k=1}^{n} u_{ik}^m \lVert x_k - v_i \rVert^2 / q$<br>$q = n(\min\{\lVert v_i - v_j \rVert^2\}), i \neq j$ |
| Fukuyama-Sugeno Index | $v_{FS}(u) = \sum_{k=1}^{N}\sum_{i=1}^{C} u_{ik}^m (\lVert x_k - v_i \rVert_A - \lVert v_i - \bar{v} \rVert)$ |

Thirdly, in order to obtain initial locations of cluster prototypes, one should implement an Agglomerative Hierarchical Clustering (AHC) algorithm as an introductory procedure to find a suitable guess for the initial locations of cluster prototypes for the FCM algorithm. For example, Ward's method [1963] could be used with unlabeled data $X=\{x_1, x_2, \ldots, x_n\}$, or any other agglomerative hierarchical clustering algorithm [Kaufman, et. al. 1990].

The results of the AHC algorithm is c hard clusters for the data, which is a good start for the fuzzy clustering procedure. With this method, one can choose the initial prototypes without any knowledge of the data *a priori*. This approach is much more efficient than exhaustive or random searches among different initial guesses. Finally, one should introduce a cluster validity index. There are several cluster validity indices, as shown in Table 5.2. Generally one is interested in determining c that minimizes one of these cluster validity indices.

It is clear that with the pair of values of "m" and "c" that would be determined with one of the variations of the FCM, one would apply the FCM algorithm, in order to find the fuzzy clusters. In particular, trace of the total scatter matrix vs the weighting exponent m for various c's and trace of the cluster validity index vs the number of cluster centers should be thoroughly analyzed in order to determine the appropriate number of rules to be included in a rule base of a system being investigated.

### *5.1.4. Interval-Valued Fuzzy Sets*

In conjunction with Type 2 fuzziness found in measurement experiments, interval-valued fuzzy sets were discovered and investigated in combination of concepts and hence approximate reasoning by Türk□en (1982-2001) where it was shown that we have to consider the disjunctive and conjunctive canonical forms FDCF and FCCF. This issues will be discussed in detail in next few chapters.

## 5.2. Type 2 Fuzziness

In modified FCM experiments which we carried out recently, we have observed that the dilemma concerning the choice of (m,c) pair, can be resolved in the following way. Instead of searching the optimal (m,c) pair, (m*,c*), we can identify a set of m's, {m} that minimize a c, say c*. Thus having identified c*, we can then choose a lower value, $mL^* = \min_m\{m, c^*\}$ $m_L^* = \min\{m, c^*\}$ and an upper value $mU^* = \max_m\{m, c^*\}$.

Thus, for different levels of fuzziness, i.e., order of fuzzy overlaps, m's, we get a scatter plot that give us a ground for Type 2 representation. This means that we can acquire Type 2 membership functions in inductive, unsupervised, learning experiments with training data (Figure 5.7)

It should be recalled that there are essential two types of Type 2 fuzziness: (i) Interval-valued Type 2 and (ii) Full Type 2.

(i) Interval-Valued Type 2 fuzziness is a special Type 2 fuzziness where the upper and lower bounds of membership are identified and the spread of membership, distribution is ignored with the assumption that membership values between upper and lower values are uniformly distributed or scattered with membership value of "1" on the $\mu(\mu(.))$ axis as shown in Figure 5.8. Thus, the upper and lower bounds of interval-valued Type 2 fuzziness specify the range of uncertainty about the membership values.

(ii) Full Type 2 fuzziness identifies upper and lower membership values as well as the spread of membership values between these bounds. That is there is a probabilistic or possibilistic distribution of membership values of the membership values that are between upper and lower bounds of membership values in the $\mu(\mu(.))$ axis as shown in Figure 5.9. Thus, we obtain a graded distribution of uncertainty between the bounds.

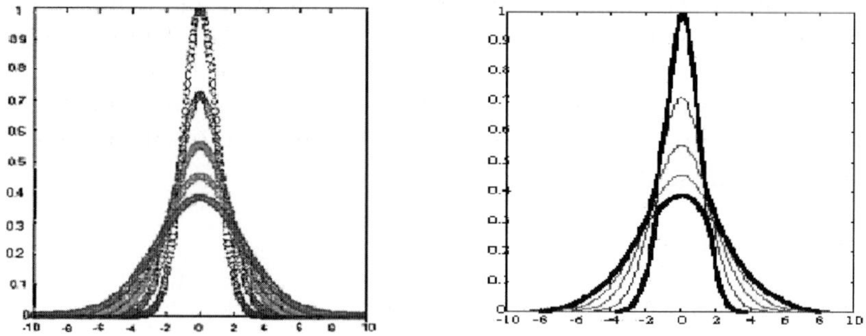

**Figure 5.7.** Type 2 sets: (a) a set of membership functions of a cluster, (b) interval-valued membership functions of a cluster.

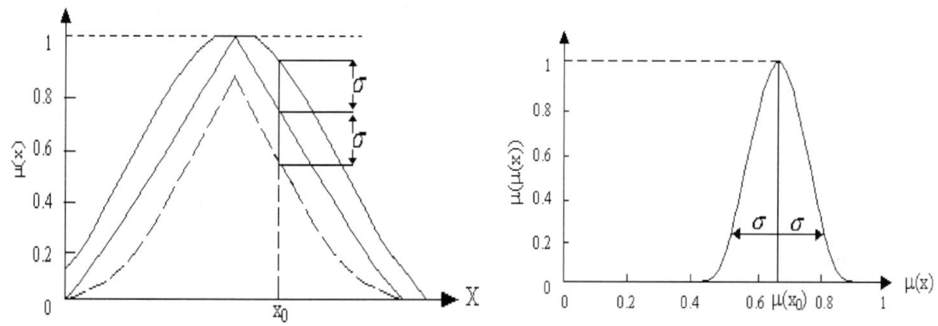


**Figure 5.8.** Full Type 2

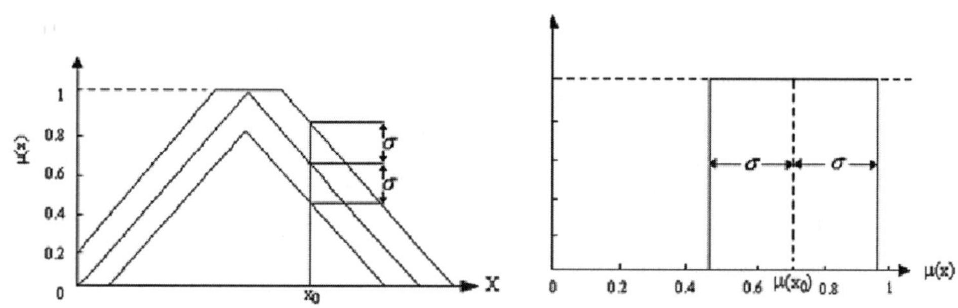


**Figure 5.9**. Interval-Valued Type 2

## 5.3. Curve Fitting to Membership Values

For Type 1 and Type 2 membership function schemas, one needs to curve fit with the usual curve fitting techniques such as least square estimates, robust curve fitting, etc.
For example , Type 1 curve fitting is based on a unique (m*,c*) pair of fuzzy clustering algorithm. The type of curves could be (a) Gaussian and (b) trapezoidal, etc., as show in Figure 5.10 (a) and (b).

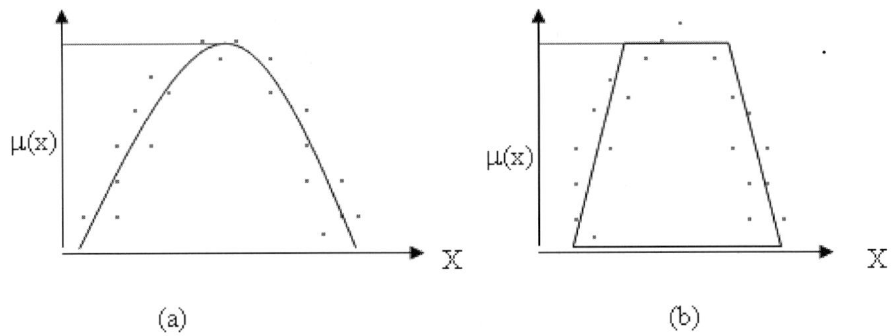


**Figure 5.10**. Type 1 curve fitted to the scatter plot of a variable X, (a) Gaussian (b) trapezoidal.

On the other hand, Type 2 curve fitting is based on either extending Type 1 curve, μ, to Type 2 representation with $2\sigma$ limit around Type 1 curve; i.e., $\mu \pm \sigma$ where $\sigma$ is the standard deviation of the scatter points around Type 1 curve, or (ii) directly determining a Type 2 curve around the unique (m*, c*) pair with the upper membership value of $m_U = \max_m(m, c^*)$ and the lower membership value of $m_L = \min_m(m, c^*)$ keeping c* constant in fuzzy clustering. These alternatives are shown in Figures 5.11 and 5.12 as an example of the exponential membership functions as follows:

$$\bar{\mu} = \frac{1}{ND}\sum \mu \qquad \sigma_\mu = \frac{\sum(\bar{\mu} - \mu)^2}{ND}$$

$$\mu_U(x') = \bar{\mu} + \sigma_\mu$$

$$\mu_L(x') = \bar{\mu} - \sigma_\mu$$

where ND is the number of data vectors.

That is for the "interval-valued Type2", all values between upper and lower values $\mu_U(x')$ and $\mu_L(x')$ have the same membership value of "1" in $\mu(\mu(x))$ dimension as shown in Figure 5.13. Whereas for the "Full Type 2", the values between $\mu_U(x')$ and $\mu_L(x')$ have different membership values in [0, 1] on $\mu(\mu(x))$ dimension as shown in Figure 5.14.

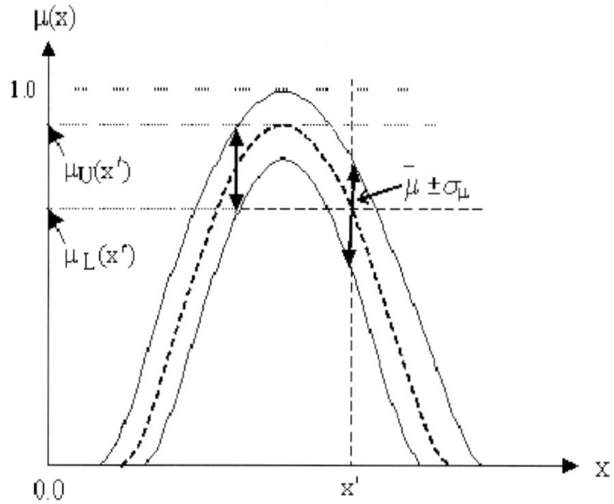


**Figure 5.11.** Gaussian Interval-Valued Type 2 membership function such that

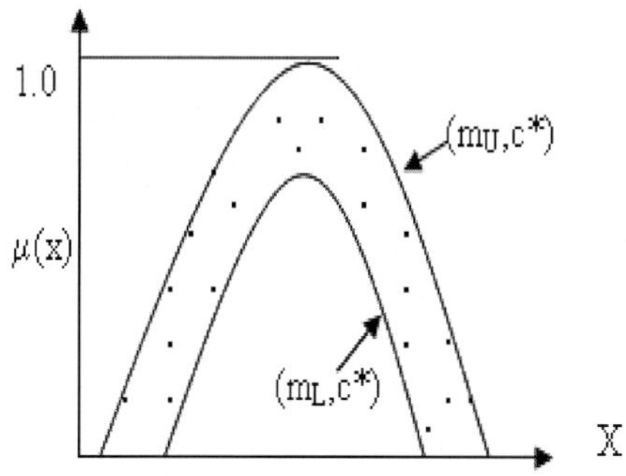


**Figure 5.12.** Type 2 membership function obtained as upper membership with $mU = \max_m(m, c^*)$ and lower membership $mL = \min_m(m, c^*)$ around (m*, c*) pair by keeping c* constant obtained from trS and CVI analysis in Fuzzy Clustering.

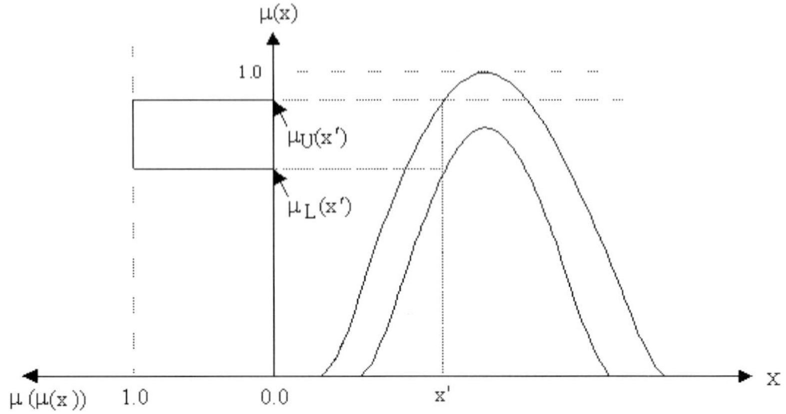


**Figure 5.13**. Gaussian Interval-Valued Type 2

Other types of curve fitting for trapezoidal, S and Π curves are done in a similar manner.Note that different m's expose different levels of fuzziness. Hence this is a full Type 2 membership representation.

The distinction between these two versions of Type 2 fuzziness are shown below:
Interval-Valued Type 2 Figure 5.13 and Full Type 2 Figure 5.14 is as follows.

As indicated before, we have other choices of curve fitting beyond exponential, trapezoidal and triangular. One example of such a curve is known as the S-curve and Π-curve, where Π-curve is defined as the aggregation of a left S-curve, $S_L$, and a right S-curve, $S_R$, around μ(x)=1 range as shown in Figure 5.15 for Type 1 membership representation.

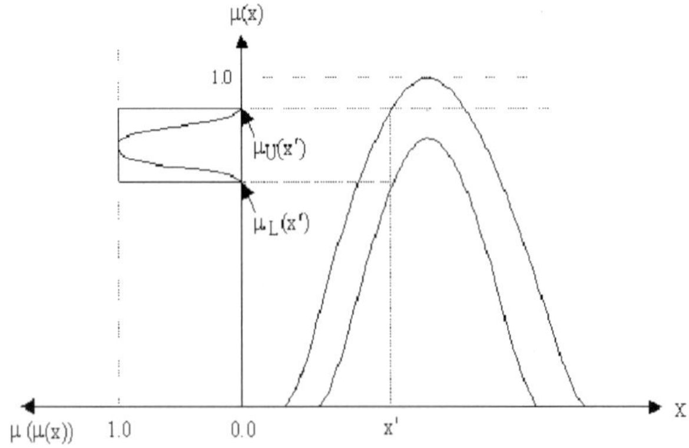


**Figure 5.14.** Gaussian Full Type 2

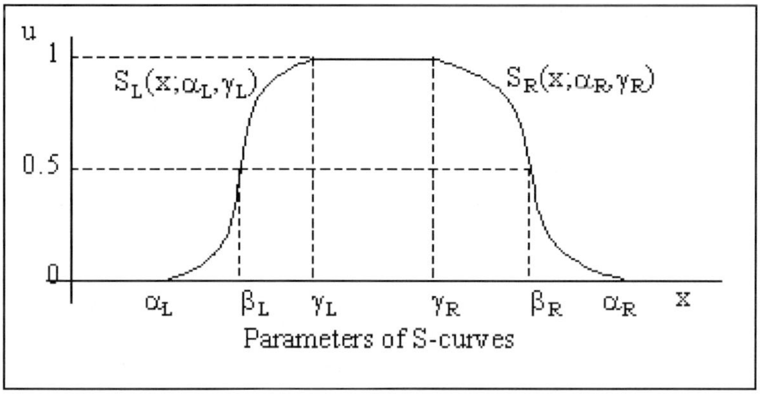


**Figure 5.15**. S-curve is defined by three parameters, α, β, γ.

Where $\beta = (\alpha + \gamma)/2$, for both left and right sides. That is the left side of the Π-curve, $S_L$ is defined as:

The right side of the curve is defined by a symmetrical S-curve with different parameters:

$$S_R(x;\ \alpha_R,\gamma_R) = \begin{cases} 0, & \text{for } x \geq \alpha_R \\ 2\left(\dfrac{x-\alpha_R}{\gamma_R-\alpha_R}\right)^2, & \text{for } \alpha_R \geq x \geq \beta_R \\ 1-2\left(\dfrac{x-\gamma_R}{\gamma_R-\alpha_R}\right)^2, & \text{for } \beta_R \geq x \geq \gamma_R \\ 1, & \text{for } \gamma_L < x \leq \gamma_R \end{cases}$$

where $\beta_R = (\alpha_R + \gamma_R)/2$ ;

Note that if the right-side S-curve has $\gamma_R = \gamma_L$ , and $\beta_R = \beta_L$; then we get a Π-curve.

Otherwise on their own $S_L$ and $S_R$ may define a right side or a left side fuzzy set, respectively.

The S and Π curves can be extended to determine either "Interval valued Type 2" or "Full Type 2" membership functions by applying the procedures described in extending Type 1 membership of exponential and/or trapezoidal membership functions.

## 5.4. Neural-fuzzy Technique

With the advance in neural network models and their applications to machine learning a considerable interest has been on unifying neural networks and fuzzy set theory (Rocha 1982, Takagi 1990, Kosko 1991; Ishibuchi, Morioka, Türkşen, 1995; Ichihashi, Türkşen, 1993). In this section, three such studies are mentioned where neural networks have been utilized to come up with membership functions from a given set of data.

For example, Yamakawa and Furukawa (1992) present an algorithm for learning membership functions using a model of the fuzzy neuron. Their method uses example-based learning and optimization of cross-detection lines. They assign trapezoidal membership functions and automatically come up with its parameters. The context is handwriting recognition. They also report some computational results for their algorithm.

On the experimental side, Erickson, Lorenzo and Woodbury (1994) claim that fuzzy membership functions and fuzzy set theory better explain the classification of taste responses in brain stems. They analyze previously published data and allow each neuron to belong to several classifications to a degree. This degree is measured by the neuron's response to the stimuli. They show that their model based on fuzzy set theory explains the data better than other statistical models.

Learning by fuzzified neural networks is investigated by Ishibuchi, Morioka and Türkşen, (1995). A general learning algorithm is implemented for the training a fuzzified feed forward neural network that has fuzzy inputs, fuzzy targets, and fuzzy connection weights. They compare, the learning algorithm results with other approaches that use real number weights and symmetric triangular fuzzy weights. They conclude that the algorithm is applicable to the adjustment of fuzzy weights with any shape if its membership function is specified by a finite number of parameters which is satisfied by trapezoidal and triangular membership functions.

## References

1. Bezdek, J.C. (1981), Pattern Recognition with Fuzzy Objective Function Algorithms, NY, *Plenum Press*.
2. Bezdek, J.C., M.P. Windham and R. Ehrlich (1980), "Statistical Parameters of Cluster Validity Functionals", *International Journal of Comput. Inf. Sci.*, 9(4), 324-336.
3. Bezdek, J.C., J.D. Harris (1979), "Convex Decompositions of Fuzzy Partitions", *Journal of Mathematical Analysis and Applications*, 6, 490-512.
4. Dunn, J.C. (1974) "A Fuzzy Relative of the ISODATA Process", *Cybernetics*, 32-57.
5. Emami, M.R., I.B. Türkşen, A.A. Goldenberg (1997), "Fuzzy Clustering and Fuzzy Modeling", IEEE-Trans on Fuzzy Systems, (to appear).

6. Erickson, R.P., P.M.D. Lorenzo and M.A. Woodbury (1994), "Classification of Taste Responses in Brain Stem: Membership in Fuzzy Sets", *Journal of Neurophysiology,* 71(6): 2139-2150.
7. Ishibuchi, H., K. Morioka and I.B. Türkşen (1995), "Learning of Fuzzified Neural Networks", *IJAR*, 13, 387-358.
8. Kandel, A. (1982), Fuzzy Techniques in Pattern Recognition, *Wiley*, New York.
9. Kaufman, L., P.J. Rousseeuw (1990), Finding Groups in Data, *Wiley*, New York.
10. Linde, Y., et. al. (1980), "An Algorithm for Vector Quantizer Design", *IEEE Trans. Commun.*, COM-28, 84-95.
11. Nakanish, H., I.B. Türkşen and M. Sugeno (1993), "A Review and Comparison of Six Reasoning Methods, Fuzzy Sets and Systems 57, 257-294.
12. Sugeno, M., T. Yasukawa (1993), "A Fuzzy Logic Based Approach to Qualitative Modelling", *IEEE Transactions on Fuzzy Systems*, 1(1), 1-24.
13. Türkşen I.B. (1999), "Type I and Type II Fuzzy System Modeling", *FSS*, 106, 11-34.
14. Younis, K.S., et. al. (1996), "Vector Quantization Based on Dynamic Adjustment of Mahalanobis Distance", *IEEE National Aerospace and Electronics Conference (NAE-CON)*, 2, 858-862.

# Chapter 6

# CLASSES OF FUZZY SET AND LOGIC THEORIES

History of human knowledge acquisition and communication has progressed from meta-languages, to speech, to writing, to mathematics, and to computer languages. In analogy to M. Mc Luhan (1995) medium is the message. It can be said that natural languages are the carries of our knowledge.

In this sense human knowledge is affectively expressed in terms of "linguistic expressions" of our natural languages. This suggests strongly that we should develop methods based on "computing with words, CWW (Zadeh, 1995). It is surmized that when CWW are developed extensively, this could form the sixth stage of development in the progression of human languages.

In CWW, we transform linguistic expressions first into "meta linguistic" expressions and then to "propositional" expressions and finally into "computational" expression, where the computational expressions receive numeric membership values for the purposes of evaluation (see Table 6.1). However, the three stages of transformations with further and further restrictions cause computational (numeric) expressions to take on a rather special structures depending on the axiomatic foundations we are willing to accept along the transformation route.

As it can be observed, each category of expressions, i.e. linguistic, meta-linguistic, propositional, and computational, have their own version of neural-networks, pattern recognition and intelligence components. In each category, we observe that I, intelligence, contains, PR, pattern recognition, and that inturn contains, NN, neural network construction, i.e., $NN \subset PR \subset I$, as well as a hierarchy of containment between the categories, i.e., Computational, Propositional, Meta-Linguistic and Linguistic, $C \subset P \subset M \subset L$.

For example, depending on the set of axioms (Table 6.2), we are willing to accept, there are four subclasses of fuzzy set theories when we limit ourselves to two-valued {T,F} logic for the derivation of the canonical expressions. Before we go into the subclassification analysis, it is worthwhile to review linguistic, meta-linguistic, propositional and computational expressions with an example.

## 6.1. Linguistic Expression

It is known that generally one needs to investigate the structure of the sixteen basic combined linguistic expressions of a natural language for any two predicates A and B (Table 6.3) at the foundation of any study of set and logic theory. For example, in the context of temperature and pressure, and fuzzy linguistic terms "high" and "low", respectively, together with crisp truth qualifiers "true" and "false", we would state:

**Table 6.1.** Human Representation of Knowledge*

| TYPE | COMPLEXITY | LEVEL |
|---|---|---|
| HUMAN KNOWLEDGE NATURAL LANGUAGE | $LNN \subset LPR \subset LI$ | L-Linguistic Expression |
| | $\cup \quad \cup \quad \cup$ | |
| SYMBOLIC KNOWLEDGE NOTATIONS-SHORTHAND | $MNN \subset MPR \subset MI$ | M-Meta-Linguistic Expression |
| | $\cup \quad \cup \quad \cup$ | |
| PROPOSITIONAL KNOWLEDGE NOTATIONS | $PNN \subset PPR \subset PI$ | P-Propositional Expresssions |
| | $\cup \quad \cup \quad \cup$ | |
| COMPUTATIONAL FORMULAE | $CNN \subset CPR \subset CI$ | C-Computational Expressions |

* Türkşen's restructuring (1994) of Bezdek's model (1994):

| | | |
|---|---|---|
| LNN | : | Linguistic Neural Networks, |
| LPR | : | Linguistic Pattern Recognition, |
| LI | : | Linguistic Intelligence, |
| MNN | : | Meta-Linguistic Neural Networks, |
| MPR | : | Meta-Linguistic Pattern Recognition, |
| MI | : | Meta-Linguistic Intelligence, |
| PNN | : | Propositional Neural Networks, |
| PPR | : | Propositional Pattern Recognition, |
| PI | : | Propositional Intelligence, |
| CNN | : | Computational Neural Networks, |
| CPR | : | Computational Pattern Recognition, |
| CI | : | Computational Intelligence |

1. It is the case that "the temperature is high (is true or false) or the temperature is not high (is true or false), and the pressure is low (is true or false) or the pressure is not high (is true or false);
2. It is the case that "the temperature is high (is true or false) and the temperature is not high (is true of false), or the pressure is low (is true or false) and the pressure is not low (is true of false)";
3. It is the case that "the temperature is high (is true or false) or the pressure is low (is true or false)";

**Table 6.2.** Axioms of Classical Two-Valued Set and Two-Valued Logic Theory.

| | |
|---|---|
| Involution | $\mathbf{c(c(A)) = A}$ |
| Commutativity | $A \cup B = B \cup A$ |
| | $A \cap B = B \cap A$ |
| Associativity | $(A \cup B) \cup C = A \cup (B \cup C)$ |
| | $(A \cap B) \cap C = A \cap (B \cap C)$ |
| Distributivity | $A \cap (B \cup C) = (A \cap B) \cup (A \cap C)$ |
| | $A \cup (B \cap C) = (A \cup B) \cap (A \cup C)$ |
| Idempotence | $A \cup A = A$ |
| | $A \cap A = A$ |
| Absorption | $A \cup (A \cap B) = A$ |
| | $A \cap (A \cup B) = A$ |
| Absorption by X and ∅ | $A \cup X = X$ |
| | $A \cap \varnothing = \varnothing$ |
| Identity | $A \cup \varnothing = A$ |
| | $A \cap X = A$ |
| Law of contradiction | $A \cap c(A) = \varnothing$ |
| Law of excluded middle | $A \cup c(A) = X$ |
| De Morgan's laws | $c(A \cap B) = c(A) \cup c(B)$ |
| | $c(A \cup B) = c(A) \cap c(B)$ |

4. It is the case that "the temperature is not high (is true of false) and the pressure is not low (is true or false)";
5. It is the case that "the temperature is not high (is true of false) or the pressure is not low (is true or false)";

6. It is the case that "the temperature is high (is true of false) and the pressure is low (is true or false)";
7. It is the case that "if the temperature is high (is true of false), then the pressure is low (is true or false)";
8. It is the case that "the temperature is high (is true of false) and the pressure is not low (is true or false)"
9. It is the case that "the temperature is high (is true of false) or the pressure is not low (is true or false)";
10. It is the case that "the temperature is not high (is true of false) and the pressure is low (is true or false)";
11. It is the case that "the temperature is high (is true of false) if and only if the pressure is low (is true or false)";
12. It is the case that "either the temperature is high (is true of false) or the pressure is low (is true or false) but not both";
13. "It is the case that the temperature is high (is true or false)";
14. "It is not the case that the temperature is high (is true or false)";
15. "It is the case that the pressure is low (is true or false)";
16. "It is not the case that the pressure is low (is true or false)";

where "temperature", and "pressure" are nouns and "high", and "low" are adjustives. Nouns are known as "linguistic variables" and adjectives are know as "linguistic values (terms)" in the terminology of fuzzy set theory [Zadeh, 1975], and "and", "or", "not", "if ... then", etc., are "linguistic connectives" which do not map directly one-to-one to a t-norm or t-conorm, as it will be explained later in more detail. It should be noted that, for further clarity, we have added explicitly in parenthesis the verification expression that are usually implicit. Usually, it is implicitly assumed that all are "true". But, if they were all taken to be implicitly "false", then we would obtain a complementary set of expressions. In the same numeric order, the meta-linguistic equivalents of these sixteen linguistic expressions are shown in Table 6.3 and their propositional equivalents are shown in Table 6.4 for the classical two-valued set and logic theory as well as fuzzy set and two-valued logic theory. In the sequel, it will be shown that these propositional expressions known as DNF, CNF and FDCF and FCCF respectively are equivalent *in form only* but not in content. In two-valued logic, for these sixteen linguistic expressions given above, (1)-(16), the adjectives "high", "low" are assigned crisp values in the lattice {0,1} and the connectives are the well defined constants "$\wedge$", "$\vee$", "-" assigned to "and", "or" and "not", respectively, in a one-to-one correspondence, in agreement with the reductionist view of set and logic paradigm.

In two-valued logic, the sixteen basic linguistic concepts (1)-(16) can be represented with propositional expressions by two canonical forms known as Disjunctive and Conjunctive Normal forms, DNF and CNF, respectively, as shown in Table 6.4. It is known that $DNF_i(\bullet) \equiv CNF_i(\bullet)$, i=1,...,16, in two-valued set and logic because of the axioms of two-valued set and logic theory (Table 6.2). In particular, the equivalence is obtained with the application of idempotency, distributivity, and the law of excluded middle or its dual the law of contradiction. For example, DNF(A AND B)=A∩B≡(A∪B)∩(c(A)∪B)∩(A∪c(B))=CNF(A AND B) in Row 6 of Table 6.4 since there is the equivalence of these two canonical forms, in general, we use the shortest of these two normal forms, i.e., A∩B, in most investigations. Thus no misrepresentation occurs in their use in two-valued set and logic theory. However, we need to be cautious in using the shortest one of the normal forms in fuzzy set and logic theories. This causes a "myopic" development of fuzzy set combinations and inference formulas. **It will be shown that this particular equivalence breaks down in fuzzy and logic theory with important consequences.**

In generalizing two-valued theory to fuzzy valued theory, we ought to make sure that both normal forms of two-valued logic are generalized appropriately. In particular, we ought to fuzzify both DNF and CNF. That is, we ought not arbitrarily select for one concept combination its CNF expression, i.e., A∪B, as we do, for example, in the case of "A OR B" and for some other concept combination its DNF expression, i.e., A∩B, as we do for example in the case of "A AND B" in generalizing the two-valued set and logic theories to the fuzzy set and logic theory. However, it appears that most researchers are making such arbitrary selections in current literature of fuzzy set and logic theory. When such arbitrary selections are made one is capturing only a part of the knowledge representation formulae, i.e., not the whole impact of such a generalization. In fact such arbitrary choices lead to a "myopic" representation with a loss of information due to the gap generated between fuzzified normal forms. But we are getting ahead of our intent. Let us now return to the next stage, i.e., the meta-linguistic expressions.

## 6.2. Meta-Linguistic Expression

Meta-linguistic expressions represent linguistic expressions in a symbolic language. The most often used form of the meta-linguistic expressions for the sixteen basic combined concepts are shown in Table 6.3. However, they are not always one-to-one. For example, there are at least four alternate forms that represent the linguistic expression (7). These are:

(7)-(1) $X_1$ isr A (is T or F) IMPLIES $X_2$ isr B (is T or F)

(7)-(2) $X_1$ is NOT A (is T or F) OR $X_2$ isr B (is T or F)

(7)-(3) NOT($X_1$ isr A (is T or F) AND NOT ($X_2$ isr B (is T or F)))

(7)-(4) LARGEST Y SUCH THAT "$X_1$ isr A (is T or F) AND Y is SOME VALUE that is CONTAINED IN $X_2$ isr B (is T or F)"

**Table 6.3.** Meta-Linguistic Expressions of Combined Concepts for any A and B.

| Number | Meta-Linguistic Expressions |
|---|---|
| 1 | UNIVERSE - X |
| 2 | EMPTY SET - ∅ |
| 3 | A OR B |
| 4 | NOT A AND NOT B |
| 5 | NOT A OR NOT B |
| 6 | A AND B |
| 7 | A IMPLIES B |
| 8 | A AND NOT B |
| 9 | A OR NOT B |
| 10 | NOT A AND B |
| 11 | A IF AND ONLY IF B |
| 12 | A EXNLUSIVE OR B |
| 13 | A |
| 14 | NOT A |
| 15 | B |
| 16 | NOT B |

where $X_1$, $X_2$ and Y are the meta-linguistic representations of the "linguistic variables" that are taken as the base axes, "A", "B" are the "meta-linguistic values"(linguistic terms) associated with $X_1$, $X_2$ and Y, where the values of Y are consequent, respectively, "isr" means "in relation to a degree" which indicates that, say, $X_1$ belongs to fuzzy set A with a membership value, where as "is" specifies the two-valued memberships, and "AND", "OR", "NOT", "IMPLIES" are "fuzzy meta-linguistic connectives" and "LARGEST", "SUCH THAT", "CONTAINED IN" have specific well defined meanings in this context. In short form, these meta-linguistic expressions are represented as:

**Table 6.4.** Classical Disjunctive Normal Form, DNF, and Fuzzy Disjunctive Canonical Form, FDCF, and Classic Conjunctive Normal, CNF, and Fuzzy Conjunctive Canonical Forms, FCCF, where $\cap$ is a conjunction, $\cup$ is a disjunction and c is a complementation operator. **(a)** Top Section, i.e., FDCF and DNF, **(b)** Bottom Section, i.e., FCCF and CNF

| No. | Fuzzy Disjunctive Canonical Forms/Classical Disjunctive Normal Forms |
|---|---|
| 1 | $(A\cap B) \cup (A\cap c(B)) \cup (c(A) \cap B) \cup (c(A) \cap c(B))$ |
| 2 | $\varnothing$ |
| 3 | $(A\cap B) \cup (A\cap c(B)) \cup (c(A) \cap B)$ |
| 4 | $(c(A) \cap c(B))$ |
| 5 | $(A\cap c(B)) \cup (c(A) \cap B) \cup (c(A) \cap c(B))$ |
| 6 | $(A\cap B)$ |
| 7 | $(A\cap B) \cup (c(A) \cap B) \cup (c(A) \cap c(B))$ |
| 8 | $(A\cap c(B))$ |
| 9 | $(A\cap B) \cup (A\cap c(B)) \cup (c(A) \cap c(B))$ |
| 10 | $(c(A) \cap B)$ |
| 11 | $(A\cap B) \cup (c(A) \cap c(B))$ |
| 12 | $(A\cap c(B)) \cup (c(A) \cap B)$ |
| 13 | $(A\cap B) \cup (A\cap c(B))$ |
| 14 | $(c(A) \cap B) \cup (c(A) \cap c(B))$ |
| 15 | $(A\cap B) \cup (c(A) \cap B)$ |
| 16 | $(A\cap c(B)) \cup (c(A) \cap c(B))$ |

| No | Fuzzy Conjunctive Canonical Forms/Classical Conjunctive Normal Forms |
|---|---|
| 1 | I |
| 2 | $(A\cup B) \cap (A\cup c(B)) \cap (c(A)\cup B) \cap c(A)\cup c(B))$ |
| 3 | $(A\cup B)$ |
| 4 | $(A\cup c(B)) \cap (c(A)\cup B) \cap c(A)\cup c(B))$ |
| 5 | $(c(A)\cup c(B))$ |
| 6 | $(A\cup B) \cap (A\cup c(B)) \cap (c(A)\cup B)$ |
| 7 | $(c(A)\cup B)$ |
| 8 | $(A\cup B) \cap (A\cup c(B)) \cap (c(A)\cup c(B))$ |
| 9 | $(A\cup c(B))$ |
| 10 | $(A\cup B) \cap (c(A)\cup B) \cap (c(A)\cup c(B))$ |
| 11 | $(A\cup c(B)) \cap (c(A)\cup B)$ |
| 12 | $(A\cup B) \cap (c(A)\cup c(B))$ |
| 13 | $(A\cup B) \cap (A\cup c(B))$ |
| 14 | $(c(A)\cup B) \cap (c(A)\cup c(B))$ |
| 15 | $(A\cup B) \cap (c(A)\cup B)$ |
| 16 | $(A\cup c(B)) \cap (c(A)\cup c(B))$ |

| | |
|---|---|
| (7')-(1') | A(is T or F)$\rightarrow$ B (is T or F), |
| (7')-(2') | NOT A (is T or F) OR B (is T or F), |
| (7')-(3') | NOT(A (is T or F) AND NOT B (is T or F)), |
| (7')-(4') | SUPREMUM {Y\| A (is T or F) AND Y$\subseteq$ B (is T or F)}, etc. |

Again for the purpose of explicitly showing the implicit truth qualifiers, T=true, F=false, we have added (is T or F) to all expressions. As stated above, therefore, the mapping from a linguistic expression to meta-linguistic expressions is not unique. That is, the transformation is, in general, one-to many from "linguistic" to "meta-linguistic" expressions. This is the reason why, for example, there are R-implications which are based on form (7')-(4') and S-implications which are based on form (7')-(2'). It is to be noted that (7')-(3') is the meta-linguistic De Morgan equivalent of (7')-(2') when the "negation" is involutive. Whereas the form (7')-(1') is generally mapped to (7')-(2') or (7')-(4') among others.

### 6.3. Propositional Expressions

As indicated earlier, in two-valued set and two-valued logic, where the descriptive and veristic assignments are two-valued, i.e., $\{ D_{\{0,1\}}, V_{\{0,1\}} \}$, there are two propositional expressions known as "Disjunctive" and "Conjunctive" normal forms, DNF and CNF, respectively, that correspond to every meta-linguistic expression. For example, DNF and CNF expressions for "A $\rightarrow$ B" are stated in form (7')-(2') as:

$$DNF(A\rightarrow B,T)=[(A\cap B)\cup(c(A)\cap B)\cup(c(A)\cap c(B)),T], \text{ and} \tag{6.1}$$
$$CNF(A\rightarrow B,T)=[c(A)\cup B,T] \tag{6.2}$$

(The definitions of DNF and CNF and their derivation will be discussed in detail later.)

where the symbols A and c(A) correspond to the affirmation and the denial, i.e., the complementation, of the propositional symbol A, respectively and similarly B and c(B); and symbols, $\cap$, $\cup$, are the propositional connectives known as conjunction and disjunction, respectively. For the sake of emphasis, recall that in two-valued set and logic theory, i.e., $\{ D_{\{0,1\}}, V_{\{0,1\}} \}$, these connectives are "constants", i.e., $\wedge$, $\vee$, - whereas in fuzzy set theories they are no longer constant connectives. In the Type 1 fuzzy set theory, these connectives are interpreted to be crisp but variable connectives. They are chosen from the three subclasses of connectives known as pseudo t-norms and co-norms, t-norms and conorms, and Min-Max, etc., shown in Diagram 6.1. Again for the purpose of clarity, we have explicitly shown that DNF and CNF expressions correspond to

T-truth evaluations. This will be explained more clearly when we discuss later Truth Table based derivation of these formulas.

Finally, within the proposed perspective of this work, the linguistic connectives are interpreted to be “fuzzy”, “imprecise’, connectives in the combination of “fuzzy”, “imprecise”, “fuzzy” linguistic values that cause the generation of “the interval-valued Type 2 fuzzy sets” via the construction of Fuzzy Disjunctive and Conjunctive Canonical Forms, FDCF and FCCF, to be discussed shortly, by a selection of crisp but variable connectives that correspond to $\cap$, $\cup$, in both of the fuzzy propositional expressions. More explicitly, it is proposed that the representation of the combined linguistic expression of the linguistic values with linguistic connectives should not just be mapped arbitrarily to one of the two fuzzy canonical forms but to the interval generated by both fuzzy canonical forms. This is required because the equivalence between DNF(.)≡CNF(.) breaks down in fuzzy theory, i.e., FDCF(.)W FCCF(.). This generates one-to-many mapping for each linguistic connective. Later in the sequel, we will also show how these canonical forms can be generated with a modified and restricted multi-valued mappings as a special case of Demspter's (1967) and Pawlak's (1991) formalism. Naturally, each specific interval is the distance between the values generated by FCNF and FCCF expressions for a, b $\in$ (0,1) with the condition that FDCF be equal to FCCF when a, b $\in$ {0,1} in order to comply with the corresponding DNF and CNF expressions in two-valued set and logic theory. It should be noted that FDCF and FCCF expressions will be obtained in two ways: (1) by directly fuzzifying the $\{D_{\{0,1\}}, V_{\{0,1\}}\}$ theory based DNF and CNF expressions, and (2) by deriving FDNF and FCNF from a new construction of Fuzzy Truth Tables. These results are further verified by a unification and modified and restricted interpretation of Dempster-Pawlak Schemas to obtain upper and lower set approximations (Türkşen, 2001).

Recall that the two canonical (normal) forms are equivalent in two-valued set and logic theory, i.e., $DNF_i(\bullet) \equiv CNF_i(\bullet)$, for all of the sixteen basic combination of linguistic concepts (see Tables 6.3 and 6.4). This will be reviewed shortly. As indicated before, because $DNF_i(\bullet) \equiv CNF_i(\bullet)$, i=1.2.,,,.16, in two-valued set and logic theory, most researchers are used to studying the shortest of these two normal forms in their discussions and writings of fuzzy set and two-valued logic theory and in most of their investigations in this domain. That is the reason why, for example, “A$\cap$B” is chosen for the representation of “A AND B” which is its FDCF; but “A$\cup$B” is chosen for “A OR B” which is its FCCF. Furthermore, “c(A)$\cup$B” is chosen most often for “A$\rightarrow$ B” which is its FCCF, and so forth (see Table 6.4). (Thus, the habit of choosing the shortest form of the two normal(canonical) forms in two-valued set and logic theory is carried over to fuzzy theory arbitrarily. This caused the Type 1 representation of any combination of two fuzzy sets in most of current practices, which are “myopic” approximations from two fuzzy normal forms to one which is the shortest form of the two-valued set and logic formula.

It has been shown and will be discussed in the sequel that fuzzy normal forms are not equivalent in general. That is $FDCF_i(\bullet) \subseteq FCCF_i(\bullet)$ in fuzzy set and two-valued logic. This was first shown by a straight forward fuzzification of the classical normal forms [Türkşen, 1986].

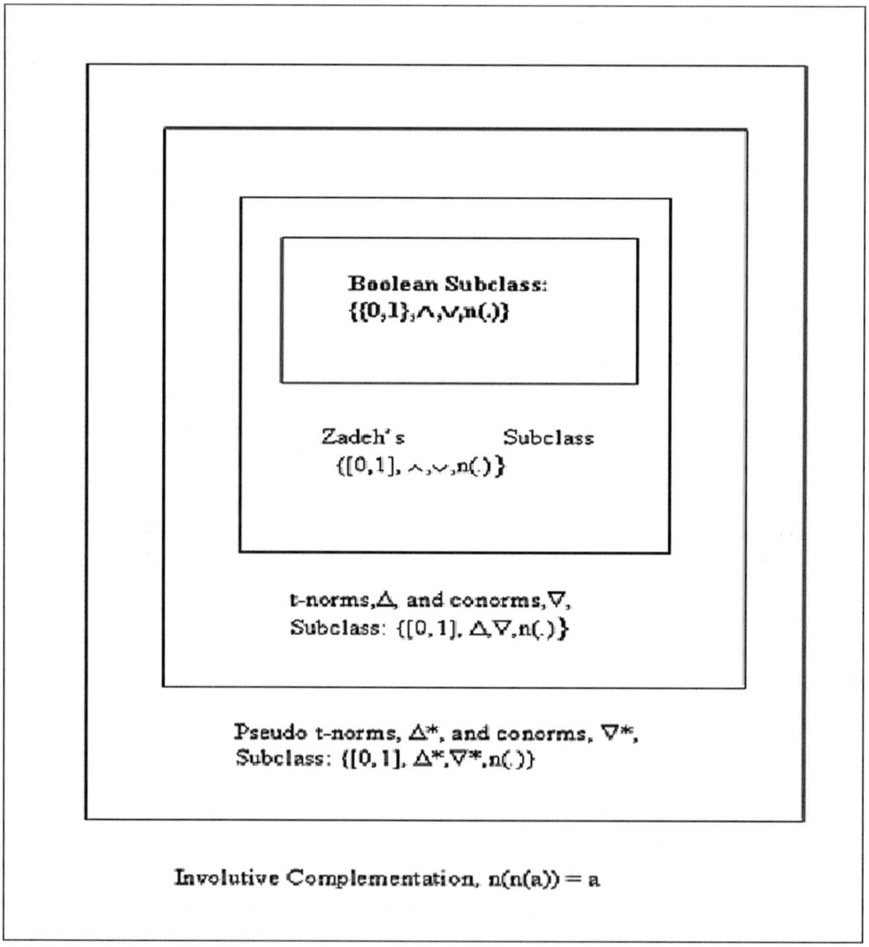


**Diagram 6.1.** The five Subclasses of Fuzzy Set Theory and two valued {t, F} logic. Computational Expressions

Propositional expressions are statements about the sets with set labels, such as A,B, etc., and address all the elements included in a set. But these expressions do not reveal the membership degree of an individual element, say, "x∈X isr A". The computational expressions specify clearly the membership degree of an element x∈X with $\mu_A$ (x)∈[0,1]. Thus at this level of specificity, it is usual to write, for example, FDCF and FCCF or more appropriately FDCF and FCCF, expressions as:

$$\mu_{\underset{(A\to B,T)}{FDCF}}(x,y) = \left[(A(x)\Delta B(y))\nabla(n(A(x))\Delta B(y))\nabla(n(A(x))\Delta n(B(y))), T\right]$$

$$\mu_{\underset{(A\to B,T)}{FCCF}}(x,y) = \left[n(A(x))\nabla B(y), T\right]$$

where A(x), B(y) are membership values of x in A and y in B, respectively, which are indicated at tiems as $\mu_A(x)$ and $\mu_B(y)$, respectively, and Δ, ∇ stand for a t-norm and a t-conorm, respectively, and n(A(x))=1-A(x)), n(B(y))=1-B(y). Furthermore, T in these expressions, indicates explicitly that these expressions are derived from the truthood of the given combination from a fuzzy truth table analysis.

## 6.4. Classes of Fuzzy Sets and Two-Valued Logic

As demonstrated in Diagram 6.1, [Türkşen, 1994], there are at least four subclasses of fuzzy sets when two-valued sets are generalized to fuzzy sets . These four subclasses of fuzzy sets are identified with the axiomatic restrictions put on the derivation of combined propositional expression derived with the use of four subclasses of crisp connectives, known as t-norms, Δ, and t-conorms, ∇.

i) If we assume that A and B are two fuzzy sets, i.e., $\mu_A$:X →[0,1], $\mu_B$:Y →[0,1] and the axioms

(1) I∩A=A, (1')∅∪A=A, Boundary,

(2) A∩B⊆C∩D, (2') A∪B⊆C∪D when A⊆C, B⊆D, Monotonicity,

(3) c(c(A))=A, (3') c(c(B))=B, Involutive Complementation,

(4) c(A∩B)=c(A) ∪c(B), ) (4') c(A∪B)=c(A) ∩c(B),De Morgan,

hold, then we get a fuzzy set and two-valued logic theory, i.e., these axioms hold to be absolutely "true" with a De Morgan Triple of conjunction, ∩, disjunction ∪, and involutive complementation, c, that are non-commutative, non-associative, and non-idempotent. There are many examples of such ordered semantics of linguistic connectives. For example, "he died severely and was buried quickly" versus "he was buried quickly and died severely" have different meanings and they need to be represented differently in the special order. These classes of crisp connectives are known as Pseudo-t-norms and co-norms and are studied by Fodor (1993) and Fodor and Keresztfalvi (1995).

ii) Next, if we assume $\mu_A:X \rightarrow[0,1]$, $\mu_B:Y \rightarrow[0,1]$ with the axioms (1), (2), (3), (4) stated above in i) together with the axioms

(5) $(A\cap B) \cap C = A\cap(B \cap C)$, (5') $(A\cup B) \cup C = A\cup (B \cup C)$, Associativity,

(6) $A\cap B = B \cap A$, (6') $A\cup B = B \cup A$, Commutativity,

hold, then we have a fuzzy set and two-valued logic theory, i.e., these axioms hold to be absolutely "true", with a De Morgan Triple of conjunction, $\cap$, disjunction $\cup$, and involutive complementation, c, that are commutative and associative, but non-idempotent. Again in natural language, there are many expressions that fall into this class. For example "he had a somewhat sweet coffee and read a good paper" versus "he read a good paper and had a somewhat sweet coffee" are symmetric and unordered and thus both representations have the same meaning. But "he put in one lump of sweet and then another lump of sweet into his drink" is different from "he put in one lump of sweet into his drink". That is algebraic structures that require "Archimedian", i.e., finiteness, axiom are not idempotent, since they require additivity and/or additivity and multiplicativity in the computational domain such as "bold intersection and union" or "algebraic sum and product" operators. This class of crisp connectives are known as t-norms and t-co-norms [Schweizer and Sklar, 1983].

iii) Next, if we assume that $\mu_A:X \rightarrow[0,1]$, $\mu_B:Y \rightarrow[0,1]$ and that the connectives are idempotent, i.e.,

(7) $A\cap A = A$, (7') $A\cup A = A$, Idempotency,

as well as being restricted by (i) (1)-(4) and (ii) (5) – (6) then we get a fuzzy set and two-valued logic theory, i.e., these axioms hold to be absolutely "true", with the original, Zadeh (1965) definitions. The connectives of this class have the dominance property. A typical example of this class of connectives appear in preference analysis and choice theory. For example, "object x is at least as good as object y and (but) object y is not as good as object x". That is "x is preferred to y" meaning some property or properties of x dominates that of those of y. This is the well known class of Max-Min, Standard Complement De

Morgan Triple. It should be recalled that idempotency implies absorption and in turn it implies distributivity (Weber, 1983).

iv) Finally, in addition to axioms 1-7, if we restrict sets to be crisp sets, i.e., two-valued sets, $\mu_A:X \rightarrow\{0,1\}$and $\mu_B:Y\rightarrow\{0,1\}$ together with the constant connective "and", $\wedge$, "or", $\vee$, and "complementation", -, then we get the classical two-valued set and two-valued logic theory, which admit the Law of Excluded Middle, LEM, and the Law of Contradiction, LC (see Table 6.2 for the complete set of axioms of two-valued set and two-valued logic theory.) Thus, it is to be observed that two-valued set and two-valued logic theory is a very special class of fuzzy set and two-valued logic theory.

### 6.5. Sub-Sub Classes of t-Norms

There are many sub-sub classes of t-norms that generate sub-sub classes of crisp set conjunction in two-valued logic. Three of the well know subclasses are:

1) Zadeh-Conjunction (special case)
   $[A(x) \text{ (is T) } \cap B(y) \text{ (is T)}]=\min[A(x), B(y)] \text{ (is T)}$
2) Algebraic-Product-Conjunction
   $[A(x) \text{ (is T) } \cap B(y) \text{ (is T)}]=[A(x)\bullet B(y)] \text{ (is T)}$
3) Lukasiewicz (or Bold)-Conjunction
   $[A(x) \text{ (is T) } \cap B(y) \text{ (is T)}]=\max[A(x)+B(y)-1,0] \text{ (is T)}$

where A(x), B(y) means the membership values of $x\in X$ in the fuzzy set A and $y\in Y$ in fuzzy set B, respectively. Again it is noted that (is T) are added in parenthesis to emphasize that these are two-valued truth-functional definitions, i.e., fuzzy descriptions, that are operated on with the two-valued logic axioms of the crisp t-norm operators that are used for the definition, i.e., description, as to how two fuzzy sets are to be conjuncted.

With the propositional (set) symbols, we have in general Triangular Norms Axioms

$$[A \text{ (is T) } \cap B \text{ (is T)}]=[A\cap B] \text{ (is T)}$$

whereas in the predicate (membership) domain symbols, we have:

1) $\Delta$ : $[0,1] \times [0,1] \rightarrow [0,1]$,
2) Commutativity: $(a \text{ (is T) } \Delta\, b \text{ (is T)})=(b \text{ (is T) } \Delta\, a \text{ (is T)})$,
3) Associativity : $(a \text{ (is T) } \Delta\, b \text{ (is T)}) \,\Delta\, c \text{ (is T)}=(a \text{ (is T)}) \,\Delta\, (b \text{ (is T) } \Delta\, c \text{ (is T)})$,
4) Monotonicity : if $a\geq a'$ (is T), $b\geq b'$ (is T), then $(a\Delta b) \geq(a'\Delta b')$ (is T),
5) $a\Delta 1$ (is T)=a (is T), $a\Delta 0$ (is T)=0 (is T),

### 6.6. Sub-Sub Classes of t-Conorms

Naturally, there are also many sub-sub classes of t-conorms that generate sub-sub classes of fuzzy set disjunction in two-valued logic. The three of the well known sub sub-classes are:

1) Zadeh-disjunction
$[A(x) \text{ (is T)} \cup B(y) \text{ (is T)}] = \max [A(x), B(y)] \text{ (is T)}$

2) Algebraic Sum-disjunction
$[A(x) \text{ (is T)} \cup B(y) \text{ (is T)}] = [A(x)+B(y)-A(y)B(y)] \text{ (is T)}$

3) Lukasiewicz (Bold)-Disjunction
$[A(x) \text{ (is T)} \cup B(y) \text{ (is T)}]=\min [A(x)+B(y), 1] \text{ (is T)}$

where A(x), B(y) are the membership values of $x \in X$ in the fuzzy set A and $y \in Y$ in the fuzzy set B, respectively. Again it is noted that (is T) are added in parenthesis to emphasize that these are two-valued truth-functional definitions that are executable under the axioms of crisp t-conorm, $\nabla$, operators that are used as disjunction definition for disjuncted fuzzy sets.

With the propositional set symbols, we have in general:

$$[A \text{ (is T)} \cup B \text{ (is T)}] = [A \cup B] \text{ (is T)}$$

whereas in the predicate (membership) domain symbols, we have:

1) $\nabla$ : $[0,1] \times [0,1] \rightarrow [0,1]$,

2) Commutativity: $(a \text{ (is T)} \nabla b \text{ (is T)})=(b \text{ (is T)} \nabla a \text{ (is T)})$,

3) Associativity : $(a \text{ (is T)} \nabla b \text{ (is T)}) \nabla c \text{ (is T)}=(a \text{ (is T)}) \nabla (b \text{ (is T)} \nabla c \text{ (is T)})$,

4) Monotonicity : if $a \geq a'$ (is T), $b \geq b'$ (is T), then $(a \nabla b) \geq (a' \nabla b')$ (is T),

5) $a \nabla 1$ (is T)=a (is T), $a \nabla 0$ (is T)=a (is T),

where a=A(x), b=B(y) are the membership values of $x \in X$ in the fuzzy set A and $y \in Y$ in the fuzzy set B, respectively, a, $b \in [0,1]$ and T is the truth qualifier for each symbol and/or expression.

## 6.7. Fuzzy-Set Complements

There are several alternative definitions of fuzzy set complement. Complement functions n are monotonic nonincreasing, and satisfy the boundary conditions:

$$n(0)=1,\ n(1)=0$$

The three well known fuzzy set complements are:

i) Standard Complement : $n(\alpha)=1-\alpha$

ii) Sugeno class of complements : $n_{\lambda}(\alpha)=\dfrac{1-\alpha}{1+\lambda\alpha}$, $\lambda>-1$

iii) Yager class of complements : $n_w(\alpha)=(1-\alpha^w)^{1/w}$, $w>0$

It should be noted that t-norms, Δ, and conorms, ∇, form De Morgan Triples with the standard complement. (hence the standard complement is the most used fuzzy complement in current literature). Furthermore, the derivation of disjunctive and conjunctive normal(canonical) forms require an involutive complementation operator. For this reason in the derivations the standard complement is used.

## 6.8. De Morgan Triples

De Morgan Laws for conventional sets are:

$$c(A\cap B) = c(A)\cup c(B),$$
$$c(A\cup B) = c(A)\cap c(B),$$

where $c(\cup)=\cap$ and $c(\cap)=\cup$.

Triples consisting of a set conjunction operator, ∩, a set disjunction operator, ∪, and the standard complement c(.) that generalize the De Morgan Laws are De Morgan Triples.

Well known examples of De Morgan Triples in fuzzy set theory are:

Zadeh: (min(a,b), max(a,b), 1-a) = (∧, ∨, -)

Algebraic: (ab, a+b-ab, 1-a) = ( ˙ , ¦ , n(a))

Lukasiewicz: (max(a+b-1, 0), min(a+b,1), n(a)) = (L· (a,b), L¦ (a,b), n(a)),

which is also known as

Bold (B· (a,b) = max (a+b,0), B| (a,b) = min (a+b,1),n(a))

Drastic

$$D_{\cdot}(a,b) = \begin{cases} a \wedge b & \text{if } a \vee b = 1 \\ 0 & \text{otherwise} \end{cases}$$

$$D_{\oplus}(a,b) = \begin{cases} a \vee b & \text{if } a \wedge b = 1 \\ 0 & \text{otherwise} \end{cases}$$

## 6.9. Parametric t-norms and t-conorms

It should be recalled that there are several classes of parametric t-norms, **Δ,** and t-conroms, **∇,** (Klir and Bo, 1995, pp. 74 and 82) some of these are shown in Table 6.5 as well their parametric ranges are shown in Table 6.6. We just list one such pair here, namely the Schweizer and Sklar (1961, 1983) parametric t-norm and t-conorm which we have applied extensively in our system modeling investigations for robotics as well as scheduling systems.(Emami, et.al., 1999-2000) (Türkşen, et.al., 1998)

$\Delta(a,b) = 1-[(1-a)^p+(1-b)^p-(1-a)^p(1-b)^p]^{1/p}, \; p > 0$

$\nabla(a,b) = [a^p + b^p + a^p b^p]^{1/p}, \; p > 0$

where Δ(•) and ∇(•) are t-norm and t-conorm, respectively.

The n-ary ∇(•) is written as:

$$\nabla(a_1, a_2, \ldots, a_n) = [\sum_{i=1}^{n} a_i^p - \sum_{i=1}^{n}\sum_{\substack{j=1 \\ j\neq i}}^{n} a_i^p a_j^p + \ldots \pm \prod_{i=1}^{n} a_i^p]^{1/p}, p > 0$$

which has a computational complexity of $O(2^n)$.

For computation speed, these t-norms and t-conorms are transforms as follows:

$\nabla(a_1, a_2, \ldots, a_n) =$

$$[a_1^p + (1-a_1^p)[a_2^p + (1-a_2^p)[\ldots[a_{n-2}^p + (1-a_{n-2}^p)[a_{n-1}^p + (1-a_{n-1}^p)a_n^p]]\ldots]]]^{1/p}$$

and $\nabla(a_1, a_2, \ldots, a_n) = 1-\nabla((1-a_1), (1-a_2), \ldots, (1-a_n))$, n>2 with complexitiy of O(n).

This parametric t-norm has a decreasing generator $f(a)=[1-(1-a)^p]^{1/P}$. It converges to D· (a,b) as p→0 to ˙ (a,b) as p→1 or –1, and to a∧b as p→∞.

A list of other parametric tnorms and t-conorms are shown in Table 6.5 and Table 6.6. Certain basic properties of negations and t-norms are included in Appendix 6.

## 6.10. Fundamental Phrases and Clauses

Let the membership of every $x \in X$ belonging to a (propositional) set A be specified by "a", a predicate membership symbol, at times written as A(x) or $\mu_A(x)$, then an atomic set $A = \{(x,a) \mid a \in \{0,1\}$, or $a \in [0,1], x \in X\}$, depending on whether A is a two-valued set or a fuzzy set such that its complement is $c(A) = \{(x,n(a)) \mid n(a) = 1\text{-}a, x \in X\}$.

It should be noted that in the propositional expressions c denotes the complement in set domain, whereas in predicate expressions n denotes the complement in the numeric domain.

**Definition 6.1.** The 1-variable-1-dimensional phrase space, where atomic propostions are written, is defined as $\text{AND}_1^1 = \{A,c(A)\}$ such that A and c(A) are atomic (propositional) set symbols as defined above.

**Definition 6.2.** The 1-variable-1-dimensional phrase space, where an atomic predicate is written, is defined as $\mathbf{and}_1^1 = \{a,n(a)\}$ such that a, n(a) are membership value symbols as defined above. It should be noted that this means only one of "a or n(a)" is taken from $\mathbf{and}_1^1 = \{a,n(a)\}$ in a 1-dimensional fundamental phrase of an atomic predicate expression, but not both! In turn, a fundamental phrase of an atomic propositional expression has either A or c(A), but not both!

**Definition 6.3.** The 1-variable-m-dimensional fundamental phrase space where compound propositions are written, is defined as:

$$\text{AND}_1^m = \{Y_1 \cap Y_2 \cap \ldots \cap Y_m \mid Y_1, Y_2, \ldots, Y_m \in \text{AND}_1^1, m \geq 2\}'$$

such that $\cap$ defines a set conjunction in the propositional domain.

**Definition 6.4.** The 1-variable-m-dimensional fundamental phrase where compound predicate expressions are written, is defined as:

$$\text{and}_1^m = \{y_1 \Delta y_2 \Delta \ldots \Delta y_m \mid y_1, y_2, \ldots, y_m \in \mathbf{and}_1^1, m \geq 2\},$$

such that $\Delta$ represents a crisp conjunction operator in the (numerical) computational domain, i.e., $\Delta$ is either Min or any one of the more general t-norm or pseudo t-norm operators . This means that m-elements are taken from $\mathbf{and}_1^1 = \{a,n(a)\}$ with repetition whether it is two-valued or infinite valued and are conjuncted with a crisp conjunction operator $\Delta$. In turn, a fundamental phrase of a compound propositional expression contains m set symbols some that are A's others c(A)'s.

For example, the following are 1-variable-m-dimensional propositional phrases:
$A \cap A \cap A \ldots \cap A$, $A \cap c(A) \cap \ldots \cap A$, $c(A) \cap c(A) \cap \ldots \cap c(A)$, etc.
where $\cap$ represents a set intersection and the total number of A's and/or c(A)'s is m.

**Table 6.5.** A list of some parametric t-conorms, $\nabla$, and t-norms, $\Delta$.

| Union, $\nabla$ | Intersection, $\Delta$ | Range | Author |
|---|---|---|---|
| $[a^p + b^p - a^p b^p]^{1/p}$ | $1-[(1-a)^p+(1-b)^p$ $-(1-a)^p(1-b)^p]^{1/p}$ | p>0 | Schweizer & Sklar [63] |
| $1-\mathrm{Max}[0,((1-a)^{-p}$ $+(1-b)^{-p}-1)]^{1/p}$ | $\mathrm{Max}(0, a^{-p}+b^{-p}-1)^{-1/p}$ | $p\in(-\infty,\infty)$ | Schweizer & Sklar [69] |
| $\dfrac{a+b-(2-\gamma)ab}{1-(1-\gamma)ab}$ | $ab/\left(\gamma+(1-\gamma)(a+b+-ab)\right)$ | $\lambda\in(0, \infty)$ | Hamacher [70] |
| $1-\log_s\left[1+\dfrac{(S^{1-a}-1)(S^{1-b}-1)}{S-1}\right]$ | $\log_s\left[1+\dfrac{(S^a-1)(S^b-1)}{S-1}\right]$ | $S\in(0, \infty)$ | Frank [71] |
| $\mathrm{Min}[1,(a^w+b^w)^{1/w}]$ | $1\text{-}\mathrm{Min}[1,((1\text{-}a)^w$ $+(1-b)^w)^{1/w}]$ | $\omega\in(0, \infty)$ | Yager [72] |
| $1-\dfrac{(1-a)(1-b)}{\max(1-a,1-b,\alpha)}$ | $\dfrac{ab}{\max(a,b,\alpha)}$ | $\alpha\in(0, \infty)$ | Dubois & Prade [67] |
| $\dfrac{1}{1+\left[\begin{array}{l}\left((1/a)-1\right)^{-\lambda}\\+\left((1/b)-1\right)^{-\lambda}\end{array}\right]^{-1/\lambda}}$ | $\dfrac{1}{1+\left[\begin{array}{l}\left((1/a)-1\right)^{\lambda}\\+\left((1/b)-1\right)^{\lambda}\end{array}\right]^{1/\lambda}}$ | $\lambda\in(0, \infty)$ | Dombi [74] |

**Definition 6.5.** The 1-variable-1-dimensional fundamental clause space, where atomic sets as propositions are written, is defined as $OR_1^1 = \{A, c(A)\}$ such that A and c(A) are atomic defined above.

**Definition 6.6.** The 1-variable-1-dimensional fundamental clause space, where atomic predicates are written, is defined as: $\mathbf{or}_1^1 = \{a, n(a)\}$. Again it should be noted that this means only one of "a or n(a)" is taken from $\mathbf{or}_1^1 = \{a, n(a)\}$ in 1-dimensional fundamental clause, but not both! In turn, a 1-dimensional fundamental clause of an atomic proposition has only one of the set symbols, either A or c(A), but not both.

**Definition 6.7.** The1-vairable-m-dimensional fundamental clause space, where compound propositions are written, is defined as:

**Table 6.6.** Limits and ranges of parametric t-norms (intersections) and t-conorms (unions).

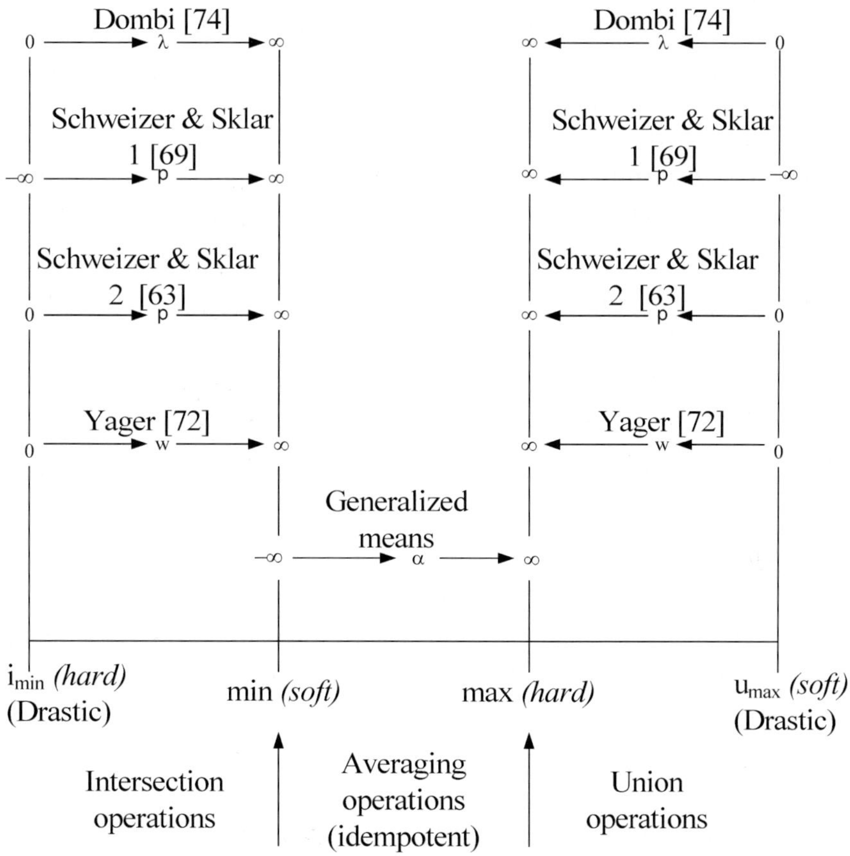


$OR_1^m = \{Y_1 \cup Y_2 \cup \ldots \cup Y_m | Y_1, Y_2, \ldots, Y_m \in AND_1^1, m \geq 2\}$
such that $\cup$ defines a set disjunction in the propositional domain.

**Definition 6.8.** The 1-variable-m-dimensional fundamental clause space, where compound predicate expressions are written, is defined as:

$$OR_m^1 = \{y_1 \nabla y_2 \nabla \ldots \nabla y_m | y_1, y_2, \ldots, y_m \in or_1^1, m \geq 2\}$$

such that $\nabla$ represents a crisp disjunction operator, in the (numerical) computational domain, i.e., $\nabla$ is either Max or any one of the more general t-conorm or pseudo-t-conorm operators. Again this means that m-elements are taken from $\mathbf{or}_1^1 = \{a, n(a)\}$ with repetition and are disjuncted with a crisp disjunction operator. In turn, a fundamental clause of a compound propositional expression contains m-set symbols some that are A's and others c(A)'s. For example: the following are 1-variable-m-dimensional propositional cluases: $A \cup A \cup A \ldots \cup A$, $A \cup c(A) \cup \ldots \cup A$, $c(A) \cup c(A) \cup \ldots \cup c(A)$, etc., where $\cup$ represents a set disjunction and the total number of A's and/or c(A)'s is m.

**Definition 6.9.** A formula W, a propositional expression, is said to be a 1-variable-m-dimensional (Fuzzy) Conjunctive Canonical Form, (F)CCF, if $(F)CCF_1^m(W) = C_1 \cap C_2 \cap \ldots \cap C_k$, where $C_i$, i = 1, 2, ..., k, is a z propositional fundamental clause such that $C_i \in OR_1^m$ and $\mu(x, C_i) = c_i \in \mathbf{or}_1^m$, m≥1, for $x \in X$.

For example, it will be shown that in two-valued logic constructed over fuzzy sets with t-norm-conorm and standard complement operators, the Fuzzy Conjunctive Canonical Form, $FCCF_1^2(W)$ for the meta-linguistic expression "W=A AND c(A)" is:

$$FCCF_1^2(A\text{ AND }c(A)) = (c(A) \cup c(A)) \cap (A \cup A) \cap (A \cup c(A))$$

where "AND" on the left hand side specifies a linguistic connective, and each of $(A \cup c(A))$, $(A \cup A)$, $(c(A) \cup c(A))$ are propositional clauses in $OR_1^2$ such that $C_1 = c(A) \cup c(A)$, $C_2 = (A \cup A)$, $C_3 = A \cup c(A)$. The corresponding predicate expression is :

$$\mu[FCCF_1^2(x, A\text{ AND }c(A)] = (n(a)\nabla n(a))\Delta(a\nabla a)\Delta(a\nabla n(a))$$ such that $c_1 = n(a)\nabla n(a)$, , $c_2 = a\nabla a$, $c_3 = a\nabla n(a)$, i.e., $c_i \in \mathbf{or}_1^2$, i =1,2,3.

In general for two propositional variables, A and B, the (Fuzzy) Conjunctive Normal Form, (F)CCF, of "W=A AND B", is written as: (F)CCF(A AND B)=(c(A)∪B)∩(A∪B) ∩(A∪c(B)) which will be discussed in more detail in the sequel. Again it should be noted that "AND" in W and on the left hand side of the FCCF(.) represents a linguistic connective.

**Definition 6.10.** A formula W, a propositional expression, is said to be a 1-variable-m-dimensional (F)DCF if $(F)DCF_1^m(W)=P_1 \cup P_2 \cup \ldots \cup P_k$ k>1, where Pi, i = 1, 2, ...k is a propositional fundamental phrase such that $P_i \in AND_1^m$ and $\mu(x,P_i)=p_i \in \mathbf{and}_1^m$ $m \geq 1$, for $x \in X$.For example, again it will be shown that in two-valued logic constructed over fuzzy sets with t-norm-conorm and standard negation operators, the Fuzzy Disjunctive Canonical Form, $FDCF_1^2$ (W) for the meta-linguistic expression "W=A OR c(A)" is ,:

$$FDCF_1^2(A\ OR\ c(A)) = (A \cap c(A)) \cup (A \cap A) \cup (c(A) \cap c(A)),$$

where "OR" in the W and on the left hand side of the FDCF(.) represents a linguistic connective, and each of $(A \cap c(A))$, $(A \cap A)$, $(c(A) \cap c(A))$ are propositional clauses in $AND_1^2$ such that $P_1 = A \cap c(A)$, $P_2 = A \cap A$, $P_3 = c(A) \cap c(A)$. The corresponding predicate expression is:

$$\mu[FDNF_1^2(x, A\ \ OR\ c(A))] = (a\Delta n(a))\nabla(a\Delta a)\nabla(n(a)\Delta n(a))$$

such that $p_1 = a\Delta n(a)$, $p_2 = a\Delta a$, $p_3 = n(a)\Delta n(a)$, i.e., $p_i \in \mathbf{and}_1^2$, i=1,2,3.

Again, in general for two propositional variables, A and B, the (Fuzzy) Disjunctive Canonical Form, (F)DCF of "W=A OR B", is written as:

$(F)DCF(A\ OR\ B)=(A\cap B)\cup(A\cap c(B)) \cup (c(A) \cap B)$

which will be discussed in more detail later.

With these preliminaries, we will next review the derivation of normal forms for the two-valued set and two-valued logic and discuss the consequences of this classical theory in order to form the foundation of comparisons when we extend these concepts to infinite (fuzzy)-valued sets and two-valued logic paradigm.

## Appendix 6.1. Basic Properties of negations and t-norms

### A-6.1. Negations

A function n: [0, 1] → [0, 1] is called a negation, if n is (i) continuous and (ii) satisfies n(0) = 1 and n(1) = 0. A negation n is called **strict**, if n is (iii) strictly decreasing, and a strict negation is called **strong**, if it satisfies (iv) $n^2(x) = n(n(x)) = x$

### A-6.2. Properties

The summary of properties of t-norms are as follows:
First, we have, for $\forall x \in [0, 1]$:

$$\Delta^{(1)}x = x$$
$$\Delta^{(n+1)}x = \Delta(\Delta^{(n)}x)$$

The following properties will be needed in Chapter 9.

A-6.2.1. Archimedean and strictness

(i) Strict monotonicity:

$\Delta$ is called strictly monotone, for $\forall x \in [0, 1]$,
if $y < z$ then $x\Delta y < x\Delta z$

(ii) Cancellation law:

satisfies the cancellation law, if
$x\Delta y = x\Delta z$ and $x>0$ imply $y=z$.

(iii) Strictness:
$\Delta$ is called strict, if it is continuous and strictly monotone.

(iv) Archimedean property:
$\Delta$ has the Archimedean property, for $\forall x \in ]0, 1[$ if there exists a natural number $n$ such that

$$\Delta^{(n)}\mathbf{x} < y.$$

(v) Limit property:

$\Delta$ has the limit property, for $\forall x \in ]0, 1[$,
if $\lim_{n\to\infty} \Delta^{(n)}x = 0$.

(vi) Diagonal property.

$\Delta$ has the diagonal property, for $\forall x \in ]0, 1[$,
if $x\, \Delta x < x$.

The following relations can be proved between the properties states above.
(a) (i) strict monotonicity is equivalent to (ii) cancellation law

(b) (iv) Archimedean property is equivalent to (v) limit property
(c) (iii) strictness implies (i) strict monotonicity = (ii) cancellation law (by the definition)
(d) (vi) diagonal property implies (i) strict monotonicity = (ii) cancellation law
(e) (iv) Archimedean property = (v) limit property implies (vi) diagonal property
And under the condition that Δ is continuous,
(f) (i) strict monotonicity = (ii) cancellation law implies (iii) strictness (therefore they are equivalent)
(g) (i) strict monotonicity = (ii) cancellation law implies (iv) Archimedean property = (v) limit property
(h) because of (f) and (g), (iii) strictness implies (iv) Archimedean property = (v) limit property
(j) (vi) diagonal property implies (iv) Archimedean property = (v) limit property (therefore they are equivalent).

### A-6.3. Nilpotency

(vii) Nilpotent element:

> $x \in ]0, 1[$ is called a nilpotent element, if there exists n such that $\Delta^{(n)}x = 0$.

(viii) Nilpotent t-norm:

> A t-norm is called a nilpotent norm, if any $x \in ]0, 1[$ is a nilpotent element.

(ix) Zero divisor:

> $x \in ]0, 1[$ is called a zero divisor, if there exists $y \in ]0, 1[$ such that $x \Delta y = 0$.

There are two theorems on continuous t-norms.

**Theorem 1**. A continuous t-norm Δ is one of the following:

> (1) $\Delta = \Delta_M (= \min)$,
> (2) Archimedean, or
> (3) an ordinal sum of t-norms.

**Theorem 2.** A continuous Archimedean t-norm is either (1) nilpotent, or (2) strict.

### A-6.4. Representation of t-norms

Archimedean t-norms and t-conorms, if continuous, can be represented by their generating functions. That is, a continuous Archimedean t-norm Δ can be expressed as

$$x \Delta y = f^{(-1)}(f(x) + f(y)).$$

Here, $f$: $[0,1] \rightarrow [0, -\infty]$ is a continuous decreasing function and $f^{(-1)}$ is a pseudo inverse function of $f$, defined by

$$f^{(-1)}(t) = \begin{cases} 1 & \text{if } 0 \leq t \leq f(1) \\ f^{(-1)}(t) & \text{if } f(1) \leq t \leq f(0) \\ 0 & \text{if } f(0) \leq t \leq +\infty. \end{cases}$$

$f$ is called an (additive) generating function of Δ.

### A-6.5. Representation of negation

For a strong negation n, there exists a strictly increasing generating function $g : [0, 1] \rightarrow [0, +\infty)$, $g(0) = 0$, and, it is expressed as:

$$n(x) = g^{-1}(g(l) - g(x)).$$

### A-6.6. De Morgan's triplet

There are two ways to define t-conorms, ∇. One way is the manner in which t-norms and t-conorms are defined mutually independently as stated in Section 6.6 and 6.7. And another way is that a t-conorm ∇ is defined by

$$\nabla(x, y) = 1-\Delta(1-x, 1-y),$$

where Δ is a t-norm. In this case, Δ and ∇ are duals (conjoints) of each other with the (standard) negation $n(x) = 1- x$.

If a continuous t-norm Δ and a continuous t-conorm ∇ are duals of each other with a strong negation n, if they satisfy the following De Morgan's law:

$$n(x \Delta y) = n(x) \nabla n(y),$$

or equivalently

$$n(x \nabla y) = n(x) \Delta n(y),$$

then a triplet $(\Delta, \nabla, n)$ is called a De Morgan triplet. De Morgan triplets now play an important role in fuzzy set theory and fuzzy logic.

### A-6.7. Other Properties of t-norms and conorms

Drastic operators $D_\cdot$ and $D_\oplus$ are not continuous.

It is well known that $\Delta=\wedge$ is the strongest t-norm while $\Delta= D_\cdot$ is the weakest. Similarly $\nabla=\vee$ is the weakest t-conorm and $\nabla= D_\oplus$ is the strongest. i.e., $D_\cdot(a,b)\leq \Delta(a,b) \leq \wedge(a,b)$ and $\vee(a,b) \leq \nabla(a,b) \leq D_\oplus(a,b)$. There is the following ordering between the six t-norms and t-conorms considered above:

$$D_\cdot \leq L_\cdot \leq \cdot \leq \wedge$$
$$D_\oplus \geq L_\oplus \geq \oplus \geq \vee$$

Of course, due to $\wedge \leq \vee$, we have:

$$D_\cdot \leq L_\cdot \leq \cdot \leq \wedge \leq \vee \leq \oplus \leq L_\oplus \leq D_\oplus$$

There are also the following distributive properties of +, -, $\wedge$ and $\vee$ operators over $\wedge$ and $\vee$ (Note that $\wedge$ is not distributive over +, etc.)

$$Di(+,\wedge) = a + (b\wedge c) = a+b \wedge a+c;$$
$$Di(+,\vee) = a + (b\vee c) = a+b \vee a+c;$$
$$Di(-,\wedge) = a - (b\wedge c) = a-b \vee a-c;$$
$$Di(-,\vee) = a - (b\vee c) = a-b \wedge a-c;$$
$$Di(\vee, \wedge) = a\vee (b\wedge c) = a\vee b \wedge a\vee c;$$
$$Di(\vee,\vee) = a\vee (b\vee c) = a\vee b \vee a\vee c;$$
$$Di(\wedge,\wedge) = a\wedge (b\wedge c) = a\wedge b \wedge a\wedge c;$$
$$Di(\wedge,\vee) = a\wedge (b\vee c) = a\wedge b \vee a\wedge c;$$

There are more sophisticated gradations of conjunctions and disjunctions which were introduced as parameterized families of t-norms and t-conorms in Section 6.11.

## References


1. Bezdek, J.C. (1994), "What is computational intelligence", in J.M. Zurada, R.J. Marks II, C.J. Robinson (eds), *Computational Intelligence* - Imitating Life, IEEE Press, New York, 1-12.
2. Dempster A.P (1967), "Upper and Lower Probabilities Induced by a Multivalued Mapping", In: *Annals of Mathematical Statistics*, 38, 325-339.
3. Fodor, J.C. (1993), "A New Look at Fuzzy Connectives", *Fuzzy Sets and Systems*, 57: 141-148.

4. Fodor, J.C., T. Keresztfalvi (1995), "Nonstandard Conjunctions and Implications in Fuzzy Logic", *International Journal of Approximate Reasoning*, 12: 69-84.
5. Klir, G., Bo Yuan (1995), Fuzzy Sets and Fuzzy Logic, *Prentice Hall.*
6. Mc Luhan, M. (1995), Essential Mc Luhan, E. Mc Luhan and F. Zingrone (eds.) Basic Books, New York.
7. Pawlak Z. (1991), Rough Sets, *Kluwer*, Netherlands.
8. Schweizer, B., A. Sklar (1983), *Probabilistic Metric Spaces,* North Holland, Amsterdam.
9. Schweizer, B., A. Sklar (1961), "Associative Functions and Statistical Triangle Inequalities", *Publ. Math. Debrecen,* 8, 169-186.
10. Türkşen I.B. (2001), "Computing with Descriptive and Veristic Words: Knowledge Representation and Reasoning", *in: Computing With Words*, P.P. Wang(ed.), Chapter 10, Wiley, New York, 297-328.
11. Türkşen, I.B. (2001), "Sources, Measurements and Models of Type 2 Fuzziness in the New Millennium", *in: Fuzziness in the New Millennium*, V. Dimitrov (ed), Springer Verlag (in press).
12. Türkşen I.B. (1994),"Interval-Valued Fuzzy Sets and Compensatory AND", *FSS,* 87-100.
13. Türkşen, I.B. (1994), "Fuzzy Normal Forms", *Fuzzy Sets and Systems,* 253-266.
14. Türkşen, I.B, (1994), "Interval-Valued Fuzzy Sets and 'Compensatory AND' ", *FSS*, 295-307.
15. Türkşen I.B. (1986), "Interval-Valued Fuzzy Sets based on Normal Forms", *Fuzzy Sets and Systems,* 191-210.
16. Weber, S. (1983), "A General Concept of Fuzzy Connectives, Negations and Implications Based on t-norms and t-conorms, *Fuzzy Sets and Systems* 11, 115-134.
17. Zadeh, L.A. (1995), Keynote address, *IFSA Congress*, Sao Paulo, Brazil.
18. Zadeh, L.A. (1965), "Fuzzy Sets", *Information and Control Systems*, Vol.8, 338-353.

# Chapter 7

# EQUIVALENCES IN TWO-VALUED LOGIC

Two-valued logic (verification) schema can be constructed either on the two-valued set descriptions or on the fuzzy-valued set descriptions. That is we have

(a) $\{D_{\{0,1\}}, V_{\{0,1\}}\}$ (Berztiss, 1971; Copy, 1982), or
(b) $\{D_{[0,1]}, V_{\{0,1\}}\}$ (Türkşen, 1986)

as indicated before in Section 2.2. There are a set of equivalences that exits in two-valued set and logic theory which break down in fuzzy-valued set and two-valued logic theory. Next, we treat each one separately.

### 7.1. Two-Valued Set(Description) and Two-Valued Logic(Verification)

Let us review two-valued set and logic formulas for the combination of two linguistic concepts A and B. In this theory, the descriptive membership values A(x)=a and B(x)=b, a, b$\in\{0,1\}$, corresponding to the set symbols A and B and the truth assignments {T,F} of the verity, i.e., verification, of the descriptive membership values, are all mapped to the lattice {0,1}.

Let us consider the linguistic combination "W= A **OR** B", a basic expression, from amongst the 16 possible meta-linguistic expressions, basic expressions, listed in Table 6.3. As an example, we demonstrate the derivation of Conjunctive and Disjunctive Normal Forms, CNF and DNF in general and for W in particular. In accordance with the discussion presented in Chapter 6, let us first write the descriptive set assignments together with the veristic, truth, assignments for the concepts A and B in predicate form.

In general for any two linguistic concepts A and B we start with:

"x$\in$X isr A, with a=1, is T" (7.1)
"x$\in$X isr A, with a=0, is F" (7.2)
"y$\in$Y isr B, with b=1, is T" (7.3)
"y$\in$Y isr B, with b=0, is F" (7.4)

Next we form the linguistic "**OR**" combination, i.e., W= A **OR** B and write the predicate expressions for the four possible combinations:

["x∈X isr A, with a=1, is T" **OR** "y∈Y isr B, with b=1, is T"] is T.
["x∈X isr A, with a=1, is T" **OR** "y∈Y isr B, with b=0, is F"] is T.
["x∈X isr A, with a=0, is F" **OR** "y∈Y isr B, with b=1, is F"] is T.
["x∈X isr A, with a=0, is F" **OR** "y∈Y isr B, with b=0, is F"] is F.

These predicate expressions form the foundation of the "Truth Table" shown in Table 7.1. It is to be emphasized that in this classical "Truth Table" the descriptive membership values for sets A and B are not stated explicitly but are assumed to be {0,1} implicitly.

With these and the application of the normal (canonical) form derivation algorithm stated Section 7.2.1 below, one can derive the DNF and CNF expressions for the two-valued set and two-valued logic paradigm.

**Table 7.1.** Classical truth table interpretations of "A OR B"

| **Truth assignments to classical meta-linguistic variables** | | **Truth assignments to the meta-linguistic expression** | **Primary conjunctions** |
|---|---|---|---|
| A | B | "A OR B" | |
| T(A) | T(B) | T(A OR B) | A ∩ B |
| T(A) | F(B) | T(A OR B) | A ∩ c(B) |
| F(A) | T(B) | T(A OR B) | c(A) ∩ B |
| F(A) | F(B) | F(A OR B) | c(A) ∩ c(B) |

## 7.2. (Canonical) Normal Form Derivation

With the "Truth Table", Table 7.1, one derives the (canonical) normal form expressions in general and in particular for "A OR B" with **normal form derivation algorithm** as follows:

### *7.2.1. (Canonical) Normal Form Derivation Algorithm*

Step 1. To derive the Disjunctive Normal Form Expression, DNF, one does the following:

For each row of the "Truth Table", if a "T" appears in the column for the combined Meta-Linguistic Expression, in our example, "A OR B" column, then:

1(a) Construct the primary conjunction, $\cap$, of the meta-linguistic symbols, e.g., "A" and "B" with the requirement that (i) if "T" appears in the truth column of that symbol, the symbol is taken as it appears in the column heading, i.e., in this case, and usually, in its affirmative form; but (ii) if "F" appears in the truth column of that symbol, the complement of the symbol is taken.

1(b) Next, combine all primary conjunctions obtained in (a) above with the disjunction operator, $\cup$.

For our example "A OR B", we get three phrases in two-variable two dimensional space in STEP 1 (a) as:

$A\cap B$; $A\cap c(B)$; $c(A)\cap B$.

In STEP 1(b), we get one complex clause

$(A\cap B)\cup(A\cap c(B)\cup(c(A)\cap B)$

in two-variable two dimensional space. Hence, we get:

$DNF(A\ OR\ B)= (A\cap B)\cup(A\cap c(B)\cup(c(A)\cap B)$

As shown in top section of Table 6.4, Table 6.4.(a) row 3.

STEP 2. To derive the Conjunctive Normal Form Expression, CNF, one does the following:

For each row of the "Truth Table", if an "F" appears in the column for the combined Meta-Linguistic Expression, in our example, "A **OR** B" column, then:

2(a) construct the primary conjunction, $\cap$, of the symbols, e.g., "A" and "B", with the same requirement as stated in STEP 1(a).

2(b) Next, combine all conjunctions obtained in (a) above with the disjunction operator, $\cup$.

For our example, "A **OR** B", we get one phrase in two-variable two-dimensional space in STEP2(a) as:

$$c(A) \cap c(B)$$

In STEP 2(b), we get the same expression, i.e., the phrase $c(A) \cap c(B)$, since there are no other phrases to combine to obtain a complex clause.

*7.2.1.1. Discussion*

Next note that the expressions obtained in STEP 2(a) and (b) are constructed for the "F" entries under the combined Meta-Linguistic Expression, e.g., "A **OR** B", thus we must take the compliment of the expression obtained in STEP 2(b) to get the Conjunctive Normal Form, CNF.

Thus, for our example, we get

$$\begin{aligned} CNF(A\ OR\ B) &= c[c(A) \cap c(B)] \\ &= A \cup B \end{aligned}$$

which is shown in the lower section of Table 6.4, Table 6.4.(b) row 3.

The Normal Form Derivation Algorithm stated above can be used to derive all DNF and CNF expression shown in Table 6.4, for all 16 possible combination of concepts shown in Table 6.3. As well, it will be used in the derivation of FDCF and FCCF, Fuzzy Disjunctive and Fuzzy Conjunctive Canonical Forms from the construction of Fuzzy Truth Tables to be discussed later in chapter 8.

## 7.3. Equivalence of Normal Forms

In Section 6.3, it was stated that $DNF_i(\bullet) \equiv CNF_i\ (\bullet)$, i=1,2,...,16, for all the possible combination of the linguistic concepts that are shown in Tables 6.3 and 6.4, when the two-valued set and two-valued logic axioms listed in Table 6.2 hold. This is the case of the fourth, boundary-valued, class of fuzzy sets discussed above in (iv) of Section 6.5.

It will be shown in this chapter and later in Chapter 8 that these 16 equivalences break down in fuzzy set and two valued logic theory.

*7.3.1. Equivalence of DNF and CNF of "IMPLICATION"*

In order to demonstrate the equivalence of the Disjunctive and Conjunctive Normal Forms, DNF and CNF, let us discuss the case of the meta-linguistic expressions:

"A IMPLIES B is true, **T**", i.e., ($A \rightarrow B$, **T**), as an example.
Let us rewrite the formulas (6.1) and (6.2). We have:

$$DNF(A \to B, \mathbf{T}) = [(A \cap B) \cup (c(A) \cap B) \cup (c(A) \cap c(B)), \mathbf{T}]$$

$$CNF(A \to B, \mathbf{T}) = [(c(A) \cup B), \mathbf{T}]$$

where **T** designates that the meta-linguistic statements on the left hand side and the propositional statement on the right hand side are "true", **T**, and A and B are crisp sets, $\cap$ and $\cup$ are crisp set conjunction and disjunction symbols, respectively.

Next let us apply the idempotency, ID, the distributivity, D, the law of excluded middle, LEM, the identity, I, and the commutativity, C, in that order, to the original expression of DNF(A→B, **T**) on the right hand side to see that we derive CNF(A→B, **T**) as follows:

$$\begin{aligned} DNF(A \to B, \mathbf{T}) &= [(A \cap B) \cup (c(A) \cap B) \cup (c(A) \cap c(B)), \mathbf{T}] & \\ &= [(A \cap B) \cup (c(A) \cap B) \cup (c(A) \cap B) \cup (c(A) \cap c(B), \mathbf{T}] & \text{ID} \\ &= [(A \cup c(A)) \cap B) \cup (c(A) \cap (B \cup c(B)), \mathbf{T}] & \text{D} \\ &= [I \cap B \cup c(A) \cap I, \mathbf{T}] & \text{LEM} \\ &= [B \cup c(A), \mathbf{T}] & \text{I} \\ &= [c(A) \cup B), \mathbf{T}] & \text{C} \\ &= CNF(A \to B, \mathbf{T}) & \end{aligned}$$

*7.3.2. Equivalence of DNF and CNF of "OR" Connective*

For another example, let us review the equivalence of DNF and CNF expressions of "OR" connective, i.e., "A OR B". We have

$$DNF(A \text{ OR } B) = (A \cap B) \cup (A \cap c(B)) \cup (c(A) \cap B)$$
$$CNF(A \text{ OR } B) = A \cup B$$

Applying distributivity and idempotency axioms as shown in Table 6.2 in general and in particular, adding $A \cap B$ due to idempotency into the DNF expression, we get

$$DNF(A \text{ OR } B) = (A \cap (B \cup c(B))) \cup (B \cap (A \cup c(A)))$$

with additional application of commutativity to the second phrase. Next with the application of the law of "Excluded Middle" we have $A \cup c(A) = I$ and $B \cup c(B) = I$. With the application of the identity axiom, we get

$$DNF(A \text{ OR } B) = A \cup B = CNF(A \text{ OR } B)$$

In a similar manner, with the appropriate applications of the axioms of two-valued set and logic paradigm, it is clear , as is well known, that

$$DNF(\bullet)_i \equiv CNF(\bullet)_i, \qquad i=1,\ldots,16.$$

Thus, the following theorem is stated.

**Theorem 7.1.** In two-valued set and two-valued logic theory, $DNF_i(\bullet) \equiv CNF_i(\bullet)$, i=1,2,…,16. A similar proof can be given for all the sixteen possible combination of concepts applying the axioms of Table 6.2.

It is for this reason and for this reason alone, in two-valued set and logic studies, one always use the shorter form either the DNF(•) or CNF(•). Thus, we use

$$DNF(A \text{ AND } B) = A \cap B,$$

But, we use

$$CNF(A \text{ OR } B) = A \cup B$$

and, we use

$$CNF(A \rightarrow B) = c(A) \cup B \text{ etc.}$$

When we generalize the descriptive set from two-valued to infinite valued fuzzy sets, but keep the verification to be two valued, i.e., when we investigate $\{D_{[0,1]}, V_{\{0,1\}}\}$, we have to ask whether these equivalences between DNF and CNF expressions hold. In fact, we find that these equivalences break down as it will be demonstrated next.

## 7.4. Direct Fuzzification of DNF and CNF Expressions

In Section 6.3, it was stated that directly fuzzified Disjunctive and Conjunctive Normal Forms and in turn Fuzzy Disjunctive and Fuzzy Conjunctive Canonical Forms, are not equivalent to each other, i.e., $DNF_i(\bullet) \not\equiv CNF_i(\bullet)$, and $FDCF(.) \not\equiv FCCF(.)$, i=1,2,…,16. We treat this in two parts as indicated before. In part one which is discussed here in this section, we will assume that "Fuzzified $DNF_i(\bullet) = FDCF_i(\bullet)$ and "Fuzzified $CNF_i(\bullet) = FCCF_i(\bullet)$, i=1,2,…,16, *in form only*, without inquiring into the derivation of FDCF and FCCF, from the Fuzzy Truth Tables, which will be treated later (Türkşen, 1986). It should be noted that we use FDCF, Fuzzy Disjunctive Canonical Form, and FCCF, Fuzzy Conjunctive Canonical Form, as a signal that they are not equivalent to each other. Even though, they are *equivalent in form only* to their two valued descriptive set versions, i.e., DNF and CNF, respectively.

Here, "Fuzzified $DNF_i(\bullet) = FDCF_i(\bullet)$" in the sense that we take $DNF_i(\bullet)$ of the two-valued set expressions and plug in fuzzy-valued descriptive sets in place of two-valued descriptive sets and try to find out what happens. And we do the same for "Fuzzified $CNF_i(\bullet)=FCCF_i(\bullet)$". Note at this time that these are surrogate expressions.

In part two, treated in Chapter 8, we will discuss the derivation of FDCF and FCCF expressions directly from Fuzzy Truth Tables and revisit the concern of non-equivalence, i.e., break down of equivalences. The discussion will treat the cases of (i), (ii) and (iii) of section 6.5. Again, we demonstrate the case of the meta-linguistic expression: "A IMPLIES B is True, **T**", i.e., (A→B, **T**). But in this demonstration let us rewrite the predicate version of the set expressions of (6.1) and (6.2) in t-norm, Δ, t-conorm, ∇, and standard negation, n, in general as:

$$\mu_{DNF(A\to B,\mathbf{T})}(x,y) = \left[ (A(x)\Delta B(y))\nabla(n(A(x))\Delta b(y))\nabla(n(A(x)\Delta n(B(y))),\mathbf{T} \right]$$

$$\mu_{CNF(A\to B,\mathbf{T})}(x,y) = \left[ n(A(x))\nabla B(y),\mathbf{T} \right]$$

where A(x) and B(y) designate the membership values of x in the fuzzy sets A and y in the fuzzy set B, and n(A(x))=1-A(x) is the standard negation, i.e., complement operation in the membership domain, i.e., in the computational expression domain. The following predicate expressions are all true, hence, we do not write **T** after each one of them on the right or left hand sides. But we are indicating the axioms that are used, such as Boundary condition, B, and Monotonicity, M, that all t-norms and t-conorms, including pseudo-t-norms and co-norms, are subject to:

### *7.4.1. Conjecture*

To show the non-equivalence, i.e., break down of the equivalence, we conjecture that if

$$\nabla[\Delta(A(x), B(y)), \Delta(n(A(x)), B(y))] \le B(y), \qquad \text{M,B}$$

holds and since in general we have

$$\Delta[n(A(x)), n(B(y)) \le \Delta[n(A(x)),1] = n(A(x)), \qquad \text{M,B}$$

Then, we get:

$$\nabla[\nabla[\Delta(A(x), B(y)), \Delta(n(A(x)), B(y))], \Delta[n(A(x)), n(B(y))]] \le \nabla(n(A(x)), B(y)).$$

It should be noted that, the proof rests on a conjecture as well as, monotonicity and boundary axioms. The conjecture will be shown to hold for a class of t-norms and t-conorms later. In particular, the conjecture holds for Archimedean t-norms that are strict and nilpotent, as well as special cases of such as, (∨, ∧, n(.)), (⊕, ⊙, n(.)), ($L_\oplus$, $L_\odot$, n(.)), etc.

Therefore, we have DNF(A→B, **T**)⊆ CNF(A→B, **T**) whenever, we take two valued normal form expressions and directly substitute fuzzy valued sets to them.

Thus, we state the following theorem. This is the theorem that essentially states the break down of the equivalences that hold in two-valued set and two valued logic paradigm.

**Theorem 7.2.** "Fuzzified $DNF_i$(•)"≢"Fuzzified $DNF_i$(•)", i= 1,2,.., 16, for all t-norms and t-conorms including pseudo-t-norms and t-conorms for which the conjecture, 7.4.1, holds.

A similar proof can be given for all the sixteen possible combination of concepts applying the axioms of t-norms and t-conorms, and pseudo-t-norms and t-conorms, recall that all that was needed monotonicity and boundary axioms as well as the conjecture. A more general treatment will be given later with generating functions of t-norms and conorms.

In Figures 7.1.1, 7.1.2; 7.2.1, 7.2.2; 7.3.1, 7.3.2; and 7.4.1, 7.4.2 we have shown the containment of DNF and CNF for two De Morgan Triples, i.e., (∨, ∧, n(.)) and (⊕, ⊙,n(.)) for "A OR B" and "A AND B", respectively in Appendix 7, together with the computer programs and graphs. A more general proof the containment of FDCF(.) ⊆ FCCF(.) beyond non-equivalence, i.e., FDCF(.)≢FCCF(.) will be treated later in Chapter 9.

## 7.5. Consequences of $D_{\{0,1\}}V_{\{0,1\}}$

In two-valued set and logic paradigm mapping the set elements and truth elements both to {0,1} simplifies the knowledge representation but creates certain paradoxes due to the loss of information by imposing an arbitrary cut off to the graded membership values of the elements of the set. In general membership values less than 0.5 are mapped to "0" and those that are greater than or equal to 0.5 are mapped to "1". Naturally other thresholds could and have been used for the purpose of certain applications. At any rate certain paradoxes are created. A well known example is called Russell's paradox or Barber's paradox, etc., under various interpretations in various contests. We state six of these paradoxes this section (Curry, 1963).

### *7.5.1. Russell' s Paradox*

Our intuition tells us that we can consider classes of objects as forming new objects. Thus, we can consider the class of all chairs in this room, the class of all men, of all houses, of natural numbers. Likewise, we can consider classes of classes, and even such notions as the class of all classes, or the class of all ideas. Among these classes, there will be two sorts, which we shall call proper and improper classes. Proper classes are those, like men, houses, numbers, which are not members of themselves; improper classes are those which, like the class of all classes, or the class of all class of all ideas, are members of themselves.

Now, let $R$ (the Russell class) be the class of all proper classes. If $R$ is a proper class, then, since $R$ is the class of all such classes, $R$ is a member of $R$, and hence $R$ is not a proper class. On the other hand, if $R$ is not a proper class, then $R$ is not a member of $R$, and therefore $R$ is a proper class. Either assumption leads to a contradiction.

#### *7.5.1.1. Formal Treatment of Russell' s Paradox*

It is instructive to express the paradox in symbols. Let the statement that x is a member of the class y be symbolized by the notation

$$x \in y$$

'x' and 'y' being variables for which names of arbitrary notions can be substituted, and let $\neg$ and $\Leftrightarrow$ be symbols for negation and logical equivalence, respectively. Then, by definition of $R$, we have for arbitrary x,

$$x \in R \Leftrightarrow \neg (x \in x)$$

and hence,

$$R \in R \Leftrightarrow \neg (R \in R)$$

Thus, the statement that $R \in R$ is equivalent to its own falsehood, and hence, if it is true, it is also false, and vice versa.

### *7.5.2. Barber's Pseudo paradox*

The council of a certain village is said to have given orders that the village barber (supposedly unique) was to shave all the men in the village who did not shave themselves, and only those men. Who shaved the barber?

### *7.5.3. Catalogue Pseudo paradox*

A certain library undertook to compile a bibliographic catalogue listing all bibliographic catalogues, and only those catalogues, which did not list themselves. Did the catalogue list itself?

These arguments are here called pseudo paradoxes, because there is no actual contradiction. In the first case, the village barber could not obey the law, which was therefore ridiculous, like that said to have been passed by an American state legislature to the effect that, when two trains approach a crossing at right angles, each one must wait until the other one passes by. Likewise, the library simply could not make a catalogue satisfying the stated requirements.

However, we come to realize that how certain orders given by kings, emperors, sultans, dictators, etc., did create ridiculous conflicts and difficulties in the past. But, such explanations as these, do not apply to the Russell Paradox. In terms of logic as it was known in the nineteenth century, the situation is simply inexplicable. This is true in spite of the fact that in the greater sophistication of the present time one may see, or think he sees, wherein the fallacy consists.

### *7.5.4. Burali-Forti Paradox*

This, the first of the mathematical paradoxes to be published, is of a more technical nature. It involves the theory of transfinite ordinal numbers. For readers acquainted with that theory, the paradox can be stated as follows. It is shown in that theory that (1) every well-ordered set has a (unique) ordinal number; (2) every segment of ordinals (i.e., any set of ordinals arranged in natural order, which contains all predecessors of each of its elements) has an ordinal number, which is greater than any ordinal in the segment; and (3) the set $B$ of all ordinals in natural order is well ordered. Then, by statements (3) and (1), $B$ has an ordinal $\beta$; since $\beta$ is in $B$, we have $\beta < \beta$ by statement (2), which is a contradiction.

### *7.5.5. Cantor's paradox*

This paradox, although not published until 1932, was known to Cantor as early as 1899; it had a great influence in leading Russell to construct his paradox, in fact rather more than did the earlier paradox of Burali-Forti. It is based on the theory of cardinal numbers. According to that theory the set of all subsets of a set $M$ has a cardinal number higher than that of $M$. This is a contradiction, if $M$ is the set of all sets.

### *7.5.6. Liar's paradox*

This paradox takes several forms. The simplest is that of the man, who says "I am lying"; if he lies, he is speaking the truth, and vice versa. Another version is that of Epimenides, the Cretan, who is alleged to have stated that all statements made by Cretans were lies, it being understood that all other statements made by Cretans were certainly false. Modern versions give a statement to the effect that a proposition described in such and such a way is false, the description being constructed so as to apply uniquely to that statement itself. The paradox caused a great commotion in antiquity, and is said to have cause the death of a certain Philites of Cos. The first-mentioned form of the paradox seems to be due to Eubilides of Miletus; any actual connection of Epimenides with any form of it is doubtful.

In barber paradox, we run into a paradox of not knowing how the barber gets his shaving. That is if he does not shave himself, he must be shaved by the barber, i.e., himself, and vice versa, if he shaves himself he cannot be shaved by the barber, i.e., himself.

Let A be the case that the barber shaves himself, and thus c(A) is the case that the barber does not shave himself. Therefore, in classical paradigm, we have

$$A \cap c(A) = \varnothing$$

Thus nobody can say how the barber gets shaved.

$A \cap c(A) = \varnothing$ is known as the "Law of Contradiction" which is used in many theorem proving exercises in mathematics that takes its bases as the two-valued set and logic theory. The dual of this is known as the "Law of Excluded Middle", i.e., $A \cup c(A) = I$ which require an arbitrary threshold to determine the $\{0,1\}$ assignments of the two-valued paradigm.

It will be shown later that in fuzzy theory, we have to replace LEM with "Fuzzy Middle" and contradiction with "Fuzzy Contradiction". They both hold only to a degree.

This is yet another example where the classical laws break down but they are upheld in a matter of degree in fuzzy set and two valued logic theory.

The difficulty with these paradoxes can in part be resolved by a separation of descriptive and veristic statements as we discussed earlier as an extension of Tarski's explanation of the nature of truth which we had discussed in Chapter 1.

## 7.6. Symbols, Propositions and Predicates

A good example of arbitrary cut off can be demonstrated for the "day" and "night" concepts in everyday life which we had discussed in Chapter 2. Let, for example, the set symbol A stand for our mental state that is linguistically expressed as "the daylight hours of a twenty-four-hour day". And hence, its complement, the set symbol, c(A), stands for our mental state that is linguistically expressed as "not the daylight hours of a twenty-four-hour day". Let also X denote the twenty four hours of a day X = [0:00 a.m., 12:00 p.m. (midnight)], and X' = [6:00 a.m., 6:00 p.m.], be the "daylight hours, and X" = [6:01 p.m., 5:59 a.m.] be not the daylight hours" such that $X = X' \cup X''$ where $\cup$ is the set disjunction operator.

In studies of logic, we investigate propositions of such linguistic expressions with the symbols of their set representations. It should be recalled that a proposition is "an expression in a language that is either true or false". For the example case stated above, the propositions of the linguistic expressions with their symbols defined above are stated below:

(1.1) X' is in the set of "the daylight hours", A, is true, T, i.e., X' isr A is T;
(1.2) X' is not in the set of "the daylight hours", A, is false, F, i.e., X' isr not A is F;
(2.1) X' is not in the set of "not the daylight hours", c(A), is true, T, i.e., X' isr not c(A) is T;
(2.2) X' is in the set of "not the daylight hours", c(A), is false, F; i.e., X' isr c(A) is F;

where "isr" is a short hand notation that stands for "is in", "belongs to", "compatible with", etc., depending on context in fuzzy theory whereas "is" is used in its usual two-valued sense of the classical set theory.

At times, we need to specify a particular time of the day with a predicate. Again it should be recalled that a predicate is "something that is affirmed or denied of the subject in a proposition in logic".

For example, the predicates of the two propositions (1.1) and (1.2) stated above are:

(1.1.1) the particular time of day, say x=6:30 a.m., or any $x \in X'$, in the set of hours of this day, X, is in the set of "the daylight hours", A, with 100% membership value, i.e., $\mu(x,A)=a=1$, is true, T; i.e., $x \in X'$ isr A, with a=1, is T;

(1.2.1) the particular time of day, say x=6:30 a.m., or any $x \in X'$, in the set of hours of this day, X, is in the set of "the daylight hours", A, with the 0% membership value, i.e., $\mu(x,A)=a=0$, is false, F, i.e., $x \in X'$ isr A, with a=0, is F.

It should be clear, there is a hidden, but rather arbitrary assumption in these statements. That is, we have defined "the daylight hours", A, crisply, to be equivalent to the time interval of [6:00 a.m., 6:00 p.m.], i.e., X' = [6:00 a.m., 6:00 p.m.] and $A=\{(x,a) \mid a=1, x \in X'\}$, and the set of "not the daylight hours", c(A), to be equivalent to [6:01 p.m., 5:59 a.m.], i.e., X" = [6.01 p.m., 5:59 a.m.], and $c(A)=\{(x,n(a)) \mid n(a)=1, x \in X''\}$ with the further assumption that our watch or clock is only able to show us the hours and minutes, i.e., a digital clock or watch, of a given day (Figure 2.4).

The examples of the predicates of the two propositions (2.1) and (2.2) stated above are:

(2.1.1) the particular time of day, say x = 6.30 a.m., or any $x \in X'$, in the set of hours of this day, X, is in the set of "not the daylight hours", c(A), with 0% membership value, i.e., $\mu(x, c(A)) = n(a) = 0$, is true, T; i.e., $x \in X'$ isr c(A), with n(a) = 0, is T.

(2.2.1) the particular time of day, say x = 6.30 a.m., or any $x \in X'$, in the set of hours of this day, X, is in the set of "not the daylight hours", c(A), with the 100% membership value, i.e., $\mu(x,c(A)) = n(a) = 1$ is false, F; i.e., $x \in X'$ isr c(A) with n(a) = 1, is F. (Figure 2.4)

It should be noted that in these examples, i.e., (1.1.1), (1.2.1), (2.1.1) and (2.2.1), $a > n(a)$ for the affirmations, T, and $a < n(a)$ for the negations, F, where $n(a)=1-a$, and $a, n(a) \in \{0,1\}$.

Furthermore, it is clear that for the other case shown in Figure 2.4, i.e., for x = 5:30 a.m., or any $x \in X''$, there are alternate and analogous propositional and predicate expressions. The examples of these alternate and analogous predicate expressions are (in short form):

(1.1.1)' $x \in X''$ isr A with a=0 is T.
(1.2.1)' $x \in X''$ isr A with a=1 is F.
(2.1.1)' $x \in X''$ isr C(A) with n(a)=1 is T.
(2.2.1)' $x \in X''$ isr C(A) with n(a)=0 is F.

It should be noted, however, that in these examples, i.e., (1.1.1)', (1.2.1)', (2.1.1)' and (2.2.1)', $a < n(a)$ for the affirmations, T, and $a > n(a)$ for the negations, F.

Naturally, in agreement with our arbitrarily assumed two-valued, crisp, and dichotomous partitions of a given day, it was also assumed that membership in these partitions are to be all or none, i.e., 100% or 0% by explicitly stating a, $n(a) \in \{0,1\}$. But these arbitrary crisp partitions disregard the periods of the sunset and the dawn which have partial darkness, i.e., degrees of light or degrees of darkness.

These concern lead us next to discuss fuzzy (graded) membership in describing sets. These will be discussed in Chapter 8.

## Appendix 7.1. Computer Programs and Graphs.

```
"Figures 7.1. (A OR B) for Max-Min"}
Clear[a, b, cnf, dnf, tN, tC];  tN[ a_,b_] = Min[a, b]; tC[a_, b_] = Max[a, b];
c[a_] = 1- a; c[b_] = 1- b; cnf = tC[a, b];
dnf = tC[tC[tN[a, b], tN[c[a], b]], tN[a, c[b]]];
Plot3D[dnf, {a, 0, 1}, {b, 0, 1},       AxesLabel -> {"a", "b", "DNF        "}];
{"Figure 7.1.1. DNF(A OR B) for Max-Min"}
Plot3D[cnf, {a, 0,1}, {b,0,1},      AxesLabel -> {"a", "b", "CNF        "}];
{"Figure 7.1.2. CNF(A OR B) for Max-Min"}
```

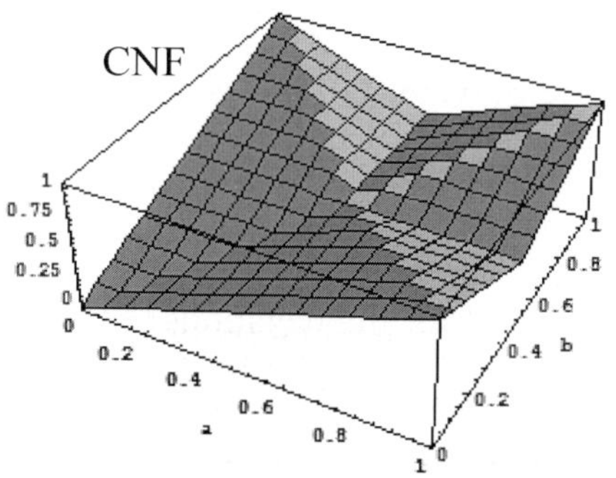


**Figures 7.1.1**. CNF(A OR B) for Max-Min

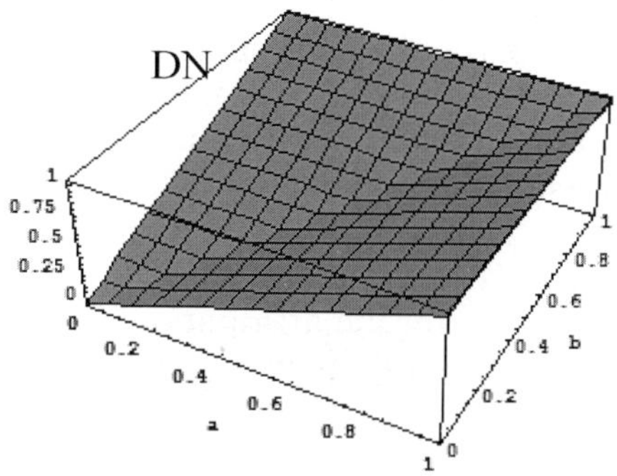


Figures 7.1.2. DNF(A OR B) for Max-Min

**Figures 7.1**. (A OR B) for Max-Min

```
Clear[a, b, cnf, dnf, tN, tC]; tN[a_,b_] = Min[a, b]; tC[a_, b_] = Max[a, b];
c[a_] = 1- a;  c[b_] = 1- b; cnf = tN[a, b];
dnf = tN[tN[tC[a, b], tC[c[a], b]], tC[a, c[b]]];
Plot3D[dnf, {a, 0, 1}, {b, 0, 1},  AxesLabel -> {"a", "b", "DNF        "}];
{"Figure 7.2.1. DNF(A AND B) for Max-Min"}
Plot3D[cnf, {a, 0,1}, {b,0,1},  AxesLabel -> {"a", "b", "CNF        "}];
{"Figure 7.2.2. CNF(A AND B) for Max-Min"}
```

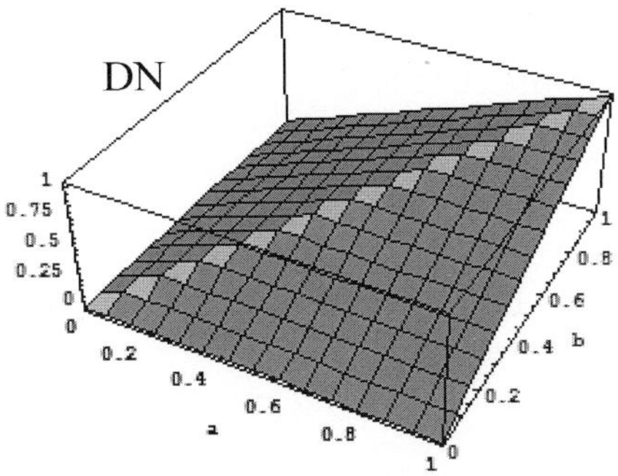


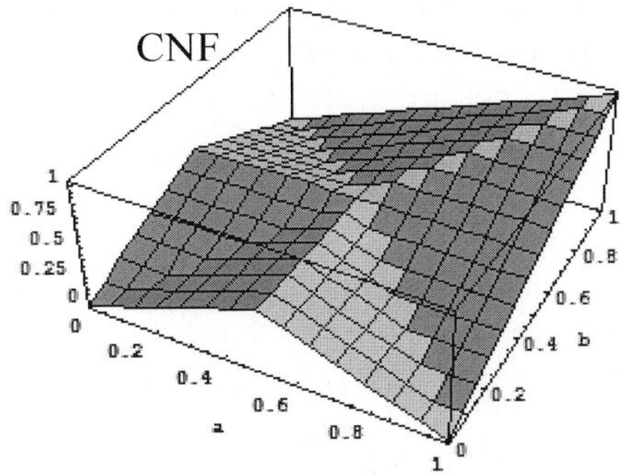


Figure 7.2.1.DNF(AANDB) for Max-Min Figure 7.2.2. CNF(AANDB) for Max-Min

Figures 7.2. (A AND B) for Max-Min

```
{"Figures 7.3. (A AND B) for Algebraic"}
Clear[a, b, cnf, dnf, tN, tC]; tN[a_,b_] = ab;  tC[a_, b_] = a + b - ab;
c[a_] = 1- a;  c[b_] = 1- b;
cnf = tN[tN[tC[a, b], tC[c[a], b]], tC[a, c[b]]]; dnf = tN[a, b];
Plot3D[dnf, {a, 0, 1}, {b, 0, 1},     AxesLabel -> {"a", "b", "DNF          "}];
{"Figure 7.3.1. DNF(A AND B) for Algebraic"}
Plot3D[cnf, {a, 0,1}, {b,0,1},     AxesLabel -> {"a", "b", "CNF          "}];
{"Figure 7.3.2. CNF(A AND B) for Algebraic"}
```

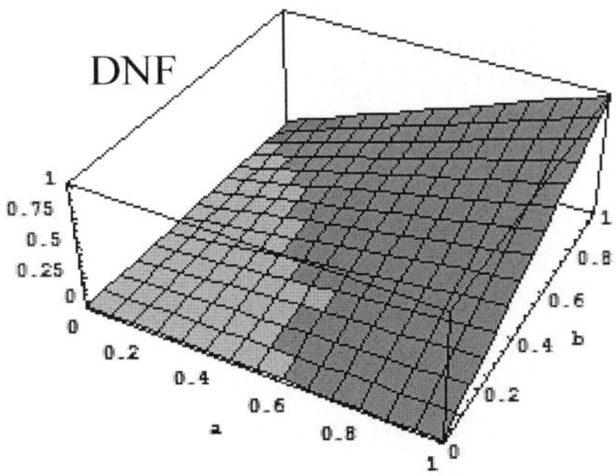


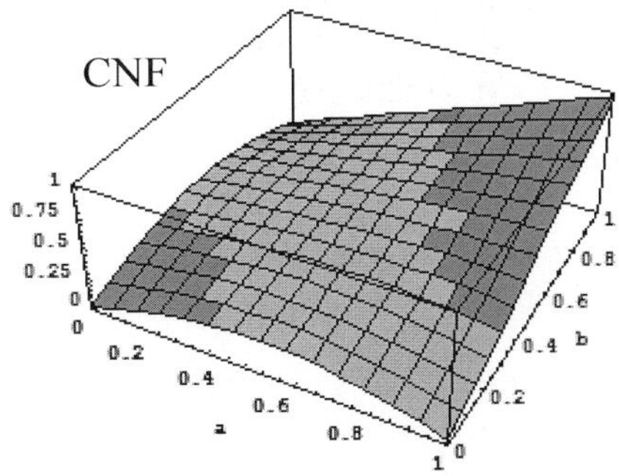


**Figure 7.3.1.** DNF(AANDB) for Algebraic **Figure 7.3.2.** CNF(AANDB) for Algebraic

**Figures 7.3**. (A AND B) for Algebraic

```
{"Figures 7.4. (A OR B) for Algebraic"}
Clear[a, b, cnf, dnf, tN, tC];
tN[a_,b_] = ab;
tC[a_, b_] = a + b - ab;
c[a_] = 1- a;
c[b_] = 1- b;
dnf = tC[tC[tN[a, b], tN[c[a], b]], tN[a, c[b]]];
cnf = tC[a, b];
Plot3D[dnf, {a, 0, 1}, {b, 0, 1},
    AxesLabel -> {"a", "b", "DNF        "}];
{"Figure 7.4.1. DNF(A OR B) for Algebraic"}
Plot3D[cnf, {a, 0,1}, {b,0,1},     AxesLabel -> {"a", "b", "CNF        "}];
{"Figure 7.4.2. CNF(A OR B) for Algebraic"}
```

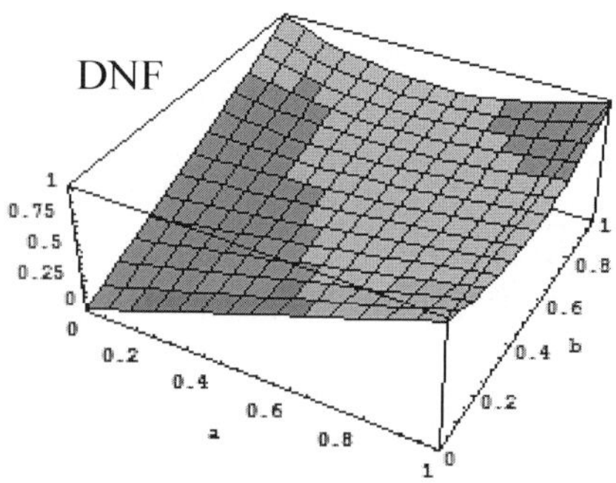


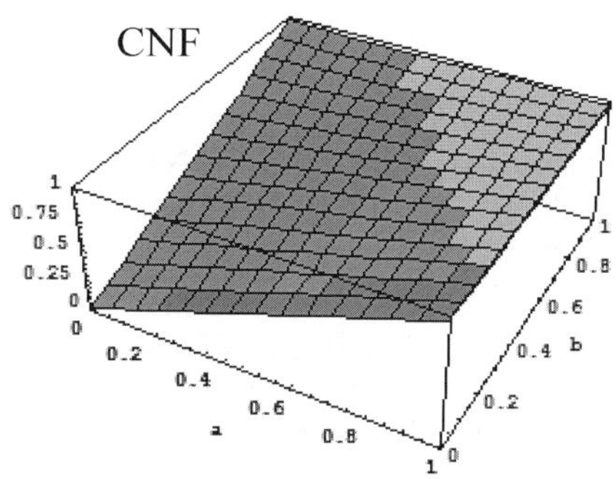


**Figure 7.4.1.** DNF(AORB) for Algebraic **Figure 7.4.2. CNF**(AORB) for Algebraic

**Figures 7.4.** (A OR B) for Algebraic

## References

1. Copy, I.M. (1982), Introduction to Logic, Mac Milland, New York.
2. Türkşen, I.B. (1986), "Interval-Valued Fuzzy Sets Based on Normal Forms", Fuzzy Sets and Systems, 191-210.
3. Curry, H.B. (1963), Foundations of Mathematical Logic, Mc Graw Hill, New York.
4. Berztiss, A.T. (1971), Data Structure, Theory and Practice, Academic Press, New York.

# Chapter 8

# FUZZY-VALUED SET AND TWO-VALUED LOGIC

In order to develop this theory and its canonical forms which are called Fuzzy Disjunctive and Fuzzy Conjunctive Canonical Forms, FDCF and FCCF, respectively, one must ask two related questions: (i) whether or not "Fuzzy Truth Tables" can be constructed, and if so (ii) can one derive the canonical, (normal), forms, i.e., FDCF and FCCF, from such tables?

These two questions were answered in the affirmative manner in Türkşen (1994, 1999) and later treated from two different unique perspectives in Türkşen, et al (1998, 1999). Various investigations attempted to show the impact of these FDCF and FCCF in fuzzy system modeling (1994, 1995, 1996, 1998). The details of these developments may be found in the articles referred to above. However, we give a brief description of the derivation of the fuzzy canonical forms here.

## 8.1. New Construction of the Truth Tables

Truth Tables can be newly constructed for the derivation of Fuzzy Normal Forms as explained in Türkşen(1994, 1999) and Türkşen et al. (1998, 1999) where three related but different explanations are given for this purpose. Here we present a modified and improved exposition of this new construction.

A newly constructed Truth Table is shown for the derivation of FDCF and FCCF of "A AND B" in Table 8.1. The new construction of the Truth Table for this purpose is based on the Preciasiated Natural Language, PNL, expressions as shown below. Let us consider the two fuzzy sets A and B defined over the base axis $X$ and a generic element $x \in X$ where $X = X_1 \cup X_2$. Let the $A(x) = a$ and $B(x) = b$, $a, b \in [0, 1]$, as shown in Figure 8.1, where $a \geq b$ representing all cases of $x_1 \in X_1$ and $a < b$ representing all cases $x_2 \in X_2$.

Now the following linguistic expressions in PNL can be stated in order to cover all possible cases where $a,b \in [0,1]$ and $\tau \in \{F,T\}$. For all $x \in X_1$:

(1.1.1) $x_1 \in X$ is in the infinite-valued descriptive(fuzzy) set A with a membership degree $a \geq b$ is verified to be true, (V = “true”) ($\tau = T$); i.e.,

' "$x_1 \in X_1$ isr A, $a \geq b$", isv V, $\tau = T$ '.

(1.2.1) $x_1 \in X_1$ is in the infinite-valued (fuzzy) set A with a membership degree $a < b$ is verified to be false, ($c$ (V) = "false") ($\tau = F$); i.e.,

' "$x_1 \in X_1$ isr A, $a < b$", isv V, $\tau = F$ '.

(2.1.1) $x_1 \in X_1$ is in the infinite-valued (fuzzy) set B with a membership degree $a \geq b$ is verified to be true, (V = "true") ($\tau = T$); i.e.,

' "$x_1 \in X_1$ isr B, $a \geq b$", isv V, $\tau = T$ '.

(2.2.1) $x_1 \in X_1$ is in the infinite-valued (fuzzy) set B with a membership degree $a < b$ is verified to be false, ($c$ (V) = "false") ($\tau = F$); i.e.,

' "$x_1 \in X_1$ isr B, $a < b$", isv V , $\tau = F$ '.

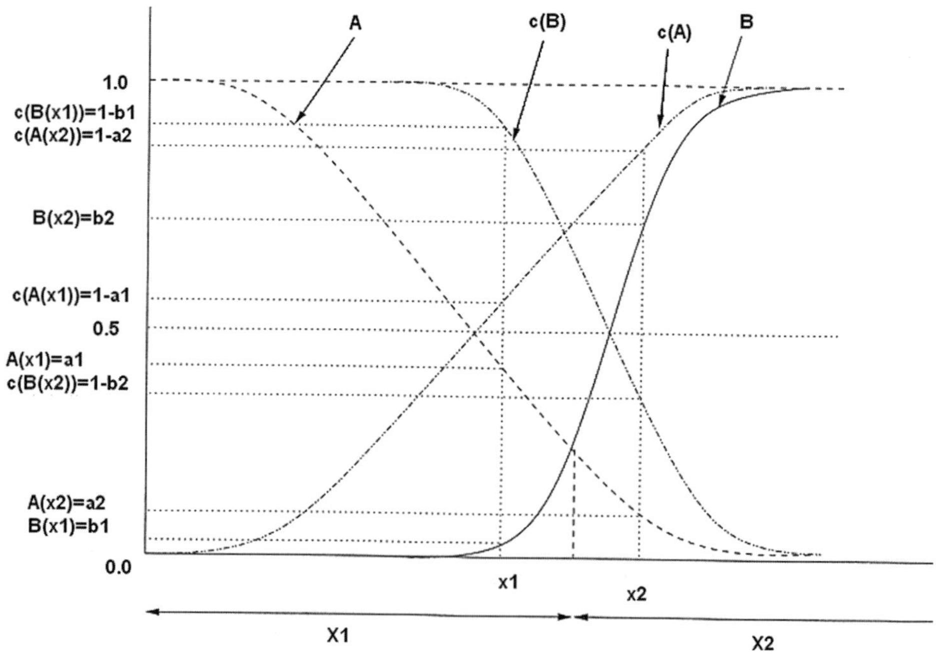


**Figure 8.1.** $x_1 \in X_1$ for $a \geq b$ and $x_2 \in X_2$ for $a<b$ such that $X = X_1 \cup X_2$

For all $x_2 \in X_2$, analogous expressions in PNL can be written, but let us just write the short hand notation only. That is, we have for all $x_2 \in X_2$:

' "$x_2 \in X_2$ isr A, a < b", isv V, $\tau$ = T '.
' "$x_2 \in X_2$ isr is A, a ≥ b", isv $\overline{V}$, $\tau$ = F '.
' "$x_2 \in X_2$ isr B, a < b", isv V, $\tau$ = T '.
' "$x_2 \in X_2$ isr B a ≥ b", isv $\overline{V}$, $\tau$ = F '.

These two sets of linguistic expressions form the bases upon which one can newly construct the Truth Table for the definition of the combined fuzzy set "A AND B" in $\{D_{[0,1]}\ V_{\{0,1\}}\}$ theory as shown in Table 8.1.

**Table 8.1.** The "Fuzzy Truth Table" of "A AND B" where a, b ∈[0, 1].

| A | | | B | | | A AND B |
|---|---|---|---|---|---|---|
| (1.1.1) | a ≥ b | T | (2.1.1) | a ≥ b | T | T |
| (1.1.1) | a ≥ b | T | (2.2.1) | a < b | F | F |
| (1.2.1) | a < b | F | (2.1.1) | a ≥ b | T | F |
| (1.2.1) | a < b | F | (2.2.1) | a < b | F | F |
| (1.1.2) | a < b | T | (2.1.2) | a < b | T | T |
| (1.1.2) | a < b | T | (2.2.2) | a ≥ b | F | F |
| (1.2.2) | a ≥ b | F | (2.1.2) | a < b | T | F |
| (1.2.2) | a ≥ b | F | (2.2.2) | a ≥ b | F | F |

The combination of the predicate expressions (1.1.1), (1.2.1), (2.1.1), (2.2.1) and (1.1.2), (1.2.2), (2.1.2), (2.2.2) together form the basis of rows of Table 8.1 as follows row by row from top to bottom. (The emphasis with ' " " ' marks are dropped in following expressions)

(1.1.1) AND (2.1.1): [$x_1 \in X_1$ isr A, a ≥ b, isv V, $\tau$ = T
AND $x_1 \in X_1$ isr B, a ≥ b, isv V, $\tau$ = T] is T;

(1.1.1) AND (2.2.1): $[x_1 \in X_1$ isr A, $a \geq b$, isv V, $\tau = T$
AND $x_1 \in X_1$ isr B, $a < b$, isv V, $\tau = F]$ is F;

(1.2.1) AND (2.1.1): $[x_1 \in X_1$ isr A, $a < b$, isv V, $\tau = F$
AND $x_1 \in X_1$ isr B, $a \geq b$, isv V, $\tau = T]$ is F;

(1.2.1) AND (2.2.1): $[x_1 \in X_1$ isr A, $a < b$, isv V, $\tau = F$
AND $x_1 \in X_1$ isr B, $a < b$, isv V, $\tau = T]$ is F.

(1.1.2) AND (2.1.2): $[x_2 \in X_2$ isr A, $a < b$, isv V, $\tau = T$
AND $x_2 \in X_2$ isr B, $a < b$, isv V, $\tau = F]$ is T;

(1.1.2) AND (2.2.2): $[x_2 \in X_2$ isr A, $a < b$, isv V, $\tau = T$
AND $x_2 \in X_2$ isr B, $a \geq b$, isv V, $\tau = F]$ is F;

(1.2.2) AND (2.1.2): $[x_2 \in X_2$ isr A, $a \geq b$, isv V, $\tau = F$
AND $x_2 \in X_2$ isr B, $a < b$, isv V, $\tau = T]$ is F;

(1.2.2) AND (2.2.2): $[x_2 \in X_2$ isr A, $a \geq b$, isv V, $\tau = F$
AND $x_2 \in X_2$ isr B, $a \geq b$, isv V, $\tau = F]$ is F;

In this way, the eight rows of the new construction of the Fuzzy Truth Table is formed in most explicit manner for the derivation of Fuzzy Canonical Forms. Now the right most column of the Table 8.1 is constructed for "A AND B". Other columns also may be constructed on the right most column for "A OR B", "A→B", etc.

The application of the Canonical Form Derivation Algorithm stated in Section 7.2.1, which is included in this chapter as Appendix 8.1, give us the derivation of FDCF and FCCF expressions for "A AND B" from the newly constructed Truth Table, Table 8.1, as follows:

FDCF (A ANDB) = $(A \cap B)$, $A \supseteq B$ or $A \subset B$ is T,

FCCF (A AND B) = $(A \cup B) \cap (c(A) \cup B) \cap (A \cup c(B))$, $A \supseteq B$ or $A \subset B$ is T,

for all t-norm, t-conorm and the corresponding complement that forms a De Morgan Triple.

It is to be noted that these are the same as classical DNF and CNF expressions, respectively. But the sameness is *in form only*. It should be noted that in previous writings (1994, 1998, 1999) a clear distinction between the two cases, i.e., $A \supseteq B$ or $A \subset B$, were not made explicit. This distinction is needed because either $x_1 \in X_1$ or $x_2 \in X_2$ but not both, i.e., it is an "Exclusive OR" case on the base axis partition. An alternate derivation is shown in Appendix 8.2 of this chapter.

It should also be noted that in Table 8.1, we have explicitly shown descriptive set membership assignments and the truth, verity, membership assignments separately.

In an analogous manner, one can derive the FDCF and FCCF expressions for the combined fuzzy set of "A OR B" to be:

$$\text{FDCF(A OR B)} = (A \cap B) \cup (c(A) \cap B) \cup (A \cap c(B)), A \supseteq B \text{ or } A \subset B \text{ is } T,$$

$$\text{FCCF(A OR B)} = (A \cup B),\ A \supseteq B \text{ or } A \subset B \text{ is T}.$$

Hence for all the remaining fourteen expressions as shown in Table 6.3 which are shown here as Table 8.2 their FDCF and FCCF expression can be obtained to be the same as the classical DNF and CNF expressions, *in form only,* as shown in Table 6.4 which is shown here as Table 8.3.

It should be further noted that derivation of these normal forms are based on *De Morgan Laws with the involutive negation and with the assumptions of monotonicity and boundary* conditions only. Hence, they are applicable to pseudo *t*-norms and co-norms as well as *t*-norm and co-norms in general. Therefore, it is straight forward to show that:

(i) First, these expressions further reduce to classical expressions when A and B are crisp, two-valued, sets with the particular characteristic relation that DNF (.)≡ CNF (.) as shown in Section 7.2. It should be noted that we are using "normal" form for the Boolean Case and "canonical" forms for the fuzzy case.

(ii) Secondly, these expressions appear to be the same in *form only* to the classical expressions shown in Table 8.3 (Table 6.4) for all t-norms and t-conorms with a *unique differentiation*, i.e., FDCF (.) T FCCF (.) and with FDCF(.) $\subseteq$ FCCF(.) for strict and nilpotent Archimedean t-norms and t-conorms.

This is the break down of the equivalences which we have indicated at the start of our book.

Therefore, the FDCF and FCCF expressions derived from newly constructed "Truth Table" for the $\{D_{[0,1]}\ V_{\{0,1\}}\}$ theory are generalizations of the classical normal forms that take into account both of the two cases $A \subset B$ or $B \subseteq A$ that are realized in infinite-valued (fuzzy) set membership assignment while keeping the two-valued verity assignment of truthood.

The consequences of the break down of these equivalences will be discussed in the sequel.

## 8.2. Dempster-Pawlak Unification

In this section, we provide a different perspective to show that in fact FDCF and FCCF expressions correspond to lower and upper set formulas derived with Dempster and Pawlak formalism.

For this purpose, we first reinterpret and modify Dempster and Pawlak formalism with information granules to identify upper and lower sets that determine upper and lower set formulas for both crisp and fuzzy sets. The aim

**Table 8.2.** Meta-Linguistic Expressions, i.e., Basic Expressions, of Combined Concepts for any A and B which may be crisp or fuzzy.

| Number | Meta-Linguistic Expressions |
|---|---|
| 1 | UNIVERSE |
| 2 | EMPTY SET |
| 3 | A OR B |
| 4 | NOT A AND NOT B |
| 5 | NOT A OR NOT B |
| 6 | A AND B |
| 7 | A IMPLIES B |
| 8 | A AND NOT B |
| 9 | A OR NOT B |
| 10 | NOT A AND B |
| 11 | A IF AND ONLY IF B |
| 12 | A EXCLUSIVE OR B |
| 13 | A |
| 14 | NOT A |
| 15 | B |
| 16 | NOT B |

is to provide an answer to a question that most skeptics of fuzzy theory had been asking for a long time.

One of the questions that non-fuzzy researchers usually ask: "why are the two valued formulas applied in fuzzy theory without any modification in the combination of concepts when propositions and their denotations are fuzzy". That is they suspect that there ought to be a reassessment of the formulas in combination of fuzzy sets with "AND", "OR", "IMP", etc.

Stated differently, they wonder why there are no new formulas since the axiomatic foundation of the fuzzy set theory relaxes the Law of Excluded Middle, LEM, $A \cap c(A) = I$, to be $A \cap c(A) \subseteq I$ and its dual Law of Contradiction, LC, $A \cup c(A) = \varnothing$, to be $A \cup c(A) \neq \varnothing$, where c(A) is the complement of A. Furthemore, it is known that the axioms of absorption, idempotency and distributivity are no longer applicable for the general class of t-norms and t-conorms and the corresponding complements that form De Morgan Triples. Thus they question the use of the two-valued formulas and ask that they be reexamined since LEM, LC and axioms such as distributivity, idempotency, and absorption are no longer applicable for the general class of connectives known as t-norms, and t-conorms.

A related question is "what happens to the equivalence between Disjunctive and Conjunctive normal forms, i.e., is DNF≡CNF, if the formulas of the two-valued theory are fuzzified by a direct substitution of fuzzy values in these formulas"?

**It is in response to these critical questions, we emphasize the break down of the basic equivalences of the classical theory. This break down is an important distinction of fuzzy set theories from the classical theory. In turn this break down generates new formulas in fuzzy theory.**

*8.2.1. Brief Review*

In classical, two-valued, set theory, we always compute the formula for "A AND B"= $A \cup B$ which is DNF(A AND B), and for "A OR B"= $A \cap B$ which is CNF(A OR B). They are the usual formulas that are used in applications. There are, however, other formulas as

"A AND B"= $(A \cap B) \cup (c(A) \cap B) \cup (A \cap c(B))$, which is CNF (A AND B)
"A OR B"= $(A \cup B) \cap (c(A) \cup B) \cap (A \cup c(B))$, which is DNF (A OR B).

The shorter forms are used because it is known and easy to show that

$A \cup B \equiv (A \cap B) \cup (c(A) \cap B) \cup (A \cap c(B))$ and
$A \cap B \equiv (A \cup B) \cap (c(A) \cup B) \cap (A \cup c(B))$.

**Table 8.3.** Classical Disjunctive Normal and Fuzzy Disjunctive Canonical Forms, DNF and FDCNF and Classic Conjunctive Normal and Fuzzy Conjunctive Canonical Forms, CNF and FCCF, where $\cap$ is a conjunction, $\cup$ is a disjunction and c is a complementation operator.

| No. | Fuzzy Disjunctive Canonical Forms/Disjunctive Normal Forms |
|---|---|
| 1 | $(A\cap B) \cup (A\cap c(B)) \cup (c(A) \cap B) \cup (c(A) \cap c(B))$ |
| 2 | . $\phi$ |
| 3 | $(A\cap B) \cup (A\cap c(B)) \cup (c(A) \cap B)$ |
| 4 | $(c(A) \cap c(B))$ |
| 5 | $(A\cap c(B)) \cup (c(A) \cap B) \cup (c(A) \cap c(B))$ |
| 6 | $(A\cap B)$ |
| 7 | $(A\cap B) \cup (c(A) \cap B) \cup (c(A) \cap c(B))$ |
| 8 | $(A\cap c(B))$ |
| 9 | $(A\cap B) \cup (A\cap c(B)) \cup (c(A) \cap c(B))$ |
| 10 | $(c(A) \cap B)$ |
| 11 | $(A\cap B) \cup (c(A) \cap c(B))$ |
| 12 | $(A\cap c(B)) \cup (c(A) \cap B)$ |
| 13 | $(A\cap B) \cup (A\cap c(B))$ |
| 14 | $(c(A) \cap B) \cup (c(A) \cap c(B))$ |
| 15 | $(A\cap B) \cup (c(A) \cap B)$ |
| 16 | $(A\cap c(B)) \cup (c(A) \cap c(B))$ |

| No. | Fuzzy Conjunctive Canonical Forms/Conjunctive Normal Forms |
|---|---|
| 1 | I |
| 2 | $(A\cup B) \cap (A\cup c(B)) \cap (c(A)\cup B) \cap c(A)\cup c(B))$ |
| 3 | $(A\cup B)$ |
| 4 | $(A\cup c(B)) \cap (c(A)\cup B) \cap c(A)\cup c(B))$ |
| 5 | $(c(A)\cup c(B))$ |
| 6 | $(A\cup B) \cap (A\cup c(B)) \cap (c(A)\cup B)$ |
| 7 | $(c(A)\cup B)$ |
| 8 | $(A\cup B) \cap (A\cup c(B)) \cap (c(A)\cup c(B))$ |
| 9 | $(A\cup c(B))$ |
| 10 | $(A\cup B) \cap (c(A)\cup B) \cap (c(A)\cup c(B))$ |
| 11 | $(A\cup c(B)) \cap (c(A)\cup B)$ |
| 12 | $(A\cup B) \cap (c(A)\cup c(B))$ |
| 13 | $(A\cup B) \cap (A\cup c(B))$ |
| 14 | $(c(A)\cup B) \cap (c(A)\cup c(B))$ |
| 15 | $(A\cup B) \cap (c(A)\cup B)$ |
| 16 | $(A\cup c(B)) \cap (c(A)\cup c(B))$ |

These equivalences are essentially due to the fact that the classical, two-valued, set theory axioms contain LEM, and its dual LC; together with idempotency, distributivity and absorption.

For example, in the demonstration of:

$$A \cup B \overset{?}{\equiv} (A \cap B) \cup (c(A) \cap B) \cup (A \cap c(B)).$$

As discussed previously, we first apply idempotency and get $(A\cap B) \cup (A\cap B) \cup (c(A) \cap B) \cup (A\cap c(B))$; next, we apply commutativity and distributivity and get $(A\cup c(A)) \cap B$ from the first and third terms and get $(B\cup c(B)) \cap A$ from second and fourth terms. As a result we obtain:

$$[(A \cup c(A)) \cap B] \cup [(B \cup c(B)) \cap A]$$

Next with the application of LC, $A \cup c(A)= \varnothing$ and $B \cup c(B)= \varnothing$, we get $B \cup A$; and then with the application of commutativity, we finally get $A\cup B \equiv A\cup B$. Because of this equivalence, the shorter form amongst the DNF and CNF is used in computations and applications. Clearly, since axioms of distributivity, idempotency, absorption and LEM and/or LC are no longer applicable in fuzzy theory, we suspect that $A \cup B$ T $(A \cap B) \cup (c(A) \cap B) \cup (A\cap c(B))$, in general, in fuzzy theory. In fact, it was shown (Türkşen, 1986) that DNF(A AND B) $\subseteq$ CNF(A AND B) where DNF(A AND B) = $A\cup B$, CNF(A AND B) = $(A \cap B) \cup (c(A) \cap B) \cup (A\cap c(B))$ for some t-norms and t-conorms.

As well, this is shown to be true four certain well known t-norms and t-conorms of fuzzy theory. (Türkşen, 1999, 2001)

It should be recalled that all of the 16 combinations of any two crisp sets have DNF and CNF, Disjunctive Normal Form and Conjunctive Normal Form, respectively, as shown in Table 8.3. Furthermore it is well known that we have $\text{DNF}(\cdot)\equiv\text{CNF}(\cdot)$ for all the sixteen possible combinations shown in Table 8.2 for the two-valued set and logic theory whose axioms are shown in Table 8.4.

There are Truth Table derivations of DNF and CNF expression for all of the 16 combination of concepts with the application of Normal Form (Canonical Form) Derivation Algorithm (Appendix 8.1)(as well stated in Section 7.2.1). As an example of the Truth Table for the "A AND B" is shown in Table 8.5.

The early investigations (Türkşen, 1986) had shown that the substitution of fuzzy values in [0,1] interval into the classical DNF and CNF formulas, instead of crisp values of {0,1} lattice, lead to the result that $\text{DNF}(\cdot)\subseteq \text{CNF}(\cdot)$ for some

**Table 8.4.** Axioms of Classical & Set & Logic Operations

| **Involution** | **c(c(A)) = A** |
|---|---|
| Commutativity | A∪B = B∪A |
| | A∩B = B∩A |
| Associativity | (A∪B)# ∪C = A∪ (B∪C) |
| | (A∩B) ∩C = A∩ (B∩C) |
| Distributivity | A∪(B∩C) =(A∪ B) ∩ (A∪C) |
| | A∩(B∪C) =(A∩ B) ∪ (A∩C) |
| Idempotence | A∪A = A |
| | A∩A = A |
| Absorption | A∪ (A∩B) = A |
| | A∩ (A ∪B) = A |
| Absorption by X and p | A∪X = X |
| | A∩ϕ = ϕ# |
| Identity | A∪ϕ = A |
| | A∩X = A |
| Law of contradiction | A∩c(A) = ϕ |
| Law of excluded middle | A∪c(A) = X |
| De Morgan's laws | c(A∩B) = c(A)# ∪# c(B) |
| | c(A∪B) = c(A) ∩ c(B) |

well known class of t-norms and t-conorms. However, there remained a crucial question:

"Could we drive, Fuzzy Disjunctive Canonical Forms, FDCF, and Fuzzy Conjunctive Canonical Forms, FCCF, directly from Fuzzy Truth Tables in a manner analogous to classical Truth Table derivation of DNF and CNF" using the Normal Form (Canonical Form) Derivation Algorithm stated in Section 7.2.1. (Appendix 8)

This concern was investigated in several papers (Türkşen, 1999, 2001) with a positive constructive result. It was shown that DNF is equivalent *in form only* FDCF and CNF is equivalent *in form only* to FCCF as shown in Table 8.3 for all the 16 possible combination concepts shown in Table 8.2.

In this chapter, in order to support and generalize these earlier results we first show that there is a connection with both the Dempster's upper and lower set formulas (Dempster, 1967) and Pawlak's upper and lower approximations on

**Table 8.5**. Classical truth table interpretations of "A AND B"

| **Truth assignments to classical meta-linguistic variables** | | **Truth assignments to the meta-linguistic Expression** | **Primary conjunctions** |
|---|---|---|---|
| A | B | "A AND B" | |
| T(A) | T(B) | T(A AND B) | **A ∩ B** |
| T(A) | F(B) | F(A AND B) | **A ∩ c(B)** |
| F(A) | T(B) | F(A ANDB) | **c(A) ∩ B** |
| F(A) | F(B) | F(A AND B) | **c(A) ∩ c(B)** |

sets (Pawlak, 1991). Secondly, we modify Dempster-Pawlak construction schema and restrict the subsets of the power set to a particular subset of information granules in the power set.

Thirdly, the formation of FDCF and FCCF from the lower and upper set approximations, respectively, of Dempster-Pawlak constructs

Recall that, the containment relationship, i.e., FDCF⊆FCCF, is shown to hold only for the well known t-norms and t-conorms, such as (Max, Min, –), i.e., (∨, ∧, –) (algebraic sum, product, –), i.e., (⊕, ⊙, –), and Lukasiewicz operators or bold operators, i.e., ($B_\oplus$, $B_\odot$, –), where "–" stands for standard complementation. In Chapter 9, it will be shown that the containment holds as well for strict and nilpotent Archimedean t-norms and t-conorms that are duals of each other with a proper negation (Bilgic, 1995) and some ordinal sums with non-trivial idempotent elements (Kilic, 2002).

With this brief review, we next turn our investigation toward Dempster's and Pawlak's constructs in order to provide an alternate approach to the structure of these containments.

## 8.3. DEMPSTER and PAWLAK Formulations

In order to handle uncertainty associated with imprecise sources of information in the identification of sets, Dempster and Pawlak propose alternate schemas with perspectives and contents that are fitting to their own agenda. With a multi-valued mapping, Dempster(1967) proposes the construction of upper and lower probabilities. In order to formulate upper and lower probabilities, he first proposes a multi-valued mapping from a space to another and proposes upper and lower set formulas. On the other hand, Pawlak(1991) approaches the same concern from the perspective of Rough Sets. Briefly, we will first summarize Dempster's and Pawlak's proposals; and then show a correspondence between these two construction schemas. Afterwards, we will modify and restrict Dempsters formalism and call it T-Formalism to derive FDCF and FCCF within this perspective.

*8.3.1. Dempster Construction*

Let X and S be two spaces, and a multi-valued mapping $\Gamma$, $\Gamma$: X$\to \mathcal{P}$(S), where $\mathcal{P}$(S) is the power set of S, i.e., $\mathcal{P}(S)=2^{|S|}$. Next let a target set be T, $\to T\subseteq S$, i.e., $T\in\mathcal{P}(S)$. Then the upper and lower sets of this multi-valued mapping, $\Gamma$, are defined for any target set T as:

$$T^* = \{x \mid x\in X, \Gamma(x) \cup T\neq\varnothing\}, \text{ and} \tag{8.1}$$
$$T_* = \{x \mid x\in X, \Gamma(x) \subseteq T, \Gamma(x) \neq \varnothing\}, \tag{8.2}$$

where, for a target set T, we are interested in determining the upper and lower sets, $T^*$ and $T_*$, respectively, under the multi-valued mapping $\Gamma$ (Dempster, 1967).
Clearly, it is the case that $T_*\subseteq T^*$ from the construction. Dempster's construction is motivated with a concern for the transformation of probabilities from a space X with a probability measure to a space S, via multi-valued mapping in order to define upper and lower probability estimates for an unsharp information source. However, prior to these probabilities, Dempster's upper and lower sets $T^*$ and $T_*$ identify upper and lower sets for the target set T constructed with the subsets of $\mathcal{P}$(S) identified by $\Gamma$(x).

For further clarity of the expressions (8.1) and (8.2), we provide a graphical representation of the expressions (8.1) and (8.2) in Figures 8.2 and 8.3.

*8.3.2. Pawlak Construction*

Let X be a given universal set and R an equivalence relation on X. The relation R induces a partition on X. Let $[x]_R$ be the equivalence class containing $x\in X$. It is to be noted that each $[x]_R$ is simply a subset of X. That is if x is a typical element of X, then "the equivalence class of x" is [x] = {t: tRx}. Given a set X and an equivalence relation R, we define a new set X/R, the set of equivalence classes of X modulo R.

Thus, the set of all $[x]_R$ will form a partition of X, denoted by X/R. Next let $A\subseteq X$, then upper and lower estimates of A, based on X/R are defined as follows:

$A^*_R = \cap\{[x]_R \mid [x]_R \cup A \neq \emptyset\}$, and $A_{*R} = \cap\{[x]_R \mid [x]_R \subseteq A\}$.

Clearly it is the case that $A_{*R} \subseteq A^*_R$. The Pawlak's rough set constructs $A_{*R}$ and $\Lambda^*_R$ identify an approximate cover of A with the information granules of X/R. (Pawlak, 1991).

**ONE-TO-MANY VALUED MAPPING**

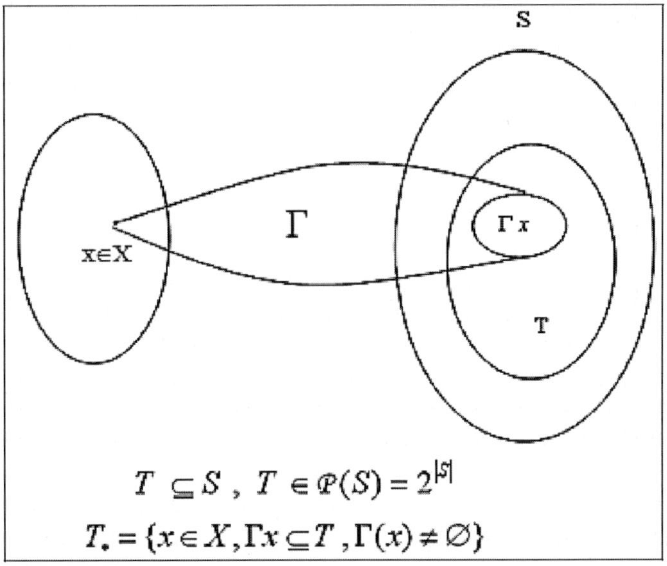


**Figure 8.2**. Lower set representation of Dempster's formalism.

### *8.3.3. Restatement of Dempster and Pawlak Constructs*

Both Demspter and Pawlak constructions rely on set containment for the lower estimate and non-empty intersection for the upper estimate.
In fact the two approaches are analogous and each can be expressed in the language of the syntax and semantics of the other. However, it is to be noted that Dempster's formalism is more general than Pawlak's.

### *8.3.4. Pawlak to Dempster Transformation*

Consider Pawlak construct with the universe of discourse U, and the equivalence relation R on U. In Dempster notation, let X be U/R and S be U. Let Γ of Demspter be the multi-valued mapping that causes the transformation

## ONE-TO-MANY VALUED MAPPING

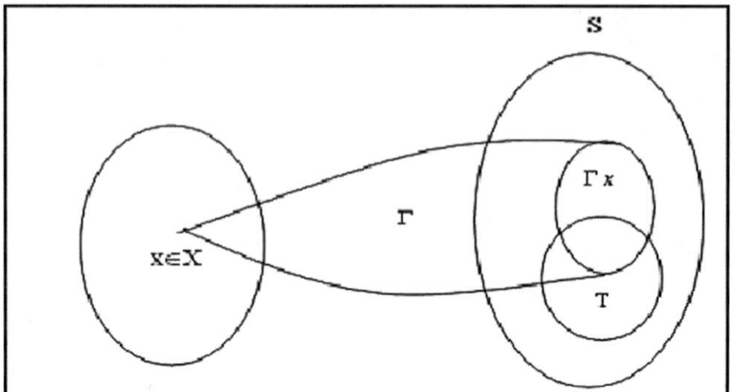


$$T \subseteq S\ ,\ T \in \mathcal{P}(S) = 2^{|S|}$$

$$T^{*} = \{x \in X, \Gamma x \cap T \neq \varnothing\}$$

**Figure 8.3**. Upper set representation of Dempster's formalism.

of the equivalence classes to the corresponding set; i.e., Γ maps each equivalence class or member of the partition to all the elements that are in it. Thus we have:
$\Gamma: U/R \rightarrow \mathcal{P}(U)$, and
$\Gamma\ ([u]_R) = \{t \mid tRu,\ t \in U\} \in \mathcal{P}(U)$.

Let $A \subseteq U$, then $A_{*R} = \cap\{[x]_R \mid [x]_R \subseteq A\}$ in the Pawlak notation, and $A_* = \{[x]_R \mid [x]_R \in U/R,\ \Gamma\ ([x]_R) \subseteq A\}$ in Dempster notation. Note that $\Gamma\ ([x]_R)$ is just the set $[x]_R$. Thus $A_{*R}$ and $A_*$ are composed of the same information granules from U/R, and capture the same information. A similar argument can be written for the correspondence between the upper estimates.

The minor structural difference between these two approaches is that the union of these equivalence classes is taken in the Pawlak's approach, whereas the set of these classes are collected in the Dempster's approach. Note that in Dempster's approach, probability measures are computed and *added* over the upper and lower sets in order to arrive at the upper and lower probabilities.Transformation of Dempster to Pawlak
Now let as recall the two spaces X and S and the multi-valued mapping $\Gamma$: $X \rightarrow \mathcal{P}(S)$ in Dempster's approach. Let a relation R on X be defined as xRy if and only if $\Gamma\ (x) = \Gamma\ (y)$. That is, two elements $x,y \in X$ are considered equivalent if they map to exactly to the same set of values in S. Naturally, this is an

equivalence relation on X and induces the partition X/R. Let $T \subseteq S$. In Dempster's approach, we have $T_* = \{x \mid x \in X, \Gamma(x) \subseteq T, \Gamma(x) \neq \emptyset\}$. Note that if $x \in T_*$ and yRx, then $y \in T_*$, because $\Gamma(x) = \Gamma(y)$. Thus $T_*$ can be written as the union of partiton members or classes in X/R, and thus can be expressed in Pawlak's notation. Naturally, one can state a similar argument for the correspondence of the upper estimates.

Since, it is shown that there is a natural correspondence between Pawlak's and Dempster's notations, we will refer to them as Dempster-Pawlak notation as D-P schema in deriving the upper and lower set formulas in the combination of concepts whether they are represented by crisp or fuzzy sets.

## 8.4. Sets and Logic Constructs

Let a proposition P be expressed by a predicate A which could be either crisp or fuzzy set depending on how A is obtained or defined. In previous writings (Türkşen, 1999, 2001), such propositions and their predicates are labeled as "descriptive" propositions and words, respectively in PNL.

Suppose values x are taken from a universe of discourse X. Let for a given value $x \in X$ with the membership value $\mu_A(x)$ be true, T, in the Type 1 fuzzy theory. In this way the truth of a descriptive proposition P is equated with the set A being true with the membership degree of $\mu_A(x)$.

Next let two linguistic concepts, $P_1$, and $P_2$, i.e., propositions, or their expressions be represented by two sets A and B. These two propositions and hence their set representations can be combined in 16 possible meta-linguistic expressions. These 16 expressions are shown in Table 8.6 together with their usual target set expressions written in set notation (crisp or fuzzy).

### *8.4.1. Restriction and Modification, T-Formalism*

We propose a restriction and a modification to D-P formalism and call it T-formalism:

> (1) Restrict the subset of the power set to the family of information granules **G** where
> $\mathbf{G}_n = \{A_1 \cup A_2 \cup \ldots \cup A_n, \ldots, c(A_1) \cap c(A_2) \cap \ldots \cap c(A_n)\}$
>
> for all possible combinations of n subsets $A_i$, i=1,...,n, in a universe of discourse by $\cup$, $\cap$, and c(.).

(2) Modify the condition for the upper set identification from "non-empty set intersection" to "set containment". (This is shown in Figure 8.4)

It is to be noted that:

(1) The restriction of the subsets of the power set to the family **G** that is formed by the primitive information granules are all directly identified by the set operators on subsets in a universe of discourse and their combinations $\Updownarrow \circ\neg \odot \neg$ the set of all meta-linguistic expressions, i.e., basic protoforms in PNL for both crisp or fuzzy cases.

(2) The modification, however, strengthens the upper set condition to "set containment" from "non-empty set intersection" which provides a conceptual and intuitive appeal.

That is, let T be the target set of a meta-linguistic expression. Then **G** covers T from below, because there exists at least one $G \in \mathbf{G}$, such that $G \subseteq T$. Because T must contain as much information granules that are contained in it, i.e., $G \subseteq T$, it makes semantic sense that T contains as much information granules that are true as the *disjunction* of all such G. Thus the disjunction of those information granules form a lower set formula for T. This is in agreement with Dempster's formalism but require disjunctive aggregation of the information granules so obtained in Pawlak's formalism.

**G** also covers T from above, because there is at least one $G \in \mathbf{G}$, such that $T \subseteq G$. Furthermore, because any such G contains T, it makes semantic sense that we take the *conjunction* of all such G that contains T. Thus, the conjunction of these information granules form an upper set formula for T. It is to be noted that this is a further modification to D-P formalism. That is we take the conjunctive aggregation of the information granules so obtained after the proposed change from "non-empty set intersection" to "set containment".

In summary, the disjunction of the proposed collection of information granules $G \subseteq T$ will give us a lower set formula for T, and the conjunction of the proposed collection of information granules G, $T \subseteq G$ will give us an upper set formula for T with the modified D-P formalism of T. It should be noted that both the upper and the lower set formulas so far are defined with containment, i.e., "$\subseteq$", this creates anamolies in the construction schema for the identification of granules $G \in \mathbf{G}$. This issue will be resolved in the next section for it depends on particular classes of five meta-linguistic expressions, i.e., basic expressions in PNL dependent on linguistic operators, "AND", and five meta-linguistic expressions, i.e., basic expressions in PNL dependent on linguistic operator "OR" and the remaining six special cases that includes bi-conditional, exclusive-or and confirmation and negation of two singletons.

**Table 8.6**. Sixteen Meta-Linguistic Expressions together with their "Usual Target Set Expressions" (crisp or fuzzy).

| Number | Meta-Linguistic Expressions | Usual Target Set Expression |
|---|---|---|
| 1 | Universe | I |
| 2 | Empty Set | $\varnothing$ |
| 3 | A OR B | $A \cap B$ |
| 4 | NOT A AND NOT B | $c(A) \cup c(B)$ |
| 5 | NOT A OR NOT B | $c(A) \cap c(B)$ |
| 6 | A AND B | $A \cup B$ |
| 7 | A IMPLIES B | $c(A) \cap B$ |
| 8 | A AND NOT B | $A \cup c(B)$ |
| 9 | A OR NOT B | $A \cap c(B)$ |
| 10 | NOT A AND B | $c(A) \cup B$ |
| 11 | A IF AND ONLY IF B | $(A \cup B) \cap (c(A) \cup c(B))$ |
| 12 | A EXCLUSIVE OR B | $(A \cap B) \cup (c(A) \cap c(B))$ |
| 13 | A | A |
| 14 | NOT A | c(A) |
| 15 | B | B |
| 16 | NOT B | c(B) |

### *8.4.2. Upper and Lower Canonical Forms*

Let us now consider again two linguistic concepts A and B, i.e., the two predicates, (words) A and B and the 16 meta-linguistic expression that can be generated with them. We start with the usual target set expressions shown in Table 8.6 as the target sets in the sense of the Dempster-Pawlak and the sets whose upper and lower set formula to be determined by the proposed T-formalism.

Our aim is to demonstrate that we could generate FDCF and FCCF expression directly with the use of information granules and with the proposed T-formalism stated above in sub-section 8.4.1 without resorting to Fuzzy Truth

Table derivation that was discussed in our previous papers. (Türkşen, 1999, 2001) and the previous sections.

**ONE-TO-MANY VALUED MAPPING**

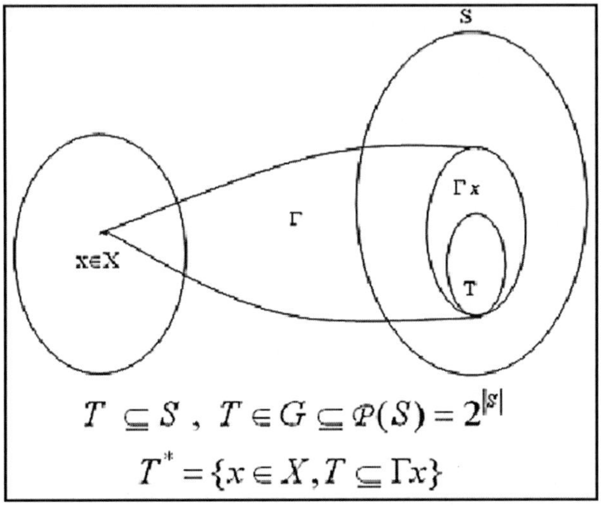


**Figure 8.4**. Modification of Dempster's formalism.

Let T be any one of these 16 expressions and $\mathbf{G}_2$ be the family of information granules that are the 8 possible combinations of two predicates A and B under conjunction, disjunction and complementation operations:

$\mathbf{G}_2 = \{G_1, G_2, ..., G_8\}$, where $G_1 = A \cup B, ..., G_8 = c(A) \cap c(B)$, such that

$\mathbf{G}_2 = \{A \cup B,\ c(A) \cup B,\ A \cup c(B),\ c(A) \cup c(B),\ A \cap B,\ c(A) \cap B,\ A \cap c(B),\ c(A) \cap c(B)\}$

It should be noted that the first four of these information granules, $G_1, ... G_4$, i.e., $A \cup B$, $c(A) \cup B$, $A \cup c(B)$, and $c(A) \cup c(B)$, form a disjoint partition of the universe, whereas the last four of these information granules, $(G_5, ..., G_8)$, i.e., $A \cap B$, $c(A) \cap B$, $A \cap c(B)$, and $c(A) \cap c(B)$ have overlaps.

As a result of our discussions and the proposed T-formalism stated in sub-section 8.3.1 above for each target set T, we define the upper and lower subset formulas to be one of the information granules as follows:

$l(T) = \{G \mid G \in \mathbf{G}, G \subseteq T\}$, and (8,3)
$u(T) = \{G \mid G \in \mathbf{G}, T \subseteq G\}$. (8.4)

It is to be observed that each element of $l(\mathrm{T})$ comes from the subsets of the conjunctive information granules, i.e., $A \cup B$, $c(A) \cup B$, $A \cup c(B)$, and $c(A) \cup c(B)$ whereas each element of $u(\mathrm{T})$ comes from the disjunctive information granules, i.e., $A \cap B$, $c(A) \cap B$, $A \cap c(B)$, and $c(A) \cap c(B)$.

With these lower and upper subsets, we determine the lower and upper set formulas of the target set T in the proposed T-formalism as:

$$L(T) = \cap l(T) \text{ and } U(T) = \cup u(T) \tag{8.5}$$

It is to be noted that the disjunction of $l(\mathrm{T})$'s are taken to form the lower set L(T) since they are all contained in the target set and they are disjoint among themselves. Thus L(T) forms the greatest lower bound.

But the conjunction of $u(\mathrm{T})$'s are taken to form the upper set, U(T), since they all contain the target set and they are not disjointed. Thus U(T) forms the least upper bound.

At first glance, it appears that $L(T) \subseteq T \subseteq U(T)$ by the construction schema. However, the containment relation $L(T) \subseteq U(T)$ requires a further investigation in t-norm and conorm space. It can be shown that $L(T) \subseteq U(T)$ for strict and nilpotent Archimedean t-norms and t-conorms that are duals of each other with a corresponding negation as well as (Max, Min, Stn) (Bilgic, 1995) (See Chapter 9) and some ordinal sums with non-trivial idempotent elements (Kilic, 2002).

Since L(T) turns out to be equal to FDCF and U(T) turns out to be equal to FCCF, by the construction, we have thus determined that $FDCF \subseteq FCCF$ not only for the well known specific t-norms and t-conorms as well as for strict and nilpotent Archimedian t-norms and t-conorms.

Now let us return to the task at hand. That is, we are to drive the upper and lower set formulas that represent each of the 16 meta-linguistic expressions. In each case, the selections of information granules are taken from **G**. It should be noted that the selection of each information granule that makes up the sets $l(\mathrm{T})$'s and $u(\mathrm{T})$'s are made with definitions given in equ.(8.3) and (8.4), i.e., $l(\mathrm{T})$ is the set of information granules that are contained in T and $u(\mathrm{T})$ is the set of T's that are contained by information granules $G \in \mathbf{G}$ where T is the usual target set in common use. Furthermore, we determine the lower set formula by the disjunctive aggregation of $l(\mathrm{T})$'s, i.e., $L(T) = \cap L(T)$ and the upper set formula by the conjunctive aggregation of $u(\mathrm{T})$'s, i.e., $U(T) = \cup u(T)$ as indicated in equ.(8.5) above.

However, there is yet another issue, we need to clarify as pointed out earlier. That is in equations (8.3) and (8.4), we observe that $G \subseteq T$ for $l(T)$ and $T \subseteq G$ for $u(T)$ for every $G \in \mathbf{G}$. Thus there are possible G's that may belong to $l(T)$ and $u(T)$ both. That is, we need to identify which G's are taken for the equality and which are taken for containment with respect to linguistic operators, "AND", "OR" and other special cases. This will be sorted out and clarified in Sections 8.3.3, 8.3.4, and 8.3.5 below because they depend whether a meta-linguistic concept is formed by "AND" or "OR" or other schemas. Thus we next develop the formulas for the cases of "AND" and "OR", and other schemas, and then we tabulate the results of all 16 expressions in Table 8.7.

### *8.4.3. Five Meta-Linguistic Expression that have "AND" Composition*

The usual, commonly used, target set T is "$A \cup B$" for "A AND B" (see Table 8.6, row 6). With the discussion and the proposed T-formalism stated in Section 8.3.1 above, we identify the set of information granules that are contained in T as:

$l(T) = \{A \cup B\}$ and thus the lower set formula is: $L(T) = \{A \cup B\} = l(T)$.
The set of information granules that contain T are

$u(T) = \{A \cap B,\ c(A) \cap B,\ A \cap c(B)\}$

and thus the upper set formula is: $U(T) = \cup u(T) = (A \cap B) \cup (c(A) \cap B) \cup (A \cap c(B))$.

It is to be observed that $l(T) = \{A \cup B\}$ which is the target set itself and thus $L(T) = l(T)$, i.e., $L(T) = \{G \mid G \in \mathbf{G},\ G = T\}$. We generalize this for all the five out of sixteen combination of concepts, i.e., the meta-linguistic expressions 2, 4, 6, 8 and 10 in Table 8.6, that admit the linguistic "AND" connective in their meta-linguistic combination, respectively. Therefore, the rule is $l(T) = \{G|,\ G = T\}$ for the targets sets $\emptyset$, $c(A) \cup c(B)$, $A \cup B$, $A \cup c(B)$, $c(A) \cup B$, where the usual target set itself forms the greatest lower bound.

This in turn entails, the rule for $u(T)$ to be $U(T) = \cup\{G \mid G \in \mathbf{G},\ T \subset G\}$ for these five out of sixteen combination of concepts that admit the linguistic "AND" connective in their meta-linguistic concept combination, i.e., meta-linguistic expressions 2, 4, 6, 8 and 10 in Table 8.6. This clarification resolves the anomaly generated by equations (8.3) and (8.4) for the cases of the linguistic

"AND" connective.
It is to be noted that
$L(T) = \text{FDCF (A AND B)}$, and
$U(T) = \text{FCCF (A AND B)}$

and therefore we FDCF (A AND B) $\subseteq$ FCCF (A AND B) from the construction which will be further investigated in Chapter 9 from the perspective of t-norms and t-conorms.

This fact holds to be true for both the crisp and fuzzy sets and for all the t-norms and t-conorms due to construction of L(T) $\subseteq$ U(T).

It is to be noted that if we apply Law of contradiction after the axiom of commutativity and distributivity, we get FDCF (A AND B) = FCCF (A AND B), as demonstrated in Section 7.2.

Thus we obtain
DNF (A AND B) = CNF (A AND B) = A$\cup$B
in the two-valued set and logic theory. Whereas in fuzzy set and logic theory FDCF and FCCF provide lower and upper set formulas, respectively, for the "AND" combination of two fuzzy concepts, A and B, i.e., fuzzy predicates, i.e., That is we get FDCF(A AND B) $\subseteq$ FCCF ( A AND B).

**Which is an example of the break down of the classical equivalence.**

*8.4.4. Five Meta-Linguistic Expression that have "OR" Composition*

The usual, commonly used, target set T is "A$\cap$B" for "A OR B" (see Table 8.6, row 3). Again with the discussion and the proposed T-Formalism, we identify the set of information granules that are contained in T as:
$l$ (T) = {A$\cup$B, c(A) $\cup$B, A$\cup$c(B)},
and thus the lower set formula is:
L(T) = $\cap l$ (T) = (A$\cup$B) $\cap$ (c(A) $\cup$B) $\cap$ (A$\cup$c(B)).
The set of information granules that contain T are:
$u$(T) = {A$\cap$B}, and thus the upper set formula is: U(T) = {A$\cap$B}=$u$(T).
In an analogous manner, we observe that $u$(T) = {A$\cap$B} which is the target set itself and thus U(T) = {G | G$\in$**G**, T=G}. We also generalize this for all the five cases out of sixteen combination of concepts that admit the linguistic "OR" connective in their combination. Therefore the rule is $u$(T) = {G | G$\in$**G**, T=G}

for the target sets I, A$\cap$B, c(A) $\cap$c(B), c(A) $\cap$B, A$\cap$c(B), i.e., the meta-linguistic expressions 1, 3, 5, 7, and 9 in Table 8.6, where the usual target set forms the least upper band.

This in turn entailes, the rule for L(T) to be L(T) = $\cap${G | G$\in$**G**, G$\subset$T} for these five out of sixteen cases that admit "OR" connective in their meta-linguistic expressions, i.e., expressions 1, 3, 5, 7, and 9 in Table 8.6. Again, this clarification resolves the anomaly introduced by equations (8.3) and (8.4) for the cases of "OR" connective.

Again it is to be noted that
L(T) = FDDCF(A OR B), and U(T) = FCCF(A OR B)

and therefore, we have
FDCF(A OR B) $\subseteq$ FCCF(A OR B) from the construction.

This fact again holds to be true for fuzzy sets and for all the strict and nilpotent Archimedian and t-norms and t-conorms due to the construction of L(T) $\subseteq$ U(T). This will be further investigated in Chapter 9.

Again it is to be noted that if we apply the commutativity and distributivity, first and then this time the Law of Excluded Middle, LEM, we get FDCF(A OR B) = FCCF(A OR B) and thus, we obtain:

DNF(A OR B) = CNF(A OR B) = A$\cap$B

in two-valued set and logic theory. Whereas in fuzzy set and logic theory FDCF and FCCF provide lower and upper set formulas, respectively, for the "OR" combination of two fuzzy concepts, A and B, i.e., fuzzy predicates. That is we get FDCF(A OR B) $\subseteq$ FCCF(A OR B).

This is yet another example of the break down of the equivalence in classical theory.

*8.4.5. Other Six Meta-Linguistic Expressions*

The remaining six basic protoforms in PNL, meta-linguistic expressions, i.e., 11, 12, 13, 14, 15, and 16 in Table 8.6 are treated in a slightly different manner. Let us investigate, for example, the meta-linguistic expression 11 in Table 8.6, i.e., the bi-conditional, "A IF AND ONLY IF B". Its usual target set is symbolically "A□B". It is clear that

$l$(T) = {A$\cup$B, c(A) $\cup$c(B)}

with the property that $l$(T) = {G | G$\in$**G**, G$\subset$T}.
Thus we have L(T) = $\cap l$(T) = (A$\cup$B) $\cap$ (c(A) $\cup$c(B)).
As well, it is clear that $u$(T) = {c(A) $\cap$B, A$\cap$c(B)},
with the property that $u$(T) = {G | G$\in$**G**, T$\subset$G}.
Thus, we have U(T) = $\cup u$(T) = (c(A) $\cap$B) $\cup$ (A$\cap$c(B)).

Therefore the rules for this remaining six meta-linguistic expressions, i.e., 11, 12, 13, 14, 15, and 16 in Table 8.6 are that
$L(T) = \cap\{G \mid G\in \mathbf{G}, G\subset T\}$ and $U(T) = \cup \{G \mid G\in \mathbf{G}, T\subset G\}$.
Thus, in this category we have upper and lower set formulas for the special cases of, A□B, A XOR B, A, c(A), B, and c(B).

*8.4.6. Results*

As a result of T-formalism we obtain the upper and lower set formulas as shown in Table 8.7 for meta linguistic concepts that are form with linguistic "AND", linguistic "OR" and those that are formed by both the linguistic "AND" and "OR".

## 8.5. Generalization

The schema developed for the determination of upper and lower set definitions for any two sets A and B, crisp or fuzzy, can be generalized to n sets.

Suppose there are n concepts that are represented by n predicates $A_1,\ldots,A_n$, crisp or fuzzy. We can write $\mathbf{G}_n$ with the formation of $2^{n+1}$ primitives, i.e., information granules, derived from conjunction, disjunction and complementation of these n concepts as:

$$G_n = \{A_1, \cup A_2 \cup \ldots \cup A_n, \ldots, c(A_1) \cap c(A_2) \cap \ldots \cap c(A_n)\}$$

Then we can apply the same method developed in Sections 8.4.2 as shown in its application to two sets A and B in Sections 8.4.3, 8.4.4 and 8.4.5 to determine the upper and lower set formulas of any meta-linguistic expression made up of these n concepts.

It is clear that with **G**, we can drive upper and lower set formulas for any meta linguistic expression, $\mathcal{MLE}$., i.e., basic expressions in PLN. But these formulas must be developed carefully in three categories, i.e., 8.4.3 those that are formed with the linguistic "AND" connective, 8.4.4 these that are formed with the linguistic "OR" connective, and 8.4.5 others that are more complex and that are singletons. Thus the lower set formula will be FDCF($\mathcal{MLE}$.) and the upper set formula will be FCCF($\mathcal{MLE}$.).

## 8.6. Interval-Valued Type 2 Fuzzy Empty and Universal Sets

The fuzzy empty and fuzzy universal sets in interval-valued Type 2 fuzzy set can be generated when we start out with the Type 1 fuzzy sets with the application of FDCF and FCCF expression. In this section we investigate some of the basic properties of these Interval-valued Type 2 fuzzy empty and universal sets. In the following, we start out first with the classical empty set related properties, next look at Type 1 empty set and its properties and finally investigate the properties of the interval-valued Type 2 fuzzy sets generated by FDCF and FCCF expressions.

### *8.6.1. Classical Empty Set*

An empty set in classical sets, is defined as a set which includes no element. The crisp empty set is denoted by the symbol $\varnothing$ or $\{\}$ and it has some properties in set theoretical operations as follows:

Assume that a set X is a universal set.

(a) Union and intersection with $\varnothing$ and X(identity laws) are:

$A \cup \varnothing = A, \; A \cap X = A$

$A \cup X = X, \; A \cap \varnothing = \varnothing$

(b) The Laws Excluded Middle and Contradiction are:

$A \cup c(A) = X, \; A \cap c(A) = \varnothing$

(c) Complement of: $\varnothing$ and X are:

$c(X) = \varnothing, \; c(\varnothing) = X$

### *8.6.2. Type 1 Fuzzy Empty Set*

In Type 1 fuzzy sets, the empty set is defined using the membership function $\mu_\varnothing$, such that

$\forall x \in X, \; \mu_\varnothing(x) = 0$

Similarly, the universal set X can be denoted by $\forall x \in X, \; \mu_X(x) = 1$.
Both empty sets in classical and fuzzy sets are exactly the same, i.e., a set which has no element is the same as a set in which the membership value of each element is o. Therefore, the set operations with the empty set in Type 1 fuzzy sets are defined as follows.

(a) Union and intersection with $\varnothing$ and X(identity laws) for ($\vee$, $\wedge$, -) are:
$A \cup \varnothing = A, \; A \cap X = A$

i.e. $\forall x \in X, \quad \mu_{A\cup\varnothing}(x) = \mu_A(x) \vee \mu_\varnothing(x) = \mu_A(x) \vee 0 = \mu_A(x)$,
$\mu_{A\cap X}(x) = \mu_A(x) \wedge \mu_X(x) = \mu_A(x) \wedge 1 = \mu_A(x)$, and $A \cup X = X, \; A \cap \varnothing = \varnothing$.
i.e. $\forall x \in X, \quad \mu_{A\cup X}(x) = \mu_A(x) \vee \mu_X(x) = \mu_A(x) \vee 1 = 1$,

$\mu_{A\cap\varnothing}(x) = \mu_A(x) \wedge \mu_\varnothing(x) = \mu_A(x) \wedge 0 = 0.$

(b) The Laws of Excluded Middle and Contradiction are not satisfied. Instead, we have:

$$A\cup c(A) \subseteq X,\ A\cap c(A) \supseteq \varnothing$$

*8.6.3. Type 2 Fuzzy Sets*

The concept of Type 2 fuzzy sets was initially proposed by Zadeh (1975) , Mizimoto and Tanaka(1976, 1981) provided certain definition. Later Türkşen(1986-2002) provided various definitions and expressions for interval-valued Type 2 fuzzy sets. Recently, Karnik and Mendel(1998), Karnik et.al.(1999), Liang and Mendel(2000) and Mendel (2001) have expressed various definitions of Type 2 fuzzy sets.

**Definition 8.6.1.** A Type 2 fuzzy set on a crisp set X is a fuzzy set, i.e., a second order fuzzy set, on a Type 1 fuzzy set which is charactrized by a membership function on membership functions of Type 1 fuzzy sets:

$$\mu_U(\mu_A) : X \rightarrow [0,1]^{[0,1]}$$

such that

$$\mu_A(x,u) = \int_{x\in X} [\int_{u\in[\underline{\mu}_A(x),\overline{\mu}_A(x)]} \mu(u)/u]/x$$

where $u\in[0,1]$ is a membership value, $\mu(u)$ is the membership of the original membership values of x, i.e., $\mu_A(x) = u\in[0,1]$ $\underline{\mu}_A(x)$ and $\overline{\mu}_A(x)$ are the lower and the upper bounds of $u = \mu_A(x)$.

**Definition 8.6.3.** A Type 2 fuzzy empty set on a crisp set X is characterized by a fuzzy membership function, such that

$$\mu_\varnothing(x,0) = \int_{x\in X} [\int_{u\in[\underline{\mu}_A(x),\overline{\mu}_A(x)]} 1/0]/x$$

This means that the Type 2 fuzzy empty set has the membership value of 1 only when its membership degree of each element $x\in X$ is zero.

**Table 8.7.** Target, Lower and Upper Set Expressions.
*** ID Number of Meta-Linguistic Expression.**

| ID* | Usual<br>Target Set Expression T | Lower Set Expression L(T) = FDCF<br>Fuzzy Disjunctive Canonical Form |
|---|---|---|
| 1 | I | (A∪B) ∩ (c(A) ∪ B) ∩ (A ∪ c(B)) ∩ (c(A) ∪ c(B)) |
| 2 | ∅ | ∅ |
| 3 | A ∩ B | (A ∪ B) ∩ (c(A) ∪ B) ∩ (A ∪ c(B)) |
| 4 | c(A) ∪ c(B) | c(A) ∪ c(B) |
| 5 | c(A) ∩ c(B) | (c(A) ∪ B) ∩ (A ∪ c(B)) ∩ (c(A) ∪ c(B)) |
| 6 | A ∪ B | A ∪ B |
| 7 | c(A) ∩ B | (A ∪ B) ∩ (c(A) ∪ B) ∩ (c(A) ∪ c(B)) |
| 8 | A ∪ c(B) | A ∪ c(B) |
| 9 | A ∩ c(B) | (A ∪ B) ∩ (c(A) ∪ B) ∩ (c(A) ∪ c(B)) |
| 10 | c(A) ∪ B | c(A) ∪ B |
| 11 | (A ∪ B) ∩ (c(A) ∪ c(B)) | (A ∪ B) ∩ (c(A) ∪ c(B)) |
| 12 | (A ∩ B) ∪ (c(A) ∩ c(B)) | (c(A) ∪ B) ∩ (A ∪ c(B)) |
| 13 | A | (A ∪ B) ∩ (A ∪ c(B)) |
| 14 | c(A) | (c(A) ∪ B) ∩ (c(A) ∪ c(B)) |
| 15 | B | (A ∪ B) ∩ (c(A) ∪ B) |
| 16 | c(B) | (A ∪ c(B)) ∩ (c(A) ∪ c(B)) |

| ID* | Usual<br>Target Set Expression T | Upper Set Expression U(T) = FCCF<br>Fuzzy Conjunctive Canonical Form |
|---|---|---|
| 1 | I | I |
| 2 | ∅ | (A∩B) ∪ (c(A) ∩ B) ∪ (A ∩ c(B)) ∪ (c(A) ∩ c(B)) |
| 3 | A ∩ B | A ∩ B |
| 4 | c(A) ∪ c(B) | (c(A) ∩ B) ∪ (A ∩ c(B)) ∪ (c(A) ∩ c(B)) |
| 5 | c(A) ∩ c(B) | c(A) ∩ c(B) |
| 6 | A ∪ B | (A ∩ B) ∪ (c(A) ∩ B) ∪ (A ∩ c(B)) |
| 7 | c(A) ∩ B | c(A) ∩ B |
| 8 | A ∪ c(B) | (A ∩ B) ∪ (A ∩ c(B)) ∪ (c(A) ∩ c(B)) |
| 9 | A ∩ c(B) | A ∩ c(B) |
| 10 | c(A) ∪ B | (A ∩ B) ∪ (c(A) ∩ B) ∪ (c(A) ∩ c(B)) |
| 11 | (A ∪ B) ∩ (c(A) ∪ c(B)) | (c(A) ∩ B) ∪ (A ∩ c(B)) |
| 12 | (A ∩ B) ∪ (c(A) ∩ c(B)) | (A ∩ B) ∪ (c(A) ∩ c(B)) |
| 13 | A | (A ∩ B) ∪ (A ∩ c(B)) |
| 14 | c(A) | (c(A) ∩ B) ∪ (c(A) ∩ c(B)) |
| 15 | B | (A ∩ B) ∪ (c(A) ∩ B) |
| 16 | c(B) | (A ∩ c(B)) ∪ (c(A) ∩ c(B)) |

**Definition 8.6.4.** A Type 2 fuzzy universal set is characterized by a fuzzy membership function, such that

$$\mu_X(x,0) = \int_{x\in X} \left[ \int_{u\in[\underline{\mu}_A(x),\bar{\mu}_A(x)]} 1/1 \right] / x$$

This means that the Type 2 fuzzy universal set has the membership value 1 only when its membership degree of each element $x\in X$ is one.

*8.6.4. Combination of Type 1 Fuzzy Sets with ∅ and X*

The combination of a Type 1 fuzzy set A with ∅ and X and with linguistic connectives "AND", "OR" generate an interval-valued Type 2 fuzzy sets by FDCF and FCCF expressions.

(a) Some identity Laws hold:

(i) A OR ∅ = A, (ii) A AND X = A

while others do not necessary hold. That is some hold.

(iii) (a) A OR X ⊆ X, (iv) A AND ∅ = ∅.

We have say, for (∨, ∧, -) FDCF(A OR ∅) = FCCF(A OR ∅) = A, i.e.,

$(a\wedge 0) \vee ((1\text{-}a) \wedge 0) \vee (a\wedge 1) = (a\vee 0) = a$

(ii) FDCF(A AND X) = FCCF(A AND X) = A, i.e.,
$(a\wedge 1) = (a\vee 1) \wedge ((1\text{-}a) \vee 1) \wedge (a\vee 0) = a$

Thus these identity laws hold!
Whereas, the other two classical identity laws do not hold.

(iii) (b) FDCF(A OR X) ⊆ FCCF(A OR X) = X, i.e.,
$(a\wedge 1) \vee ((1\text{-}a) \wedge 1) \vee (a\wedge 0) \leq a\vee 1 = 1$, i.e., $a\vee(1\text{-}a) \leq 1$

(iv) ∅ = FDCF(A AND ∅) ⊆ FCCF(A AND ∅), i.e.,
$(a\wedge 0) \leq (a\vee 0) \wedge ((1\text{-}a) \vee 0) \wedge (a\vee 1) = a\wedge(1\text{-}a)$, i.e., $0 \leq a\wedge(1\text{-}a)$

(b) As well, it will be shown in the Chapter 10 that the Laws of Excluded Middle and Contradiction give way to the Laws of "Fuzzy Middle" and "Fuzzy Contradiction". That is we have:

A OR c(A) $\subseteq$ X and A AND c(A) $\supseteq \emptyset$

Note that:
FDCF(A OR c(A)) $\subseteq$ FCCF(A OR c(A))

for $(\vee, \wedge, -)$

$(a\wedge(1-a) \vee((1-a) \wedge(1-a)) \vee (a\wedge a) \leq a\vee (1-a) \leq 1$
But for $(\oplus, \odot, -)$, we have

FDCF(A OR c(A)) $\subseteq$ FCCF(A OR c(A)) $\subseteq$ X
$(a\odot(1-a)) \oplus ((1-a) \odot(1-a)) \oplus (a\odot a) \leq (a\oplus (1-a) \leq 1$

## Appendix 8.1. Fuzzy Canonical Form Derivation Algorithm

Step 1. To derive the Fuzzy Disjunctive Canonical Form Expression, FDCF, one does the following:

For each row of the "Truth Table", if a "T" appears in the column for the combined Meta-Linguistic Expression, in our example, "A OR B" column, then:

1(a) Construct the primary conjunction, $\cap$, of the meta-linguistic symbols, e.g., "A" and "B" with the requirement that (i) if "T" appears in the truth column of that symbol, the symbol is taken as it appears in the column heading, i.e., in this case, and usually, in its affirmative form; but (ii) if "F" appears in the truth column of that symbol, the complement of the symbol is taken.

1(b) Next, combine all primary conjunctions obtained in (a) above with the disjunction operator, $\cup$.
For our example "A OR B", we get three phrases in two-variable two dimensional space in STEP 1 (a) as:

$A\cap B$; $A\cap c(B)$; $c(A)\cap B$.

In STEP 1(b), we get one complex clause

$(A\cap B)\cup(A\cap c(B)\cup(c(A)\cap B)$

in two-variable two dimensional space. Hence, we get:

$$FDCF(A \text{ OR } B) = (A \cap B) \cup (A \cap c(B) \cup (c(A) \cap B)$$

As shown in top section of Table 8.3 row 3.(first part)

STEP 2. To derive the Fuzzy Conjunctive Canonical Form Expression, FCCF, one does the following:
For each row of the "Truth Table", if an "F" appears in the column for the combined Meta-Linguistic Expression, in our example, "A **OR** B" column, then:

2(a) construct the primary conjunction, $\cap$, of the symbols, e.g., "A" and "B", with the same requirement as stated in STEP 1(a).

2(b) Next, combine all conjunctions obtained in (a) above with the disjunction operator, $\cup$.

For our example, "A **OR** B", we get one phrase in two-variable two-dimensional space in STEP2(a) as:

$$A \cup B$$

In STEP 2(b), we get the same expression, i.e., the phrase $A \cup B$, since there are no other phrases to combine to obtain a complex clause. Thus, we obtain FCCF(A OR B) = $A \cup B$ as shown in Table 8.3 row 3(second part).

**Appendix 8.2. Canonical Form Derivation with Set Membership Consideration**

Let us for simplicity use $\mu_A(x) = a$, $\mu_B(x) = b$, and a, b ε [0.1].
For $x_1$ ε $X_1$, where a. b, we say "x belongs to the combined fuzzy set $A \cap B$ with membership aΔ*b" where Δ* is a pseudo t-norm. Thus FDCF (A AND B) = $A \cap B$ for $A \supseteq B$. It is to be noted when we say "x belongs to ....", we are in fact stating implicitly what the logicians explicitly say "it is the case that x belongs to .... ", etc. When Δ=Min=∧, we get a ∧ b as depicted in Figure 8.1.

Similarly, for $x_2$ ε $X_2$, where a < b, we say "x belongs to the combined fuzzy set $A \cap B$ with membership aΔ*b", where Δ* is a pseudo t-norm. Thus FDCF(A AND B) = $A \cap B$ for A ⊂. B. Again, when Δ = Min = ∧, we get a ∧ b as depicted in Figure 8.1. It should be recalled that we are assuming, boundary, monotonicity, involutive negation, and De Morgan Laws. Hence, we leave a. b and a < b to be two distinct cases. Thus the results are applicable pseudo t-norms and t-conorms that are non-associative and non-commutative. Hence for both cases, the formulas turn out to be the same. In particular, we have $\Delta$ = Min = $\wedge$, we get a $\wedge$ b as depicted in Figure 8.1.

Next, let us consider $x_1 \varepsilon X_1$ where again, a$\gtrsim$ b. Now we say, "x does not belong to the combined fuzzy set of $(c(A) \cap c(B)) \cup (A \cap c(B)) \cup (c(A) \cap B)$ with membership of $(n(a)\Delta^*n(b))\ \nabla^*(a\Delta^*n(b))\ \nabla\ (n(a)\Delta^*b)$", where $\Delta$ *is a pseudo t-norm and $\nabla$* is a pseudo t-conorm. Again, it should be noted that when we say "x does not belong to ....." in fact we are stating implicitly what the logicians explicitly say "it is not the case that x belongs to ...", etc. When $\Delta$ = Min = $\wedge$ and $\nabla$ = Max = $\vee$ and n(a) = 1 - a, the standard complement, we get $(n(a) \wedge n(b)) \vee (a \wedge n(b)) \vee (n(a) \wedge b)$ as depicted in Figure 8.1. Furthermore, because we say "x does not belong ...." we then have to take the complement of the given expressions in both the set and membership domains.

Thus with De Morgan Laws we get, for $A \supseteq B$:

$$\begin{aligned}\text{FCCF(A AND B)} &= c[(c(A) \cap c(b)) \cup (A \cap c(B)) \cup (c(A) \cap B)] \\ &= (A \cup B) \cap (c(A) \cup B) \cap (A \cup c(B)).\end{aligned}$$

And for $A \subset B$, we get similarly

$$\begin{aligned}\text{FCCF(A AND B)} &= c[(c(A) \cap c(B)) \cup (A \dot{\cup} c(B) \cup (c(A) \dot{\cap} B))] \\ &= (A \cup B) \cap (c(A) \dot{\cup} B) \cap (A \cup c(B)).\end{aligned}$$

as expected. Again the results are also applicable to regular t-norms and t-conorms. Hence for both cases, the formulas turn out to be the same. In particular, again for example, when $\nabla$ = Max = $\vee$, $\Delta$ = Min = $\vee$ and n(a) = 1 - a, the Standard Complement, we get:

FCCF(A AND B) = $(a \vee b) \wedge (n(a) \vee b) \wedge (a \vee n(b))$ for both $a \geq b$ and $a < b$, etc., as depicted in Figure 8.1.

## References

1. Bilgic, T. (1995), Measurement-Theoretic Frameworks for Fuzzy Set Theory with Applications to Preference Modelling, PhD thesis, University of Toronto, Dept. of Industrial Engineering Toronto, Ontario M5S 1A4 Canada.(supervisor, I.B. Türkşen)
2. Dempster A.P (1967), "Upper and Lower Probabilities Induced by a Multivalued Mapping", In: *Annals of Mathematical Statistics*, 38, 325-339.
3. Karnik, N.N., J.M. Mendel and Q. Liang (1999), "Type 2 Fuzzy Logic System", *IEEE Trans. Fuzzy Systems.* Vol. 7, No. 5, Oct.
4. Karnik, N.N., J.M. Mendel (1998), Introduction to Type 2 Fuzzy Logic Systems, *Presented at1998 IEEE FUZZ Conference,* May, Anchorage, AK.

5. Liang, Q., J.M. Mendel (2000), "Decision Feedback Equalizers for Non-linear Time Varying Channels using Type 2 Fuzzy Adaptive Filters", *Proceedings of the 9th IEEE International Conference on Fuzzy Systems*, 7-10 May, San Antonio, Texas, 883-888.
6. Pawlak Z. (1991), Rough Sets, *Kluwer*, Netherlands.
7. Türkşen I.B. (2002), "Type 2 Representation and Reasoning for CWW", *FSS*, 127, 17-36.
8. Türkşen I.B. (2002), "Interval-valued Type 2 Fuzzy Sets, Multi-valued Maps and Rough Sets", (in) A Grmela and N.E.Mastorakis (eds.), Advances in Intelligent Systems, *Fuzzy Sets*, Evolutionary Computation, WSEAS, 142-146.
9. Türkşen I.B. (2001), "Computing with Descriptive and Veristic Words: Knowledge Representation and Reasoning", *in: Computing With Words*, P.P. Wang(ed.), Chapter 10, Wiley, New York, 297-328.
10. Türkşen, I.B. (2001), "Sources, Measurements and Models of Type 2 Fuzziness in the New Millennium", *in: Fuzziness in the New Millennium*, V. Dimitrov (ed), Springer Verlag (in press).
11. Türkşen, I.B. (1999), "Computing with Descriptive and Veristic Words, *Proceedings of NAFIPS'99* (Special Invited Presentation), New York, June 10-12, 13-17.
12. Türkşen, I.B. (1999), "Theories of Set and Logic with Crisp or Fuzzy Information Granules", *J.of Advanced Computational Intelligence*, 3,4, 264-273.
13. Türkşen I.B. (1999), "Type I and Type II Fuzzy System Modeling", *FSS*, 106, 11-34.
14. Türkşen I.B. (1994),"Interval-Valued Fuzzy Sets and Compensatory AND", *FSS,* 87-100.
15. Türkşen, I.B. (1994), "Fuzzy Normal Forms", *Fuzzy Sets and Systems,* 253-266.
16. Türkşen, I.B, (1994), "Interval-Valued Fuzzy Sets and 'Compensatory AND' ", *FSS*, 295-307.
17. Türkşen I.B. (1986), "Interval-Valued Fuzzy Sets based on Normal Forms", *Fuzzy Sets and Systems,* 191-210.
18. Zadeh, L.A. (1965), "Fuzzy Sets", *Information and Control Systems*, Vol.8, 338-353.

# Chapter 9

# CONTAINMENT OF FDCF IN FCCF

On the basis of ontological and epistemological assumption, in previous chapters, it was shown that the equivalence between canonical forms break down in fuzzy set and logic theory, i.e., FDCF $\subseteq$ FFCF.

However, we have stated that for particular classes of t-norms and t-conorms FDCF(.)$\subseteq$FCCF(.) holds for all sixteen combination of concepts that are formed with De Morgan triples. This is now developed in this chapter in detail with the application of generating functions for strict and nilpotent Archimedean t-norms and t-conorms. For this purpose, we next review the generators of continuous t-norms and t-conorms and other relevant concepts in addition to the properties of t-norms and t-conorms stated in Appendix 6.

## 9.1. Generators of Continuous Archimedean Norms

Archimedean t-norms and t-conorms admit an additive representation on the extended non-negative real numbers, $\overline{\Re}^+=[0, +\infty]$.The following is a representative theorem of Ling (1965), Schweizer and Sklar(1983).

**Theorem 9.1.** A t-norm $\Delta$ is continuous and Archimedean if and only if there exists a continuous and strictly decreasing function $f:[0,1]\rightarrow \overline{\Re}^+$ with $f(1)=0$ and

$$\Delta(x,y)=f^{-1}(\min\{f(x) + f(y), f(0)\}). \tag{9.1}$$

Similarity, a t-conorm $\nabla$ is continuous and Archimedean if and only if there exists a continuous and strictly increasing function $g:[0,1]\rightarrow \overline{\Re}^+$ with $g(0) = 0$ and

$$\nabla(x,y)=g^{-1}(\min\{g(x) + g(y), g(1)\}). \tag{9.2}$$

Furthermore, f and g are unique up to a positive constant (ratio scale), i.e., if f' and g' are two other such representations then $f'=\lambda f$ and $g'=\lambda' g$, for $\lambda, \lambda' > 0$.

**Remark:** Actually, the theorem is valid with a strictly monotonic function. But for convenience, the generator functions of t-norms are taken to be strictly decreasing and the generator function of t-conorms, strictly increasing.

The functions f and g from Theorem 9.1 are called the additive generators of t-norm, Δ, and t-conorm, ∇, respectively. The following result is about the representation of continuous Archimedean strict t-norms.

**Corollary 9.1.** A t-norm Δ and a t-conorm ∇ are strict if and only if they admit the following representations

$$\Delta(x,y) = f^{-1}(f(x) + f(y)) \tag{9.3}$$
$$\nabla(x,y) = g^{-1}(g(x) + g(y)) \tag{9.4}$$

In terms of Theorem 9.1, a t-norm is strict if and only if $f(0) = +\infty$ and has zero divisors if and only if $f(0)<+\infty$.

**Examples**: Algebraic product, ⊙ is additively generated by $g_{\odot}(x) = -\log x$. The corresponding t-conorm is given by $\oplus(x,y)=x+y-xy$. The multiplicative generator is $h_{\odot}(x)=z$.

Lukasiewicz t-norm, $L_{\odot}$, (at times it is also referred to as Bold intersection, $B_{\odot}$) is additively generated by $f_L(x) = \max\{1 - z,0\}$ for all $x \in \Re^+$ and $g_L(x) =1 - x$ for all $x \in[0,1]$. Multiplicative generators are given by $h_L(x) = \max\{1 + \log x, 0\}$ and $k_L(x) = e^{x-1}$.

Since all continuous t-norms and t-conorms are in [0,1] it is interesting to find representations for continuous Archimedean t-norms in terms of the automorphism of the unit interval. The following is proven in (Schweizer and Sklar, 1983).

**Theorem 9.2.** Any continuous, strict t-norm Δ can be represented as a φ-transform of ⊙ as

$$\odot(x,y) = \varphi^{-1}(\varphi(x)\varphi(y)) \tag{9.5}$$

and any continuous, strict t-conorm can be represented as a φ-transform of ⊕ as

$$\oplus(x,y) = \varphi^{-1}(\varphi(x) + \varphi(y) - \varphi(x)\varphi(y)) \tag{9.6}$$

for all $x, y\in[0,1]$, where φ is an automorphism of the unit interval.This result shows that any continuous, strict t-norm is isomorphic to ⊙ and any continuous, strict t-conorm is isomorphic to ⊕.The following are similar results for t-norms with zero divisors and nilpotent t-conorms (Ovchinikov and Roubens, 1991).

**Theorem 9.3.** Any continuous t-norm Δ having zero divisors can be represented as a φ-transform of $L_{\odot}$ as

$$L_{\odot}(x,y) = \varphi^{-1}(\max\{\varphi(x) + \varphi(y) - 1, 0\})$$

and any continuous, nilpotent t-conorm $\nabla$, can be represented as a φ-transform of $L_{\oplus}$ as

$$L_{\oplus}(x,y) = \varphi^{-1}(\min\{\varphi(x) + \varphi(y), 1\})$$

where φ is an automorphism of the unit interval.

This shows that any continuous t-norm with zero divisors is isomorphic to $L_{\odot}$ and any continuous, nilponent t-conorm is isomorphic to $L_{\oplus}$ .The following result shows that the uniqueness of the representation in term of automorphisms of the unit interval is stronger.

**Corollary 9.2.** If φ is the automorphism of the unit interval that generates a t-norm with zero divisers or a nilpotent t-conorm then φ' is another such generator if and only if φ' = φ.

**Proof**: For a t-norm with zero divisors, let $\varphi(x) = 1 - f(x)/f(0)$ where f is the additive generator of the t-norm. Such a φ is an automorphism of the unit interval (Ovchinikov and Roubens, 1991). But f is unique up to a positive constant (Theorem 9.1). Define, φ' as another such generator from:

$$\varphi'(x) = 1 - \frac{\lambda f(x)}{\lambda f(0)} = 1 - f(x)/f(0) = \varphi(x),$$

for all $x \in [0,1]$ and a $\lambda > 0$.

For a nilpotent t-conorm, define $\varphi(x) = 1 - g(x)/g(1)$ where g is the additive generator of the t-conorm. The proof follows exactly as for a t-norm because g is also unique up to a positive constant.

## 9.2. Non Archimedean Triangular Norms and Conorms

The functions, $\wedge$ and $\vee$ defined as: $\wedge$=min and $\vee$=max are the t-norm and the t-conorm, respectively. However, min and max are not Archimedean since min{x, x} =max{x, x,} = x (i.e., they are idempotent).

Therefore, min and max do not admit an additive representation of the form given in Theorem 9.1.

Moreover, it is well known that min and max are the extreme t-norm and t-conorm in the sense that for any t-norm, $\Delta$ and t-conorm, $\nabla$ we have

$$\Delta < \min \leq \max < \nabla$$

In order to have min and max as the only t-norms and t-conorms that follow, one forms an independent set of axioms. It is necessary and sufficient to add an axioms for distributivity to the definition of t-norms and t-conorms (Alsina at al. 1983, Weber 1983). Hence $\Delta$ =min and $\nabla$ = max are the unique solutions if and only if the two place functions $\Delta$ and $\nabla$ on the unit interval are monotonic, associative, have boundaries and are distributive.

It should be noticed that commutativity (symmetry) need not be explicitly assumed since it follows from others. More can be said about the characteristics of the min function (for max there are dual arguments) (Schweizer and Sklar 1983). That is, there does not exist continuous (not necessarily monotonic) functions f, g and h such that $\min\{x,y\} = f(g(x) + h(y))$. On the other hand, there does not exist strictly monotonic (not necessarily continuous) functions, f and g such that $\min\{x,y\} = f^{[-1]}(g(x)+g(y))$, where $f^{[-1]}$ is the pseudo-inverse of f.

Among all the t-norms and t-conorms, only min and max distribute over each other. For others, weaker versions of distributive laws hold (Gottwald 1998):

**Proposition 9.1.** For any (not necessarily continuous) t-norm, $\Delta$, and t-conorm, $\nabla$, the following distributive laws hold for all x, y, z $\in[0,1]$:

(i) $\Delta(x, \min(y,z)) = \min(\Delta(x,y), \Delta(x,z))$
(ii) $\Delta(x, \max(y,z)) = \max(\Delta(x,y), \Delta(x,z))$
(iii) $\nabla(x, \min(y,z)) = \min(\nabla(x,y), \Delta(x,z))$
(iv) $\nabla(x, \max(y,z)) = \max(\nabla(x,y), \nabla(x,z))$

as well as the following sub distributive laws, for all x, y, z $\in[0,1]$:

(v) $\min(x, \Delta(y,z)) \geq \Delta(\min(x,y), \min(x,z))$
(vi) $\max(x, \Delta(y,z)) \geq \Delta(\max(x,y), \max(x,z))$
(vii) $\min(x, \nabla(y,z)) \leq \nabla(\min(x,y), \min(x,z))$
(viii) $\max(x, \nabla(y,z)) \leq \nabla(\max(x,y), \max(x,z))$

### 9.3. Ordinal Sums

The following representation of continuous t-norms by means of ordinal sums can be derived from results in (Mostert and Shields, 1957) in the context of I-semigroups. (originally proved in Schweizer and Sklar, 1983). (more recently Klement, et.al. 2000)

**Theorem 9.4.** For a function $\Delta$:[0,1] the following are equivalent:

(i) $\Delta$ is a continuous t-norm.
(ii) $\nabla$ is uniquely representable as an ordinal sum of continuous Archimedean t-norms, i.e., there exits a uniquely determined (finite or countably infinite) index set A, a family of uniquely determined pairwise disjoint open subintervals $(]a_\alpha,e_\alpha[)_{a\in A}$ of [0,1] and a family of uniquely determined continuous Archimedean t-norms $(\Delta_\alpha)_{a\in A}$ such that $\Delta = (<a_\alpha,e_\alpha,\Delta_\alpha>)_{a\in A}$ .

**Proof**: (See Klement, et.al. 2000)

## 9.4. De Morgan Triples

If a t-norm, $\Delta$, a t-conorm, $\nabla$ and a negation function, n are related to each other, then such a triple is called a De Morgan Triple.

**Definition 9.1.** If $\Delta$ is a continuous t-norm, $\nabla$ is a continuous t-conorm, n is a strict negation and De Morgan Laws hold, i.e., $n(a\Delta b) = n(a)\nabla n(b)$, and $n(a\nabla b) = n(a)\Delta n(b)$, then the triple $(\Delta,\nabla,n)$ is called a De Morgan triple. According to properties of t-norms, t-conorms and the negation, De Morgan triples can be further classified.

**Definition 9.2.** If $(\Delta,\nabla,n)$ is a De Morgan triple such that $\Delta$ and $\nabla$ are strict, n is a strong negation and both are generated by the same automorphism $\varphi$, then $(\Delta,\nabla,n)$ is called a strict De Morgan triple. In this case,

$$\Delta(x,y) = \varphi^{-1}(\varphi(x)\,\varphi(y)) \tag{9.7}$$
$$\nabla(x,y) = \varphi^{-1}(\varphi(x) + \varphi(y) - \varphi(x)\,\varphi(y)) \tag{9.8}$$
$$n(x) = \varphi^{-1}(1 - \varphi(x)) \tag{9.9}$$

**Definition 9.3.** If $(\Delta,\nabla,n)$ is a De Morgan triple such that $\Delta$ has zero divisors, n is a strong negation and both are generated by the same automorphism $\varphi$, then $(\Delta,\nabla,n)$ is called a strong or Lukasiewicz like De Morgan triple. In this case,

$$\Delta(x,y) = \varphi^{-1}(\max\{\varphi(x) + \varphi(y) - 1, 0\}) \tag{9.10}$$
$$\nabla(x,y) = \varphi^{-1}(\min\{\varphi(x) + \varphi(y), 1\}) \tag{9.11}$$
$$n(x) = \varphi^{-1}(1 - \varphi(x)) \tag{9.12}$$

A strong or Lukasiewicz like De Morgan triple can also be characterized by a continuous, strictly decreasing function $g:[0,1]\to \overline{\Re}^+$ with $g(1)=0$, $g(0) < +\infty$ as follows(Schweizer and Sklar, 1983) (Ling, 1965):

$$\Delta(x,y) = g^{-1}(\min\{g(x) + g(y), g(0)\}) \tag{9.13}$$
$$n(x) = g^{-1}(g(0 - g(x)) \tag{9.14}$$

$\nabla(x,y) = g^{-1}(\max\{g(x) + g(y) - g(0), 0\})$ (9.15)
Zadeh's ($\vee$, $\wedge$, n) triple given by:
$\wedge(x,y) = \min\{x,y\}$ (9.16)
$\vee(x,y) = \max\{x,y\}$ (9.17)
$n(x) = 1 - x$ (9.18)
is a bona fide De Morgan triple but Zadeh triple is neither strict nor strong.

It is worthwhile to note also that Zadeh triple has the following properties a part from the properties embedded in the definition of t-norms (associativity, monotonicity and boundary conditions).

**Definition 9.4.**

| | | |
|---|---|---|
| Idempotency | $\wedge(x,x) = x$, | |
| | $\vee(x,x) = x$, | $\forall x \in [0,1]$ |
| Absorption | $\wedge(\vee(x,y),x) = x$, | |
| | $\vee(\wedge(x,y),x) = x$, | $\forall x,y \in [0,1]$ |
| Distributivity | $\vee(x,\wedge(y,z)) = \wedge(\vee(x,y),\vee(x,z))$, | |
| | $\wedge(x,\wedge(y,z)) = \vee(\wedge(x,y),\wedge(x,z))$, | $\forall x,y,z \in [0,1]$ |
| Fuzzy Contradiction | $\wedge(x,n(x)) \geq 0$ , | |
| Fuzzy Middle | $\vee(x,n(x)) \leq 1$, | $\forall x \in [0,1]$ |

In Chapter 10, fuzzy middle and fuzzy contradiction will be treated in general for strict and nilpotent Archimedian t-norms and conorms.

For the two-valued De Morgan triple, we have:
(Crisp) Contradiction $\wedge(x,n(x) = 0, \forall x \in \{0,1\}$
(Crisp) Excluded Middle $\vee(x, n(x)) = 1, \forall x \in \{0,1\}$.

## 9.5. Basic Protoforms: FDCF and FCCF

The basic protoforms of the basic combination of concepts (Table 6.3) which we had introduced in Table 6.4 of Chapter 6 are restarted here as Table 9.1 As well, their t-norm and t-conorm version in the membership domain are given in Table 9.2 where $x=\mu_x(.)$.

## 9.6. Preliminary Observations

It should be observed that for a given De Morgan triple, the sixteen representations given in Table 9.2 can be partitioned as {1,2}, {3,...,10},

{11,12}, {13,...,16}, i.e., (1) and (2) are equivalent, (3)–(10) are equivalent, (11) and (12) are equivalent and (13)-(16) are equivalent to each other in form in Table 9.2.

This fact is first realized in (Piaget, 1949) for two-valued logic. Dubois and Prade (1980) discuss it for fuzzy sets without going into normal forms. Türkşen(1984) establishes the equivalence of (1) and (2), (3)-(6) and (13)-(14). Bo, Wangming and Yin (1992) mention the equivalence as given in this study.

The first question to answer is whether FDCF(.)=FCCF(.) for any De Morgan triple. This question is answered with the following results: (Bilgic, 1995)

**Proposition 9.2.** If (Δ,∇,n) is a De Morgan triple such that Δ is t-norm which has zero divisors, ∇ is a nilpotent t-conorm and n is a negation function for which n(0)=1 and n(1)=0, then FDCF(.)=FCCF(.) whenever a=1 and $b\in[0,1]$ or $a\in[0,1]$ and b=1.(Bilgic, 1995)

**Proof:** The result is established by showing it for rows (1), (3), (11) and (16) of Table 9.2. Let a=1 then the case for b=1 and $a\in[0,1]$ is similar.

It should be noted that, the specific requirements on t-norm, Δ and t-conorm,∇ of Proposition 9.2 are that Δ has zero divisors and is nilpotent. No continuity is assumed. Similarly nothing needs to be assumed for the negation function, n, except that n(0) = 1 and n(1) = 0 to prove Proposition 9.2.

This result states that having zero divisors (for the t-norm) and being nilpotent (for the t-conorm) are necessary conditions for FDCF(.) to be equal to FCCF(.). Therefore, this leaves out both Zadeh triples and strict De Morgan triples from further consideration. The only continuous candidate is a strong De Morgan triple. However, the following result states another necessary condition for De Morgan triples that already satisfy the conditions of Proposition 9.2.

**Proposition 9.3.** If (Δ,∇,n) is a De Morgan triple such that Δ has zero divisors, then in order for FDCF(.)=FCCF(.), Δ and ∇ must be idempotent. (Bilgic, 1995)

**Proof**: The result is established by considering row (3) of Table 9.2(for example) and setting a = b:

$$\nabla[\Delta(a,b),\Delta(a,n(b)),\Delta(n(a),b)] = \nabla(a,b)$$

$$\nabla[\Delta(a,a),\Delta(a,n(a)),\Delta(n(a),a)] = \nabla(a,a)$$

**Table 9.1.** Set and set connective representation of FDCF and FCCF

| No. | Fuzzy Disjunctive Canonical Forms |
|---|---|
| 1 | $(A\cap B)\cup(A\cap c(B))\cup(c(A)\cap B)\cup(c(A)\cap c(B))$ |
| 2 | $\phi$ |
| 3 | $(A\cap B)\cup(A\cap c(B))\cup(c(A)\cap B)$ |
| 4 | $(c(A)\cap c(B))$ |
| 5 | $(A\cap c(B))\cup(c(A)\cap B)\cup(c(A)\cap c(B))$ |
| 6 | $(A\cap B)$ |
| 7 | $(A\cap B)\cup$ $(c(A)\cap B)\cup(c(A)\cap c(B))$ |
| 8 | $(A\cap c(B))$ |
| 9 | $(A\cap B)\cup(A\cap c(B))$ $(c(A)\cap c(B))$ |
| 10 | $(c(A)\cap B)$ |
| 11 | $(A\cap B)\cup$ $(c(A)\cap c(B))$ |
| 12 | $(A\cap c(B))\cup(c(A)\cap B)$ |
| 13 | $(A\cap B)\cup(A\cap c(B))$ |
| 14 | $(c(A)\cap B)\cup(c(A)\cap c(B))$ |
| 15 | $(A\cap B)\cup$ $(c(A)\cap B)$ |
| 16 | $(A\cap c(B))\cup$ $(c(A)\cap c(B))$ |
| **No.** | **Fuzzy Conjunctive Canonical Forms** |
| 1 | I |
| 2 | $(A\cup B)\cap(A\cup c(B))\cap(c(A)\cup B)\cap c(A)\cup c(B))$ |
| 3 | $(A\cup B)$ |
| 4 | $(A\cup c(B))\cap(c(A)\cup B)\cap c(A)\cup c(B))$ |
| 5 | $(c(A)\cup c(B))$ |
| 6 | $(A\cup B)\cap(A\cup c(B))\cap(c(A)\cup B)$ |
| 7 | $(c(A)\cup B)$ |
| 8 | $(A\cup B)\cap(A\cup c(B))\cap$ $(c(A)\cup c(B))$ |
| 9 | $(A\cup c(B))$ |
| 10 | $(A\cup B)\cap$ $(c(A)\cup B)\cap(c(A)\cup c(B))$ |
| 11 | $(A\cup c(B))\cap(c(A)\cup B)$ |
| 12 | $(A\cup B)\cap$ $(c(A)\cup c(B))$ |
| 13 | $(A\cup B)\cap(A\cup c(B))$ |
| 14 | $(c(A)\cup B)\cap(c(A)\cup c(B))$ |
| 15 | $(A\cup B)\cap$ $(c(A)\cup B)$ |
| 16 | $(A\cup c(B))\cap$ $(c(A)\cup c(B))$ |

Since has zero divisors this equation reduces to $\Delta(a,a)=\nabla(a,a)$. Since $\Delta(a,a)\le a\le\nabla(a,a)$ by definition of t-norms and t-conorms, this can only happen at $\Delta(a,a)=a=\nabla(a,a)$, i.e., when $\Delta$ and $\nabla$ are idempotent.

**Table 9.2.** $\mu_{FDCF}$ and $\mu_{FCCF}$ formulas for membership values and membership connectives, i.e., t-norms, $\Delta$ and t-conorms, $\nabla$.

| No. | $\mu_{FDCF}$ | $\mu_{FCCF}$ |
|---|---|---|
| 1 | $\nabla[\Delta(a,b),\Delta(a,n(b)),\Delta(n(a),b),\Delta(n(a),n(b))]$ | 1 |
| 2 | 0 | $\Delta[\nabla(a,b),\nabla(a,n(b)),\nabla(n(a),b),\nabla(n(a),n(b))]$ |
| 3 | $\nabla[\Delta(a,b),\Delta(a,n(b)),\Delta(n(a),b)]$ | $\nabla(a,b)$ |
| 4 | $\Delta(n(a),n(b))$ | $\Delta[\nabla(a,n(b)),\nabla(n(a),b),\nabla(n(a),n(b))]$ |
| 5 | $\nabla[\Delta(a,n(b)),\Delta(n(a),b),\Delta(n(a),n(b))]$ | $\nabla(n(a),n(b))$ |
| 6 | $\Delta(a,b)]$ | $\Delta[\nabla(a,b),\nabla(a,n(b)),\nabla(n(a),b)]$ |
| 7 | $\nabla[\Delta(a,b)$ $\Delta(n(a),b),\Delta(n(a),n(b))]$ | $\nabla(n(a),b)$ |
| 8 | $\Delta(a,n(b))$ | $\Delta[\nabla(a,b),\nabla(a,n(b)),\nabla(n(a),n(b))]$ |
| 9 | $\nabla[\Delta(a,b),\Delta(a,n(b)),$ $\Delta(n(a),n(b))]$ | $\nabla(a,n(b))$ |
| 10 | $\Delta(n(a),b)$ | $\Delta[\nabla(a,b),\nabla(n(a),b),\nabla(n(a),n(b))]$ |
| 11 | $\nabla[\Delta(a,b)$ $\Delta(n(a),n(b))]$ | $\Delta[\nabla(a,n(b)),\nabla(n(a),b)]$ |
| 12 | $\nabla[\Delta(a,n(b)),\Delta(n(a),b)]$ | $\Delta[\nabla(a,b),\nabla(n(a),n(b))]$ |
| 13 | $\nabla[\Delta(a,b),\ \Delta(a,n(b))]$ | $\Delta[\nabla(a,b),\nabla(a,n(b))]$ |
| 14 | $\nabla[\Delta(n(a),b),\Delta(n(a),n(b))]$ | $\Delta[\nabla(n(a),b),\nabla(n(a),n(b))]$ |
| 15 | $\nabla[\Delta(a,b),$ $\Delta(n(a),b)]$ | $\Delta[\nabla(a,b),\nabla(n(a),b)]$ |
| 16 | $\nabla[\Delta(a,n(b)),$ $\Delta(n(a),n(b))]$ | $\Delta[\nabla(a,n(b)),\nabla(n(a),n(b))]$ |

$$\nabla[\Delta(a,b),\Delta(a,n(b)),\Delta(n(a),b),\Delta(n(a),n(b))] = 1 \quad (1)$$
$$\nabla[(b,n(b)),0,0) = 1, \text{ since } \nabla \text{ is nilpotent.}$$
$$\nabla[\Delta(a,b),\Delta(a,n(b)),\Delta(n(a),b)] = \nabla(a,b) \quad (3)$$
$$\nabla[(b,n(b)),0) = 1= \nabla(1,b)$$
$$\nabla[\Delta(a,b),\Delta(n(a),n(b))] = \Delta[\nabla(a,n(b)),\nabla(n(a),b)] \quad (11)$$
$$\nabla(b,0)] = b=\Delta(1,b)$$
$$\nabla[\Delta(a,n(b)),\Delta(n(a),n(b))] = \Delta(\nabla(a,n(b)),\nabla(n(a),n(b)) \quad (16)$$
$$\nabla(n(b),0) = n(b)=\Delta(1,n(b)$$

Proposition 9.3 identifies a second necessary condition for the equality of FDCF and FCCF, i.e., idempotency. Recall, the only idempotent t-norm and t-conorm are min and max, respectively and neither min has zero divisors nor max is nilpotent. Therefore, the following impossibility result can be stated.

**Carollary 9.3.** If $(\Delta,\nabla,n)$ is a De Morgan triple such that $\Delta$ is a t-norm, $\nabla$ is a t-conorm and n is a negation satisfying $n(0)=1$ and $n(1)=0$, then FDCF(.) cannot be equal to FCCF(.) (Bilgic, 1995)

**Proof:** Follows from the fact that there does not exist a t-norm that has zero divisors and idempotent and from Propositions 9.2 and 9.3.

Corollary 9.3 verifies Türkşen(1986)'s work by giving a formal proof for the non-equivalence of two Boolean normal forms in fuzzy logics. The result is very general and applies to all (not necessarily continuous) t-norms and t-conorms with any negation function that satisfies boundary conditions. This is again an important results that shows the break down of Boolean equivalence in fuzzy set and logic theory.

Once it is established that the equality cannot be obtained, whether there is containment or not is investigated in the next subsection for continuous Archimedean t-norms.

## 9.7. Containment for continuous Archimedean t-norms

In order to show FDCF(.)$\subseteq$FCCF(.) for all sixteen combined concepts with a particular De Morgan triple, it is sufficient to show that the following are satisfied:

$$\nabla[\Delta(a,b),\Delta(a,n(b)),\Delta(n(a),b))] \leq \nabla(a,b) \tag{9.19}$$
$$\nabla[\Delta(a,b),\Delta(n(a),n(b))] \leq \Delta[\nabla(a,n(b)),\nabla(n(a),b))] \tag{9.20}$$
$$\nabla[\Delta(a,b),\Delta(a,n(b))] \leq \Delta[\nabla(a,b),\nabla(a,n(b))] \tag{9.21}$$

Note that for rows (1) and (2) of Table 9.2 $\mu_{FDCF}(.) \geq \mu_{FCCF}(.)$ is always satisfied for all t-norms and it is sufficient to show that either one of

$$\nabla(\Delta(a,b),\Delta(a,n(b))) \leq a \tag{9.22}$$

$$\nabla(\Delta(a,b),\Delta(n(a),n(b))) \leq a \tag{9.23}$$
$$\nabla(\Delta(a,n(b)),\Delta(n(a),b)) \leq a \tag{9.24}$$

holds for all a, b$\in$[0,1] in order 9.19 to be satisfied. Condition 9.22 is stated by Türkşen (1986).

**Theorem 9.5.** If $(\Delta,\nabla,n)$ is a strong De Morgan Triple then FDCF(.)$\subseteq$FCCF(.) for the sixteen combined concepts. (Bilgic, 1995)
The proof is straightforward but tedious and is therefore given in Appendix 9. An immediate consequence is the following corollary.

**Corollary 9.4.** If $\varphi(z) = z$ for all $z \in[0,1]$ then FDCF(.)$\subseteq$FCCF(.) for Lukasiewicz triples.
The same result can be shown for strict De Morgan triples.

**Theorem 9.6.** If $(\Delta,\nabla,n)$ is a strict De Morgan Triple then FDCF(.)$\subseteq$FCCF(.) for the sixteen combined concepts. (Bilgic, 1995)
The proof is given in Appendix 9. An immediate consequence is the following corollary.

**Corollary 9.5.** If $\varphi(z) = z$ for all $z \in [0,1]$ then FDCF(.)$\subseteq$FCCF(.) for Algebraic triples.The results of Corollaries 9.4 and 9.5 are also obtained in Türkşen (1986) as special cases of families of operators. Note that one does not have to resort to twice diffrentiability in order to prove Theorem 9.6 as is required in Türkşen (1986).

Since a valued relation can formally be defined as a fuzzy subset of the Cartesian product AxA, an interval-valued relation can be defined as an interval-valued fuzzy subset of AxA. This concept is central to some of the developments in interval-valued Type 2 theory.

**Definition 9.5.** If a valued relation is constructed from other valued relations by means of linguistic connectives (as "AND", "OR", "NOT" constructs as well as the other thirteen constructs), then it is defined as an interval-valued fuzzy relation (IVFR) as

IVFR(.) = [FDCF(.), FCCF(.)], (Türkşen, 1986-2002)

**Remark:** IVFR(.)'s are interval-valued Type 2 Fuzzy sets.

## 9.8. Combination of More Than Two Propositions

We next investigate, the containment of Fuzzy Canonical forms for any number of propositions. As an example consider A AND B AND C.

$$\text{FDCF(A AND B AND C)} = A \cap B \cap C \tag{9.25}$$

FCCF (A AND B AND C)

$$= (A \cup B \cup C) \cap (\bar{A} \cup B \cup C) \cap (A \cup \bar{B} \cup C) \cap (A \cup B \cup \bar{C}) \cap (\bar{A} \cup \bar{B} \cup C) \cap (\bar{A} \cup B \cup \bar{C}) \cap (A \cup \bar{B} \cup \bar{C}) \tag{9.26}$$

For this purpose, it is constructive first to show the equivalence of normal forms in the two-valued case (Bilgic, 1995). We start with A and B for which it is known that

$$\text{DNF(A AND B)} = A \cap B = (A \cup B) \cap (\bar{A} \cup B) \cap (A \cup \bar{B}) = \text{CNF(A AND B)} \tag{9.27}$$

Apply the following two steps to 9.27:

**Step 1.** Disjunct $(C \cap \bar{C})$ to each term in the right hand side. The equality remains valid since $(C \cap \bar{C}) = \varnothing$.

**Step 2.** Now conjunct $(C \cup \bar{A} \cup \bar{B})$ to both sides. Using the distributive law, the right hand side becomes the CNF of $(A \cap B \cap \bar{A})$ :

$$(A \cup B \cup C) \cap (\bar{A} \cup B \cup C) \cap (A \cup \bar{B} \cup C) \cap (A \cup B \cup \bar{C})$$
$$\cap (\bar{A} \cup \bar{B} \cup C) \cap (\bar{A} \cup B \cup \bar{C}) \cap (A \cup \bar{B} \cup \bar{C})$$

Again, using the distributive law, the left hand side becomes:

$$(A \cap B) \cap (C \cup \bar{A} \cup \bar{B}) \equiv (A \cap B \cap C) \cup (A \cap B \cap \bar{A}) \cup (A \cap B \cap \bar{B})$$
$$\equiv (A \cap B \cap C)$$

which is the DNF of $(A \cap B \cap C)$.

Therefore, we have DNF(A AND B AND C) = CNF(A AND B AND C).
Our concern is whether the containment of FDCF in FCCF still holds in the fuzzy logic under a similar construction. If the connectives $(\cap,\cup,-)$ are modelled by a De Morgan triple $(\Delta,\nabla,n)$ it is known that (at least for the triples utilized in Theorems 9.5, 9.6 and the ones given in (Türkşen, 1986)):

$$\Delta(a,b) \le \Delta[\nabla(a,b),\nabla(n(a),b),\nabla(a,n(b))].$$

Disjuncting anything to the right hand side does not decrease the membership value (by the definition of a t-conorm). Therefore, in order to show containment between $\mu_{FDCF}$ and $\mu_{FCCF}$, conjuct a $c \in [0,1]$ to both sides and disjunct $\Delta(c,n(c))$ to each of the first three terms of the right hand side and disjunct $\nabla(n(a),n(b))$ to the last one, i.e., to c.

We set:

$$\Delta(a,b,c) = \mu_{FDCF(A \text{ AND } B \text{ AND } C)}(a,b,c)$$

$$= \Delta[\nabla(a,b),\nabla(n(a),b),\nabla(a,n(b)),c]$$
$$\le \Delta[\nabla(\nabla(a,b),\Delta(z,n(c))),\nabla(\nabla(n(a),b),\Delta(c,n(c))),$$
$$\nabla(\nabla(a,n(b)),\Delta(c,n(c))),\nabla(c,\nabla(n(a),n(b)))]$$

$$\le \Delta[\nabla(\nabla(a,b,c),\nabla(a,b,n(c)),\nabla(n(a),b,c),\nabla(n(a),b,n(c)),$$
$$\nabla(a,n(b),c),\nabla(a,n(a),n(c)),\nabla(n(a),n(b),c)]$$

$$= \mu_{FCCF(A \text{ AND } B \text{ AND } C)}(a,b,c)$$

where the first and second inequalities follows from $\nabla(p,q) \ge \max(p,q)$ and $\Delta(p,n(p) \ge 0$ for all $p,q \in [0,1]$. Notice that one does not have to resort to distributivity in order to show the containment of FDCF in FCCF.
In fact, this construction can be inductively increased to any number of propositions:

**Theorem 9.7.** If $(\Delta,\nabla,n)$ is a strict De Morgan Triple constructed from a continuous Archimedean t-norm, t-conorm and a strong negation then
$FDCF(.) \subseteq FCCF(.)$
for any number of propositions.

**Proof:** By construction as discussed in the text.
This result is important in the sense that, the interval-valued fuzzy set construction does not yield a higher order fuzziness than second order, i.e., a Type 2 fuzzy set. In other words, the interval-valued fuzzy set corresponding to complex sentences, is always interval-valued) but not lattice valued as in L-fuzzy-sets).

**Appendix 9.1 (Bilgic, 1995)**

**Theorem 9.5.** If $(\Delta,\nabla,n)$ is a strong De Morgan Triple then
$FDCF(.) \subseteq FCCF(.)$ for all the sixteen basic expressions of the basic combination of concepts introduced in Table 6.4 and which have been retstated in Tables 9.1 and 9.2 for set and membership domain expressions, respectively. For the Proof of the Theorem, we restate the generating functions for strong De Morgan Triples which are characterized by a continuous, strictly decreasing function

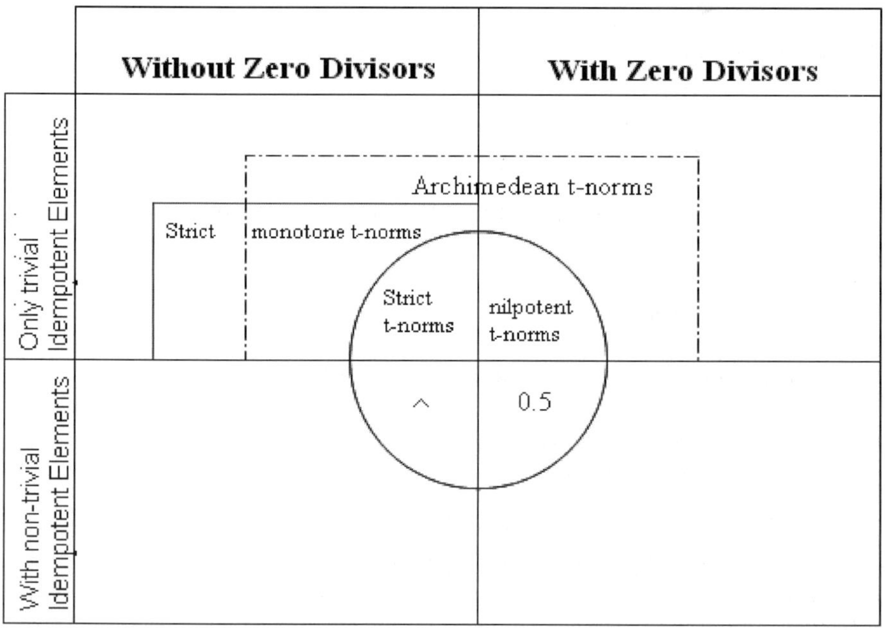


**Figure 9.1.** Different classes of t-norms, $\nabla$, where $\wedge$ is the minimum and 0.5 are the ordinal sums of t-norms (Reprinted from: E.P. Klement, R. Mesiar, E. Pap "Triangular Norms", page 35, Fig. 2.3 © Kluwer Academic Publishers, Dordrecht, Boston, London, 2000).

g: [0,1]→ $\overline{\Re}^+$ with g(1)=0, g(0)<+∞ as follows:

$$\Delta(x,y) = g^{-1}(\min\{g(x) + g(y), g(0)\}) \tag{9.13}$$

$$n(x) = g^{-1}(g(0) - g(x) \tag{9.14}$$

$$\nabla(x,y) = g^{-1}(\max\{g(x) + g(y)- g(0), 0\}) \tag{9.15}$$

We will show that (9.19), (9.20), and (9.21) are satisfied for if (9.22), (9.23) and (9.24) are satisfied. For this purpose, first we rewrite the left hand sides of (9.22) and (9.23) in terms of (9.13), (9.14), (9.15) as

$$\begin{aligned}\nabla(\Delta(x,y),\Delta(x,n(y)) &= \qquad (9.22')\\ &= g^{-1}(\max\{g(g^{-1}(\min\{g(x) + g(y), g(0)\}))\\ &\qquad +\{g(g^{-1}(\min\{g(x) + \{g(g^{-1}(g(0) - g(y))), g(0)\}))\\ &\qquad - g(0), 0\})\\ &= g^{-1}(\max\{\min\{g(x) + g(y), g(0)\}\\ &\qquad +\min\{g(x) + g(0) - g(y), g(0)\}\\ &\qquad - g(0), 0\})\end{aligned}$$

$$\begin{aligned}&\nabla(\Delta(x,y),\Delta(n(x),n(y)) \qquad (9.23')\\ &= g^{-1}(\max\{g(g^{-1}(\min\{g(x) + g(y), g(0)\}))\\ &\qquad +\{g(g^{-1}(\min\{g(g^{-1}(g(0) - g(x)))\\ &\qquad +g(g^{-1}(g(0) - g(y))), g(0)\}\\ &\qquad - g(0), 0\})\\ &= g^{-1}(\max\{\min\{g(x) + g(y), g(0)\}\\ &\qquad +\min\{g(0) - g(x) + g(0) - g(y), g(0)\}\\ &\qquad - g(0), 0\})\\ &= g^{-1}(\max\{\min\{g(x) + g(y), g(0)\}\\ &\qquad +\min\{2g(0) - g(x) - g(y), g(0)\}\\ &\qquad - g(0), 0\})\end{aligned}$$

As well, we rewrite the right hand side of (9.20) and (9.21) in terms of (9.13), (9.14) and (9.15) as

$$\begin{aligned}&\Delta(\nabla(x,n(y)),\nabla(n(x),y) \qquad (9.20')\\ &= g^{-1}(\min\{g(g^{-1}(\max\{g(x) + g(g^{-1}(g(0) - g(y)))\\ &\qquad - g(0), 0\})\\ &\qquad + g(g^{-1}(\max\{g(g^{-1}(g(0) - g(x)))\\ &\qquad + g(y) - g(0), 0\})\\ &\quad ,g(0)\})\\ &= g^{-1}(\min\{\max\{g(x) + g(0) - g(y) - g(0),0\}\\ &\quad +\max\{g(0) - g(x) + g(y) - g(0),0\}\\ &\quad ,g(0)\})\\ &= g^{-1}(\min\{\max\{g(x) - g(y),0\}\\ &\quad +\max\{g(y) - g(x),0\},g(0)\})\end{aligned}$$

$\Delta(\nabla(x,y),\nabla(x,n(y))$
(9.21′)
$= g^{-1}(\min\{g(g^{-1}(\max\{g(x) + g(y) - g(0), 0\}))$
$+g(g^{-1}(\max\{g(x) + g(g^{-1}(g(0) - g(y))$
$- g(0),0\}))$
$,g(0)\})$
$= g^{-1}(\min\{\max\{g(x) + g(y) - g(0), 0\}$
$+\max\{g(x) - g(0) - g(y) - g(0), 0\}$
$,g(0)\})$
Next we need to show that

$$\nabla(\Delta(x,y),\Delta(x,n(y)) \leq x, \tag{9.22}$$

$$\nabla(\Delta(x,y),\Delta(x,n(y)) \leq \Delta(\nabla(x,y),\nabla(x,n(y))) \tag{9.21}$$

and

$$\nabla(\Delta(x,y),\Delta(n(x),n(y)) \leq \Delta(\nabla(x,n(y)),\nabla(n(x),y)) \tag{9.20}$$

For this purpose, there are four cases, we need to consider:

(i) (a) $g(0) \geq g(x) + g(y)$, and
(b) $g(x) - g(y) > 0$
(ii) (a) $2g(0) \geq g(x) + g(y) > g(0)$, and
(b) $0 \leq g(y) - g(x) \leq g(0)$
(iii) (a) $2g(0) \geq g(x) + g(y) > g(0)$, and
$g(y) - g(x) \geq g(0)$
(iv) $g(0) \geq g(x) + g(y)$, and
$g(y) - g(x) \leq 0$
Case (i) $g(0) \geq g(x) + g(y)$ and
$g(x) - g(y) > 0$

$\nabla(\Delta(x,y),\Delta(x,n(y))$

$=(A\text{-}9\text{-}1) = g^{-1}(\max\{g(x) + g(y), 0\})$
$= g^{-1}(g(x) + g(y))$
$\leq g^{-1}(g(x))$
$= x$

Hence,

$\nabla(\Delta(x,y),\Delta(x,n(y)) \leq x$
Since we have

$\nabla(\Delta(x,y),\Delta(x,n(y)) \le x$

and therefore, we get

$\nabla(\Delta(x,y),\Delta(x,n(y)),\Delta(n(x),y)) \le \nabla(x,y)$

Hence (9.16) holds. From above since we have

$$\begin{aligned}
&\nabla(\Delta(x,y),\Delta(x,n(y)) = g^{-1}(g(x) + g(y)) \\
&\le g^{-1}(g(x)\ g(y)) \\
&= g^{-1}(\min\{g(x) - g(y), g(0)\})
\end{aligned}$$

Under this case, we also have:

$$\begin{aligned}
&\Delta(\nabla(x,y),\nabla(x,n(y)) = \text{A-9-4} \\
&= g^{-1}(\min\{g(x) - g(y), g(0)\})
\end{aligned}$$

Therefore, we get:

$$\nabla(\Delta(x,y),\Delta(x,n(y)) \le \Delta(\nabla(x,y),\nabla(x,n(y))) \tag{9.21}$$

Next, we show that

$$\nabla(\Delta(x,y),\Delta(n(x),n(y)) = \tag{9.23'}$$

$$\begin{aligned}
&= g^{-1}(\max\{g(x) + g(y), 0\}) \\
&= g^{-1}(g(x) + g(y)) \\
&\le g^{-1}(g(x) - g(y))
\end{aligned}$$

But, we have

$\Delta(\nabla(x,y),\nabla(x,n(y)) =$ (Right hand side of (9.21))

$= g^{-1}(g(x) - g(y))$ under Case (i)

Therefore, it is shown that

$$\nabla(\Delta(x,y),\Delta(n(x),n(y)) \le \Delta(\nabla(x,y),\nabla(x,n(y))) \tag{9.20}$$

Case (ii) $2g(0) \ge g(x) + g(y) > g(0)$, and $0 \le g(y) - g(x) \ge g(0)$

Under these conditions, we get
$\nabla(\Delta(x,y),\Delta(x,n(y)) =$ (Left hand side of (9.22))

$= g^{-1}(\max\{g(0) + g(x) - g(y), 0\})$
$= g^{-1}(g(0) + g(x) - g(y))$
$\leq g^{-1}(g(x))$
$= x.$

Therefore, as shown before 9.19 holds.

As well, from above, we got

$\nabla(\Delta(x,y),\Delta(x,n(y))= g^{-1}(g(0) + g(x) - g(y))$
$\leq g^{-1}(g(x) + g(y) - g(0))$
$= g^{-1}(\min\{g(x)+g(y) - g(0), g(0)\})$
$= \Delta(\nabla(x,y),\nabla(x,n(y)))$

Note that under case (ii):

$\Delta(\nabla(x,y),\nabla(x,n(y))) =$ (Right hand side of (9.21))

$= g^{-1}(\min\{g(x) + g(y) - g(0), g(0)\})$

Therefore, we get

$\nabla(\Delta(x,y),\Delta(x,n(y)) \leq \Delta(\nabla(x,y),\nabla(x,n(y)))$

Hence (9.21) holds.

Next, we investigate

$\nabla(\Delta(x,y),\Delta(n(x),n(y)) =$ (Left hand side of (9.21))
$= g^{-1}(\max\{2g(0) - g(x) - g(y), 0\})$
$= g^{-1}(2g(0) - g(x) - g(y))$
$\leq g^{-1}(g(y) - g(x)) = g^{-1}(\min\{g(y) - g(x), g(0)\})$
$= \Delta(\nabla(x,n(y)),\nabla(n(x),y))$

Note that under case (ii)

$\Delta(\nabla(x,n(y)),\nabla(n(x),y)) =$ (Right hand side of (9.20))
$= g^{-1}(\min\{g(y) - g(x), g(0)\})$

Therefore, we get

$\nabla(\Delta(x,y),\Delta(n(x),n(y)) \leq \Delta(\nabla(x,n(y)),\nabla(n(x),n(y))$
Hence (9.20) holds.

Case (iii) Under conditions:

(a) $2g(0) \geq g(x) + g(y) > g(0)$, and
(b) $g(y) - g(x) < 0$
We get

$\nabla(\Delta(x,y),\Delta(x,n(y)) =$ (Left hand side of (9.21))
$= g^{-1}(\max\{g(0), 0\})$
$= g^{-1}(g(0))$
$= 0 \leq x$

Thus again we have (9.19) holds. On the other hand,

$\Delta(\nabla(x,y),\nabla(x,n(y)) =$ (Right hand side of (9.21))
$= g^{-1}(\min\{g(x) + g(y) - g(0))$
$+ g(x) - g(y), g(0)\})$
$\geq 0$

Therefore, we get
$\nabla(\Delta(x,y),\Delta(x,n(y)) = 0$
$\leq \Delta(\nabla(x,y),\nabla(x,n(y))$.

Thus, again (9.21) holds.

Next consider

$\nabla(\Delta(x,y),\Delta(n(x),n(y)) =$ (Right hand side of (9.20))
$= g^{-1}(\max\{2g(0) - g(x) - g(y), 0\})$
$= g^{-1}(2g(0) - g(x) - g(y))$
$\leq g^{-1}(g(x) - g(y))$
$= \Delta(\nabla(x,n(y)),\nabla(n(x),y)$
$= g^{-1}(g(x) - g(y))$

Therefore, we get

$$\nabla(\Delta(x,y),\Delta(n(x),n(y)) \leq \Delta(\nabla(x,n(y)),\nabla(n(x),y)$$

Thus, (9.20) holds.

Case (iv) under the condition
(a) $g(0) \geq g(x) + g(y)$ and
(b) $g(x) - g(y) \leq 0$,
We get

$\nabla(\Delta(x,y),\Delta(x,n(y)) =$ (Left hand side of (9.22))
$= g^{-1}(2g(x)) \leq g^{-1}(g(x)) = x$

Thus we have (9.19) holds. Furthermore

$\nabla(\Delta(x,y),\Delta(x,n(y))) \leq x \leq 1 = g^{-1}(0) = \Delta(\nabla(x,y),\nabla(x,n(y)))$

Thus, we get (9.21) holds.

Next, we assess

$\nabla(\Delta(x,y),\Delta(n(x),n(y))) =$ (Left hand side of (9.20))
$= g^{-1}(g(x)) + g(y)) \leq g^{-1}(g(y) - g(x))$
$= \Delta(\nabla(x,n(y)),\nabla(n(x),y)) = g^{-1}(g(y)) - g(x))$

Thus, we have (9.20) holds. As a result, we have shown that in four possible cases, the inequalities (9.19), (9.20) and (9.21) hold. Therefore, FDCF(.)⊆FCCF(.) for all sixteen protoforms of the combination of concepts for all strong De Morgan Triples (Δ,∇,n).

## Appendix 9.2 (Bilgic, 1995)

**Theorem 9.6.** If (Δ,∇,n) is a strict De Morgan Triple, then FDCF(.) ⊆ FCCF(.) for all the sixteen basic protoforms of the basic combination of concepts introduced in Table 6.4 and which have been retstated in Tables 9.1 and 9.2 for set and membership domain expressions, respectively.

Let us recall that strict De Morgan Triples (Δ,∇,n) are generated by the automorphism,φ such that

$\Delta(x,y) = \varphi^{-1}(\varphi(x)\,\varphi(y))$ (9.7)
$\nabla(x,y) = \varphi^{-1}(\varphi(x) + \varphi(y)- \varphi(x)\,\varphi(y))$ (9.8)
$n(x) = \varphi^{-1}(1 - \varphi(x)$ (9.9)

Let us rewrite the following expressions in terms of this automorphism:

$\nabla(\Delta(x,y),\Delta(x,n(y))) = \varphi^{-1}(\varphi(x)\,\varphi(y) + \varphi(x)\,\varphi(y)) - \varphi(x)^2\,\overline{\varphi}(y))$ (A-9-5)

Note that, we have the following notational simplification

$n(y) = \varphi^{-1}(1 - \varphi(y))$
$\varphi(n(y)) = \varphi\,\varphi^{-1}(1-\varphi(y)) = 1- \varphi(y)$
$= \overline{\varphi}(y)$
$\nabla(\Delta(x,y),\Delta(n(x),n(y))) = \varphi^{-1}(\varphi(x)\,\varphi(y) + \overline{\varphi}(x)\,\overline{\varphi}(y))$

$(\overline{\varphi}(x)+\varphi(y) - \overline{\varphi}(x)\varphi(y))$ (A-9-7)

$\Delta(\nabla(x,n(y)),\nabla(n(x),y)) = \varphi^{-1}(\varphi(x)+\overline{\varphi}(y) - \varphi(x)\overline{\varphi}(y)$
$- \varphi(y)\overline{\varphi}(x))\overline{\varphi}(x)$ (A-9-6)

$\Delta(\nabla(x,y),\nabla(x,n(y))) = \varphi^{-1}((\varphi(x) +\varphi(y) - \varphi(x)\varphi(y))$
$(\varphi(x)+\overline{\varphi}(y) - \varphi(x)\overline{\varphi}(y)))$ (A-9-8)

Thus, we need to show that (A-9-5) ≤ x(A-9-5) ≤ (A-9-8) and (A-9-6) ≤(A-9-7).

To show (A-9-7) ≤ x, it is sufficient to show that
$\varphi(x)\,\varphi(y) - \varphi(x)\,\overline{\varphi}(y) - \varphi(x)^2\,\varphi(y)\,\overline{\varphi}(y) \le \varphi(x)$

or equivalentlywhich obviously holds for all x, y∈[0,1].

To show that (A-9-6) ≤ (A-9-7), let us rewrite (A-9-7) as

$(A\text{-}9\text{-}7) = \varphi^{-1}(\varphi(x)\overline{\varphi}(x)+\varphi(x)\varphi(y) - \varphi(x)\overline{\varphi}(x)\varphi(y)$
$+\overline{\varphi}(x)\,\overline{\varphi}(y)+\varphi(y)\overline{\varphi}(y) - \overline{\varphi}(x)\varphi(y)\overline{\varphi}(y)$
$- \varphi(x)\overline{\varphi}(x)\varphi(y) - \overline{\varphi}(x)\varphi(y)\overline{\varphi}(y)$
$+\varphi(x)\varphi(y) - \varphi(x)\overline{\varphi}(y))$

$(A\text{-}9\text{-}6) = \varphi^{-1}(\varphi(x)\varphi(y) + \overline{\varphi}(x)\,\overline{\varphi}(y) - \varphi(x)\,\varphi(y)\overline{\varphi}(x)\,\overline{\varphi}(y)$

If we cancel $\varphi(x)\varphi(y) + \overline{\varphi}(x)\,\overline{\varphi}(y)$ from both A-9-7 and A-9-6 and add $+\varphi(x)\varphi(y)\overline{\varphi}(x)\,\overline{\varphi}(y)$ to both, then by rearranging, we get

$0 \le \varphi(x)\overline{\varphi}(x)+\varphi(x)\varphi(y) - \varphi(x)\varphi(y)\overline{\varphi}(x)$
$+\varphi(y)\overline{\varphi}(y) - \overline{\varphi}(y)\varphi(y)\overline{\varphi}(y)$
$- \varphi(x)\overline{\varphi}(x)\,\overline{\varphi}(y) - \overline{\varphi}(y)\varphi(y)\overline{\varphi}(y)$
$+ 2\varphi(x)\varphi(y)\overline{\varphi}(x)\,\overline{\varphi}(y)$

$0 \le \varphi(x)\overline{\varphi}(x)+\varphi(x)\varphi(y)\,\overline{\varphi}(x)$
$- \varphi(x)\varphi(y)\overline{\varphi}(x)+\varphi(x)\varphi(y)\overline{\varphi}(y)$
$+ 2\varphi(x)\varphi(y)\overline{\varphi}(x)\,\overline{\varphi}(y))$

$0 \le \varphi(x)\overline{\varphi}(x)+\varphi(x)\varphi(y)\,\overline{\varphi}(y)\,\overline{\varphi}(x)$

Thus (A-9-6) ≤ (A-9-7) holds.To show that

(A-9-5) ≤ (A-9-8)

Again cancel the common terms and rearrange, then get

$0 \le \varphi(y)\overline{\varphi}(y)(1 - 2\varphi(x) + 2\varphi(x)^2)$
$= \varphi(y)\overline{\varphi}(y)[(1\text{-}\varphi(x)^2 + \varphi(x)^2]$

which holds for all x,y∈[0,1] for the automorphisim φ on the unit interval.

## References

1. Alsina C., E. Trillas and I. Valverde (1983), "On Some Logical Connectives for Fuzzy Sets Theory", *Journal of Mathematical Analysis and Applications,* 93, 15-26.
2. Bilgic, T. (1995), Measurement-Theoretic Frameworks for Fuzzy Set Theory with Applications to Preference Modelling, PhD thesis, University of Toronto, Dept. of Industrial Engineering Toronto, Ontario M5S 1A4 Canada.(supervisor, I.B. Türkşen)
3. Dubois, D., H. Prade (1980), Fuzzy Sets and Systems: Theory and Applications, A*cademic Press*, New York.
4. Ling, C.H. (1965), "Representation of Associative Functions", Publicationes Mathematicae Debrecen, 12, 189-212.
5. Ovchinnikov, S., M. Roubens (1991), "On Strict Preference Relations", *Fuzzy Sets and* Systems 43, 319-326.
6. Schweizer, B., A. Sklar (1983), *Probabilistic Metric Spaces,* North Holland, Amsterdam.
7. Türkşen I.B. (1986), "Interval-Valued Fuzzy Sets based on Normal Forms", *Fuzzy Sets and Systems,* 191-210.
8. Weber, S. (1983), "A General Concept of Fuzzy Connectives, Negations and Implications Based on t-norms and t-conorms, *Fuzzy Sets and Systems* 11, 115-134.

# Chapter 10

# CONSEQUENCES OF $\{D_{[0,1]}, V_{\{0,1\}}\}$ THEORY

There are a number consequences and new insight that come to light as a result of the Fuzzy Canonical Forms. We identify some of these in this section under the headings of:

Laws of Fuzzy Middle and Fuzzy Contradiction
Laws of Fuzzy Conservation
Fuzzy Re-Affirmation
Fuzzy Re-Negation
Fuzzy Modus Ponens
Clearly, there ought to be many other consequences. These are left for future expositions.

This chapter is a modified and improved version of Türkşen, Kandel and Zhang (1998, 1999). In general, these consequences are examples of the general break-down of the classical equivalences in fuzzy theory. In particular, it leads to our proposal for the reassessment of the axions of classical theory in terms of "Meta-Linguistis Axioms" which are discussed in Chapter 17.

## 10.1. Laws of Middle and Contradiction

Let us now investigate the meta-linguistic expressions of the Middle, i.e. "A OR c(A)" and the Contradiction, i.e. "A AND c(A)" for both the Boolean and Fuzzy Sets.

### *10.1.1. Boolean CNF' s and DNF*

In two-valued set and logic theory, the membership values {a,n(a)} of the set symbols {A,c(A)} and the truth assignments {T,F} are all mapped to the lattice {0,1}. This mapping simplifies knowledge representation. However, it creates certain paradoxes, e.g., Russel's paradox, Barber's paradox and Gödel's paradox, etc., with slightly differing conceptual models and interpretations. The digital computer that operates with binary strings of 0's and 1's stands as the ultimate emblem of the black and white theory and its triumpt over western scientific mind. The black and white theory was canonized by Aristotle (384-322, B.C.).

It is based on a reductionist philosophy that excludes the middle ground of the shades of gray rather arbitrarily.

Let us now recall the predicate expressions we have stated in Section 8.1 above. Let us now substitute c(A) in place of B in those expressions. For the combination of the predicate expressions (1.1.1), (1.2.1), (2.1.1), (2.2.1), and (1.1.2), (1.2.2), (2.1.2), (2.2.2), we can form a "Truth Table" as shown in Table 8.2 for the meta-linguistic expression known as "Excluded Middle", i.e., "A OR c(A)". It should be noted that the combined predicate expressions of the meta-linguistic expression "A OR c(A)" for the first four rows are:

"$x \in X_1$ is A, with a = 1, is T" or "$x \in X_1$ is c(A), with n(a) = 0, is T", is T; (1.1.1)
"$x \in X_1$ is A, with a = 1, is T", or "$x \in X_1$ is c(A), with n(a) = 1, is F", is T;(1.2.1)
"$x \in X_1$ is A, with a = 0, is F", or "$x \in X_1$is c(A), with n(a) = 0, is T", is T; (2.1.1)
"$x \in X_1$ is A, with a = 0, is F", or "$x \in X_1$is c(A), with n(a) = 1, is F", is F; (2.2.1)
where "isr" is replaced by "is" since $a \in \{0,1\}$.

The next four rows can be written similarly and, thus, the Table 10.1 can be formed.

**Table 10.1.** The reconstructed "Truth Table" of "A OR c(A)" where a, $n(a) \in [0, 1]$.

| A | | | c(A) | | | A OR c(A) |
|---|---|---|---|---|---|---|
| (1.1.1) | a = 1 | T | (2.1.1) | n(a) = 0 | T | T |
| (1.1.1) | a = 1 | T | (2.2.1) | n(a) = 1 | F | T |
| (1.2.1) | a = 0 | F | (2.1.1) | n(a) = 0 | T | T |
| (1.2.1) | a = 0 | F | (2.2.1) | n(a) = 1 | F | F |
| (1.1.1)′ | a = 0 | T | (2.1.1)′ | n(a) = 1 | T | T |
| (1.1.1)′ | a = 0 | T | (2.2.1)′ | n(a) = 0 | F | T |
| (1.2.1)′ | a = 1 | F | (2.1.1)′ | n(a) = 1 | T | T |
| (1.2.1)′ | a = 1 | F | (2.2.1)′ | n(a) = 0 | F | F |

(Note (1.1.1)′ is a duplicate of (1.1.1),…, (2.2.1)′ is a dublicate of (2.2.1), etc.)

*10.1.2. LEM and CONTRADICTION*

With the execution of the canonical normal, form derivation algorithm given in Section 7.2.1 and interpret as in Appendix 10.1, we can now write the DNF and CNF expression of "A OR c(A)" in prepositional form as follows:

$$DNF_1^2(A \text{ OR } c(A)) = (A \cap c(A)) \cup (A \cap A) \cup (c(A) \cap c(A))$$

(Idempotency)

$$= (A \cap c(A)) \cup A \cup (c(A)) \quad \text{(LEM)}$$
$$= (A \cap c(A)) \cup X \quad \text{(LC)}$$
$$= X$$

$$CNF_1^2(A \text{ OR } c(A)) = c\{c(A) \cap A\} \quad \text{(Involutive, Negation)}$$
$$= A \cup c(A) \quad \text{(LEM)}$$
$$= X$$

Hence, it is found as it should be $DNF_1^2(A \text{ OR } c(A)) = CNF_1^2(A \text{ OR } c(A)) = X$. It should be noted that we started with the new construction of Fuzzy Truth Table and obtained the classical result.

In a similar manner to the formation of Table 10.1, we can reconstruct the Truth Table, Table 10.2 as an extraction of Table 8.5, for the meta-linguistic expression known as "Contradiction", i.e., "A AND c(A)", and derive the DNF and CNF as follows, again eliminating the duplicate, we get:

$$DNF_1^2(A \text{ AND } c(A)) = A \cap c(A) \quad \text{(LC)}$$
$$= \varnothing$$

$$CNF_1^2 \; (A \text{ AND } c(A))$$
$$= c\{(A \cap A) \cup (c(A) \cap c(A)) \cup (c(A) \cap A)\} \quad \text{(Involutive Negation)}$$
$$= (c(A) \cup c(A)) \cap (A \cup A) \cap (A \cup c(A)) \quad \text{(Idempotency)}$$
$$= (c(A) \cap A) \cap (A \cup c(A)) \quad \text{(LEM)}$$
$$= (c(A) \cap A) \cap X \quad \text{(I)}$$
$$= (c(A) \cap A) \quad \text{(LC)}$$
$$= \varnothing$$

Hence, it is found as it should be $DNF_1^2(A \text{ AND } N(A)) = CNF_1^2(A \text{ AND } N(A)) = \varnothing$.

*10.1.3. Fuzzy DCF' s and CCF' s*

When we relax the restriction of the reductionist philosophy and allow the shades of gray to exist between the black and white, we receive the infinite (fuzzy) valued set theory, where the set symbols {A,c(A)} become values (labels) of linguistic terms such as "the daylight hours" and "not the daylight hours" as in our example of Chapter 2. This relaxation, i.e., letting a, n(a) be in [0,1] resolves the Russel' s paradox and its other varieties stated earlier.

For the infinite valued case, we should again recall the predicate expressions of Section 8.1 from (1.1.1) to (2.2.2) and substitute c(A) in place of B and combine them to form the "Truth Table" Table 10.3 first and write FCCF and FDCF expressions of "A OR c(A)" for infinite (fuzzy) valued set and two-valued logic. It is to be noted that there is a one-to-one correspondence between Table 10.2 and 10.3. The only difference is for the two-valued case, we identify the values of a, and n(a) as {0,1}, i.e., precisely, but in infinite-valued case, we could not identify the values of a and n(a) precisely as "0" or "1" but as degrees of membership values in [0,1]. If we wanted to form a Truth Table to identify the crisp values of a, $n(a) \in [0,1]$ for all possible cases, then we would have to form a Truth Table with infinitely many rows. This is an impossibility. This is why sometime ago it was stated that the construction of Fuzzy Truth Table is impossible [Marinos, 1969]. (A long this line of research Schwede and Kandel (1977) proposed the contruction of "Fuzzy Maps".) But by construction, we know that in Table 10.3, there are two rows that are for the verification of $a \geq n(a)$ to be true, T, and there are two rows where $a \geq n(a)$ is denied, i.e., falsified, as well there are two rows where $a < n(a)$ is verified as true, T, and there are two rows where $a < n(a)$ is denied, i.e., falsified, F. In this manner, infinitely many rows of an infinite Fuzzy Truth Table is reduced to a finite Fuzzy Truth Table with eight entries only.

*10.1.4. Fuzzy Middle*

For further clarity of the construction of the Tables 10.3, the combined predicate expressions of the meta-linguistic expression "A OR c(A)" are stated below as an example for the first four rows of Table 10.3: (note that either the first four expressions or the last four expressions are applicable as explained in Chapter 8)

"$x \in X_1$ isr A, with $a \geq n(a)$, is T" or "$x \in X_1$ isr c(A), with $n(a) \leq a$, is T", is T.
"$x \in X_1$ isr A, with $a \geq n(a)$, is T", or "$x \in X_1$ isr c(A), with $n(a) > a$, is F", is T.
"$x \in X_1$ isr A, with $a < n(a)$, is F", or "$x \in X_1$ isr c(A), with $n(a) \leq a$, is T", is T.
"$x \in X_1$ isr A, with $a < n(a)$, is F", or "$x \in X_1$ isr c(A), with $n(a) > a$, is F", is F.

With this reconstruction of the Fuzzy Truth Table, Table 10.3, the canonical forms are obtained to be the same with the execution of the canonical form derivation algorithm stated in Section 7.2.1. Hence we get

$$FDNF_1^2(A \text{ OR } c(A)) = (A \cap c(A)) \cup (A \cap A) \cup (c(A) \cap c(A))$$

$$FCNF_1^2(A \text{ OR } c(A)) = (A \cup (c(A)).$$

But now, it is known that the crisp operators of the fuzzy set theory, i.e., t-norms and conorms, are non-idempotent. Therefore, $FDCF_1^2$ (A OR c(A)) and $FCCF_1^2$ (A OR c(A)) can not be simplified for the general class of fuzzy sets that are combined with t-norm and conorm operators. Furthermore, the laws of excluded middle and contradiction are no longer applicable since $A \cup c(A) \subseteq X$ and $A \cap c(A) \supseteq \varnothing$. But, with the generalization of Theorem 2 to fuzzy theory, we have $FDCF_1^2$ (A OR c(A)) $\subseteq$ $FCCF_1^2$ (A OR c(A)). Thus, it is clear that fuzzy middle exist to a matter of varying degrees between FDCF and FCCF.

**Table 10.2.** The reconstructed "Truth Table" of "A AND c(A)" where a, n(a)∈[0, 1].

| A | | | c(A) | | | A AND c(A) |
|---|---|---|---|---|---|---|
| (1.1.1) | a = 1 | T | (2.1.1) | n(a) = 0 | T | T |
| (1.1.1) | a = 1 | T | (2.2.1) | n(a) = 1 | F | F |
| (1.2.1) | a = 0 | F | (2.1.1) | n(a) = 0 | T | F |
| (1.2.1) | a = 0 | F | (2.2.1) | n(a) = 1 | F | F |
| (1.1.1)' | a = 0 | T | (2.1.1)' | n(a) = 1 | T | T |
| (1.1.1)' | a = 0 | T | (2.2.1)' | n(a) = 0 | F | F |
| (1.2.1)' | a = 1 | F | (2.1.1)' | n(a) = 1 | T | F |
| (1.2.1)' | a = 1 | F | (2.2.1)' | n(a) = 0 | F | F |

### *10.1.5. Fuzzy Contradiction*

In a similar manner, we form Tables 10.4 for the meta linguistic expression known as contradiction, i.e., "A AND c(A)". Then derive the FDCF and FCCF expressions from Table 8.5 with the execution of the normal form derivation algorithm as follows:

$FDNF_1^2$ (A AND c(A))= $A \cap c(A)$

$FCNF_1^2$ (A AND c(A)) $= (c(A) \cup c(A)) \cap (A \cup A) \cap (A \cup c(A))$

Once again, t-norms and conorms are non idempotent and $A \cap c(A) \supseteq \varnothing$ and $A \cup c(A) \subseteq X$ in general in all fuzzy set theories therefore there is no simplification possible in these terms. Hence, we find that $FDCF_1^2$ (A AND c(A)) $\subseteq$ $FCCF_1^2$ (A AND c(A)) with Theorem 9.5.

**Table 10.3.** The reconstructed "Truth Table" of "A OR c(A)" where a, n(a)$\in$[0, 1].

| A | | | c(A) | | | A OR c(A) |
|---|---|---|---|---|---|---|
| (1.1.2) | $a \geq n(a)$ | T | (2.1.2) | $n(a) \leq a$ | T | T |
| (1.1.2) | $a \geq n(a)$ | T | (2.2.2) | $n(a) > a$ | F | T |
| (1.2.2) | $a < n(a)$ | F | (2.1.2) | $n(a) \leq a$ | T | T |
| (1.2.2) | $a < n(a)$ | F | (2.2.2) | $n(a) > a$ | F | F |
| (1.1.2) | $a < n(a)$ | T | (2.1.2) | $n(a) > a$ | T | T |
| (1.1.2) | $a < n(a)$ | T | (2.2.2) | $n(a) \leq a$ | F | T |
| (1.2.2) | $a \geq n(a)$ | F | (2.1.2) | $n(a) > a$ | T | T |
| (1.2.2) | $a \geq n(a)$ | F | (2.2.2) | $n(a) \leq a$ | F | F |

## 10.2. Zadehean Fuzzy Middle and Contradiction

Zadeh' s Fuzzy Logic is a special subclass of fuzzy theories where the axioms of "Distributivity", "Absorption" and "Idempotency" are applicable both in propositional and predicate domain expressions. Thus, we have the following canonical forms for the Zadeh's Fuzzy Logic and its Fuzzy Middle and contradiction:

**Table 10.4.** The reconstructed "Truth Table" of "A AND c(A)" where a, n(a)$\in$[0, 1].

| A | | | c(A) | | | A AND c(A) |
|---|---|---|---|---|---|---|
| (1.1.2) | $a \geq n(a)$ | T | (2.1.2) | $n(a) \leq a$ | T | T |
| (1.1.2) | $a \geq n(a)$ | T | (2.2.2) | $n(a) > a$ | F | F |
| (1.2.2) | $a < n(a)$ | F | (2.1.2) | $n(a) \leq a$ | T | F |
| (1.2.2) | $a < n(a)$ | F | (2.2.2) | $n(a) > a$ | F | F |
| (1.1.2) | $a < n(a)$ | T | (2.1.2) | $n(a) > a$ | T | T |
| (1.1.2) | $a < n(a)$ | T | (2.2.2) | $n(a) \leq a$ | F | F |
| (1.2.2) | $a \geq n(a)$ | F | (2.1.2) | $n(a) > a$ | T | F |
| (1.2.2) | $a \geq n(a)$ | F | (2.2.2) | $n(a) \leq a$ | F | F |

$FDNF_1^2$ (A OR c(A)) = A $\cup$ c(A) $\subseteq$ X (Commutativity, Idempotency, Max Absorption)

$FCCF_1^2$ (A OR c(A)) = A $\cup$ c(A) $\subseteq$ X (Idempotency)

$FDNF_1^2$ (A AND c(A)) = A $\cap$ c(A) $\supseteq \varnothing$ (Idempotency)

$FCCF_1^2$ (A AND c(A)) = A $\cap$ c(A) $\supseteq \varnothing$ (Commutativity, Idempotency, Min Absorption)

It is to be observed that these expressions are equivalent in *form only* to the Boolean expressions of "A OR c(A)" and "A AND c(A)" in propositional and predicate forms. But in computational, i.e., numeric domain, the graphs of these expressions appear as shown in Figure 10.1 for ($\vee$, $\wedge$, -) De Morgan Triple.

Furthermore, it is to be realized that the expression of the "Fuzzy Middle" and the "Fuzzy Contradiction" are realized as a matter of degree specified by $a \vee n(a) \in [0,1]$, and $a \wedge n(a) \in [0,1]$, respectively (Figure 10.1). This turns out to be a Type 1 fuzzy set. This is quite possible to realize when, for example two different sensors, or the same sensor at two distinct points of time give readings one that affirms A and the other that affirms c(A).

## 10.3. Fuzzy Middle and Fuzzy Contradiction with t-norms and co-norms

In the t-norm-conorm class of fuzzy theories, let us investigate two well known subclasses formed by Algebraic Product and Sum, and Bold Intersection and Union for the case of "A OR c(A)" and "A AND c(A)" where the axioms of "Distributivity", "Absorption", and "Idempotency" are not applicable either in the propositional or in the predicate domain expressions.

*10.3.1. Algebraic Product and Sum*

In the predicate domain, the well-known operators of Algebraic Product and Sum are:

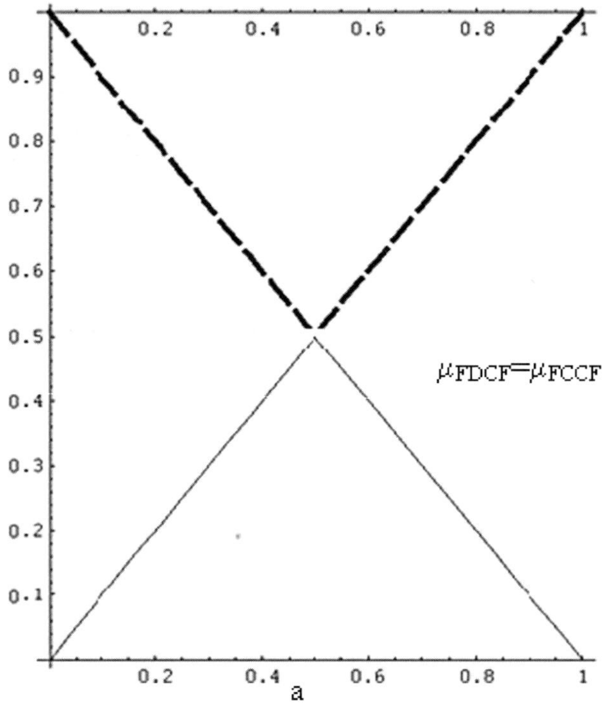


**Figure 10.1.** Type 1 fuzzy set of "A OR c(A)" (— — — ) and "A AND c(A) (———) for Zadehean connectives.

$\Delta(a,b) = a\odot b = ab$ and and $\nabla(a,b) = a \oplus b = a+b-ab$ where $a,b \in [0,1]$ are the generic membership values for every $x \in X$.

We rewrite the propositional expressions obtained in Section 8.3.2 and we get:

$\mu\,(FDCF_1^2(A \text{ OR } c(A)) = (a\odot(a)) \oplus (a\odot a) \oplus (n(a) \odot n(a))$
$\mu\,(FCCF_1^2\,(A \text{ OR } c(A)) = (a\oplus n(a))$
$\mu\,(FDCF_1^2\,(A \text{ AND } c(A)) = (a\odot n(a))$
$\mu\,(FCCF_1^2\,(A \text{ AND } c(A)) = (a\oplus n(a)) \odot (a\oplus a) \odot (n(a) \oplus n(a))$

In Figures 10.2 and 10.3, we observe the Type 2 representation of "A OR c(A)" and "A AND c(A)", respectively. The interpretation of the meta-linguistic expressions of "Fuzzy Middle" and "Contradiction" are to be stated as a relative matter of degree in the intervals defined by

$$\mu(FDCF_I^2(A \text{ OR } c(A)) \quad \mu(FCCF_I^2(A \text{ OR } c(A)) \text{ and by}$$

$$\mu(FDCF_I^2(A \text{ AND } c(A)) \subseteq \mu(FCCF_I^2(A \text{ AND } c(A))$$

, respectively. Therefore, we cannot conclude an absolute contradiction when we observe both A and c(A) in two identical but independent experiments and/or when we receive two input information one for A and the other for c(A) from two identical but independent sensors.

*10.3.2. Bold Intersection and Union*

In the predicate domain, the well known operators of Bold Intersection and Union (for which alternate names are Lukasiewic' s Intersection and Union) are:

$$L_{\odot}(a,b)=\Delta(a,b) = a\Delta\, b = \mathrm{Max}\{0, a+b-1\} = a\, L_{\odot} b$$

$$L_{\oplus}(a,b)=\nabla(a,b) = a\, \nabla\, b = \mathrm{Min}\{1, a+b\} = a\, L_{\oplus}\, b$$

where $a,b \in [0,1]$ are the generic membership values for every $x \in X$.

Again, we rewrite the propositional expressions obtained in Section 10.1.4 and 10.1.5, we get:

$$\mu\,(FDCF_1^{\,2}\,(A \text{ OR } c(A)) = (a\, L_{\odot}\, n(a))\, L_{\oplus}\, (a\, L_{\odot}\, a)\, L_{\oplus}\, (n(a)\, L_{\odot}\, n(a))$$
$$\mu\,(FCCF_1^{\,2}\,(A \text{ OR } c(A)) = (a\, L_{\oplus}\, n(a))$$
$$\mu\,(FDCF_1^{\,2}\,(A \text{ AND } c(A)) = (a\, L_{\odot}\, n(a))$$
$$\mu\,(FCCF_1^{\,2}\,(A \text{ AND } c(A)) = (a\, L_{\oplus}\, n(a))\, L_{\odot}\, (a\, L_{\oplus}\, a)\, L_{\odot}\, (n(a)\, L_{\oplus}\, n(a))$$

In Figures 10.4 and 10.5, again we observe the Type 2 representation of "A OR c(A))" and "A AND c(A))", respectively. Again the interpretations of the meta-linguistic expressions of "Fuzzy Middle" and "Fuzzy Contradiction" are

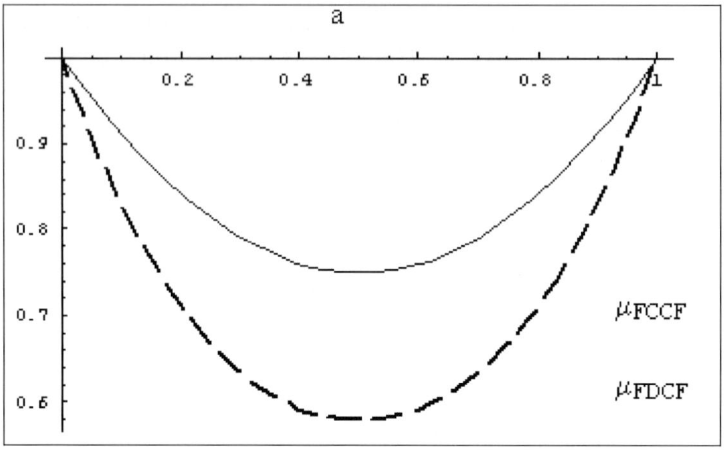


**Figure 10.2.** Type 2 fuzzy set of "A OR c(A)" for the Algebraic De Morgan Triple (□, ⊕, -), where FDCF(A OR c(A)) (— — — ) and FCCF(A OR c(A)) (————)

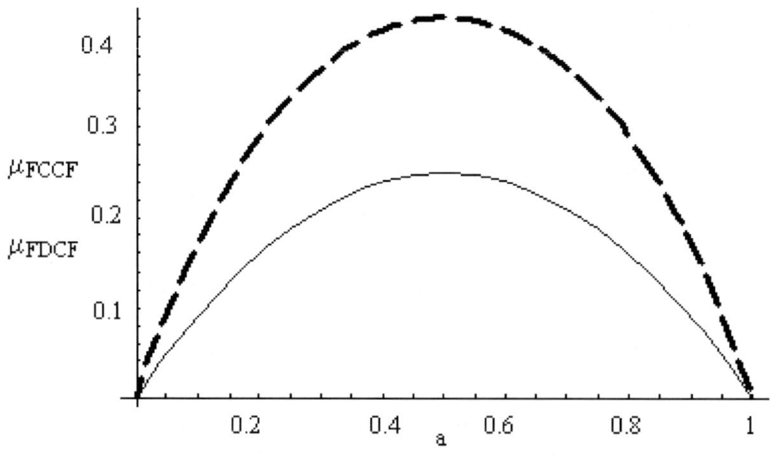


**Figure 10.3.** Type 2 fuzzy set of "A AND c(A)" for the Algebraic De Morgan Triple (⊙, ⊕, -), where FDCF(A AND c(A)) (— — —) and FCCF(A AND c(A))(————)

to be stated as a matter of degree in the intervals defined by $\mu(FDCF_2^1$ (A OR c(A)) $\subseteq \mu(FCCF_2^1$ (A OR c(A)) and by $\mu(FDCF_2^1$ (A AND c(A)) $\subseteq \mu(FCCF_2^1$ (A AND c(A)), respectively. Here, again, we cannot conclude an absolute contradiction when we observe both A and c(A) in two identical but independent experiments when we receive two input information one for A and the other c(A) from two identical but independent sensors or when we receive two consecutive information from the same sensor as A and c(A).

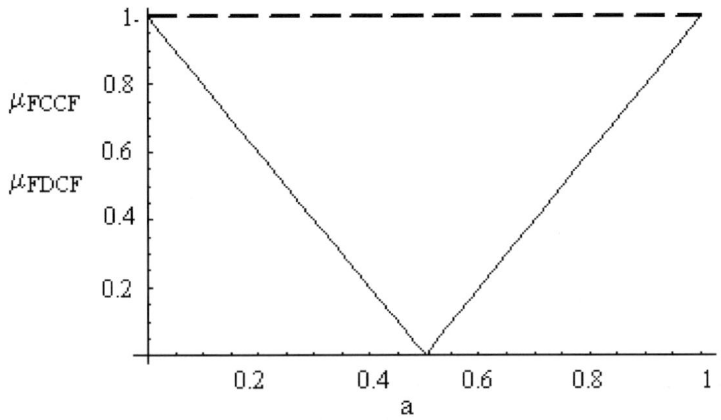


**Figure 10.4.** Type 2 fuzzy set of "A OR c(A)" for Lukasiewicz De Morgan Triple ($L_{\odot}$, $L_{\oplus}$, -), where FDCF(A OR c(A)) (———) and FCCF(A OR c(A)) ( — — — )

For emphasis and clarification, it should be pointed out that "Fuzzy Middle" expressed with the interval

$[\mu(FDCF_1^2(\text{A OR c(A)})), [\mu(FCCF_1^2(\text{A OR c(A)}))]$
and the "Fuzzy Contradiction" expressed with the interval
$[\mu(FDCF_1^2(\text{A AND c(A)})), [\mu(FCCF_1^2(\text{A AND c(A)}))]$

are Type 2 fuzzy sets. Therefore the interpretation of "Fuzzy Middle" and "Fuzzy Contradiction" is to be understood as a matter of many degrees within these intervals. That is "Fuzzy Middle" and "Fuzzy Contradiction" are subject to second order vagueness. This is fitting to our common sense understanding that the combination of two Type 1 imprecise information ought to produce a Type 2 imprecise information. Furthermore this second order vagueness is realized in the membership domain. Naturally, these affect the results of the defuzzification algorithms.

## 10.4. Laws of Fuzzy Conservation

In two-valued set and logic, we find that

$$\left\{\begin{matrix} \mu[DNF_1^2(\text{A OR c(A)}) \\ +\mu[DNF_1^2(\text{A AND c(A)}) \end{matrix}\right\} = \left\{\begin{matrix} \mu[CNF_1^2(\text{A OR c(A)})] \\ \mu[CNF_1^2(\text{A AND c(A)})] \end{matrix}\right\} = 1$$

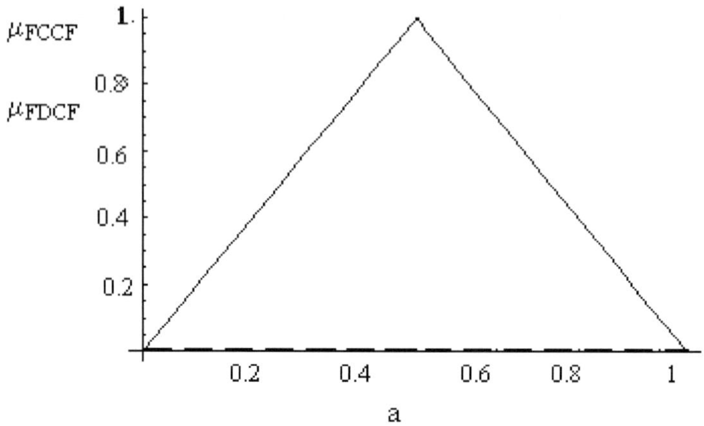


**Figure 10.5.** Type 2 fuzzy set of "A AND c(A)" for Lukasiewicz De Morgan Triple ($L_{\odot}$, $L_{\oplus}$, -), where FDCF(A AND c(A))(———) and FCCF(A AND c(A)) (– – – –)

This is known as the Law of Conservation. It is found that in Fuzzy Logic, there are also "Laws of Fuzzy Conservation" but they are formed in a slightly modified manner. It is found that, in Fuzzy Logic, the Laws of Fuzzy Conservation hold between $FDCF_1^2$ (A OR c(A)) and $FCCF_1^2$(A AND c(A)) as well as between $FDCF_1^2$ (A AND c(A)) and $FCCF_1^2$(A OR c(A)). That is, we have

$$\mu[FDCF_1^2(A \text{ OR } c(A))] + \mu[FCCF_1^2(A \text{ AND } c(A))] = 1, \text{ and}$$

$$\mu[FDCF_1^2(A \text{ AND } c(A))] + \mu[FCCF_1^2(A \text{ OR } c(A))] = 1$$

In general, it can be shown that the "Laws of Fuzzy Conservation" hold. It is straight forward to show for example that:

$$\begin{aligned}\mu[FDCF_1^2(A \text{ AND } c(A))] + \mu[FCCF_1^2(A \text{ OR } c(A))] &\\ &= [(a \,\Delta\, n(a)) + (a \,\nabla\, n(a)] \\ &= n[(a \,\nabla\, n(a))] + (a \,\nabla\, n(a)) \\ &= 1-(a \,\nabla\, n(a)) + (a \,\nabla\, n(a)) \\ &= 1\end{aligned}$$

The other case is left as an exercise for the reader. Furthermore, in specific examples, it can be shown that the Laws of Fuzzy Conservation hold for max-min as well as for Algebraic Sum and Product and for Bold Union-Intersection and when all these three operator sets form a De Morgan triple with Standard Negations.

### *10.4.1. Laws of Fuzzy Conservation for Zadeh Operators*

For Max-Min and Standard Negation, we have from Section 10.2:

$$\mu[FDCF_1^2(\text{A OR c(A)})] = a \vee n(a) \text{, and}$$
$$\mu[FCCF_1^2(\text{A AND c(A)})] = a \wedge n(a)$$

Therefore,

$$a \vee n(a) + a \wedge n(a) = 1$$

On the other hand, we have:

$$\mu[FDCF_1^2(\text{A AND c(A)})] = a \wedge n(a) \text{, and}$$
$$\mu[FCCF_1^2(\text{A OR c(A)})] = a \vee n(a)$$

and hence again, we have:

$$a \wedge n(a) + a \vee n(a) = 1$$

### *10.4.2. Laws of Fuzzy Conservation for $\Delta$ and $\nabla$ Combinations*

In the t-norm-conorm class of fuzzy theories, let us investigate the laws of conservation for the two well known subclasses formed by Algebraic Product and Sum, and Bold Intersection and Union operators for the case of "A OR c(A)" and "A AND c(A))" where the axioms of "Distributivity", "Absorption", and "Idempotency" are not applicable in either the propositional or in the predicate domain expressions.

#### *10.4.2.1. Laws of Fuzzy Conservation with Algebraic Product and Sum*

Laws of Fuzzy Conservation can be written directly with the results obtained in Section 10.3.1 for the Algebraic Product and Sum operators and Standard Negation. Rewriting the predicate expressions of these normal forms, we have:

$\mu[FCCF_1^2$ (A OR c(A))] = $(a \oplus n(a))$, and
$\mu[FDCF_1^2$ (A AND c(A))]= $(a \odot n(a)$
Therefore, it is straight forward to show that:

$$\mu[FCCF_1^2(\text{A OR c(A)})] + \mu[FDCF_1^2(\text{A AND c(A)})] = 1$$

At the same time, we have

$$\mu[FDCF_1^2(A \text{ OR } c(A))] = (a \odot n(a) \oplus (a \odot a) \oplus (n(a) \odot n(a))$$
$$\mu[FCCF_1^2(A \text{ AND } c(A))] = (a \oplus n(a)) \odot (a \oplus a) \odot (n(a) \oplus n(a))$$

Hence it can also be shown that

$$\mu[FDCF_1^2(A \text{ OR } c(A))] + \mu[FCCF_1^2(A \text{ AND } c(A))] = 1.$$

*10.4.2.2. Laws of Fuzzy Conservation with Bold Intersection and Union*

Laws of Fuzzy Conservation can be written directly with the results obtained in Section 10.3.2 for the Bold Intersection and Union operators and Standard Negation. Rewriting the predicate expression of these normal forms, we have:

$\mu[FCCF_1^2 (A \text{ OR } c(A))] = (a \nabla n(a))$, and
$\mu[FDCF_1^2 (A \text{ AND } c(A))] = (a \Delta n(a))$.

Therefore, it is straight forward to show that:

$(a \nabla n(a)) = \text{Min}\{1, a+n(a)\} = 1$
$a \Delta n(a) = \text{Max}\{0, a+n(a)-1\} = 0$

Therefore,

$$\mu[FCCF_1^2(A \text{ OR } c(A))] + \mu[FDCF_1^2(A \text{ AND } c(A))] = 1$$

In a similar manner it can be shown that

$$\mu[FDCF_1^2(A \text{ OR } c(A))] + \mu[FCCF_1^2(A \text{ AND } c(A))] = 1$$

by substituting the bold intersection and union operators where appropriate.

*10.4.2.3. Remarks on Laws of Fuzzy Conservation*

There are two observations that should be made in connection with the Laws of Fuzzy Conservation.

Observation 1. It should be noted that in showing that the Laws of Fuzzy Conservation holds for all t-norms and t-conorms the following relationships hold:

(i) c(FDCF(A AND c(A)) = FCCF(A OR c(A)) and vice-versa, and

(ii) c(FDCF(A OR c(A)) = FCCF(A AND c(A)) and vice-versa.

Therefore the Laws of Fuzzy Conservation hold as discussed above.

**Observation 2.** It should also be pointed out that the Laws of Fuzzy Conservation forms a foundation of the **invariance** in Fuzzy set and logic theory.

This principle of the **invariance** can be obtained in Full Type 2 fuzzy set and logic structure. (However, this is left for further research)

## 10.5. Canonical Forms of Re-Affirmation And Re-Negation

An affirmative expression such as "temperature is cold" may be re-affirmed by two independent sources of information, say, by two independent sensors. Similarly, negative expression such as "temperature is not cold" may also be re-negated by two independent sources of information again, say, by two independent sensors. Such expressions at times are known as sub-expression. Let us next investigate the fuzzy normal forms of such expressions.

### *10.5.1. Canonical Forms of Re-Affirmation*

At the linguistic level a re-affirmation is a combination of an affirmative statement by another affirmative statement with one of the linguistic connectives "OR", "AND".

### *10.5.2. Canonical Forms of Re-Affirmation with "OR"*

Such expressions may be stated in general in the meta linguistic form as "A OR A". However, they would be expressed in detailed linguistic form as follows:

"$x \in X$ isr $A_1$, with $a_1 \in [0,1]$, is T" or "$x \in X$ isr$A_2$, with $a_2 \in [0,1]$ is T", is T.
"$x \in X$ isr $A_1$, with $a_1 \in [0,1]$, is T"or "$x \in X$ isr $A_2$, with $a_2 \in [0,1]$ is F", is T.
"$x \in X$ isr $A_1$, with $a_1 \in [0,1]$, is F"or "$x \in X$ isr $A_2$, with $a_2 \in [0,1]$ is T", is T.
"$x \in X$ isr $A_1$, with $a_1 \in [0,1]$, is F"or "$x \in X$ isr $A_2$, with $a_2 \in [0,1]$ is F", is F.

where "$\in$" means "which is a value in" and $A_1$ and $A_2$ represent the two indicator readings that are received from the first and the second independent sources of information.

Clearly, we can reconstruct the Truth Table for such expressions and obtained the fuzzy canonical forms for the meta-linguistic expression "$A_1$ OR $A_2$" which represents detailed realizations of a re-affirmation expression, "A OR A", where in the above statements we have labeled the first A as $A_1$, and the second A as $A_2$ in order to emphasize the fact that they are received from distinct independent sources of information and/or they are received at different moments in time as t = 1, 2 from the same source, which could happen in some instrumental readings. But without re-producing a Truth Table, we can write the fuzzy disjunctive and conjunctive canonical expressions in propositional form as follows:

$$FDCF_1^2(A \text{ OR } A) = (A \cap A) \cup (A \cap c(A)) \cup (c(A) \cap A)$$

$$FCCF_1^2(A \text{ OR } A) = A \cup A$$

Clearly, if we are in two-valued set theory, we get:

$$\mu[DNF_1^2(A \text{ OR } A)] = \mu[CNF_1^2(A \text{ OR } A)] = a \in \{0,1\}$$

That is, re-affirmation is absolutely true as expected. However, if we are in Zadeh's Max-Min and Standard Negation theory, then we get:

$\mu[FDNF_1^2(A \text{ OR } A)] = \mu[FCNF_1^2(A \text{ OR } A)] = a \in [0,1]$, in form only.
But now, the re-affirmation takes on a value only to a degree! (Figure 10.6), and this turns out to be Type 1 fuzzy set.

With the Algebraic Sum and Product operators and Standard Negation for "A OR A", we obtain a Type 2 interval valued fuzzy set as shown in Figure 10.2, somewhat analogous, but not exactly as Figure 10.7. The predicate expressions of "A OR A" are:

$$\mu[FDCF_1^2(A \text{ OR } A)] = a^2 \oplus a \odot n(a) \oplus n(a)a \subseteq \mu[FCCF_1^2(A \text{ OR } A)] = 2a - a^2$$

Finally, for "A OR A" with for Bold Union and Intersection, and Standard Negation, we get:

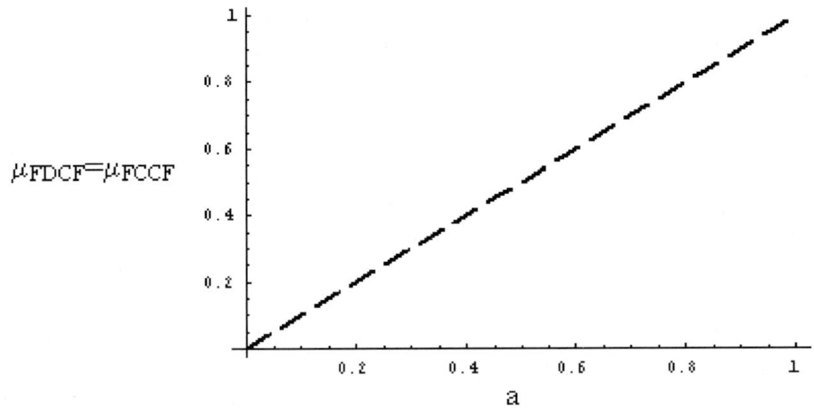


**Figure 10.6.** Type 1 fuzzy set of "A OR A" for Zadehean Connectives, where FDCF(A OR A)=FCCF(A OR A)=A( ---- ). As well, this is the Graph of FDCF(A AND A)=FCCF( A AND A)=A( ---- ).

$\mu[FDCF_1^2(A\ OR\ A)] = Max\{0, 2a\text{-}1\} \subseteq \mu[FCCF_1^2(A\ OR\ A)] = Min\{1,2a\}$

Therefore, the re-affirmation "A OR A" again turns out to be a Type 2 interval valued fuzzy set with membership values specified within the interval $[\mu[FDCF_1^2(A\ OR\ A)],\ \mu[FCCF_1^2(A\ OR\ A)]]$ (Figure 10.8).

*10.5.3. Canonical Forms of Re-Affirmation with "AND"*

On the other hand, such expressions of re-confirmation may also be stated in general in the meta-linguistic form as "A AND A". Their detailed linguistic expressions would be stated as follows:

"$x \in X$ isr $A_1$, with $a_1 \in [0,1]$, is T"and "$x \in X$ isr $A_2$, with $a_2 \in [0,1]$ is T",is T.
"$x \in X$ isr $A_1$, with $a_1 \in [0,1]$, is T" and "$x \in X$ isr $A_2$, with $a_2 \in [0,1]$ is F",is F.
"$x \in X$ isr $A_1$, with $a_1 \in [0,1]$, is F" and "$x \in X$ isr $A_2$, with $a_2 \in [0,1]$ is T",is F.
"$x \in X$ isr $A_1$, with $a_1 \in [0,1]$, is F" and "$x \in X$ isr $A_2$, with $a_2 \in [0,1]$ is F",is F.

Again, we can write the FDCF and FCCF expressions by re-constructing the Truth Table, which is again left for the reader to do. But without reproducing these tables here, we can write the fuzzy disjunctive and conjunctive expressions as follows:

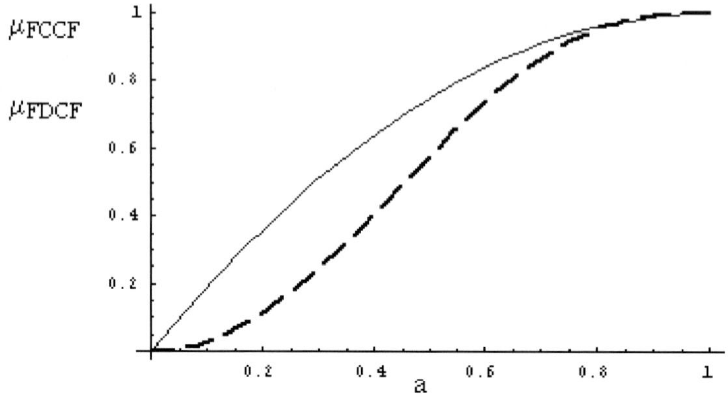


**Figure 10.7.** Type 2 Interval-valued fuzzy set of "A OR A" with the Algebraic De Morgan Triple (⊙, ⊕, -), where FCCF(A OR A)( ——— ) and FDCF(A OR A)(– – – –).

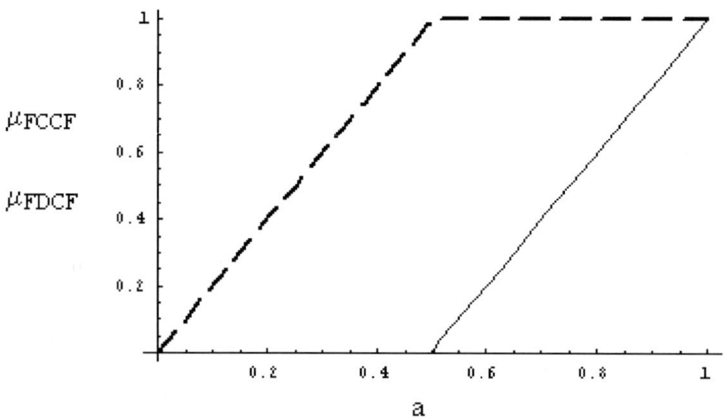


**Figure 10.8.** Type 2 Interval-valued fuzzy set of "A OR A" with the Lukasiewicz De Morgan Triple, where FDCF(A OR A)( ——— ) and FCCF(A OR A) (– – – –).

$$\mathrm{FDCF}_1^2(\mathrm{A\ AND\ A}) = \mathrm{A} \cap \mathrm{A}$$

$$\mathrm{FCCF}_1^2(\mathrm{A\ AND\ A}) = (\mathrm{A} \cup c(\mathrm{A})) \cap (\mathrm{A} \cup c(\mathrm{A})) \cap (\mathrm{A} \cup \mathrm{A})$$

Again clearly, if we are in *two-valued theory* we get:
$\mu[\mathrm{FDCF}_1^2(\mathrm{A\ AND\ A})] = \mu[\mathrm{FCCF}_1^2(\mathrm{A\ AND\ A})] = a \in \{0,1\}$
Hence, the re-affirmation takes on an absolute membership value as expected.
In Zadeh's Max-Min and Standard Negation, again we get:

$\mu[FDCF_1^2(A \text{ AND } A)] = \mu[FCCF_1^2(A \text{ AND } A)] = a \in [0,1]$

Therefore, the re-affirmation takes on a membership value to a degree in [0,1] ! Furthermore, it should be noted that this turns out to be equivalent to "A OR A". Therefore the graph is as shown in Figure 10.6.

Next, if we are in t-Norm-Conorm and Standard Negation theory, again, we would realize a separation between FDCF and FCCF expressions. We obtain a Type 2 interval valued fuzzy set with Algebraic Sum and Product operators and Standard Negation for "A AND A" is expressed as

$$\mu[FDCF_1^2(A \text{ AND } A)] = a^2$$

$$\mu[FCCF_1^2(A \text{ AND } A)] = (a \oplus a) \odot (a \oplus n(a)) \odot (n(a) \oplus a)$$

and as shown in Figure 10.9. For "A AND A" with for Bold Intersection-Union and Standard Negation, we obtain:

$$\mu[FDCF_1^2(A \text{ AND } A)] = \text{Max}\{0, 2a-1\}$$

$$\mu[FCCF_1^2(A \text{ AND } A)] = \text{Min}\{1, 2a\}\}$$

Therefore, the re-affirmation "A AND A" is true in an interval of membership degrees specified within the interval

$\mu[FDCF_1^2(A \text{ AND } A)], \mu[FCCF_1^2(A \text{ AND } A)]]$ as shown in Figure 10.10

## 10.6. Canonical Forms of Re-Negation

At the linguistic level, re-negation is a combination of a negative statement by another negative statement with one of the linguistic connectives "OR", "AND". We leave writing such detailed linguistic expressions to the reader and move directly to propositional expressions of canonical forms for re-negation.

### *10.6.1. Canonical Forms of Re-Negation with "OR"*

Without re-constructing the Truth Tables, we can write the Fuzzy Disjunctive and Conjunctive, Canonical Forms of "c(A) OR c(A)" as follows:

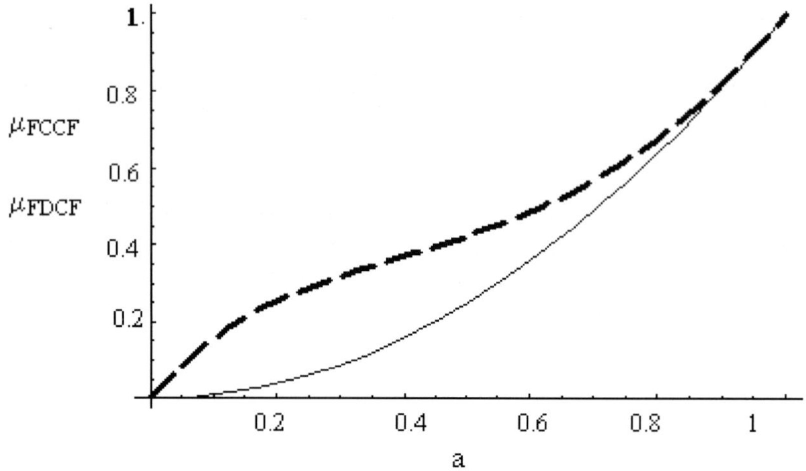


**Figure 10.9.** Type 2 Interval-valued fuzzy set of "A AND A" with Algebraic De Morgan Triple ($\odot$, $\oplus$, -), where FDCF(A AND A)( — — — — ) and FCCF(A AND A)(————).

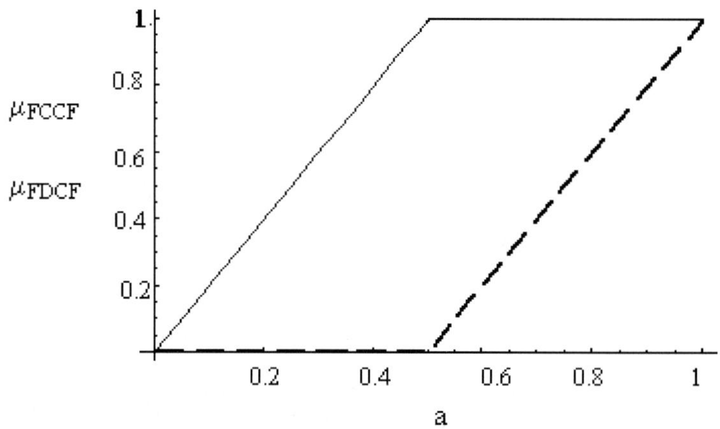


**Figure 10.10.** Type 2 Interval-valued fuzzy set of "A AND A" with the Lukasiewicz De Morgan Triple, where FDCF(A AND A)( — — — — ) and FCCF(A AND A) (————).

$$FDCF_1^2\,(c(A)\text{ OR }c(A)) = (c(A)\cap c(A))\cup(c(A)\cap A)\cup(A\cap c(A))$$

$$FCCF_1^2\,(c(A)\text{ OR }c(A)) = (c(A)\cup c(A))$$

Clearly, if we are in <u>two-valued theory</u>, we would get:

$\mu[DNF_1^2\,(c(A)\text{ OR }c(A))] = \mu[DNF_1^2\,(c(A)\text{ OR }c(A))] = n(a) \in \{0,1\}$

That is, re-negation takes on absolute values as expected.
If, however, we are in Zadeh's Max-Min Standard Negation theory, then we get:

$\mu[FDCF_1^2\,(c(A)\text{ OR }c(A))] = \mu[FCCF_1^2\,(c(A)\text{ OR }c(A))] = n(a) \in [0,1],$

But now, the re-negation takes on membership values to a degree [0,1]! (Figure 10.11); and this turns out to be a Type 1 fuzzy set.

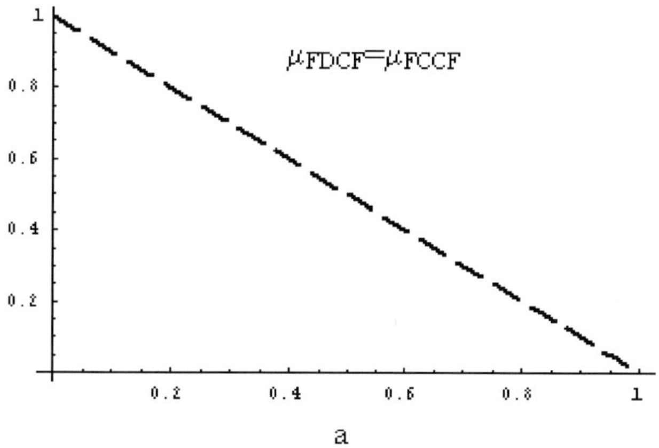


**Figure 10.11.** Type 1 fuzzy set of "c(A) OR c(A)" for Zadehean Connectives, where FDCF(c(A) OR c(A))=FCCF(c(A) OR c(A))=A(– – – – ). As well, this is the Graph of FDCF(c(A) AND c(A))=FCCF(c(A) AND c(A))=A(– – – – ).

Finally, again if we are in t-norm-conorm and standard negation theory, we would realize a separation between FDCF and FCCF expressions. We obtain a Type 2 interval valued fuzzy set with Algebraic Sum and Product operators and Standard Negation for "c(A) OR c(A)" as shown in Figure 10.12.

For "c(A) OR c(A)" with Bold Union and Intersection and Standard Negation, we obtain:

$$\mu[FDCF_1^2(c(A)\,OR\;c(A))] = (n(a)\;L_{\oplus}\;n(a))\;L_{\odot}\;(n(a)\;L_{\oplus}\;a)\;L_{\odot}\;(a\;L_{\oplus}\;n(a))$$
$$\mu[FCCF_1^2(c(A)\,OR\;c(A))] = n(a)\;L_{\oplus}\;n(a)$$

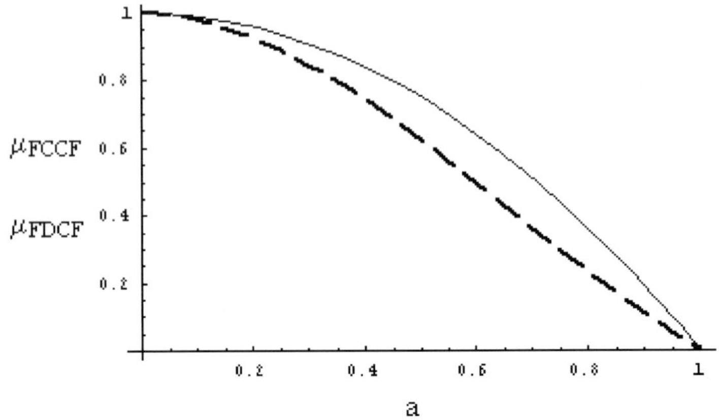


**Figure 10.12.** Type 2 fuzzy set of "c(A) OR c(A)" for Algebraic Product and Sum Operators, where FDCF(c(A) OR c(A)) (— — — —), FCCF(c(A) OR c(A)) (————).

When we simplify these expressing, we get

$$\mu[FDCF_1^2(c(A)\,OR\,c(A))] = Max\{0, 2n(a)\text{-}1\}$$
$$\subseteq \mu[FCCF_1^2(c(A)\,OR\,c(A))] = Min\{1, 2n(a)\}$$

Therefore, the re-negation "c(A) OR c(A)" takes on values in an interval of membership degrees specified within the interval $[\mu[FDCF_1^2(c(A)\,OR\,c(A))]$, $\mu[FCCF_1^2(c(A)\,OR\,c(A))]]$. Clearly, this is a Type 2 interval valued fuzzy set. (Figure 10.13)

Again, without re-constructing the Truth Tables, we can write Fuzzy Disjunctive and Conjunctive Canonical Forms of "c(A) AND c(A)" as follows:

$FDCF_1^2$ (c(A) AND c(A)) = c(A) $\cap$ c(A)
$FCCF_1^2$(c(A) AND c(A)) = (A$\cup$c(A)) $\cap$ (c(A)$\cup$A) $\cap$(c(A)$\cup$c(A))

*10.6.2. Canonical Forms of Re-Negation with "AND"*

Clearly, if we were in two-valued theory, we get:

$$\mu[FDCF_1^2(c(A)\,AND\,c(A))] = \mu[FCCF_1^2(c(A)\,AND\,c(A))] = n(a) \in \{0,1\}$$

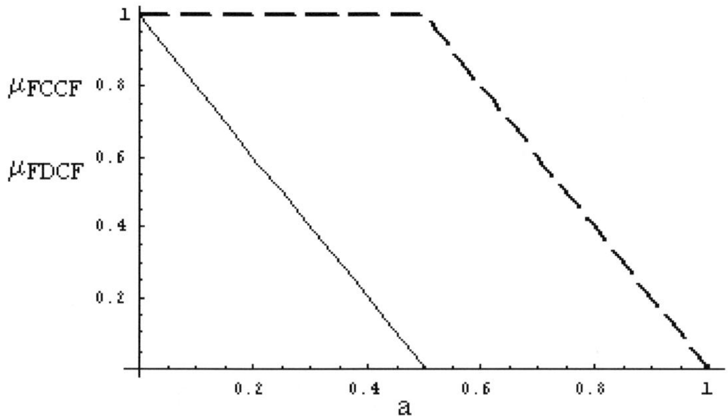


**Figure 10.13.** Type 2 Interval-Valued fuzzy set of "c(A) OR c(A)" with the Lukasiewicz De Morgan Triple ($L_\odot$, $L_\oplus$, -), where FCCF(c(A) OR c(A)) (– – – – ), FDCF(c(A) OR c(A)) (———).

That is, re-negation takes on absolute values as expected.
If however, we are in Zadeh's Max-Min Standard Negation theory, then we get:

$$\mu[FDCF_1^2(c(A)\,AND\,c(A))] = \mu[FCCF_1^2(c(A)\,AND\,c(A))] = n(a) \in [0,1]$$

But now, the re-negation takes on membership values to a degree [0,1]! (Figure 10.11). It is noted that this is equivalent to the expression of "c(A) OR c(A)". De Morgan Triple (⊙, ⊕, -), the FDCF and FCCF or "c(A) AND c(A)" turns out to be as show in Figure 10.14.
Finally, if we were in t-Norm-Conorm and Standard Negation theory, we again realize a separationbetween FDCF and FCCF expressions. For example, for Bold Union and Intersection and Standard Negation, we get:

$$\mu[FDNF_1^2(c(A)\,AND\,c(A))] = Max\{0,\, 2n(a)-1\},$$
$$\mu[CNF_1^2(c(A)\,AND\,c(A))] = Min\{1, 2n(a)\}.$$

Therefore, the re-negation "c(A) AND c(A)" takes on values in an interval of membership degrees specified within the interval

$$[\mu[FDCF_1^2(c(A)\,AND\,c(A))],\ \mu[FCCF_1^2(c(A)\,AND\,c(A))]]$$

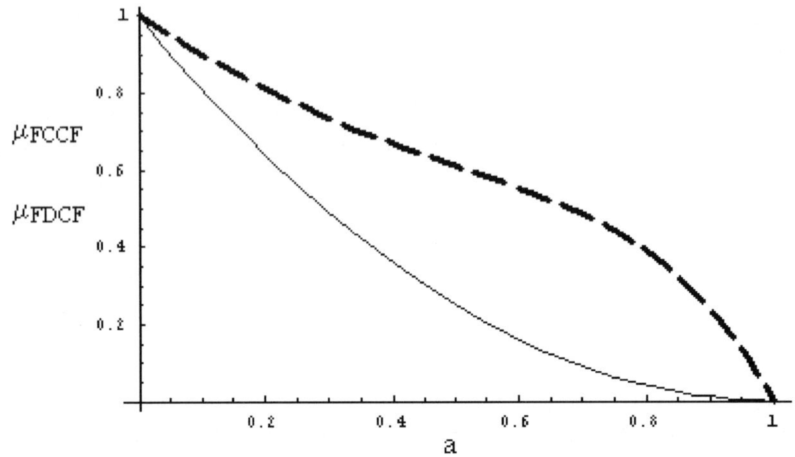


**Figure 10.14.** Type 2 Interval-Valued fuzzy set of "c(A) AND c(A)" with the Algebraic De Morgan Triple (⊙, ⊕, -), where FCCF(c(A) OR c(A)) (– – – – ), FDCF(c(A) OR c(A)) (————).

It is noted that this expression is equivalent to "c(A) OR c(A)". (Figure 10.13) Note: Recall that superscript (²) in $FDCF^2(.)$ and $FCCF^2(.)$ was used to indicate that these expressions have two arguments. In the remaining chapters we drop the superscript (²) and assume that it will be clear in context!

## 10.7. Conclusion

We have demonstrated that one should start out one' s discussion in logic initially with a conceptual linguistic statement and transform them to meta-linguistic expressions, then to propositional expressions and finally to predicate expressions before launching into numerical computations either in {0,1} or [0,1] domains of membership assignments. In computing with words, it is to be understood that we are in fact transforming a natural language expression stated in linguistic terms into meta-linguistic, then to prepositional, and then to computational formal languages. At each stage, the assumptions of the language constrain the meaning representation and leading to results expressed within that constrained language.

It was also pointed out that we must distinguish and separate membership assignments from the truth qualifications attributed to propositions and in particular to predicates in the re-construction of the Truth Tables. This distinction was not needed in two-valued theory and its associated Truth Table constructions. The results obtained by re-constructed Truth Tables produce the same well-known results of two-valued theory under the axioms applicable to that theory for fuzzy sets and two-valued logic expressions. But, the advantages of re-constructed Truth Tables are that:

1) We get more expressive information content from the "Fuzzy Middle" and the "Fuzzy Contradiction" expressions in infinite (fuzzy) valued theory as opposed to "Excluded Middle" and "Crisp Contradiction" expressions of two-valued theory.

2) Furthermore, we obtain results for sub concepts such as "Re-affirmation" and "Re-negation" but again with more expressive information content.

These results suggest that the power of expression in the infinite (fuzzy) valued theory helps us discern more information content for better systems analyses and decision-making.

Furthermore, we have shown that the break down of classical equivalences generate new formulas and new laws in fuzzy set and logic theory and demonstrate clearly the richness of fuzzy theories.

## References

1. Marinos, P.N. (1969), "Fuzzy Logic and its Applications to Switching Systems", *IEEE Trans.Comp.*, Vol 4, 343-348.
2. Türkşen, I.B., A. Kandel, Y-Q. Zhang (1998), "Universal Truth Tables and Normal Forms", *IEEE-Fuzzy Systems*, 6-2, 295-303.
3. Schwede, G.W., A. Kandel (1977), "Fuzzy Maps", *IEEE Trans. SMC* Vol 9, 669-674.
4. Türkşen, I.B., A. Kandel, Y-Q. Zhang (1999), "Normal Forms of Fuzzy Middle and Fuzzy Contradiction", *IEEE-SMC, 29-2, Part B, Cybernetics*, 237-253.

# Chapter 11

# COMPENSATORY "AND" *

In this chapter, we show that the break down of classical equivalences, i.e., FDCF $\subseteq$ FCCF, together with the containment, i.e., FDCF $\subseteq$ FCCF for strict and nilpotent Archimeidan t-norm and t-conorm provide a representation and an explanation for the "compensatory 'AND'".

It was stated earlier that linguistic operators "AND", "OR", etc., do not correspond one-to-one, isomorphically to t-norms and t-conorms. Experimental studies conducted by Zimmerman and Zysno (1980,1985) clearly validates the fact that linguistic "AND", "OR" do not correspond to the t-norms and t-conorms. This is treated in this chapter with FDCF and FCCF structures.

It is shown that Zimmerman and Zysno's $\gamma$ operator takes on values within the FDCF and FCCF bounds of Type 2 interval-valued fuzzy set representation. Zimmermann and Zysno (1980, 1985) proposed the "Compensatory AND", based on certain experimental results. Zimmermann and Zysno (1980, 1985) and Zimmermann (1986) defined the "Compensatory AND" as:

$$\mu_{A \odot B} = \mu_{A \cap B}{}^{1-\gamma} \mu_{A \cup B}{}^{\gamma} \tag{11.1}$$

where $\odot$ is the "Compensatory AND" operator, A and B are two fuzzy sets, and $\gamma \in [0,1]$ is the compensatory parameter.
A particular version of (11.1) is defined by Zimmermann (1986), for the algebraic sum-product operators, as:

$$\mu_{\odot} = \left(\prod_{i=1}^{m} \mu_i\right)^{1-\gamma} \left(1 - \prod_{i=1}^{m} (1-\mu_i)\right)^{\gamma}, \quad 0 \leq \mu_i \leq 1,\ 0 \leq \gamma \leq 1 \tag{11.2}$$

for fuzzy sets i=1,2,...,m, each with membership value $\mu_i$. Another version was also suggested:

$$\mu_{A \odot B} = (1-\gamma)\mu_{A \cap B} + \gamma \mu_{A \cup B} \tag{11.3}$$

---

* Extracted and modified from "Interval-valued Fuzzy Sets and 'Compensatory AND' ", FSS, 51 (1992) 295-307

for two fuzzy sets A and B. A particular version of (11.3) is defined as

$$\mu_{\odot}=(1-\gamma)\prod_{i=1}^{m}\mu_i+\gamma(1-\prod_{i-1}^{m}(1-\mu_i)) \tag{11.4}$$

for the algebraic sum-product operators, for m fuzzy sets with membership values $\mu_i$ i=1,2,...,m.

A generalization of the "Compensatory AND" defined by (11.1) and (11.3) may be written in terms of t-conorms $\nabla$ and t-norms $\Delta$ as

$$\mu_{A\odot B}=[\Delta(a,b)^{1-\gamma}[\nabla(a,b)]^{\gamma}, \text{ and} \tag{11.1'}$$

$$\mu_{A\odot B}=(1-\gamma)\,\Delta(a,b)+\gamma\nabla(a,b) \tag{11.3'}$$

where $a,b\in[0,1]$ are the membership values of fuzzy sets A, B, and $\gamma\in[0,1]$ is the compensatory parameter. Let us call form (11.1') the "Exponential-Compensatory AND" and form (11.3') the "Convex-Linear- Compensatory AND".

### 11.1. Exponential-Compensatory "AND"

Consider the "Exponential-Compensatory AND" as defined by (11.1'), i.e.,

$$\mu_{A\odot B}=[\Delta(a,b)^{1-\gamma}[\nabla(a,b)]^{\gamma}=f(\gamma). \tag{11.5}$$

It is clear that

$$f(\gamma=0)=\Delta(a,b) \quad \text{and} \quad f(\gamma=1)=\nabla(a,b) \tag{11.5'}$$

Furthermore, differentiating with respect to $\gamma$, we obtain

$$\partial f(\gamma)/\partial\gamma=[\Delta(a,b)^{1-\gamma}[\nabla(a,b)]^{\gamma}\ln[\nabla(a,b)/\Delta(a,b)].$$

Since, $\nabla(a,b) \geq \Delta(a,b)$ , $\partial f(\gamma)/\partial\gamma \geq 0$. Hence, $f(\gamma)$ is a monotonic non-decreasing function of $\gamma$.

Note also that, when $\Delta(a,b)=\nabla(a,b)$, $\gamma$ disappears in (11.5). This is a characteristic of the "Exponential-Compensatory AND", along with the characteristic expressed in (11.5').

## 11.2. Containment of FDCF in FCCF of "AND"

In this section, our aim is to show that in the membership domain (due to the containment of FDCF in FCCF), we have:

$$\mu_{FDCF(A\text{ AND }B)} \leq \mu_{A \odot B} \leq \mu_{FCCF(A\text{ AND }B)}$$

i.e., in set notation:

$$A \odot B \in [FDCF(A\text{ AND }B),\ FCCF(A\text{ AND }B)]$$

for certain values of the compensatory parameter $\gamma \in [0,1]$. That is, we wish to show the conditions on a, b, and $\gamma$ under which the following two inequalities hold simultaneously:

$$\Delta(a,b) \leq [\Delta(a,b)^{1-\gamma}[\nabla(a,b)]^{\gamma} \tag{11.6a}$$

$$[\Delta(a,b)]^{1-\gamma}[\nabla(a,b)]^{\gamma} \leq \Delta[\Delta(\nabla(a,b),\nabla(a,n(b)),\nabla(n(a),b)] \tag{11.6b}$$

First, let us consider the conditions on $\gamma$. On the one hand, (11.6a) is equivalent to

$$1 \leq [\nabla(a,b)/\Delta(a,b)]^{\gamma}$$

which is true only if $\gamma \geq 0$.

On the other hand, (11.6b) is equivalent to

$$[\nabla(a,b)/\Delta(a,b)]^{\gamma} \leq \mu_{FCCF(A\text{ AND }B)}/\Delta(a,b).$$

Solving for $\gamma$, we find

$$\gamma \leq \ln[\mu_{FCCF(A\text{ AND }B)}/\Delta(a,b)] \,/\, \ln[\nabla(a,b)/\Delta(a,b)] = \gamma^{*}$$

Thus, the two inequalities hold only if $\gamma \in [0,\gamma^{*}]$.
In general,

$$\mu_{FCCF(A\text{ AND }B)} \leq \nabla(a,b),\ \text{and}\ \nabla(a,b) \geq \Delta(a,b);$$

and

$$[\mu_{FCCF(A\text{ AND }B)} \,/\, \Delta(a,b)] \geq 1,\ \text{and}\ \nabla(a,b) \,/\, \Delta(a,b) \geq 1.$$

Finally, since $\mu_{FCCF(A\text{ AND }B)} \leq \nabla(a,b)$, we have

$$[\mu_{FCCF(A\ AND\ B)} / \Delta(a,b)] \le [\nabla(a,b)/\Delta(a,b)]$$

Thus, by the definition of $\gamma^*$, $0 \le \gamma^* \le 1$, except under certain conditions, which we now consider.

In order that $\gamma^* \in [0,1]$, the solution for $\gamma$ requires that: (i) $\nabla(a,b) \ne \Delta(a,b)$; and (ii) $\nabla(a,b) \ne 0$, $\Delta(a,b) \ne 0$, which imply that neither a nor b nor both be equal to 0, which guarantees that $\mu_{FCCF(A\ AND\ B)} \ne 0$. Thus, on the basis of this, the two inequalities hold when $\gamma \in [0,\gamma^*]$ only if $\nabla(a,b) \ne \Delta(a,b)$, and $a,b \in (0,1]$.

These two restrictions on the solution rest on the characteristics of "Exponential-Compensatory AND" noted in the last section.

It is clear that in the cases excluded by restriction (ii), when a, b or both are equal to 0, the inequality holds with no restriction on $\gamma$. The case excluded by restriction (i) is more complicated. The conditions on a, b that follow from the requirement that $\nabla(a,b) \ne \Delta(a,b)$ depend on the $\Delta$-$\nabla$ pair being used. For example, for max-min operators, the requirement implies that $a \ne b$. But these conditions do not affect (11.6b), since it is equivalent to

$$\Delta(a,b)[\nabla(a,b)/\Delta(a,b)]^{\gamma} \le \mu_{FCCF(A\ AND\ B)}$$

which, when $\nabla(a,b) = \Delta(a,b)$, reduces to $\Delta(a,b)] \le \mu_{FCCF(A\ AND\ B)}$, that is,

$$\mu_{FDCF(A\ AND\ B)} \le \mu_{FCCF(A\ AND\ B)},$$

which is true from the containment of FDCF and FCCF which we have established in previous chapters.

**Theorem 11.1.** The "Exponential Compensatory AND" is included in the interval defined by FDCF and FCCF of "AND", i.e.,

$$A \odot B \in [FDCF(A\ AND\ B), FCCF(A\ AND\ B)],$$

for $a,b \in [0,1]$ and $\gamma \in [0,\gamma^*]$, $0 \le \gamma^* \le 1$.

## 11.3. Compensatory "OR"

In this section, our aim is to show that in the membership domain, when inequality (11.5) holds, we have:

$$\mu_{FDCF(A\ AND\ B)} \leq \mu_{A \odot B} \leq \mu_{FCCF(A\ OR\ B)},$$

i.e., in set notation:

$$A \odot B \in [FDCF(A\ OR\ B), FCCF(A\ OR\ B)],$$

for certain values of the compensatory parameter $\gamma \in [0,1]$. That is, we wish to show the conditions on a, b, and $\gamma$ when the following two inequalities hold simultaneously

$$\nabla[\nabla[\Delta(a,b),\Delta(a,b^c)],\Delta(a^c,b)] \leq [\Delta(a,b)]^{1-\gamma}[\nabla(a,b)]^{\gamma}, \tag{11.7a}$$
$$[\Delta(a,b)^{1-\gamma}[\nabla(a,b)]^{\gamma} \leq \nabla(a,b). \tag{11.7b}$$

From (11.7a), since $\mu_{FDCF(A\ OR\ B)} \leq \nabla(a,b)$, solving for $\gamma$, we get

$$\gamma_0 = \ln[\mu_{FDCF(A\ OR\ B)}/\Delta(a,b)] \ / \ \ln[\nabla(a,b)/\Delta(a,b)] \leq \gamma \tag{11.8}$$

From (11.7b) we get

$$[\Delta(a,b)/\nabla(a,b)]^{1-\gamma} \leq 1$$

which is true only if $\gamma \leq 1$. Thus, the inequalities only hold for $\gamma \in [\gamma_0, 1]$.

Further, we know that

$$\mu_{FDCF(A\ OR\ B)} \leq \nabla(a,b), \text{ and } \nabla(a,b) \geq \Delta(a,b);$$

and

$$[\mu_{FDCF(A\ OR\ B)} / \Delta(a,b) \geq 1, \text{ and } [\nabla(a,b)/\Delta(a,b) \geq 1.$$

Finally, it is clear that;

$$[\mu_{FDCF(A\ OR\ B)}/\Delta(a,b)] \leq [\nabla(a,b)/\Delta(a,b)].$$

Thus, by the definition of $\gamma_0$, $0 \leq \gamma_0 \leq 1$, except under certain conditions. Similarly, in order that $\gamma_0 \in [0,1]$, the solution of (11.8) for $\gamma$ requires that: (i) $S(a,b) \neq T(a,b)$; and (ii) $S(a,b) \neq 0$, $T(a,b) \neq 0$, which imply that neither a nor b nor both must be equal to 0. This guarantees that $\mu_{DNF(A\ OR\ B)} \neq 0$. Thus, on the basis of (11.8) alone, the two inequalities hold when $\gamma \in [\gamma_0, 1]$ only if $S(a,b) \neq T(a,b)$, and $a,b \in (0,1]$. These restrictions, as noted before, rest on characteristics of "Exponential-Compensatory AND".

It is clear that when a and b both are equal to 0 the inequalities hold. But, if either $a = 0$ when $b \neq 0$, or $a \neq 0$ when $b = 0$, (11.7a) does not hold. However, when $S(a,b) \neq T(a,b)$, both inequalities hold with no restrictions on $\gamma$. We thus have the following theorem.

**Theorem 11.2.** The "Exponential Compensatory AND" is included in the FDCF and FCCF interval, i.e.,

$$A \odot B \in [\text{FDCF(A OR B)}, \text{FCCF(A OR B)}]$$

for $a,b \in (0,1]$ and $\gamma \in [\gamma_0, 1]$.

## 11.4. Specific Operators

In this section, for the specific cases of t-conorms $\nabla$ and t-norms $\Delta$, we investigate the details of max-min operators and briefly look at the algebraic sum-product operators.

### *11.4.1. Max-Min Operators*

The specific form of the (basic) "Exponential-Compensatory AND" for max ($\vee$) - min ($\wedge$) and the standard-complement $(n(\cdot))=1-(.)$ is:

$$\mu_{A \odot B} = (a \wedge b)^{1-\gamma} (a \vee b)^{\gamma}$$

(i) With the FDCF and FCCF of "A AND B", we have:

$$\mu_{FDCF(A\ AND\ B)} = a \wedge b, \text{ and}$$
$$\mu_{FCCF(A\ AND\ B)} = (a \vee b) \wedge (n(a) \vee b) \wedge (a \vee n(b)).$$

As a special case of the general solution discussed above, one can investigate all possible cases with the fuzzy truth table, which has eight cases (Table 11.1) in order to show that:

$$a \wedge b \leq (a \wedge b)^{1-\gamma} (a \vee b)^{\gamma} \leq (a \vee b) \wedge (n(a) \vee b) \wedge (a \vee n(b))$$

**Table 11.1.** Fuzzy Truth Table.

| |
|---|
| $a \leq b \leq n(b) \leq n(a)$ |
| $a \leq n(b) \leq b \leq n(a)$ |
| $n(a) \leq b \leq n(b) \leq a$ |
| $n(a) \leq n(b) \leq b \leq a$ |
| $b \leq a \leq n(a) \leq n(b)$ |
| $b \leq n(a) \leq a \leq n(b)$ |
| $n(b) \leq a \leq n(a) \leq b$ |

$$n(b) \leq n(a) \leq a \leq b$$

A graphic representation of four sub-cases from Table 11.1 is shown in Figure 11.1. A detailed analysis of all eight cases is included in Appendix 11.1.

(ii) For the FDCF and FCCF intervals of "A OR B", we have

$$\mu_{FDCF(A\ OR\ B)} = (a \wedge b) \vee (n(a) \wedge b) \vee (a \wedge n(b))$$

$$\mu_{FCCF(A\ OR\ B)} = a \vee b\ .$$

"A AND B": $\mu_{FDCF} = (a \wedge b)$

$\mu_{FCCF} = (a \vee b) \wedge (a \vee n(b) \wedge (n(a) \vee b)$

$\mu_\gamma = (a \wedge b)^{1-\gamma}(a \vee b)^\gamma$

Again, as a special case of the general solution, one can investigate all possible cases with the fuzzy truth table (Table 11.1) in order to show that

$$(a \wedge b) \vee (n(a) \wedge b) \vee (a \wedge n(b)) \leq (a \wedge b)^{1-\gamma}(a \vee b)^\gamma \leq (a \vee b)\ .$$

A graphic representation of four sub-cases from Table 11.1 is shown in Figure 11.2. A detailed analysis of all eight cases is included in Appendix 11.2.

In Figure 11.3, we show the relationship of FDCF and FCCF Intervals of "A AND B" and FDCF and FCCF Intervals of "A OR B", for the "Exponential-Compensatory AND", using the max-min case of t-conorm, $\nabla$, and t-norm, $\Delta$, for the specific values of $\mu_A(\bullet)=.8$ and $\mu_B(\bullet)=.6$ for all values of $\gamma \in [0,1]$. In passing, it should be noted that the general upper and lower bounds indicated in (11.5') are shown to hold in Figure 11.3.

"A OR B": $\mu_{FDCF} = (a \wedge b) \vee (a \wedge n(b)) \vee (n(a) \wedge b)$

$\mu_{FCCF} = a \vee b$

$\mu_\gamma = (a \wedge b)^{1-\gamma}(a \vee b)^\gamma$

Zimmermann [1986] states: "as 'extremal' operators we prefer the product and sum". For this reason, we will briefly turn our attention to this form. The specific form of the (basic) "Exponential-Compensatory AND" for the algebraic sum $\oplus$ and product ,operators and the standard-complement (n(a)=1-a) is:

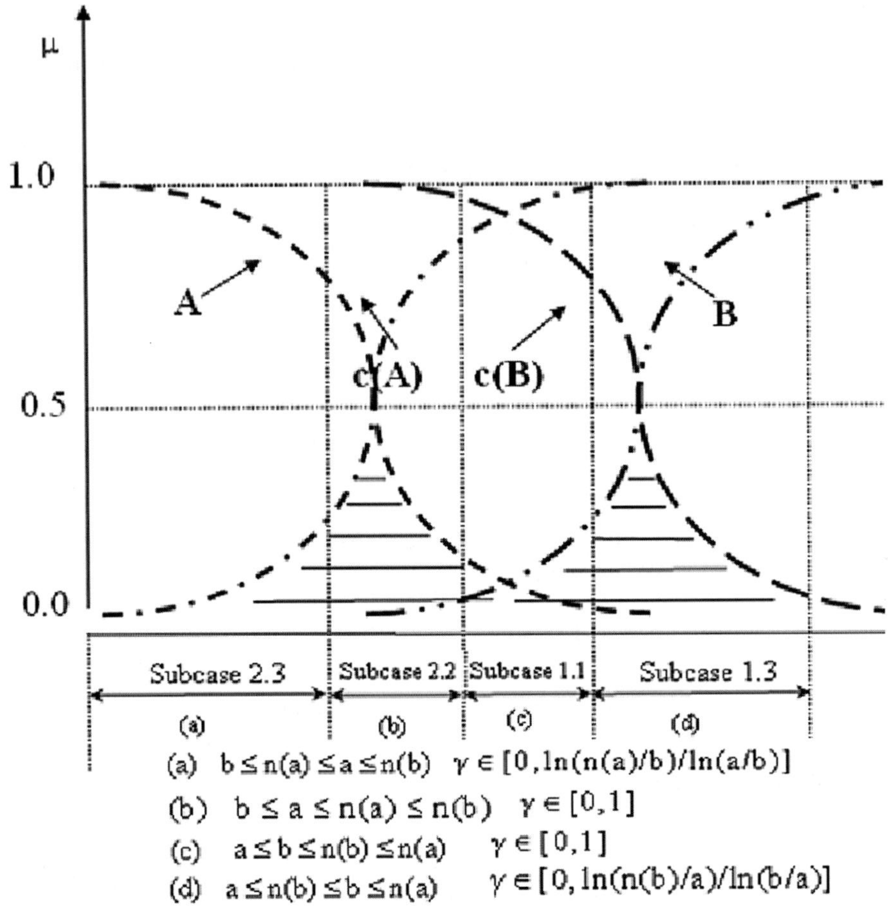


**Figure 11.1.** a Comparison of the FDCF and FCCF intervals for "A AND B" and the "Compensatory AND".

$$\mu_{A,B} = (a \odot b)^{1-\gamma} (a \oplus b)^{\gamma}$$

Naturally, for m component cases, the form can be expressed as shown in (11.2) above. However, we will be satisfied to investigate the basic form since it can be generalized to (11.2) in a natural way.

(i) For the FDCF and FCCF intervals of "A AND B", we have

$$\mu_{FDCF(A\ AND\ B)} = a \odot b$$
$$\mu_{FCCF(A\ AND\ B)} = (a \oplus b) \odot (n(a) \oplus b) \odot (a \oplus n(b))$$

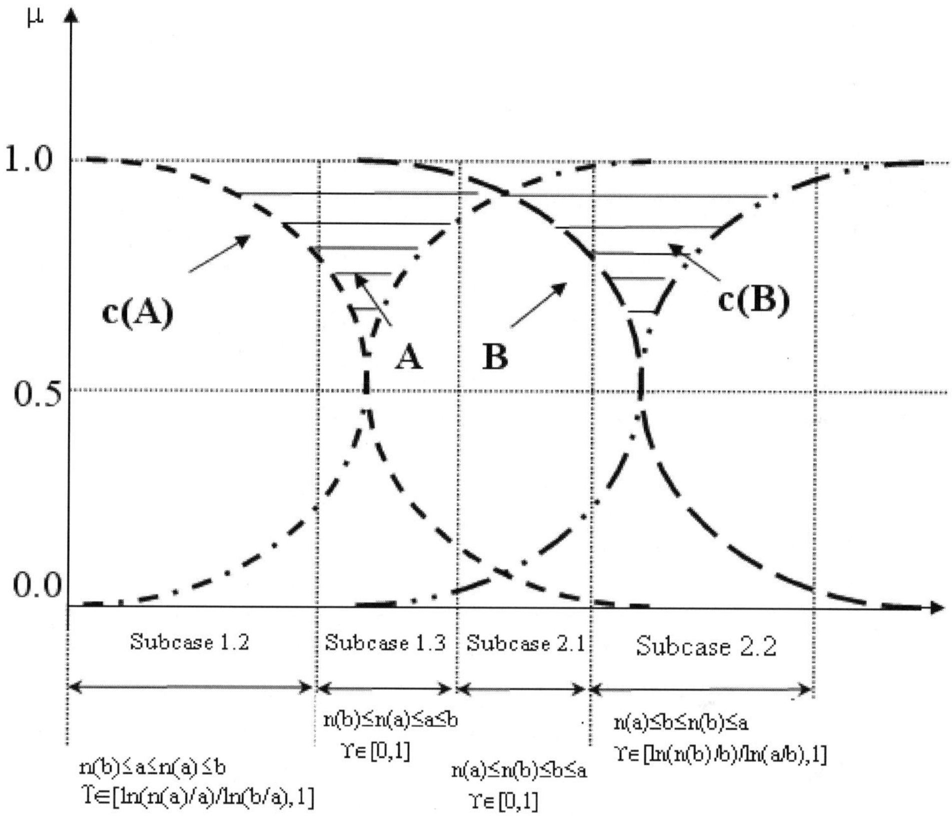


**Figure 11.2.** A Comparison of the FDCF and FCCF intervals for "A OR B" and the "Compensatory AND".Algebraic Sum-Product Operators

Here, we need to determine the values of $\gamma$ when the following inequalities hold:

$$(a\odot b) \leq (a\odot b)^{1-\gamma}\,(a\oplus b)^{\gamma} \leq (a\oplus b)\odot(n(a)\oplus b)\odot(a\oplus n(b))$$

From the left hand side, LHS, we directly obtain

$$1 \leq [(a\oplus b)/(a\odot b)]^{\gamma}$$

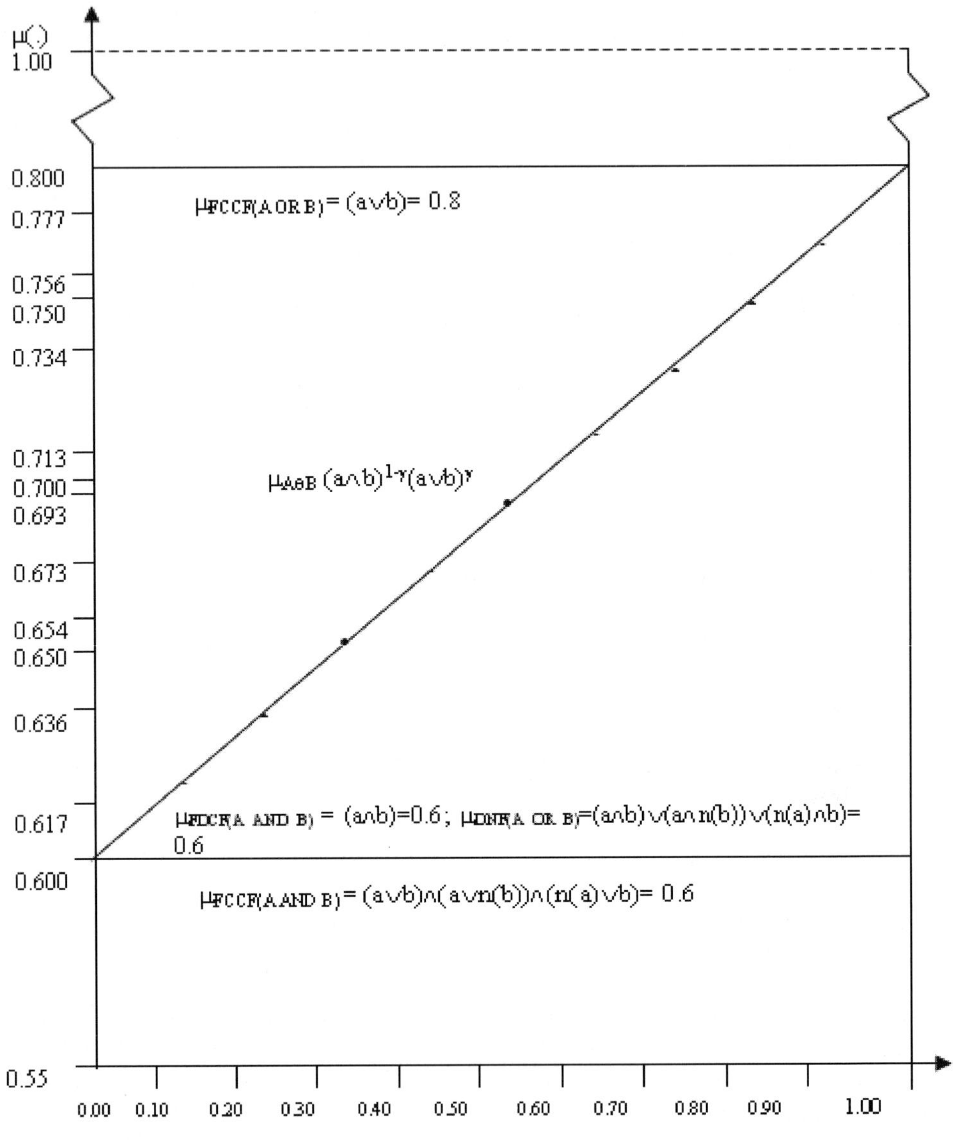


**Figure 11.3.** Intervals of "A AND B", "A OR B", expressed in FDCF and FCCF and "Exponential Compensatory AND" using Max-Min realization of t-conorm and t-norm for $\mu_A$=0.8, $\mu_B$=0.6 and $\gamma \in [0,1]$.

which is true only when $0 \leq \gamma$. From the right hand side, RHS, based on the results expressed in Theorem 11.1, for the specific case of the algebraic sum and product, we can write

$$\gamma \leq \ln[(a \oplus b) \odot (n(a) \oplus b) \odot (a \oplus n(b))/(a \odot b)]/\ln[(a \oplus b)/(a \odot b)] = \gamma^* \leq 1$$

Thus, for the case of the algebraic sum product operators and the standard-complement n(a)=1-a, we obtain the expected result when $\gamma \in [0, \gamma^*]$.

(ii) Similarly, for the FDCF and FCCF interval of "A OR B", we have:

$$\mu_{FDCF(A\ OR\ B)} = (a \odot b) \oplus (n(a) \odot b) \oplus (a \odot n(b)), \text{ and}$$

$$\mu_{FCCF(A\ OR\ B)} = a \oplus b$$

We need to determine values of $\gamma$ when the following inequalities hold:

$$(a \odot b) \oplus (n(a) \odot b) \oplus (a \odot n(b)) \leq (a \odot b)^{1-\gamma} (a \oplus b)^{\gamma} \leq a \oplus b$$

Clearly the RHS is satisfied as

$$[(a \odot b)/ (a \oplus b)]^{1-\gamma} \leq 1$$

which is true only if $\gamma \leq 1$. From the LHS, with the application of (11.8) it is easy to show that

$$\ln[(a \oplus b) \odot (n(a) \oplus b) \odot (a \oplus n(b))/(a \odot b)]/\ln[(a \oplus b)/(a \odot b)] = \gamma_0 \leq \gamma$$

Therefore, the inequality is true when $\gamma \in [\gamma_0, 1]$, provided neither a = 0 when $b \neq 0$, nor $a \neq 0$ when b = 0, i.e., $a,b \in (0,1]$.

Here, we do not need to analyze the cases of a fuzzy truth table in detail for the algebraic product form of the "IVFS (A AND B)", "IVFS (A OR B)" and "Exponential-Compensatory AND", as we did for max-min, since algebraic forms are available, as shown above. For the sake of illustration, however, we do show the relationship of these three expressions for the particular values of $\mu_A(\bullet)$=.8 and $\mu_B(\bullet)$=.6, and for all values of $\gamma \in [0,1]$ in Figure 11.4. Again, in passing, it should be noted that the general upper and lower bounds indicated by (11.5') are shown to hold in Figure 11.4.

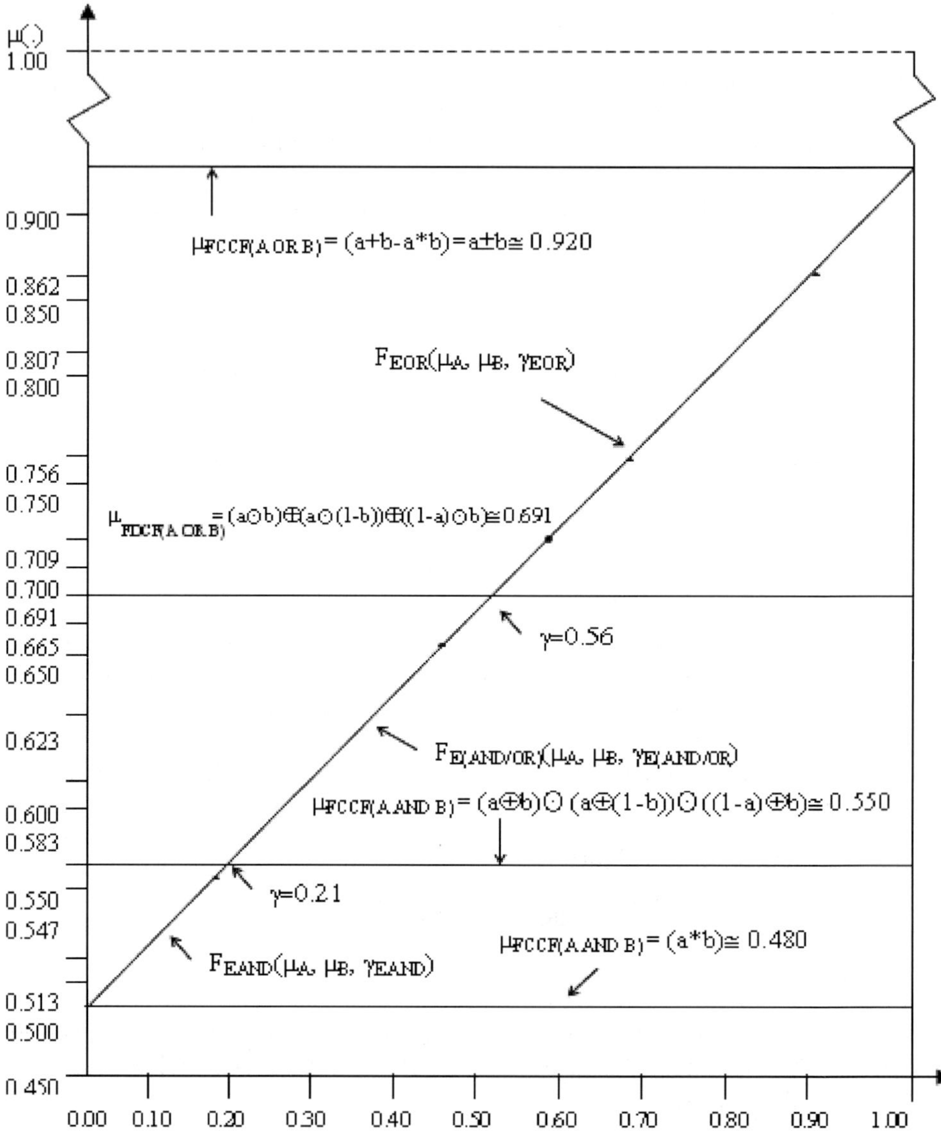


**Figure 11.4.** Intervals of "A AND B", "A OR B", expressed in FDCF and FCCF and "Exponential Compensatory AND" using algebraic Sum-Product realization of t-conorm and t-norm for $\mu_A=0.8$, $\mu_B=0.6$ and $\gamma \in [0,1]$

## 11.5. An Observation

An investigation of the FCCF(A AND B) versus FDCF(A OR B) for the special case of max-min operators and the standard-complement, under the eight possible cases of the Fuzzy Truth Table (Table 11.2), shows that

$$(a \vee b) \wedge (n(a) \vee b) \wedge (a \vee n(b)) \leq (a \wedge b) \vee (n(a) \wedge b) \vee (a \wedge n(b))$$

That is to say,

$$\mu_{FCCF(A\ AND\ B;\vee,\wedge,n(.))} \leq \mu_{FDCF(A\ OR\ B;\vee,\wedge,n(.))} \tag{11.9}$$

(see Appendix 11.3). Furthermore, with a cumbersome but detailed analysis, it can be shown that

$$(a \oplus b) \odot (n(a) \oplus b) \odot (a \oplus n(b)) \leq (a \odot b) \oplus (n(a) \odot b) \oplus (a \odot n(b)) \tag{11.10}$$

That is to say,

$$\mu_{FCCF(A\ AND\ B;(\oplus,\ddot{Y},\, n(.)))} \leq \mu_{FDCF(A\ OR\ B;(\oplus,\ddot{Y},n(.)))} \tag{11.11}$$

These results might seem counter-intuitive but are true for the two special cases. What they say is that the upper bound of "AND" is less than or equal to the lower bound of "IVFS OR", for these two conjugate pairs. In no case is the upper bound of "IVFS AND" greater than the lower bound of "OR". So, the results imply that FDCF and FCCF intervals of "AND" and "OR" have no fuzzy overlap for these operators. In Figure 11.4, we observe a special case of the inequality (11.11).

At this juncture, it is conjunctured that, in general,

$$\mu_{FCCF(A\ AND\ B;\nabla,\Delta,n(.))} \leq \mu_{FDCF(A\ OR\ B;\nabla,\Delta,n(.))}$$

The results (11.9) and (11.10) help us clarify the relations between "AND" and "OR" and "Exponential-Compensatory AND" for these special cases. Recall that $\gamma^*$ was defined by as

$$\gamma^* = \ln[\mu_{FCCF(A\ AND\ B)}/\Delta(a,b)]/ \ln[\nabla(a,b)/ \Delta(a,b)], \qquad 0 \leq \gamma^* \leq 1$$

and $\gamma_0$ was defined by (11.8) as

$$\gamma_0 = \ln[\mu_{FDCF(A\ OR\ B)}/\Delta(a,b)]/ \ln[\nabla(a,b)/ \Delta(a,b)], \qquad 0 \leq \gamma_0 \leq 1$$

Since, by the results obtained above,

$$\mu_{FCCF(A\ AND\ B)} \leq \mu_{FDCF(A\ OR\ B)}$$

for the operator sets $(\vee,\wedge,n(.))$ and , it is clear that $\gamma^* \leq \gamma_0$. Therefore, since $\gamma \in [0,1]$, there are values of $\gamma \in [\gamma^*,\gamma_0]$ for the "compensatory AND", i.e.,

$$A \Theta B \in [FCCF(A\ AND\ B),\ FDCF(A\ OR\ B)]$$

for the special case of $(\vee,\wedge,n(.))$ and $(\oplus,\odot,n(.))$.

Combined with Theorem 11.1 and Theorem 11.2, this allows us to say the following about the expressive power of "Exponential-Compensatory AND" for the operator sets $(\vee,\wedge,n(.))$ and $(\oplus,\odot,n(a))$. When $\gamma \in [0,\gamma^*]$ "Exponential-Compensatory AND" exposes the granulation of linguistic "AND" operator in its FDCF and FCCF interval. Similarly, when $\gamma \in [\gamma_0,1]$ "Exponential Compensatory AND" exposes the granulation of the linguistic "OR" operator within its FCCF and FDCF interval. The case of $\gamma \in [\gamma^*,\gamma_0]$ is examined next. In order to generalize this result one needs to show that

$$\Delta[\Delta[\nabla(a,b),\nabla(a,n(b))],\nabla(n(a),b)] \leq \nabla[\nabla[\Delta(a,b),\ \Delta(a,n(b))],\ \Delta(n(a),b)]$$

It is shown in Appendix 11.3 that

$$\mu_{FCCF(A\ AND\ B;\vee,\wedge,n(.))}(x,y) \leq \mu_{FDCF(A\ OR\ B;\vee,\wedge,n(.))}(x,y)$$

hold. It can also be shown that this inequality holds for all boundary conditions, and for all t-norms and conorms, i.e., $a,b \in \{0,1\}$. From the monotonicity of t-norms and t-conorms, it is clear that the inequality holds for all possible values of $a,b \in [0,1]$, i.e.,

$$\mu_{FCCF(A\ AND\ B;\nabla,\Delta,n(.))}(x,y) \leq \mu_{FDCF(A\ OR\ B;\nabla,\Delta,n(.))}(x,y).$$

These are graphically shown in Figures 11.3 and 11.4.It can also be observed that the LHS of the inequality gets smaller and smaller as we move from (Max, Min,n(.)) to $(\oplus,\odot,n(.))$whereas RHS gets larger and larger until we get the case of (Max, Min,n(.)) for both RHS and the left-hand side as demonstrated in Figure 11.3. This means that the interval between FCCF(AND) and FDCF(OR) expose the granulation of "AND-OR"ness of these linguistic operators.

## 11.6. "Convex-Linear-Compensatory AND"

The "Convex-Linear-Compensatory AND" suggested by Zimmermann [1986] is as indicated in (11.3). Consider the generalization (11.3'), i.e.,

$$f_1(\gamma)=(1-\gamma)\Delta(a,b)+\gamma\nabla(a,b).$$

It is clear that

$$f_1(0)=\Delta(a,b), \text{ and } f_1(1)=\nabla(a,b)$$

Furthermore, by differentiating with respect to $\gamma$, we obtain

$$\partial f_1(\mu_{FCCF(A\ AND\ B;\nabla,\Delta,n(.))}(x,y)\ \nabla\ )/\partial\gamma=-\Delta(a,b)+\nabla(a,b).$$

Since both $\Delta(a,b)$ and $\nabla(a,b)$ are monotonic and $\Delta(a,b)\le\nabla(a,b)$, $f_1(\gamma)$ is a monotonic function of $\gamma$.In this case, we need to determine the values of a, b, $\gamma$ such that the following inequalities hold:

i) $\mu_{FDCF(A\ AND\ B)} \le (1-\gamma)\Delta(a,b)+\gamma\nabla(a,b) \le \mu_{FCCF(A\ AND\ B)}$

ii) $\mu_{FDCF(A\ OR\ B)} \le (1-\gamma)\Delta(a,b)+\gamma\nabla(a,b) \le \mu_{FCCF(A\ OR\ B)}$

For the sake of illustration, let us examine the case of the algebraic sum-product. For inequality i) we need to determine the values of a, b and $\gamma$ such that the following inequalities hold:

$$(a\odot b)\le(1-\gamma)(a\odot b)+\gamma(a\oplus b) \le (a\oplus b) \odot (n(a)\oplus b) \odot (a\oplus n(b)). \qquad (11.12)$$

From the LHS of (11.12), we have $0 \le \gamma[(a\oplus b)-(a\odot b)]$

which is satisfied for $0 \le \gamma$ and $a,b \in [0,1]$.

From the RHS of (11.12), we have:

$$\gamma \le \gamma^+=\{[(a\oplus b) \odot (n(a)\oplus b) \odot (a\oplus n(b))]-(a\odot b)\}/[(a\oplus b)-(a\odot b)] \le 1. \qquad (11.13)$$

Thus, it is clear that inequality i) holds for

$\gamma \in [0,\gamma^+]$, $0 < \gamma^+ \le 1$, $a,b \in [0,1]$, provided a and b are not both equal to 0 in (11.13), i.e., $a,b\in(0,1]$. The restriction on $\gamma^+$ arises from the characteristic of "Convex-Linear-Compensatory AND", but $\gamma$ is irrelevant when a and b both = 0.

For inequality ii) we need to determine the values of a, b and $\gamma$ such that the following inequalities hold:

$$(a\odot b)\oplus(n(a)\odot b)\oplus(a\odot n(b)) \le (1-\gamma)(a\odot b)\oplus\gamma(a\oplus b) \le a \oplus b. \qquad (11.14)$$

From the RHS of (11.14), we have $(1-\gamma)(a\odot b)+\gamma(a\oplus b) \le a \oplus b$.

Solving for $\gamma$, we find $\gamma \le [(a\oplus b)-(a\odot b)] / [(a \oplus b)-(a\odot b)]=1$.

From the LHS of (11.14), we have $(a\odot b)\oplus(n(a)\odot b)\oplus(a\odot n(b)) \le (1-\gamma)(a\odot b)+\gamma(a\oplus b)$

Solving for $\gamma$, it can be shown that

$$\gamma^{++} = \{[(a\odot b)\oplus(n(a)\odot b)\oplus(a\odot n(b))]-(a\odot b)\}/[(a\oplus b)-(a\odot b)] \le \gamma,$$

i.e.,

$$\gamma^{++} = [\mu_{FDCF(A\ OR\ B)} - (a\odot b)]/[\mu_{FCCF(A\ OR\ B)} - (a\odot b)] \le \gamma,$$

Since

$[\mu_{FDCF(A\ OR\ B)} \le \mu_{FCCF(A\ OR\ B)} - (a\odot b)]$, and $(a\odot b)\le \mu_{FDCF(A\ OR\ B)}$, $0 \le \gamma^{++} \le 1$.

Thus, inequality ii) holds when

$$\gamma \in [\gamma^{++},1], 0 \le \gamma^{++} \le 1, \text{ for all } a,b \in [0,1].$$

## 11.7. An Observation

Recall that by (11.11)

$$\mu_{FCCF(A\ AND\ B;(\oplus,\ddot{Y},n(.)))} \le \mu_{FDCF(A\ OR\ B;(\oplus,\ddot{Y},n(.)))}$$

Therefore, for this operator set, we have a similar conclusion concerning values of $\gamma \in [\gamma^{+},\gamma^{++}]$ as we had for the values of $\gamma \in [\gamma^{*},\gamma_0]$, in section 11.7. When $\gamma \in [\gamma^{+},\gamma^{++}]$ "Convex-Linear-Compensatory AND" exposes the granulation of the combined linguistic operator, i.e. "AND OR"ness. When $\gamma \in [\gamma^{+},1]$ it exposes the granulation of the linguistic operator "OR"; when $\gamma \in [0,\gamma^{+}]$ it exposes the granulation of the linguistic operator "AND".

## 11.8. Conclusion

Once again it is clear that the break down of the classical equivalences generate a richer representation in fuzzy theory.

The "Compensatory AND" was discovered in the experiments conducted by Zimmermann and Zysno [1980]. Türk□en [1986, 2002] showed that FDCF and

FCCF forms of logical expressions, which are equivalent to their counterpart DNF and CNF *in form only* respectively in two valued theory, split apart in fuzzy set and logic theory as it was shown in previous chapters. Here, we have shown that the "Compensatory AND" proposed by Zimmermann and Zysno [1980] has a granulation within the interval bounds of the FDCF and FCCF expressions of the linguistic connectives "AND" and "OR". It depends on the value of the compensatory parameter. Specifically, when $\gamma \in [0,\gamma^*]$, for "Exponential-Compensatory AND", or $\gamma \in [0,\gamma^+]$, for "Convex-Linear-Compensatory AND", "Compensatory AND" exposes graded meaning representation within the FDCF and FCCF intervals of "AND". Under these conditions the operator $\gamma$ expresses the graded meaning of the linguistic operator "AND". On the other hand, when $\gamma \in [\gamma_0,1]$, for "Exponential-Compensatory AND", or $\gamma \in [\gamma^{++},1]$, for "Convex-Linear-Compensatory AND", the value of the operator $\gamma$ expresses the graded meaning of the linguistic operator "OR".

Furthermore, for the special cases of max-min and algebraic sum-product operators, "Compensatory AND" is also contained between FCCF(A AND B) and FDCF(A OR B). That is, for these operator sets, when $\gamma \in [\gamma^*,\gamma_0]$, for "Exponential-Compensatory AND", or $\gamma \in [\gamma^+,\gamma^{++}]$, for "Convex-Linear-Compensatory AND", "Compensatory AND". Furthermore, the $\gamma$ operator of "Compensatory AND" expresses graded meaning of the combined linguistic variables of "AND" and "OR", and identifies "AND-OR"ness between the upper bound of the linguistic operator "AND" and the lower bound of "OR", i.e., FDCF(OR). Under these conditions the operators can be said to express the gradation of a true mixture of meaning of "AND" and "OR".

These results clarify the expressive power of the experimental "Compensatory AND" by exhibiting its relation to the linguistic "AND" and "OR" boundaries. An examination of the actual experimental values of $\gamma$in terms of these containment relations may serve to help us understand the real-life distribution of $\gamma$, and hence to understand more clearly the human use of linguistic “AND”, “OR”, etc. It is thus clear once again that linguistic “AND”, “OR” do not correspond to t-norm and t-conorm isomorphically.

**Appendix 11.1.**

A detailed analysis of the IVFS "A AND B" and the "Exponential-Compensatory AND" for the case of max ($\vee$) - min ($\wedge$) operators.

**Case 1:** $a \le b$

**Subcase 1.1**: For the case $a \le b \le n(b) \le n(a)$,

since $\mu_{FDCF}$=a, $\mu_{FCCF}$=b we have: $a \le a^{1-\gamma} b^{\gamma} \le b$. From the left hand side, LHS, we get $1 \le (b/a)^{\gamma}$ which holds for $\gamma, a, b \in [0,1]$; from right hand side, RHS, we get $(b/a)^{\gamma} \le (b/a)$, which holds for $\gamma, a, b \in [0,1]$. Therefore, $\mu_{FDCF} \le \mu_{FCCF}$. for all $\gamma, a, b \in [0,1]$.

**Subcase 1.2**: For the cases $n(b) \le n(a) \le a \le b$,
$n(b) \le a \le n(a) \le b$,

since $\mu_{FDCF}=\mu_{FCCF}$=a , we have: $a = a^{1-\gamma} b^{\gamma}$, and thus $1 = (b/a)^{\gamma}$, which holds only for $\gamma$=0 or a=b.

Therefore, $\mu_{FDCF}=\mu_{\gamma}=\mu_{FCCF}$ only for $\gamma$=0 or a=b.

**Subcase 1.3**: For the case $a \le n(b) \le b \le n(a)$, since $\mu_{FDCF}$=a, $\mu_{FCCF}$=n(b), we have: $a \le a^{1-\gamma} b^{\gamma} \le n(b)$. From LHS, we get $1 \le (b/a)^{\gamma}$ which holds for all $\gamma, a, b \in [0,1]$. From RHS, we get $(b/a)^{\gamma} \le (n(b)/a)$, which holds only when $\gamma \le \ln(n(b)/a)/\ln(b/a) \le 1$.

Therefore $\mu_{FDCF} \le \mu_{\gamma} \le \mu_{FCCF}$ for

$\gamma \in [0, \ln(n(b)/a)/\ln(b/a)]$, $a \ne b \ne 0$, $n(b) \ne 0$.

A numerical example 1.1: For the subcase $a \le b \le n(b) \le n(a)$, consider the values: $a=.09 \le b=.25 \le n(b)=.75 \le n(a)=.91$
From $1 \le (b/a)^{\gamma} = (.25/.09)^{\gamma}$, $\gamma \in [0,1]$. Thus we have:

$\mu_{FDCF(\,.\,,\wedge,\vee,-)}(.09,.25) = .09$, $\mu_{FCCF(\,.\,,\wedge,\vee,-)}(.09,.25) = .25$, and
$\mu_{\gamma}(.09,.25) = (.09)^{1-\gamma}(0.25)^{\gamma}$ .

Therefore, $.09 \le (0.9)^{1-\gamma}(.25)^{\gamma} \le .25$.
In particular, for $\gamma$=.5, $.09 \le .15 \le .25$
A second numerical example 1.2: For the subcase $a \le n(b) \le b \le n(a)$, consider the values: $a=.09 \le n(b)=.19 \le b=.81 \le n(a)=.91$. From $1 \le (b/a)^{\gamma} = (.81/.09)^{\gamma}$, $\gamma \in [0,1]$;
and from $(b/a)^{\gamma} \le (n(b)/a)$, for $\gamma \in [0, \ln(n(b)/a)/\ln(b/a)]$
we get $(.81/.09)^{\gamma} \le (.19/.09)$; $9^{\gamma} \le 2.111$.

In particular, for example, $\gamma$=1/4 or $\gamma$=1/3 satisfies the above inequality. Therefore, we have:
$\mu_{FDCF(\,.\,,\wedge,\vee,-)}(.09,.81) = .09$,
$\mu_{FCCF(\,.\,,\wedge,\vee,-)}(.09,.81) = .19$, and

$\mu_\gamma(.09,.81) = (.09)^{1-\gamma}(0.81)^\gamma$
$.09 \le (0.9)^{1-\gamma}(.81)^\gamma \le .19$, for all $\gamma \in [0,\ln(.19/.09)/\ln(.81/.09)]$

**Case 2:** $b \le a$

**Subcase 2.1**: For the cases $n(a) \le a \le n(b) \le b$,
$n(a) \le n(b) \le b \le a$,

since $\mu_{FDCF}$=$\mu_{FCCF}$=b, we thus have: $b=b^{1-\gamma}a^\gamma$.

Therefore, $1 \le (b/a)^\gamma$ which only holds when $\gamma=0$ or $a=b$.

**Subcase 2.2**: For the case $b \le a \le n(a) \le n(b)$,

since $\mu_{FDCF}$=b, $\mu_{FCCF}$=a, we thus have: $b \le b^{1-\gamma}a^\gamma \le a$.
From LHS, we get $b \le b^{1-\gamma}a^\gamma$, and $1 \le (a/b)^\gamma$ which holds for $\gamma,a,b \in [0,1]$.

From RHS, we get $b^{1-\gamma}a^\gamma \le a$, and $(a/b)^\gamma/(a/b)$, which holds for $\gamma,a,b \in [0,1]$.

**Subcase 2.3**: For the case $b \le n(a) \le a \le n(b)$,
since $\mu_{FDCF}$=b, $\mu_{FCCF}$=n(a), we thus have: $b \le b^{1-\gamma}a^\gamma \le n(a)$.
From LHS, we get $1 \le (a/b)^\gamma$ which holds for $\gamma,a,b \in [0,1]$.
From RHS, we get $b^{1-\gamma}a^\gamma \le n(a)$, $(a/b)^\gamma \le (n(a)/b)$, which holds when $\gamma \le \ln(n(a)/b)/\ln(a/b)]$.

Therefore, $\gamma \in [0,\ln(n(a)/b)/\ln(a/b)]$, $a \ne b \ne 0$, $n(a) \ne 0$.

A numerical example 2.1: Subcase $b \le a \le n(a) \le n(b)$

For the values of $b{=}.16 \le a{=}.49 \le n(a){=}.51 \le n(b){=}.84$, the boundaries are,
$\mu_{FDCF}$=.16, $\mu_{FCCF}$=.49; we have: $\mu_\gamma{=}(.16)^{1-\gamma}(0.49)^\gamma$.

From $1 \le (.49/.16)^\gamma$, $\gamma \in [0,1]$.
From $(.49/.16)^\gamma \le (.49/.16)^\gamma$, $\gamma \in [0,1]$.
Therefore, $.16 \le (16)^{1-\gamma}(.49)^\gamma \le .49$, $\gamma \in [0,1]$.

In particular, for example, for $\gamma{=}.5$, we have:

$$.16 \le (.4)(.7) \le .49, .16 \le .28 \le .49.$$

A second numerical example 2.2: Subcase $b \le n(a) \le a \le n(b)$
For the values of $b{=}.04 \le n(a){=}.36 \le a{=}.64 \le n(b).96$ the boundaries are:

$\mu_{FDCF}$=.04, $\mu_{FCCF}$=.36; thus we have:

$1 \le (.64/.04)^{\gamma}$, $\gamma \in [0,1]$; and
$(.64/.04)\gamma \le (.36/.04)$, $\gamma \in [0, \ln(.36/.04)/\ln(.64/.04)]$

That is, the values of $\gamma \in [0, \ln 9/\ln 16]$ will produce desirable results. A graphic representation of some of these cases is shown in Figure 11.5.

**Appendix 11.2.**

For the special case of max ($\vee$) - min ($\wedge$) operators, a detailed analysis of the FDCF and CBF interval of "A OR B" and the "Exponential-Compensatory AND".

**Case 1:** $a \le b$

**Subcase 1.1**: For the cases $a \le b \le n(b) \le n(a)$
$a \le n(b) \le b \le n(a)$

since $\mu_{FDCF}$=b, $\mu_{FCCF}$=b, we have: $b = \mu^{\gamma} = a^{1-\gamma} b^{\gamma}$.

Therefore, $(b/a) = (b/a)^{\gamma}$, which holds only when $\gamma=1$ or $a=b$.

**Subcase 1.2**: For the case $n(b) \le a \le n(a) \le b$,

since $\mu_{FDCF}$=n(a), $\mu_{FCCF}$=b, we have: $n(a) \le a^{1-\gamma} b^{\gamma} \le b$

From LHS, we get $(n(a)/a) \le (b/a)^{\gamma}$, which holds when $\ln(n(a)/a)/\ln(b/a) \le \gamma$, $a \ne b \ne 0$.

From RHS, we get $(b/a)^{\gamma} \le (b/a)$..which holds for $\gamma, a, b \in [0,1]$.
Therefore $\mu_{FDCF} \le \mu_{\gamma} \le \mu_{FCCF}$ for $\gamma \in [\ln(n(a)/a)/\ln(b/a), 1]$, $a \ne b \ne 0$.

**Subcase 1.3**: For the case $n(b) \le n(a) \le a \le b$,

since $\mu_{FDCF}$=a, $\mu_{FCCF}$=b, we have: $a \le a^{1-\gamma} b^{\gamma} \le b$

From LHS, we get $1 \le (b/a)^{\gamma}$, which holds for $\gamma, a, b \in [0,1]$.

From RHS, we get $(b/a)^{\gamma} \le (b/a)$, which holds for $\gamma, a, b \in [0,1]$.
Therefore $\mu_{FDCF} \le \mu_{\gamma} \le \mu_{FCCF}$ , which holds for $\gamma, a, b \in [0,1]$.

**Case 2**: $b \le a$

**Subcase 2.1**: For the case $n(b) \le n(a) \le a \le b$,

since $\mu_{FDCF}$=b, $\mu_{FCCF}$=a, we thus have: $b \le b^{1-\gamma} a^{\gamma} \le a$

From LHS, we get $1 \le (b/a)^{\gamma}$, which holds for $\gamma, a, b \in [0,1]$.

From RHS, we get $(a/b)^{\gamma} \le (a/b)$, which holds for $\gamma, a, b \in [0,1]$.

Therefore $\mu_{FDCF} \le \mu_{\gamma} \le \mu_{FCCF}$, which holds for $\gamma, a, b \in [0,1]$.

**Subcase 2.2**: For the case $n(a) \le b \le n(b) \le a$,

since $\mu_{FDCF}$=n(b), $\mu_{FCCF}$=a, we thus have: $n(b) \le b^{1-\gamma} a^{\gamma} \le a$

From LHS, we get $(n(b)/b) \le (a/b)^{\gamma}$, which holds when $1 \ge \gamma \ge \ln(n(b)/b)/\ln(a/b) \le \gamma,\ a \ne b \ne 0$.

From RHS, we get $(a/b)^{\gamma} \le (a/b)$, which holds for $\gamma, a, b \in [0,1]$.

Therefore $\mu_{FDCF} \le \mu_{\gamma} \le \mu_{FCCF}$, for all $\gamma \in [\ln(n(b)/b)/\ln(a/b), 1],\ a \ne b \ne 0$.

**Subcase 2.3**: For the cases $b \le a \le n(a) \le n(b)$,

$b \le n(a) \le a \le n(b)$,

since $\mu_{FDCF}$=a, $\mu_{FCCF}$=a, we thus have: $a = b^{1-\gamma} a^{\gamma}$.

Therefore, $(b/a) \le (b/a)^{\gamma}$, which holds only for $\gamma=1$ or $a=b$.

## Appendix 11.3. A Truth Table Analysis of the containment relationship between $\mu_{FCCF(A\ AND\ B;\vee,\wedge,n(.))} \le \mu_{FDCF(A\ OR\ B;\vee,\wedge,n(.))}$.

$$\underbrace{(a \vee b) \wedge (a \vee n(b)) \wedge (n(a) \vee b)}_{(1)} \overset{?}{\le} \underbrace{(a \wedge b) \vee (a \wedge n(b)) \vee (n(a) \wedge b)}_{(2)}$$

| possible cases | (1) | ? | (2) |
|---|---|---|---|
| $a \le b \le n(b) \le n(a)$ | $b \wedge n(b) \wedge n(a) = b$ | (=) | $a \vee a \vee b = b$ |
| $a \le n(b) \le b \le n(a)$ | $b \wedge n(b) \wedge n(a) = n(b)$ | (≤) | $a \vee a \vee b = b$ |
| $n(a) \le b \le n(b) \le a$ | $b \wedge a \wedge b = b$ | (≤) | $b \vee n(b) \vee n(a) = n(b)$ |
| $n(a) \le n(b) \le b \le a$ | $a \wedge a \wedge b = b$ | (=) | $b \vee a \vee b = a$ |
| $b \le a \le n(a) \le n(b)$ | $a \wedge n(b) \wedge n(a) = a$ | (=) | $b \vee a \vee b = a$ |
| $b \le n(a) \le a \le n(b)$ | $a \wedge n(b) \wedge n(a) = n(a)$ | (≤) | $b \vee a \vee b = b$ |
| $n(b) \le a \le n(a) \le b$ | $b \wedge a \wedge b = a$ | (≤) | $a \vee n(b) \vee n(a) = n(a)$ |
| $n(b) \le n(a) \le a \le b$ | $b \wedge a \wedge b = a$ | (=) | $a \vee n(b) \vee n(a) = a$ |

## References


1. Türkşen I.B. (2002), "Belief, Plausibility and Probability Measures over Type 2 Fuzzy Sets", *I.J. of Applied Mathematics and Computer Science*, (to appear).
2. Türkşen I.B. (2002), "Type 2 Representation and Reasoning for CWW", *FSS*, 127, 17-36.

3. Türkşen I.B. (2002), "Interval-valued Type 2 Fuzzy Sets, Multi-valued Maps and Rough Sets", (in) A Grmela and N.E.Mastorakis (eds.), Advances in Intelligent Systems, *Fuzzy Sets*, Evolutionary Computation, WSEAS, 142-146.
4. Türkşen I.B. (2002), "Upper and Lower Set Formulas: A Modification of Dempster-Pawlak Formalism", *Special Issue of the International Journal of Applied Mathematics and Computer Science,* (submitted).
5. Türkşen, I.B. (1992), "Interval Valued Fuzzy Sets and 'Compensatory AND' ", *Fuzzy Sets and Systems*, 51, 295-307.
6. Türkşen I.B. (1986), "Interval-Valued Fuzzy Sets based on Normal Forms", *Fuzzy Sets and Systems,* 191-210.
7. Zimmermann, H.J., P. Zysno (1985), "Quantifying Vagueness in Decision Models", *European Journal of Operational Research*, 22, 148-154.
8. Zimmermann, H.J., P. Zysno (1980), "Latent Connectives in Human Decision Making", *Fuzzy Sets and Systems*, 4, 37-51.

# Chapter 12

# BELIEF, PLAUSIBILITY AND PROBABILITY MEASURES ON INTERVAL-VALUED TYPE 2 FUZZY SETS

Within last fourty years, Zadeh (1965-2004) has proposed seminal concepts for fuzzy sets, approximate reasoning, Type 2 fuzzy sets, CWW, Computing With Words, and CWP, Computing With Perceptions.

For about twenty years or so, we have shown that the combination of two or more Type 1 fuzzy sets with linguistic connectives, i.e., “AND”, “OR”, etc., generate interval-valued Type 2 fuzzy sets (Türkşen, 1982-2002) due to the break down of classical equivalences in fuzzy theory.

Shortly after the introduction of fuzzy sets Dempster (1967) proposed upper and lower probabilities induced by multi-valued maps that generate upper and lower sets. About a few years later, Shafer (1976) introduced the belief measures in response to the opponents of Bayesian theory. About twenty years ago, Pawlak (1982, 1991) proposed the rough sets.

Recently, we have proposed a T-formalism as a modification and restriction of Demspter's approach (Türkşen, 2001, 2002). Along the way, it was shown that there are correspondences between Dempster's and Pawlak's approaches and that Dempster's approach is more general than Pawlak's. (Türkşen, 1982-2002) which was discussed in Chapter 8.

Briefly, in particular, Pawlak (1991) proves that belief measures are outer measures of basic assignments if the focal elements are mutually exclusive. In a number of related papers, T.Y.Lin (1997, 1999) proves that belief measures are inner measures (1999) for the general case when focal elements overlap. In all these investigations the aim is concentrated on the belief measures.

In this chapter, we first start with Interval-valued Type 2 fuzzy sets that are defined by lower and upper membership functions generated by FDCF, Fuzzy Disjunctive Canonical Form, and FCCF, Fuzzy Conjunctive Canonical Form, expressions computed with two Type 1 fuzzy sets. We have shown previously

(Türkşen, 2001-2002) that FDCF and FCCF boundaries are naturally generated by T-Formalism which is a modified and restricted extension of Dempster's multi-valued maps. (Chapter 8) The proposed T-formalism generates FDCF and FCCF, Fuzzy Disjunctive and Fuzzy Conjunctive Canonical Forms for "AND", "OR" and "other" classes of combination of concepts.

Secondly, we review the basic axioms of belief measures and apply them to each boundary of an Interval-valued Type 2 fuzzy set defined by FDCF and FCCF expressions. It should be recalled that each FDCF and FCCF expression identifies a Type 1 fuzzy set membership function.

In this chapter, we show how the probabilities, beliefs and plausibilities can be computed on interval-valued Type 2 fuzzy sets. In turn, it is suggested that FDCF and FCCF formulas should be taken as the schemas of copcept in combinations CWW. Furthermore, these schemas are enriched by probabilities, belief and plausibility measures over interval-valued Type 2 fuzzy sets. Therefore, these schemas provide representation of uncertainties associated with fuzzy rules. Thus their interpretation cause CWW and CWP to have a more expressive power.

As well, we show once again, as we had indicated before (Türkşen, 2001), that fuzzy sets provide a "description" whereas belief, plausibility, probability, possibility measures on fuzzy set help us to deal with different types of uncertainty.

## 12.1. Belief and Plausibility over Fuzzy Sets

In classical setting, Belief, **Bel**, Plausibility, **Pl**, and Probability, **P**, measures are defined over crisp sets. In this section, we first define **Bel** and **Pl** on Type 1 fuzzy sets. Secondly, we investigate **Bel** and **Pl** on Interval-valued Type 2 fuzzy sets which are generated by combinations of Type 1 fuzzy sets where the combination operators are linguistic connectives "AND", "OR", etc.

Briefly, we first discuss what we mean by **Bel**(A) when A is a Type 1 fuzzy set. Secondly, we investigate **Bel**($A_1$ OR $A_2$) instead of **Bel**($A_1 \cup A_2$) and **Pl**($A_1$ AND $A_2$) instead of **Pl**($A_1 \cap A_2$) when both $A_1$ and $A_2$ are Type 1 fuzzy sets. But, we have to re-structure and re-state **Bel** and **Pl** over $\alpha$-level (crisp sets) of the fuzzy sets since **Bel** is a fuzzy measure over the crisp sets.

We know from our previous analysis that "$A_1$ OR $A_2$" and "$A_1$ AND $A_2$" have lower and upper bound approximations known as FDCF($A_1$ OR $A_2$), FCCF($A_1$ OR $A_2$), and FDCF($A_1$ AND $A_2$), FCCF($A_1$ AND $A_2$), respectively. (Türkşen, 1999, 2002)

Therefore, we define:

$\mathbf{Bel}(A_1 \text{ OR } A_2) \geq [\mathbf{Bel}_L(A_1 \text{ OR } A_2), \mathbf{Bel}_U(A_1 \text{ OR } A_2)]$
where $\mathbf{Bel}_L(A_1 \text{ OR } A_2) = \mathbf{Bel}[\text{FDCF}(A_1 \text{ OR } A_2)]$
$= \mathbf{Bel}[(A_1 \cap A_2) \cup (c(A_1) \cap A_2) \cup (A_1 \cap c(A_2))]$
$\mathbf{Bel}_U(A_1 \text{ OR } A_2) = \mathbf{Bel}[\text{FCCF}(A_1 \text{ OR } A_2)] = \mathbf{Bel}[A_1 \cup A_2]$

where the interval $[\mathbf{Bel}_L(A_1 \text{ OR } A_2), \mathbf{Bel}_U(A_1 \text{ OR } A_2)]$ defines the uncertainty interval of $\mathbf{Bel}(A_1 \text{ OR } A_2)$ generated by the linguistic "OR" combination of two

Type 1 fuzzy sets $A_1$ and $A_2$ and c(A) is the complement of A. As well, we define:

$\mathbf{Pl}(A_1 \text{ AND } A_2) \leq [\mathbf{Pl}_L(A_1 \text{ AND } A_2), \mathbf{Pl}_U(A_1 \text{ AND } A_2)]$
where $\mathbf{Pl}_L(A_1 \text{ AND } A_2) = \mathbf{Pl}[\text{FDCF}(A_1 \text{ AND } A_2)]$
$= \mathbf{Pl}[A_1 \cap A_2\}$

$\mathbf{Pl}_U(A_1 \text{ AND } A_2) = \mathbf{Pl}[\text{FCCF}(A_1 \text{ AND } A_2)]$
$= \mathbf{Pl}[(A_1 \cup A_2) \cap (c(A_1) \cup A_2) \cap (A_1 \cup c(A_2))]$

where the interval $[\mathbf{Pl}_L(A_1 \text{ AND } A_2), \mathbf{Pl}_U(A_1 \text{ AND } A_2)]$ defines the uncertainty interval of $\mathbf{Pl}(A_1 \text{ AND } A_2)$ generated by the linguistic "AND" combination of two Type 1 fuzzy sets $A_1$ and $A_2$.

Recall that the Interval-valued Type 2 fuzzy sets are generated by the combination of two Type 1 fuzzy sets, $A_1$ and $A_2$ with linguistic connectives "AND", "OR", etc. For this purpose, we next briefly review the basic definitions of **Bel** and **Pl** first from classical perspective and then from a fuzzy perspective. Afterwards, we determine how we can compute the **Bel** and **Pl** measures on the interval-valued Type 2 fuzzy sets.

As well, we illustrate the related concepts for upper and lower beliefs and probabilities over interval-valued Type 2 fuzzy sets.

### *12.1.1. Belief Measures*

A belief measure is a function:
$\mathbf{Bel}: \mathcal{P}(X) \rightarrow [0,1]$

that satisfies the axioms of fuzzy measures that are known as:
(1) Boundary Condition: $h(\varnothing)=0$, $h(X)=1$
(2) Monotonicity: For every $A_1, A_2 \in \mathcal{P}(X)$ if $A_1 \subseteq A_2$, then $h(A_1) \leq h(A_2)$.
(3) Continuity: For ever sequence, $A_i \in \mathcal{P}(X)$, $i \in \{1,2,...,\}$ of subsets of X, if either $A1 \subseteq A2 \subseteq ...$ or $A1 \supseteq A2 \supseteq ...$, i.e., the sequence is monotonic, then

$$\lim_{i\to\infty} h(A_i) = h(\lim_{i\to\infty} A_i).$$

The boundary axiom states that if the element in question definitely does belong to the empty set, then the degree of evidence associated with the empty set is zero. On the other hand, if the element in question definitely belongs to the universal set, then the degree of evidence associated with the universal set is one.

The monotonicity axiom states that the evidence of the membership of an element in a set must be at least as large as the evidence that the element belongs to any subset of that set.

The continuity axiom states that for every infinite sequence $A_1, A_2, \ldots$ of nested (monotonic) subsets of X that converge to the set

$$A = \lim_{i\to\infty} A_i$$

the sequence of associated fuzzy measures, $h(A_i)$, $i=1,2,\ldots$, must converge to the fuzzy measure $h(A)$.

More generally, a fuzzy measure is defined as:
$h: \mathbf{B} \to [0,1]$
where $\mathbf{B} \subset \mathcal{P}(X)$ is a family of subsets of such that:

(i) $\varnothing \in \mathbf{B}$, and $X \in \mathbf{B}$, (A-1)
(ii) If $A \in \mathbf{B}$, then $c(A) \in \mathbf{B}$, where $c(A)$ is the complement of A. (A-2)
(ii) $\mathbf{B}$ is closed under the operation of set union, i.e., if $A_1 \in \mathbf{B}$ and $A_2 \in \mathbf{B}$, then $A_1 \cup A_2 \in \mathbf{B}$. (A-3)

The set $\mathbf{B}$ is generally known as "Borel field" or "σ-field". Since we have $A_1 \cup A_2 \supseteq A_1$ and $A_1 \cup A_2 \supseteq A_2$, we get $\vee[h(A_1),h(A_2)] \leq h(A_1 \cup A_2)$ due to monotonicity of h. Similarly, since we have $A_1 \cap A_2 \subseteq A_1$ and $A_1 \cap A_2 \subseteq A_2$, we get $h(A_1 \cap A_2) \leq \wedge[h(A_1),h(A_2)]$ where $\vee$=Max and $\wedge$=Min operators.

The ***belief measure***, in addition to the axioms discussed above, satisfies the following axiom:

$$\mathbf{Bel}(A_1 \cup A_2 \cup \ldots \cup A_n) \geq \sum_i \mathbf{Bel}(A_i) - \sum_{i<j} \mathbf{Bel}(A_i \cap A_j) + \ldots + (-1)^{n+1}\mathbf{Bel}(A_1 \cap A_2 \cap \ldots \cap A_n) \quad \text{(A-4)}$$

for every $n \in N$ and every collection of subsets of X. For n=2, this axiom is expressed as:

$\mathbf{Bel}(A_1 \cup A_2) \geq \mathbf{Bel}(A_1) + \mathbf{Bel}(A_2) - \mathbf{Bel}(A_1 \cap A_2)$.

For n=3, this axiom is expressed as

$\mathbf{Bel}(A_1 \cup A_2 \cup A_3) \geq \mathbf{Bel}(A_1) + \mathbf{Bel}(A_2) + \mathbf{Bel}(A_3) - \mathbf{Bel}(A_1 \cap A_2) - \mathbf{Bel}(A_1 \cap A_3)$
$- \mathbf{Bel}(A_2 \cap A_3) + \mathbf{Bel}(A_1 \cap A_2 \cap A_3)$

If the sets $A_1, A_2, \ldots, A_n$, in the axiom discussed above, are pairwise disjoint, i.e., $(A_i \cap A_j) = \emptyset$, for all i, j $\in$N such that i$\neq$j, the axiom requires that the degree of belief associated with the union of the sets is not smaller than the sum of the degrees of belief pertaining to the individual sets. The basic axiom of belief measures is thus a weaker version of the additive axiom of probability theory. This states that probability measures are special cases of belief measures.

It can be shown that the axiom of belief measures implies the monotonicity axiom of fuzzy measures. Let $A_1 \subset A_2$ and let $A_3 = A_2 - A_1$. Thus $A_1 \cup A_3 = A_2$ and $A_1 \cap A_3 = \emptyset$. Now applying the axiom of belief, we get:

$\mathbf{Bel}(A_1 \cup A_3) = \mathbf{Bel}(A_2) \geq \mathbf{Bel}(A_1) + \mathbf{Bel}(A_3) - \mathbf{Bel}(A_1 \cap A_3) = \mathbf{Bel}(A_1) + \mathbf{Bel}(A_3)$, and consequently $\mathbf{Bel}(A_2) \geq \mathbf{Bel}(A_1)$.

In turn let $A_1 = A$, and $A_3 = c(A)$, then, we get

$\mathbf{Bel}(A) + \mathbf{Bel}(c(A)) \leq \mathbf{Bel}(A \cup c(A))$

Thus $\mathbf{Bel}(A) + \mathbf{Bel}(c(A)) \leq 1$.

Associated with each belief measure is a plausibility measure, **Pl**, defined by

$\mathbf{Pl}(A) = 1 - \mathbf{Bel}(c(A))$

for all $A \in \mathcal{P}(X)$; and similarly,

$\mathbf{Bel}(A) = 1 - \mathbf{Pl}(c(A))$.

In theory, **Bel**(A), for each $A \in \mathcal{P}(X)$, is interpreted as the "degree of belief" based on the evidence that a given element $x \in X$ belongs to the set A. Explicitly, this interpretation should be written as $\mathbf{Bel}(A, A(x), x \in X)$ where A(x)=1 is assumed, i.e., the full membership of x in A. But implicitly, we can write it as **Bel**(A(x)). That is, we mean **Bel**(A(x)), $x \in X$, where A(x)=1 is the absolute degree of belonging in the classical interpretation. Naturally, with this understanding, we need to express our belief over a fuzzy set A, interms of its $\alpha$-cuts. Thus we write

$$\textbf{Bel}(A) = \sum_{\alpha} \alpha[\textbf{Bel}(A(x))_{\alpha}] \text{ where } A = \sum_{\alpha} \alpha(A(x))_{\alpha},$$

and $(A\ (x))_{\alpha}$ is the $\alpha$-level set, i.e., the crisp set at the $\alpha$-cut, because **Bel** is a fuzzy measure over crisp sets.

It should be noted that $\alpha$ levels sets have overlaps. As shown in the examples, these overlaps expose valuable information content. The information content hidden in the overlaps are treated with the basic axiom of **belief measures** (A-4). That is all possible conjunctions of the form, $A_i \cap A_j$, $A_i \cap A_j \cap A_k$, ..., $(A_1 \cap A_2 \cap \ldots \cap A_n)$ are treated as required by the basic axiom of belief measures which are the $\alpha$-level crisp sets of a fuzzy set. In turn, the $\alpha$'s are treated with the fuzzy conjunctions, i.e., t-norms as required by the fuzzy Type 1 fuzzy theory. It will be shown how fuzzy conjunctions of $\alpha$'s affect the weights assignable to **Bel** measures computed over $\alpha$-level (crisp) sets of Type 1 fuzzy sets.

*12.1.2. Degree of Evidence*

Every belief measure and its dual plausibility measure can be expressed in terms a function

$$m: \mathcal{P}(X) \rightarrow [0,1] \tag{12.1}$$

Such that $m(\varnothing)=0$, and $\sum_{A \in P(X)} m(A) = 1$,

where m(A) is interpreted either as:

(i) the degree of evidence supporting the claim that a specific element of X belongs to a set A but not to any special subset of A, or
(ii) the degree to which an expert believes that such a claim is warranted; and it is called the *basic assignment*.

The basic assignment has the following properties:
(i) It is not required that $m(X) = 1$
(ii) It is not required that $m(A) \leq m(B)$ when $A \subset B$.
(iii) No relationship between m(A) and m(c(A)) is required.

Given a basic assignment m, a belief measure and plausibility measure are uniquely determined by formulas

$$\textbf{Bel}(A) = \sum_{B \subseteq A} m(B), \text{ and} \tag{12.2}$$

$$\mathbf{Pl}(A) = \sum_{B \cap A \neq \varnothing} m(B) \tag{12.3}$$

for all $A \in \mathcal{P}(X)$.

There are associated special meanings between m(A), **Bel**(A), and **Pl**(A). On the one hand the basic assignment m(A) characterizes the degree of evidence or belief that the element in question belongs to the set A alone, i.e., exactly to the set A.

Whereas, **Bel**(A) represents the total evidence or belief that the element belongs to A as well as to all possible subset of A. On the other hand, the plausibility measure **Pl**(A) represents not only the total evidence or belief that the element in question belongs to A or to any of its subsets but also the additional evidence or belief associated with sets that overlap with A. Therefore, we get

$$\mathbf{Pl}(A) \geq \mathbf{Bel}(A) \tag{12.4}$$

for all $A \in \mathcal{P}(X)$.

Every set $A \in \mathcal{P}(X)$ for which $m(A)>0$ is usually called a focal element of m. The pair (F,m) is called a *body of evidence* where F denote a set of *focal elements* and m is the associated basic assignment.

*Total ignorance* is expressed in terms of the basic assignment by $m(X)=1$ and $m(A)=0$ for all $A \neq X$. It is also clear from above that the expression of total ignorance in terms of the corresponding belief measure that $\mathbf{Bel}(X)=1$ and $\mathbf{Bel}(A)=0$ for all $A \neq X$.

In turn, we have $\mathbf{Pl}(\varnothing)=0$ and $\mathbf{Pl}(A)=1$ for all $A \neq \varnothing$. A basic assignment is said to be a simple support function focused, say, at B, if $m(B)=f$, $m(X)=1-f$ and $m(B)=0$ for all other sets $B \in \mathcal{P}(X)$. Furthermore, evidence obtained in the same context from two independent sources, e.g., from two experts in the field of inquiry, and expressed by two basic assignments $m_1$ and $m_2$ on some power set $\mathcal{P}(X)$ must be appropriately combined to obtain a joint basic assignment $m_{1,2}$. In general, evidence can be combined in various ways, some of which may take into consideration the reliability of the sources and other relevant aspects. The standard way of combining evidence is expressed by the formula

$$m_{1,2}(B) = (1-K)^{-1} \sum_{C \cap D = B} m_1(C).m_2(D) \tag{12.5}$$

for $B \neq \varnothing$, where

$$K = \sum_{C \cap D = \varnothing} m_1(C).m_2(D) \tag{12.6}$$

and $m_{1,2}(\varnothing)=0$.

The formula (12.5) is known as Dempster's rule of combination. According to this rule, the degree of evidence $m_1(C)$ from the first source that focuses on set $C \in \mathcal{P}(X)$ and the degree of evidence $m_2(D)$ from the second source that focuses on $D \in \mathcal{P}(X)$ are combined by taking the product $m_1(C).m_2(D)$, which focuses on the intersection $C \cap D$. This is exactly the same way in which the joint probability distribution is calculated from two independent marginal distribution and, consequently, it is justified on the same grounds.

### *12.1.3. Belief on Type 1 Combination of Fuzzy Sets*

Suppose we are concerned with the assessment of Belief on the Type 1 combination of two Type 1 fuzzy sets, say, "$x \in X$ isr $A_1$", OR "$x \in X$ isr $A_2$".
As we have indicated, a fuzzy set A is a description. We are to assess the Belief on a fuzzy set A given that we have access to basic assignments provided by two independent experts.

**Example 1:**

Suppose a crime is committed and that there are three suspects, say, $S=\{S_1, S_2, S_3\}$. There is a witness who describes the suspect to be either "tall" or "heavy" build and gives a fuzzy set definition over the three suspects as follows:

tall $= (.3, S_1) \cup (.7, S_2) \cup (1, S_3)$
heavy $= (.4, S_1) \cup (.6, S_2) \cup (1, S_3)$

Therefore, we get: A = tall OR heavy when we construct a Type 1 fuzzy set combination we get, assuming $(\vee, \wedge, -)$ De Morgan Triple:

$$A = \text{tall} \cup \text{heavy} = (.4, S_1) \cup (.7, S_2) \cup (1, S_3) \tag{12.7}$$

(Remark: at times, we write:
$A = (.4, S_1) + (.7, S_2) + (1, S_3)$
where "+" is interpreted as a set aggregation)

In addition, suppose two independent sources of evidence, i.e., two experts, state their basic assignments about their belief that these suspects have committed the crime.

In Table 12.2, we show the calculation of combined evidence based on the basic assignments obtained from two independent experts on the focal elements (Table 12.1). As well, we show how these entries of Table 12.1 are computed in Appendix 12.1.

Our concern is to combine the description given by the witness and the belief combination computed as shown in Table 12.2 in order to sort out and therefore to narrow down the potential suspects list as much as possible in the light of description given by fuzzy set combination and belief assessments provided by two independent sources of evidence. In order to determine the belief over Type 1 fuzzy set A, we first express it in terms of its α-cuts and α-level sets as:

$$A=(0.4)\{S_1 \cup S_2 \cup S_3\} \cup (0.7)\{S_2 \cup S_3\} \cup (1)\{S_3\} \qquad (12.8)$$

**Table 12.1.** Basic Assignments provided by two independent sources of evidence on the focal elements.

| **Focal elements** | **Expert 1 ($m_1$)** | **Expert 1 ($m_2$)** |
|---|---|---|
| $S_1$ | 0.05 | 0.15 |
| $S_2$ | 0.05 | 0.05 |
| $S_3$ | 0.05 | 0.05 |
| $S_1 \cup S_2$ | 0.15 | 0.05 |
| $S_1 \cup S_3$ | 0.1 | 0.2 |
| $S_2 \cup S_3$ | 0.05 | 0.05 |
| $S_1 \cup S_2 \cup S_3$ | 0.55 | 0.45 |

It should be noted that the right hand side of equ. (12.8) is derived from equ. (12.7) by the α level representation.

Recall, that for Type 1 fuzzy sets, we have, from the previous section:

$$\begin{aligned}&\mathbf{Bel}(A_1 \text{ OR } A_2) = \mathbf{Bel}(A_1 \cup A_2)\\ &= \mathbf{Bel}(A)\\ &= \sum_{\alpha} \mathrm{Bel}(\alpha(A(x))_{\alpha})\\ &= \sum_{\alpha} [\alpha, \mathrm{Bel}(A(x))_{\alpha}]\end{aligned}$$

(note that Σ=∪, i.e., it is set aggregation and α is a scalar)

Thus, we get from the fuzzy set description of A and **Bel** calculation from Table 12.2:

$$\begin{aligned}\mathbf{Bel}_{1,2}(A)=\{&[(0.4), \mathbf{Bel}_{1,2}\{S_1 \cup S_2 \cup S_3\}]\\ &\cup[(0.7), \mathbf{Bel}_{1,2}\{S_2 \cup S_3\}]\\ &\cup[(1), \mathbf{Bel}_{1,2}\{S_3\}]\}\end{aligned}$$

From Table 12.2, we have $\mathbf{Bel}_{1,2}\{S_1 \cup S_2 \cup S_3\}=1$, $\mathbf{Bel}_{1,2}\{S_2 \cup S_3\}=0.22$, and $\mathbf{Bel}_{1,2}\{S_3\}=0.09$.

Thus fuzzy set description, with its α-cuts, α-level sets, helps us to narrow down the total possibilities shown in Table 12.2. Therefore, we are able to conclude that with 0.4 degree of membership the perpetrator of the crime is in the set $\{S_1 \cup S_2 \cup S_3\}$ and the belief associated with $S=\{S_1, S_2, S_3\}$ is one as expected!

**Table 12.2.** Combination of Degrees of Evidence from Two Independent Sources, i.e., Experts, Agents, etc.

| Focal elements | Expert 1 | | Expert 2 | | Combined Evidence | |
|---|---|---|---|---|---|---|
| | $m_1$ | $Bel_1$ | $m_2$ | $Bel_2$ | $m_{1,2}$ | $Bel_{1,2}$ |
| $S_1$ | 0.05 | 0.05 | 0.15 | 0.15 | 0.21 | 0.21 |
| $S_2$ | 0.05 | 0.05 | 0.05 | 0.05 | 0.08 | 0.08 |
| $S_3$ | 0.05 | 0.05 | 0.05 | 0.05 | 0.09 | 0.09 |
| $\{S_1, S_2\}$ | 0.15 | 0.25 | 0.05 | 0.25 | 0.13 | 0.42 |
| $\{S_1, S_3\}$ | 0.1 | 0.2 | 0.2 | 0.4 | 0.18 | 0.48 |
| $\{S_2, S_3\}$ | 0.05 | 1.5 | 0.05 | 0.15 | 0.05 | 0.22 |
| $\{S_1, S_2, S_3\}$ | 0.55 | 1.0 | 0.45 | 1.0 | 0.26 | 1.0 |

Although we are uncertain which one of $\{S_2, S_3\}$ it may be, we are able to discern that the perpetrator is in the set $\{S_2, S_3\}$ with 0.7 membership and with an associated belief of 0.22 and that the perpetrator is $\{S_3\}$ with full membership but with the associated belief of 0.09. This is a good example of how we can combine the descriptive information obtained from fuzzy set representation and fuzzy measures such as belief, etc. Thus "meaning" representation provided by fuzzy set description helps us to further extract more useful information. In particular, in this example, we were able to show how the contents of Table 12.2 is reduced to three options with associated degrees of belonging!

On the other hand, by the application of the belief axiom, we are able to show that

$$
\begin{aligned}
\mathbf{Bel}(A) &= \mathbf{Bel}\{[0.4,\{S_1 \cup S_2 \cup S_3\}] \cup [(0.7),\{S_2 \cup S_3\}] \cup [(1),\{S_3\}]\} \\
&\geq \{[(0.4), \mathbf{Bel}\{S_1 \cup S_2 \cup S_3\}] \\
&+ [(0.7), \mathbf{Bel}\{S_2 \cup S_3\}] \\
&+ [(1), \mathbf{Bel}\{S_3\}] \\
&- [(0.4 \wedge 0.7), \mathbf{Bel}\{S_2 \cup S_3\}] - [(0.4 \wedge 1), \mathbf{Bel}\{S_3\}] \\
&- [(0.7 \wedge 1), \mathbf{Bel}\{S_3\}] + [(0.4 \wedge 0.7 \wedge 1), \mathbf{Bel}\{S_3\}] \\
&= [(0.4), \mathbf{Bel}\{S_1 \cup S_2 \cup S_3\}] + [(0.3), \mathbf{Bel}\{S_2 \cup S_3\}] + [(0.3), \mathbf{Bel}\{S_3\}].
\end{aligned}
$$

It is to be noted that $\{S_1 \cup S_2 \cup S_3\}$ is the 0.4-level set, $\{S_2 \cup S_3\}$ is the 0.7-level set, and $\{S_3\}$ is the 1-level set. The equality follows from the axiom of **belief measure** (A-4). It is an application of the axiom (A-4) to crisp α-level sets of Type 1 fuzzy sets.

This demonstrates how the axiom of **belief measure** is applied over the $\alpha$-level fuzzy sets. It is to be noted **Bel**(A) discussed above is shown in Figure 12.1. While the original membership values of $\alpha$-level sets $\{S_1 \cup S_2 \cup S_3\}$, $\{S_2 \cup S_3\}$, $\{S_3\}$ are 0.4, 0.7, and 1, respectively, which are shown with solid black dots, ⊙, the application of the axiom of

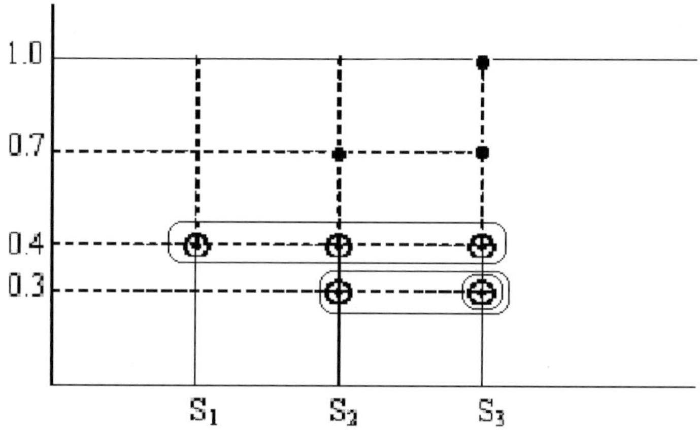


**Figure 12.1.** Crisp $\alpha$-level sets of Type 1 fuzzy set are shown.

belief on crisp sets entails a reduction on the membership values for $\{S_2 \cup S_3\}$ and $\{S_3\}$ from 0.7 to 0.3 and from 1 to 0.3, respectively, which are shown with a circle around the solid black dot, ⊙, whereas the membership of $\{S_1 \cup S_2 \cup S_3\}$ stays the same, i.e., 0.4.

This is a consequence of the intersection operation that is applied to crisp $\alpha$-level sets of Type 1 fuzzy theory. In that context it requires that the membership values of the corresponding $\alpha$-level sets be properly intersected. This generates appropriate weight for **Bel** measure over the intersections of crisp sets, i.e., the $\alpha$-level sets. Thus in the above example, we have shown the results of applying the $\wedge$-Min version of t-norms, $\Delta$. Thus in general, it is expressed as:

$$
\begin{aligned}
\mathbf{Bel}(A) = {} & \mathbf{Bel}(\alpha_1 A_1(x)_\alpha \cup \alpha_2 A_2(x)_\alpha \cup \alpha_3 A_3(x)_\alpha) \\
& \geq [\alpha_1, \mathbf{Bel}(A_1(x))_{\alpha_1}] + [\alpha_2, \mathbf{Bel}(A_2(x))_{\alpha_2}] + [\alpha_3, \mathbf{Bel}(A_3(x))_{\alpha_3}] \\
& - [(\alpha_1 \Delta \alpha_2), \mathbf{Bel}(A_1(x) \cap A_2(x))_{\alpha_1 \Delta \alpha_2}] - [(\alpha_1 \Delta \alpha_3), \mathbf{Bel}(A_1(x) \cap A_3(x))_{\alpha_1 \Delta \alpha_2}] \\
& - [(\alpha_2 \Delta \alpha_3), \mathbf{Bel}(A_2(x) \cap A_3(x))_{\alpha_2 \Delta \alpha_3}] \\
& + [(\alpha_1 \Delta \alpha_2 \Delta \alpha_3), \mathbf{Bel}(A_1(x) \cap A_2(x) \cap A_3(x))_{\alpha_1 \Delta \alpha_2 \Delta \alpha_3}]
\end{aligned}
$$

as it should be, where $\Delta$ is a t-norm. We next demonstrate the effect of other t-norms.

### *12.1.4. Belief on Type 1 Fuzzy set formed with other De Morgan Triples*

In the previous section, we have defined A=tall ∪ heavy by using (∨, ∧, -) De Morgan Triple. Let us next look at very briefly Algebraic and Lukasiewicz De Morgan Triples, i.e., (⊕,⊙, -) and ($L_\oplus$, $L_\odot$, -).

#### *12.1.4.1. Algebraic Combination*

For this case A=tall ∪ heavy gives us:

$$A=((.3+.4-.12), S_1) \cup((.7+.6-.42), S_2) \cup ((1+1-1), S_3)$$

Thus we get:

$$A= (.58, S_1) \cup(.88, S_2) \cup (1, S_3)$$

Therefore, with the α-cuts, we get

$$A= [(.58),\{S_1, S_2, S_3\}] \cup [(.88),\{S_1, S_2\}]\cup [(1),\{S_3\}].$$

Hence the interpretation is that the culprit is with membership of (.58) one of $\{S_1, S_2, S_3\}$ and the belief equal to one. Although, we are uncertain which one it is the perpetrator is one of $\{S_2, S_3\}$ with membership of (.88) with belief equal to 0.22, etc.

We note, however, that the belief values stays the same while the membership degrees of the perpetrators' class are increased.

#### *12.1.4.2. Lucasiewicz Combination*

In this case, we compute

$$A = (\mathrm{Min}(1, .3 + .4), S_1) \cup (\mathrm{Min}(1, .7 + .6), S_2) \cup (\mathrm{Min}(1, 1 + 1), S_3)$$

which has an α-cut representation as:

$$A= [(.7),\{S_1, S_2, S_3\}] \cup [(1),\{S_2, S_3\}]$$

Again, we observe an increase in the membership values for perpetrators belonging to $\{S_1, S_2, S_3\}$ or $\{S_2, S_3\}$ without a change in the associated belief. Note however, in this case, we have full membership of the culprit belonging to the set $\{S_2, S_3\}$ but again with belief (0.22).

### *12.1.5. Belief on Internal-valued Type 2 Fuzzy Sets*

In our interval-valued Type 2 theory, which recently is shown to correspond to a restricted and modified multi-valued mapping of Dempster, which we call "T-formalism"(Türkşen, 2001, 2002), the membership values of the meta-linguistic combination of "$A_1$ OR $A_2$=A" are mapped into:

(1) the upper bound set approximation

$$FCCF(A) = A_1 \cup A_2 = A_U$$

(2) the lower bound set approximation

$$FDCF(A) = (A_1 \cap A_2) \cup (c(A_1) \cap A_2) \cup (A_1 \cap c(A_2)) = A_L$$

Recall that these two representations are equal to each other in two-valued set and logic theory and hence, the usual formula of $(A_1 \cup A_2)$ is used in applications of two-valued set and logic theory. However, it is also known that $FDCF(.) \subseteq FCCF(.)$ in fuzzy set and logic theory for strict and nilpotent classes of Archimedean t-norms and t-conorms that are conjugates of each other based on a negation as well as (Max, Min, -), $(\oplus,\odot,-)$ and $(L_{\oplus},L_{\odot},-)$ De Morgan Triples. (Türkşen, 1986-2002; Bilgic, 1995).

Thus for the belief measure on the meta-linguistic expression, "$A_1$ OR $A_2$ = A", i.e.,

$\sum_{\alpha} \alpha \mathbf{Bel}[A(x)]_{\alpha}$, we naturally have to assess our belief on both the upper and lower set approximations.

It is to be noted that we first determine FDCF and FCCF then identify their α-level sets and then assess **Bel** on the α-level, crisp, sets. Thus we write

$$\sum_{\alpha} \alpha\, \mathbf{Bel}_L[A\,(x)]_{\alpha}$$

$$= \sum_{\alpha} [\alpha, \mathbf{Bel}[FDCF(A(x))]_{\alpha}] = \sum_{\alpha} [\alpha, \mathbf{Bel}[A_L\,(x)]_{\alpha}], \text{ and}$$

$$\sum_{\alpha} [\alpha, \mathbf{Bel}_U[A(x)]_{\alpha}]$$

$$= \sum_{\alpha} [\alpha, \mathbf{Bel}[FCCF(A(x))]_{\alpha}] = \sum_{\alpha} [\alpha, \mathbf{Bel}[A_U\,(x)]_{\alpha}]$$

It is clear that

$\mathbf{Bel}_U[A(x)]_\alpha$

$$= \sum_\alpha [\alpha, \mathbf{Bel}[A_U(x)]_\alpha]$$

$$= \sum_\alpha [\alpha, \mathbf{Bel}[(A_1 \cup A_2)\ (x)]_\alpha\ ]$$

which is the $\alpha$-level set representation that corresponds to the classical Belief measure, as we have shown in the previous Section 12.1.3, i.e.,

$$\sum_\alpha [\alpha, \mathbf{Bel}_U[A(x)]_\alpha]$$

$$= \sum_\alpha [\alpha, \mathbf{Bel}[(A_1 \cup A_2)\ (x)]_\alpha] \geq [\alpha_1, \mathbf{Bel}[(A_1)(x)]_\alpha]$$

$$+ [\alpha_2, \mathbf{Bel}[(A_2)(x)]_\alpha] - [(\alpha_1 \Delta \alpha_2), \mathbf{Bel}[(A_1 \cap A_2)\ (x)]_\alpha],$$

where $\Delta$ is a t-norm, with the basic axiom of Belief measures where $\cap$ is the crisp set intersection. But now, we have, in addition to the upper, a lower belief measure to compute over the meta-linguistic expression of "$A_1$ OR $A_2 = A$". Thus, now, we require that the belief measure be greater than or equal to the belief measure determined on the interval specified by its upper and lower bounds; i.e.,

$$\sum_\alpha [\alpha, \mathbf{Bel}[(A_1 OR A_2)\ (x)]_\alpha] \geq [\sum_\alpha [\alpha, \mathbf{Bel}_U[(A_1 OR A_2)\ (x)]_\alpha], \sum_\beta [\beta, \mathbf{Bel}_L[(A_1\ OR\ A_2)\ (x)]_\beta]$$

Alternately, we write $\sum_\alpha [\alpha, \mathbf{Bel}[A(x)]_\alpha] \geq [\sum_{\alpha_1} [\alpha_1, \mathbf{Bel}[A_L(x)]_{\alpha_1}],$

$$\sum_{\alpha 2} [\alpha_2, \mathbf{Bel}[A_U(x)]_{\alpha_2}]$$

Note that the $\alpha$-cuts of the upper and lower sets generally turn out to be different. Therefore, we use $\alpha_1$-cut for the one and $\alpha_2$-cut for the other to indicate this potential difference when we write them together. When we write them separately, we simply use $\alpha$-cuts in general.

This introduces a region of uncertainty between upper and lower bounds of the belief assessment which we have determined above. This means that our belief would be greater than some value in this uncertain region to the degrees identified by the upper and lower limits. This identifies possible uncertainty interval of belief measure. In a sense the uncertainty region of belief in the interval [$\mathbf{Bel}_U$, $\mathbf{Bel}_L$] identifies possible belief values, that could occur.

Since we know that the upper bound corresponds to the classical measure applied to the α-cuts of a fuzzy set as we have discussed in Section 12.1.3, we need to investigate the properties of the lower bound assessment of our belief on the α-cuts of the lower fuzzy set.

### *12.1.6. Lower Belief Assessments*

Let us next investigate the lower bound assessment of our degree of belief, i.e.,

$$\sum_{\alpha}[\alpha, \mathbf{Bel}_L[(A_1 \text{ OR } A_2)(x)]_\alpha]$$
$$= \sum_{\alpha}[\alpha, \mathbf{Bel}[\text{FDCF}(A_1 \text{ OR } A_2)\,(x)]_\alpha]$$
$$= \sum_{\alpha}[\alpha, \mathbf{Bel}[\{(A_1 \cup A_2) \cap (c(A_1) \cup A_2) \cup (A_1 \cup c(A_2))\}\,(x)]_\alpha]$$

When we interpret this with the classical axiom of belief, we get

$$\sum_{\alpha}[\alpha, \mathbf{Bel}_L[(A_1 \text{ OR } A_2)\,(x)]_\alpha] \geq [\alpha_1, \mathbf{Bel}[(A_1 \cup A_2)(x)\,]_{\alpha_1}]$$
$$+ [\alpha_2, \mathbf{Bel}[(c(A_1) \cup A_2)(x)\,]_{\alpha_2}]$$
$$+ [\alpha_3, \mathbf{Bel}[(A_1 \cup c(A_2))(x)\,]_{\alpha_3}]$$
$$- [(\alpha_1 \Delta \alpha_2), \mathbf{Bel}[(A_1 \cup A_2 \cup c(A_1) \cup A_2)(x)\,]_{\alpha_1 \Delta \alpha_2}]$$
$$- [(\alpha_1 \Delta \alpha_3), \mathbf{Bel}[(A_1 \cup A_2 \cup A_1 \cup c(A_2))\,(x)\,]_{\alpha_1 \Delta \alpha_3}]$$
$$- [(\alpha_2 \Delta \alpha_3), \mathbf{Bel}[(c(A_1) \cup A_2 \cup A_1 \cup c(A_2))\,(x)\,]_{\alpha_2 \Delta \alpha_3}]$$
$$+ [(\alpha_1 \Delta \alpha_2 \Delta \alpha_3), \mathbf{Bel}[(A_1 \cup A_2 \cup c(A_1) \cup A_2 \cup A_1 \cup c(A_2))\,(x)\,]_{\alpha_1 \Delta \alpha_2 \Delta \alpha_3}]$$

It should be noted that in classical set theory, because we have idempotecy and the law of contradiction, and $A_1$ and $A_2$ are crisp sets we obtain the classical belief measure.

In particular, we know that, in classical two-valued theory. We have

$\mathbf{Bel}(A_1 \text{ OR } A_2) = \mathbf{Bel}(\text{DNF}(A_1 \text{ OR } A_2)) \equiv \mathbf{Bel}(\text{CNF}(A_1 \text{ OR } A_2))$
$\mathbf{Bel}\{(A_1 \cup A_2) \cup (c(A_1) \cup A_2) \cup (A_1 \cup c(A_2))\} \equiv \mathbf{Bel}(A_1 \cup A_2)$
$\geq \mathbf{Bel}(A_1) + \mathbf{Bel}(A_2) - \mathbf{Bel}(A_1 \cup A_2)$

from the equivalence of DNF and CNF.

Let us briefly return to our example to assess the belief on the lower bound expression of interval-valued Type 2 representation of the combination:
If we assume ($\vee$, $\wedge$, -) De Morgan Triple, for

$$A_L = (\text{tall OR heavy})_L$$
$$= (\text{tall} \cap \text{heavy}) \cup (c(\text{tall}) \cap \text{heavy}) \cup (\text{tall} \cap c(\text{heavy}))$$

and recalling the definition of tall and heavy from before, we get

$$A_L = [[(.3)\vee(.4)\vee(.3)], S_1] \cup [[(.6)\vee(.3)\vee(.4)], S_2] \cup [[(1)\vee(0)\vee(0)], S_3]$$
$$= (.4, S_1) \cup (.6, S_2) \cup (1, S_3)$$

Therefore, we get

$$\mathbf{Bel}_{1,2}(A_L) = [(.4), \mathbf{Bel}\{S_1, S_2, S_3\}] \cup [(.6), \mathbf{Bel}\{S_2, S_3\}] \cup [(1), \mathbf{Bel}\{S_3\}]$$

Thus, recall from Section 12.1.3 and 12.1.4, we have:

$$\mathbf{Bel}_{1,2}(A_U) = [(0.4), \mathbf{Bel}\{S_1, S_2, S_3\}] \cup [(0.7), \mathbf{Bel}\{S_2, S_3\}] \cup [(1), \mathbf{Bel}\{S_3\}]$$

Therefore there is a membership uncertainty interval of [0.6, 0.7] where our belief is 0.22 over the set $\{S_2, S_3\}$ one of which is discerned to be the culprit.
If we compute $A_L$ with the Algebraic De Morgan Triple ($\oplus$,$\odot$, -), we would get

$$A_L = [[(.3)(.4) \oplus (.7)(.4) \oplus (.3)(.6)], S_1] \cup [[(.7)(.6) \oplus (.3)(.6) \oplus (.7)(.4)], S_2]$$
$$\cup [[(1)(1) \oplus (0)(1) \oplus (1)(0)], S_3] = (.48, S_1) \cup (.66, S_2) \cup (1, S_3)$$

Hence the results is:

$$A_L = (.48)\{S_1, S_2, S_3\} \cup (.66)\{S_2, S_3\} \cup (1)\{S_3\}$$

and thus

$$\mathbf{Bel}_{1,2}(A_L) = [(.48), \mathbf{Bel}\{S_1, S_2, S_3\}] \cup [(.66), \mathbf{Bel}\ \{S_2, S_3\}] \cup [(1), \mathbf{Bel}\ \{S_3\}]$$

Recall from Section 12.1.4.1, we have

$$A_U = [(0.58), \mathbf{Bel}\{S_1, S_2, S_3\}] \cup [(0.88), \mathbf{Bel}\{S_2, S_3\}] \cup [(1), \mathbf{Bel}\{S_3\}]$$

Thus, we observe an uncertain membership in the interval [.48, .58] with the belief "one" that the culprit is in the universal set $S = \{S_1, S_2, S_3\}$ and that with an uncertain membership of [.66, .88] and the belief (.22) he/she is in the set $\{S_2, S_3\}$, etc.

## 12.2. Upper and Lower Probabilities over Interval Valued Type 2 Fuzzy Sets

In this section, present an example for the computation of upper and lower probabilities over Type 2 fuzzy sets in analogy to Dempster's upper and lower probabilities. Consider a three element set $S=\{S_1, S_2, S_3\}$, say, three persons. Suppose we are interested how "big" they would be if we know how "tall" they are and how "heavy" they are if "big"::="tall" "AND" "heavy".

Let "tall" = $(.3, S_1) + (.7, S_2) + (1, S_3)$
and heavy = $(.4, S_1) + (.6, S_2) + (1, S_3)$ same as our previous example.From Interval-valued Type 2 Fuzzy sets, we have

$FDCF_{big}(S_i \in S) = (tall) \cap (heavy)$
$FCCF_{big}(S_i \in S) = (tall \cup heavy) \cap (c(tall) \cup heavy) \cap (tall \cup c(heavy))$

For $(\vee, \wedge, -)$, we get, the lower band to be:

$FDCF_{big}(S_i \in S) = (.3, S_1) + (.6, S_2) + (1, S_3)$, and

the upper band to be:

$FCCF_{big}(S_1 \in S) = (.4, S_1) \wedge (.7, S_1) \wedge (.6, S_1) = (.4, S_1)$
$FCCF_{big}(S_2 \in S) = (.7, S_2) \wedge (.6, S_2) \wedge (.7, S_2) = (.6, S_2)$
$FCCF_{big}(S_3 \in S) = (1, S_3) \wedge (1, S_3) \wedge (1, S_3) = (1, S_3)$
$\therefore FCCF_{big}(S_i \in S) = (.4, S_1) + (.6, S_2) + (1, S_3)$

Thus, the $\alpha$-cut representations would be:

$FDCF_{big}(S_i \in S)=(.3)\{S_1, S_2, S_3\}+(.6)\{S_2, S_3\} + (1)\{S_3\}$
$FCCF_{big}(S_i \in S)=(.4)\{S_1, S_2, S_3\}+(.6)\{S_2, S_3\} + (1)\{S_3\}$

The calculation of upper and lower probabilities are shown in Tables 12.3, 12.4, and 12.5 for each $\alpha$-cut. Once again, we ought to observe that there are advantages in combining the results of the descriptive information obtained from fuzzy set representation with the probability measure. Thus, the only valuable information in these tables are identified as three entries in Table 12.3, i.e., $\{S_1, S_2, S_3\}$, $\{S_2, S_3\}$ and $\{S_3\}$ at $\alpha$=0.3, there are three entries in Table 12.4, i.e., $\{S_2\}$, $\{S_3\}$ and $\{S_2, S_3\}$; and there is only one entry in Table 12.5.

**Table 12.3.** Upper and lower probabilities when we have (.3) $\{S_1, S_2, S_3\}$;we also have the same upper and lower probabilities when we have (.4) $\{S_1, S_2, S_3\}$.

| $T_{(.3)}$ | $P^*(T_{(.3)})$ | $P_*(T_{(.3)})$ |
|---|---|---|
| ∅ | 0 | 0 |
| $\{S_1\}_{(.3)}$ | $(.3)[P_{100}+P_{110}+P_{101}+P_{111})/(1-P_{000})]$ | $(.3)[P_{100}/(1-P_{000})]$ |
| $\{S_2\}_{(.3)}$ | $(.3)[P_{010}+P_{110}+P_{011}+P_{111})/(1-P_{000})]$ | $(.3)[P_{010}/(1-P_{000})]$ |
| $\{S_3\}_{(.3)}$ | $(.3)[P_{001}+P_{101}+P_{011}+P_{111})/(1-P_{000})]$ | $(.3)[P_{001}/(1-P_{000})]$ |
| $\{S_1,S_2\}_{(.3)}$ | $(.3)[P_{100}+P_{010}+P_{110}+P_{101}+P_{011}+P_{111})/(1-P_{000})]$ | $(.3)[P_{100}+P_{010}+(P_{110})/((1-P_{000})]$ |
| $\{S_1,S_3\}_{(.3)}$ | $(.3)[P_{100}+P_{001}+P_{110}+P_{101}+P_{011}+P_{111})/(1-P_{000})]$ | $(.3)[P_{100}+P_{001}+P_{101})/((1-P_{000})]$ |
| $\{S_2,S_3\}_{(.3)}$ | $(.3)[P_{010}+P_{001}+P_{110}+P_{101}+P_{011}+P_{111})/(1-P_{000})]$ | $(.3)[P_{010}+P_{001}+P_{011})/((1-P_{000})]$ |
| $\{S_1,S_2,S_3\}_{(.3)}$ | (.3) 1 | (.3) 1 |

**Table 12.4.** Upper and lower probabilities when we have {S2}, {S3} and {S2, S3}, at $\alpha$=(.6). Note: For two and single element sets their upper and lower sets.

| $T_{(.6)}$ | $P^*(T_{(.6)})$ | $P_*(T_{(.6)})$ |
|---|---|---|
| $\varnothing_{(.6)}=\{\varnothing,S_1\}$ | 0 | 0 |
| $\{S'_1\}=\{S_2\}_{(.6)}$ | $(.6)\left(\frac{P_{010}+P_{110}+P_{011}+P_{111}}{1-P_{000}-P_{100}}\right)$ (.6) | $(.6)\left(\frac{P_{010}}{1-P_{000}-P_{100}}\right)$ |
| $\{S'_2\}=\{S_3\}_{(.6)}$ | $(.6)\left(\frac{P_{001}+P_{101}+P_{011}+P_{111}}{1-P_{000}-P_{100}}\right)$ | $(.6)\left(\frac{P_{001}}{1-P_{000}-P_{100}}\right)$ |
| $\{S'_1,S'_2\}=\{S_2,S_3\}_{(.6)}$ | (.6)(1) | (.6)(1) |

**Table 12.5. Upper and lower probabilities at $\alpha$=1, (1){S3}.**

| $T_{(1)}$ | $P^*(T_{(1)})$ | $P_*(T_{(1)})$ |
|---|---|---|
| $\varnothing_{(1)}$ | 0 | 0 |
| $\{S_3\}_{(1)}$ | 1 | 1 |

## 12.3. Interval-Valued Type 2 Fuzzy Sets and Fuzzy Beliefs

In the previous sections, we have discussed the case where Belief measures were obtaioned from Basic Assigments as shown in Table 12.1. In Computing With Words within the context of fuzzy sets, our approach is that every thing is a matter of degree and is given as fuzzy set. In this regard, we next look at the case where Belief Measures are given as fuzzy sets such as "low Beliefs", "moderate Beliefs", "high Beliefs".For this purpose, we re-state the example as follows.

**Example 2**. Suppose a crime is committed and there are three suspects, say, $S=\{S_1,S_2,S_3\}$ as before. As well, there is a withness who describes the suspect to be either "tall" or "heavy" build and gives a fuzzy set definition over the three suspects as follows:

tall = $(.3, S_1) \cup (.7, S_2) \cup (1, S_3\}$

heavy = $(.4, S_1) \cup (.6, S_2) \cup (1, S_3\}$

Furthermore, there are two experts who describe their Belief with a fuzzy description to be "low" and "somewhat high" such that, as shown in Figure 12.2.

$$\text{Low(Belief)} = (0.05,S_1) \cup (0.05,S_2) \cup (0.05,S_3) \cup (0.15,\{S_1 \cup S_2\}) \cup (0.15, \{S_1 \cup S_3\}) \cup (0.15,\{S_2 \cup S_3\}) \cup (0.55, \{S_1 \cup S_2 \cup S_3\})$$

$$\text{Somewhat High(Belief)} = (0.15,S_1) \cup (0.15,S_2) \cup (0.15,S_3) \cup (0.20,\{S_1 \cup S_2\}) \cup (0.25, \{S_1 \cup S_3\}) \cup (0.30,\{S_2 \cup S_3\}) \cup (0.65, \{S_1 \cup S_2 \cup S_3\})$$

For the description of the witness we get

A = tall OR heavy

In Interval-Valued Type 2 fuzzy sets with the $(\vee,\wedge,-)$ De Morgan Triple
$A_L$(tall OR heavy)

$= (0.4,S_1) \cup ([0.6\text{-}0.7],S_2) \cup (1,S_3)$

$=(0.4,\{S_1,S_2,S_3\}) \cup ([0.6\text{-}0.7],\{S_2 \cup S_3\}) \cup (1,S_3)$

From the two experts descriptions of their belief, we set

Low(Belief) AND Somewhat high(Belief) = Concensus(Belief)

In Interval-Valued Type 2 fuzzy sets. We next determine upper and lower belief limits for the consensus.

Consensus(Belief)$_L$
$= (0.05,S_1) \cup (0.05,S_2) \cup (0.05,S_3)$
$\cup (0.15,\{S_1 \cup S_2\}) \cup (0.15, \{S_1 \cup S_3\}) \cup (0.15,\{S_2 \cup S_3\}) \cup (0.55, \{S_1 \cup S_2 \cup S_3\})$

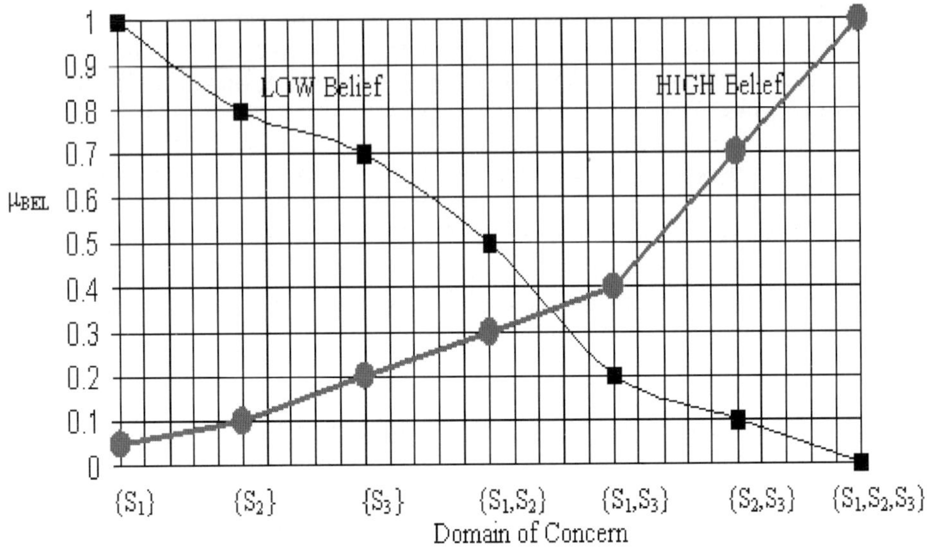


**Figure 12.2.** Fuzzy Belief Sets

Consensus(Belief)$_U$

$$
\begin{aligned}
= \; & [(0.15) \vee (0.95) \vee (0.85)] = (0.15, S_1) \\
& \cup [(0.15) \vee (0.95) \vee (0.85)] = (0.15, S_2) \\
& \cup [(0.15) \vee (0.95) \vee (0.85)] = (0.15, S_3) \\
& \cup [(0.2) \vee (0.85) \vee (0.8)] = \cup (0.2, \{S_1 \cup S_2\}) \\
& \qquad = \cup (0.25, \{S_1 \cup S_3\}) \\
& \qquad = \cup (0.3, \{S_2 \cup S_3\}) \\
& \qquad = \cup (0.55, \{S_1 \cup S_2 \cup S_3\})
\end{aligned}
$$

Thus, we have the upper and lower belief limits of the consensus between the two experts as:

Consensus(Belief) = B

$$
\begin{aligned}
& = ([0.05\text{-}0.15], S_1) \cup ([0.05\text{-}0.15], S_2) \cup ([0.05\text{-}0.15], S_3) \\
& \quad \cup ([0.15\text{-}0.2], \{S_1, S_2\}) \cup ([0.15\text{-}0.25], \{S_1, S_3\}) \\
& \quad \cup ([0.15\text{-}0.3], \{S_2, S_3\}) \cup ([0.55], \{S_1, S_2, S_3\})
\end{aligned}
$$

Next, we combine this belief concensus with the witness description, that the culprit is

$A = (0.4, \{S_1, S_2, S_3\}) \cup (0.6, \{S_2, S_3\}) \cup (1, \{S_3\})$

As a result, we obtain that the culprit is $A > B =$

$$
\begin{aligned}
& = (0.4, [0.55 - 0.65]\{S_1, S_2, S_3\}) \\
& \quad \cup ([0.6\text{-}0.7], [0.15\text{-}0.3], \{S_2, S_3\})
\end{aligned}
$$

$\cup$ (1, [0.05-0.15].{ $S_3$})

The interpretation is that The culprit being any one of $S_1$ or $S_2$ or $S_3$ has a membership degree of 0.4 in the class of "tall OR heavy" people, and the belief of the expert's concensus for this case, is in an interval of [0.55-0.65], alternately $S_2$ or $S_3$ could be the calprite with an interval-valued membership degree of [0.6-0.7] being "tall or heavy" people with the belief the experts' concensus, for this case is an interval of [0.15-0.3] and finally, $S_3$ could be the culprite with a full membership in the class of "tall or heavy" people but with a reduced degree of the concensus belief being in the interval [0.05-0.15].

## 12.4. Conclusions

It is shown that we can compute upper and lower beliefs and thus plausibilities and probabilities over interval-valued Type 2 fuzzy sets when we take FDCF and FCCF as basic protoforms in determining combination of concepts for the development of CWW. These computations are naturally carried over the $\alpha$-cuts of the fuzzy sets since fuzzy measures are over the crisp sets.
We have also shown that belief can be taken to be point value as in the classical approach. Alternately we can take belief expressed linguistically as "low", "somewhat low", "moderate", etc., belief and we can use the interval-valued Type 2 representation to represant the uncertainty associated both in description and in belief.

Appendix 12.1. Computation of the entries of Table 12.2.

$K = m_1(S_1).m_2(S_2)+m_1(S_1).m_2(S_3)$
$+ m_1(S_1).m_2(S_2{\cup}S_3)+m_1(S_2).m_2(S_1)$
$+ m_1(S_2).m_2(S_3)+m_1(S_1).m_2(S_1{\cup} S_3)$
$+ m_1(S_3).m_2(S_1)+m_1(S_3).m_2(S_2)$
$+ m_1(S_3).m_2(S_1{\cup}S_2)$
$+ m_1(S_1{\cup}S_2).m_2(S_3)$
$+ m_1(S_1{\cup}S_3).m_2(S_2) + m_1(S_2{\cup}S_3).m_2(S_1)$
$K = (.05)(.05) + (.05)(.05) + (.05)(.05) +(.05)(.15)$
$+ (.05)(.05) + (.05)(.2) +(.05)(.15) + (.05)(.05)$
$+ (.05)(.05) +(.15)(.05) + (.1)(.05)$
$+ (.05)(.15)$
$1-K = 1 - .06 = .94$
$m_{1,2}(S_1) = [m_1(S_1)\ m_2(S_2)+m_1(S_1)\ m_2(S_1{\cup}\ S_2)$
$+ m_1(S_1)\ m_2(S_1{\cup}S_3)+m_1(S_1)\ m_2(S_1{\cup}S_2{\cup}S_3)$
$+ m_1(S_1{\cup}\ S_2)\ m_2(S_1)+m_1(S_1{\cup}S_2)\ m_2(S_1{\cup}S_3)$
$+ m_1(S_1{\cup}S_3)\ m_2(S_1)+m_1(S_1{\cup}S_3)\ m_2(S_1{\cup}S_2)$
$+ m_1(S_1{\cup}S_2{\cup}S_3)\ m_2(S_1)]\ /\ .94$
$= [(.05)(.15) + (.05)(.05) + (.05)(.2)$
$+ (.05)(.45) + (.15)(.12) +(.15)(.2)$

$+ (.1)(.15) + (.1)(.05) + (.55)(.15)] / .94$
$= .21$

$m_{1,2}(S_2) = [m_1(S_2)\, m_2(S_2) + m_1(S_2)\, m_2(S_1 \cup S_2)$
$+ m_1(S_2)\, m_2(S_2 \cup S_3) + m_1(S_2)\, m_2(S_1 \cup S_2 \cup S_3)$
$+ m_1(S_1 \cup S_2)\, m_2(S_2) + m_1(S_1 \cup S_2)\, m_2(S_1 \cup S_3)$
$+ m_1(S_2 \cup S_3)\, m_2(S_2) + m_1(S_2 \cup S_3)\, m_2(S_1 \cup S_2)$
$+ m_1(S_1 \cup S_2 \cup S_3)\, m_2(S_2)] / .94$
$= [(.05)(.05) + (.05)(.05) + (.05)(.05)$
$+ (.05)(.45) + (.15)(.12) + (.15)(.05)$
$+ (.05)(.05) + (.05)(.05) + (.55)(.05)] / .94$
$= .082$

$m_{1,2}(S_3) = [m_1(S_3)\, m_2(S_3) + m_1(S_3)\, m_2(S_1 \cup S_3)$
$+ m_1(S_3)\, m_2(S_2 \cup S_3) + m_1(S_3)\, m_2(S_1 \cup S_2 \cup S_3)$
$+ m_1(S_1 \cup S_3)\, m_2(S_3) + m_1(S_1 \cup S_3)\, m_2(S_2 \cup S_3)$
$+ m_1(S_2 \cup S_3)\, m_2(S_3) + m_1(S_2 \cup S_3)\, m_2(S_1 \cup S_3)$
$+ m_1(S_1 \cup S_2 \cup S_3)\, m_2(S_3)] / .94$
$= [(.05)(.05) + (.05)(.2) + (.05)(.05)$
$+ (.05)(.45) + (.1)(.05) + (.1)(.05)$
$+ (.05)(.05) + (.05)(.2) + (.55)(.05)] / .94$
$= .093$

$m_{1,2}(S_1 \cup S_2) = [m_1(S_1 \cup S_2)\, m_2(S_1 \cup S_2) + m_1(S_1 \cup S_2)\, m_2(S_1 \cup S_2 \cup S_3)$
$+ m_1(S_1 \cup S_2 \cup S_3)\, m_2(S_1 \cup S_2)] / .94$
$= [(.15)(.05) + (.15)(.45) + (.55)(.05)] / .94 = .132$

$m_{1,2}(S_2 \cup S_3) = [m_1(S_2 \cup S_3)\, m_2(S_2 \cup S_3) + m_1(S_2 \cup S_3)\, m_2(S_1 \cup S_2 \cup S_3)$
$+ m_1(S_1 \cup S_2 \cup S_3)\, m_2(S_2 \cup S_3)] / .94$
$= [(.05)(.05) + (.05)(.45) + (.55)(.05)] / .94 = .055$

$m_{1,2}(S_1 \cup S_3) = [m_1(S_1 \cup S_3)\, m_2(S_1 \cup S_3) + m_1(S_1 \cup S_3)\, m_2(S_1 \cup S_2 \cup S_3)$
$+ m_1(S_1 \cup S_2 \cup S_3)\, m_2(S_1 \cup S_3)] / .94$
$= [(.1)(.2) + (.1)(.45) + (.55)(.2)] / .94 = .186$

$m_{1,2}(S_1 \cup S_2 \cup S_3) = [m_1(S_1 \cup S_2 \cup S_3)\, m_2(S_1 \cup S_2 \cup S_3)] / .94$
$= [(.55)(.45)] / .94 = .26$

$\mathbf{Bel}_1(S_1 \cup S_2) = m_1(S_1) + m_1(S_2) + m_1(S_1 \cup S_2)$
$= .05 + .05 + .15 = .25$

$\mathbf{Bel}_1(S_1 \cup S_3) = m_1(S_1) + m_1(S_3) + m_1(S_1 \cup S_3)$
$= .05 + .05 + .1 = .2$

$\mathbf{Bel}_1(S_2 \cup S_3) = m_1(S_2) + m_1(S_3) + m_1(S_2 \cup S_3)$
$= .05 + .05 + .05 = .15$

$\mathbf{Bel}_1(S_1 \cup S_2 \cup S_3) = m_1(S_1) + m_1(S_2) + m_1(S_3)$
$+ m_1(S_1 \cup S_2) + m_1(S_1 \cup S_3) + m_1(S_2 \cup S_3)$

$+ m_1(S_1 \cup S_2 \cup S_3)$
$= .05 + .05 + .05 + .15 + .1 + .05 + .55 = 1$

$\mathbf{Bel}_2(S_1 \cup S_2) = m_2(S_1) + m_2(S_2) + m_2(S_1 \cup S_2)$
$= .15 + .05 + .05 = .25$

$\mathbf{Bel}_2(S_1 \cup S_3) = m_2(S_1) + m_2(S_3) + m_2(S_1 \cup S_3)$
$= .15 + .05 + .2 = .4$

$\mathbf{Bel}_2(S_2 \cup S_3) = m_2(S_2) + m_2(S_3) + m_2(S_2 \cup S_3)$
$= .05 + .05 + .05 = .15$

$\mathbf{Bel}_2(S_1 \cup S_2 \cup S_3) = m_2(S_1) + m_2(S_2) + m_2(S_3)$
$+ m_2(S_1 \cup S_2) + m_2(S_1 \cup S_3) + m_2(S_2 \cup S_3)$
$+ m_2(S_1 \cup S_2 \cup S_3)$
$= .15 + .05 + .05 + .05 + .2 + .05 + .45 = 1$

$\mathbf{Bel}_{1,2}(S_1 \cup S_2) = m_{1,2}(S_1) + m_{1,2}(S_2)+m_{1,2}(S_1 \cup S_2)$
$= (.21) + (.08) + (.13) = .42$

$\mathbf{Bel}_{1,2}(S_1 \cup S_3) = m_{1,2}(S_1) + m_{1,2}(S_3)+m_{1,2}(S_1 \cup S_3)$
$= (.21) + (.09) + (.18) = .48$

$\mathbf{Bel}_{1,2}(S_2 \cup S_3) = m_{1,2}(S_2) + m_{1,2}(S_3)+m_{1,2}(S_2 \cup S_3)$
$= (.08) + (.09) + (.05) = .22$

$\mathbf{Bel}_{1,2}(S_1 \cup S_2 \cup S_3) = m_{1,2}(S_1) + m_{1,2}(S_2) + m_{1,2}(S_3) +$
$+ m_{1,2}(S_1 \cup S_2) + m_{1,2}(S_1 \cup S_3)$
$+ m_{1,2}(S_2 \cup S_3) + m_{1,2}(S_1 \cup S_2 \cup S_3)$
$= .21 + .08 + .09 + .13 + .18 + .05 + .26 = 1$

## References


1. Bilgic, T. (1995), Measurement-Theoretic Frameworks for Fuzzy Set Theory with Applications to Preference Modelling, PhD thesis, University of Toronto, Dept. of Industrial Engineering Toronto, Ontario M5S 1A4 Canada.(supervisor, I.B. Türkşen)
2. Dempster A.P (1967), "Upper and Lower Probabilities Induced by a Multivalued Mapping", In: *Annals of Mathematical Statistics*, 38, 325-339.
3. Pawlak Z. (1991), Rough Sets, *Kluwer*, Netherlands.
4. Pawlak Z. (1982), "Rough Sets", *Int. J. of Computer and Information Sciences*, 11, 341-356.
5. Shafer, G. (1976), A Mathematical Theory of Evidence, *Princeton University Press.*
6. Türkşen I.B. (2001), "Computing with Descriptive and Veristic Words: Knowledge Representation and Reasoning", *in: Computing With Words*, P.P. Wang(ed.), Chapter 10, Wiley, New York, 297-328.

7. Türkşen, I.B. (2001), "Sources, Measurements and Models of Type 2 Fuzziness in the New Millennium", *in: Fuzziness in the New Millennium*, V. Dimitrov (ed), Springer Verlag (in press).
8. Zadeh, L.A. (1965), "Fuzzy Sets", *Information and Control Systems*, Vol.8, 338-353.

# Chapter 13

# VERISTIC FUZZY SETS OF TRUTHOODS[*]

It should be recalled that starting with the discussion of Philosophical Grounding and through out the text, we have identified two distinct treatments in fuzzy set theory: (i) Descriptive fuzzy sets and (ii) Veristic fuzzy sets, i.e., fuzzy sets of truthoods.

Most of the previous chapters have dealt with essentially Descriptive fuzzy sets and the formation of Interval-Valued Type 2 fuzzy sets generated by the combination of two Type 1 fuzzy sets.
In this chapter, we treat the Veristic Fuzzy sets of Truthoods. Both the "Fuzzy Truth Table" and T-formalism (a modified and restricted multi-valued mapping of Dempster) derivations are applicable for this purpose. Thus we obtain FDCF(.) and FCCF(.) expressions when we replace, in the Fuzzy Truth Tables, the values of Descriptive set memberships for sets A and B with Veristic set membership for sets A and B.

Similarly in T-formalism the information granules of Descriptive fuzzy sets are replaced with the Veristic fuzzy sets to obtain the upper and lower set expressions. As well the break down of the equivalences, i.e., FDCF(.)FCCF(.) and the containment of FDCF(.)$\subseteq$FCCF(.) holds for all possible sixteen expression for the class of Archimedean t-norms and t-conorms that are duals of each other with a required negation provided they are strict and nilpotent.

However, in this chapter, we provide an alternate approach based on modal logic and the meta-theory of fuzzy truthoods (Resconi and Türkşen, 2001)

---

[*] extracted and modified from Resconi and Türkşen (2001).

## 13.1. Modal Logic

There is a wide-spread acceptance that Kripke models (Chellas, 1980) offer the fundamental semantics for modal logic. Further there are several generalisations of Kripke models and, among them, the so-called Scott-Montague models (Chellas, 1980) have the most general characteristics of such models. In this section, we give a brief description of Kripke and Scott-Montague models.

### *13.1.1. Kripke Modal Framework*

Let ATOMS, be elements of a denumerable set of *atomic sentences.* Let $L_{ML}$ be a language for modal logic formed from ATOMS using propositional connectives such as $\neg$ (negation), $\wedge$ (conjunction), $\vee$ (disjunction), $\rightarrow$ (material implication), and $\leftrightarrow$ (equivalence) and modal operators such as $\Box$ (necessity) and $\Diamond$ (possibility).

A *Kripke model* is defined as the following three-tuple:
$M = < W, R, V >$,
where W is a non-empty set of possible worlds, $R \subseteq W \times W$ is an accessibility relation on W, and V is a valuation operator for atomic sentences at each possible world, $w \in W$ :
$V: ATOM \times W \rightarrow \{T, F\}$
The valuation operator is extended in the usual way and, in particular, for modal operators, $\Box$ - necessity and $\Diamond$ - possibility:

$$V(\Box p, w) = T \Leftrightarrow \forall\, w'(wRw' \Rightarrow V(p, w')) = T \Leftrightarrow \Xi_w \subseteq ||p||^M,$$
$$V(\Diamond p, w) = T \Leftrightarrow \exists w'\, (wRw' \text{ and } V(p, w')) = T \Leftrightarrow \Xi_w \cap ||p||^M \neq \varnothing$$

where $\Xi_w = \{w' \mid wRw'\}$ and $||p||^M = \{w \mid V(p, w) = T\}$. Note that $||p||^M$ is called the truth set associated with the proposition of an atomic sentence p in a model M in the literature of philosophical logic. In what follows, we omit the superscript $^{'M'}$ unless confusion arises. It is well-known that different conditions on accessibility relations correspond to different axiom schemas. Some of them are shown in Table 13.1. For emphasis, it should be noted that the valuation V of an atomic sentence assigns {T, F} to that atomic sentence but does not specify how the atomic sentence itself is to be represented by a "descriptive" set notation. The "descriptive" words of atomic sentences may be represented either by two-valued or infinite valued set representations. This is dealt with in previous chapters.

Within the context of "measurement of membership functions (Türksen, 1991, 1999), it should also be pointed out that the “pooled” estimate exercises extract membership values and functions of "fuzzy truthood". In such experiments, individuals (worlds) respond to a question, i.e., an atomic sentence, say, "Is this person tall". The answer from the individual (world) is a "Yes/No" response that only verifies the atomic sentence by assigning a "Yes", i.e., it is "true", or "No", i.e., it is "false". Thus in this context, such pooling exercises are experimental models that are implicitly operating on a modal logic framework described above, i.e., Kripke model.

A *system* of modal logic is, in general, a set of sentences in $L_{ML}$. In particular, a system $\Sigma$ is said to be *normal* if it contains all tautologies, e.g., **Df** ◊, and **K**. A system $\Sigma$ is closed under the following two inferential rules:

**Table 13.1.** Accessibility relation and axiom schemas.

| | |
|---|---|
| No condition | **Df** $\Diamond . \Diamond p \leftrightarrow \neg \Box \neg p$ |
| No condition | **K.** $\Box (p \rightarrow q) \rightarrow (\Box p \rightarrow \Box q)$ |
| Serial: $\forall w \exists w'$ (w R w') | **D.** $\Box p \rightarrow \Diamond p$ |
| Reflexive $\forall w$ (w R w) | **T.** $\Box p \rightarrow \Diamond p$ |
| Symmetric $\forall w \forall w'$ (w R w' $\Rightarrow$ w' R w) | **B.** $p \rightarrow \Box \Diamond p$ |
| Transitive $\forall w \forall w' \forall w''$ (w R w' and w' R w'' $\Rightarrow$ w R w'') | **S4.** $\Box p \rightarrow \Box \Box p$ |
| Euclidean $\forall w \forall w' \forall w''$ (w' R w and w' R w'' $\Rightarrow$ w R w'') | **S5.** $\Diamond p \rightarrow \Box \Diamond p$ |

Modus Ponens: $p, p \rightarrow q \in \Sigma \Rightarrow q \in \Sigma$

Necessitation: $p \in \Sigma \Rightarrow \Box p \in \Sigma$.

## 13.2. Meta-Theory Based On Modal Logic

In a series of papers Klir, *et al.*, (1994, 1995; Resconi, *et al.*, 1992-2000) a meta-theory was developed. The approach is based on Kripke model. A Kripke model is given by the structure:

$$< W, R, V >.$$

Resconi, *et al.* (1992-1996) suggested to adjoin a function

$$\Psi : W \rightarrow R$$

where R is the set of real numbers assigned to the model in a way to obtain the new model

$$S1 = < W, R, V, \Psi > \qquad (13.1)$$

For every world, there is an associated real number that is assigned to it. With the model S1, we can build a *hierarchical meta-theory* where we can calculate

the expression for the membership function of truthood in a fuzzy set theory to verify a given sentence via a computational method based on {1,0} values corresponding to the truth values {T , F} assigned to a given sentence as the response of a world, a person, an agent or a sensor, etc.
At this point, we should ask: "what are the linkages between the concepts of a population of observers, a population of possible worlds and the algebra of fuzzy subsets of well defined universe of discourse?" To replay at this question, we need to point out that fuzzy set was introduced for the representation of imprecision in natural languages. However, imprecision means that a word representing an "entity" (temperature , velocity ...) cannot have a crisp logic evaluation. The meaning of a word in a proposition may and usually does evaluate in different ways for different assessments of an entity by different agents, i.e. worlds. An important principle is:

"*we cannot separate the assessments of the entity without some loss property in the representation of the entity itself*"

Different and in some cases conflicting evaluations for the same proposition would come up for the same entity. The presence of conflicting properties within the entity itself is the principal source of the imprecision in meaning representation of the entity. For example, suppose the entity is a particular temperature of a room, and we ask for the property *cold*, when we have no instrument to measure that property. The meaning of the entity, "temperature", is composed of assessments that are the *opinions* of a population of observers that evaluate the predicate "cold". Without the population of observers, and their assessments, we cannot have the entity "temperature" and the predicate "cold". When we move from crisp set to the fuzzy set, we move from atomic *elements*, i.e., individual assessments to non-atomic *entities*, i.e., aggregate assessments. In an abstract way, the population of the assessments of the entity becomes the population of the worlds that we associate the set of propositions with crisp logic evaluation. *Perception* based data embedded in a context generate imprecise properties. With the *perception* based data, as assessments of an entity, in context, we evaluate the properties of an entity. The evaluation can be conflicting. In such cases, the world population assessment composed of individual assessments give the context structure of the perception based data. If we know only the name of a person (entity) we cannot know if s/he is "old". Additional observation-based information that s/he is married, and is a manager, and that s/he plays with toys are conflicting information of the proposition that associates her/him with the predicate "old", etc.

The aim of this chapter is to show that with a model of a perception based imprecision generated by worlds, i.e., agents, we can on the one hand simplify the definitions of the operations in fuzzy logic and on the other hand expose and explain deeper issues embedded within the fuzzy theory.

Consider a sentence such as: "John is tall" where $x \in X$ is a person, "John", in a population of peoplc, X, and "tall" is a linguistic term of a linguistic variable, the height of people. Let this linguistic term be represented by the fuzzy set A, then the meta-linguistic expression that represents the proposition "John is tall" written in fuzzy set canonical form as:
" $x \in X$ isr A",
where "isr" means " $x \in X$ is in relation to a fuzzy information granule A". Next consider the sentence that is uttered by a world $wk \in W$ which evaluates the atomic sentence "John is tall" in the sense of the Kripke model as: " 'John is tall' is true", say, in a world $wk \in W$. That is a world wk, a person, or a sensor, responds either as "true" or "false" when presented with the sentence "John is tall". It should be noted that the world, i.e., the person or the sensor, does not say anything about the qualification, i.e., descriptive gradation of "John's being tall", but just verifies on the basis of a valuation schema V. Then we can write for short: pA(x) such that V(pA(x), wk) evaluates to T, true. Next let us assign $\eta$(wk , x , A) = 1 if $V(p_A(x), w_k) = T$ or $\eta$(wk , x , A ) = 0 if $V(p_A(x), w_k) = F$. With this background, we next define the membership expression of truthood of a given atomic sentence in a finite set of worlds wk $\in$ W as follows:

$$\mu_{p_A}(x) = \frac{(set\ of\ worlds \text{ where } p_A(x) \text{ is true})}{|W(x)|} = \frac{\sum_k \eta(w_k, x, A)}{|W(x)|} \tag{13.2}$$

where $p_A$ the proposition is the atomic sentence "John is tall", for x = "John", and A = "tall" such that $V(p_A(x)), w_k) = T$; and |W| is the cardinality of the set of worlds in our domain of concern. *Recall once again that these worlds, $w_k$ $\in W$, may be agents, sensors, persons,* etc. Let us define the subset of worlds $W_A = \{w_k \in W \mid V(p_A(x), w_k) = T\}$. We can write, expression (12.2) as follows:

$$\mu_{p_A}(x) = \frac{|W_A|}{|W|} \tag{13.2'}$$

With the understanding that $W_A$ represent the subset of the worlds W where the valuation of $p_A(x)$ is "true" in the Kripke sense. In Figure 13.1, we show the set $W_A$ in the universal set W .

For the special case, where the relation R in the Kripke model is

$w_k \text{ R } w_k$

at any world $w_k$ only itself is accessible ( any world $w_k$ is isolated from the others ) the membership expression is computed as the value of $\Psi$ in S1 stated in (13.1) above. It is computed by the expression

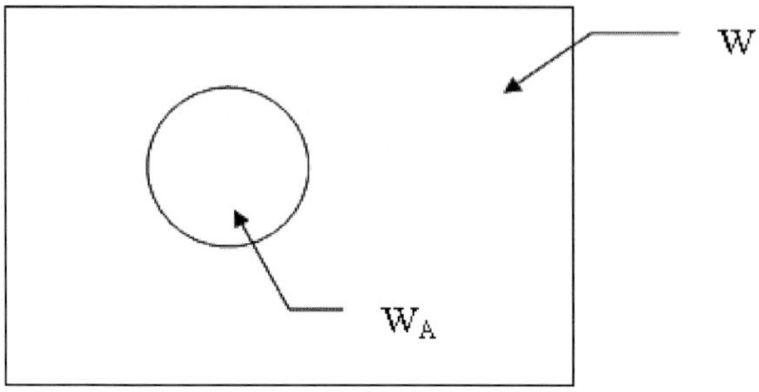


**Figure.13.1**. Crisp set of the worlds W and subset of the worlds WA where

$V(p_A(x),w_k)=T$

$$\Psi = \frac{1}{|W|}$$

for any (single) world w in W. Thus we can write the expression (13.2) as follows:

$$\mu_{p_A}(x) = \sum_k \eta(w_k)\Psi_k$$

It should be noted that the valuation operator V assigns {T, F} to every atomic sentence $p_A$ for any world; but it does not assess how the atomic sentence itself is to be represented by a "descriptive" set notation, i.e., the descriptive set may be represented either with a crisp or fuzzy set. Therefore it should be clear that only the membership value of truthood for the "descriptive" words of atomic sentences is computed by $\Psi$ for a given (single) world w in W.

We remark that the same $\mu_{p_A}$ can be obtained by a different $W_A$, where each one may be obtained from the other by a suitable transformation T on W that changes the configuration of the worlds. For any transformation T it is clear that

$$|W_A| = |T\,W_A|,$$

i.e., $|W_A|$ is invariant for any transformation T.

For example when we ask to a group of people, if "John is tall", then a sub-set of the people in the group, i.e., $W_A$, reply Yes and the complementary sub-set reply No, such that Yes means true and No means false. Any one person can be considered as one world. When the opinion of the persons change by a transformation T with the same number of Yes's and No's, then we get the cardinality $| T W_A |$ or the number of person that reply Yes to be the same as that before the transformation. In conclusion for each modal logic model such as S1, there exist one "descriptive" set notation and one membership function computation but the reverse is not true. That is for one membership value, there exist many equivalent modal logic models S1.(see Figure.13.2)

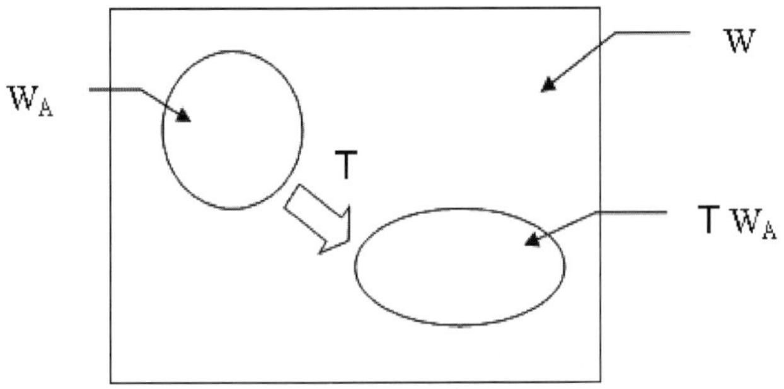


**Figure.13.2.** Subsets of the worlds $W_A$ and T $W_A$ with the same membership function μ

*13.2.1. Consonance and Dissonance of Propositions*

For any proposition p, we associate a set $\| p \|$ ::= W of the worlds where the valuation of the proposition p is true. Given two propositions $p_1$ and $p_2$ each containing a linguistic term A or B, respectively, we have two sets of worlds, $W_1$ and $W_2$, where $p_1$ and $p_2$ are "true", respectively. That is all the worlds where the proposition that contains A is true make up $W_1$ and all the worlds where the proposition that contains B is true make up $W_2$.

For the assessment of the truthood sets of any two propositions $p_1$ and $p_2$, we have either

$$cW_1 \cap W_2 \neq \varnothing \ \text{ or } \ cW_2 \cap W_1 \neq \varnothing$$

$$cW_1 \cap W_2 = \varnothing \ \text{ or } \ cW_2 \cap W_1 = \varnothing$$

where cW is the complement set of W. In this setting, any two propositions $p_1$ and $p_2$ are dissonant when the conditions (a) are applicable or they are consonant when the conditions (b) are applicable.

### *13.2.2. Dissonant Propositions*

The propositions $p_2$ is dissonant respect to $p_1$ when

$$\| \neg p_1 \wedge p_2 \| ::= cW_1 \cap W_2 \neq \varnothing \tag{13.3}$$

where cW is the set complement of W. Note that in (13.3), the left hand expression is the intersection of propositions. Whereas the right hand expression is the intersection of worlds that give true responses to the propositions, i.e., they are sets of the worlds that represent responding agents. In this case, $cW_1 \cap W_2$ is the part of $W_{2,}$ which disagrees with $W_1$.(see Figure.13.3)

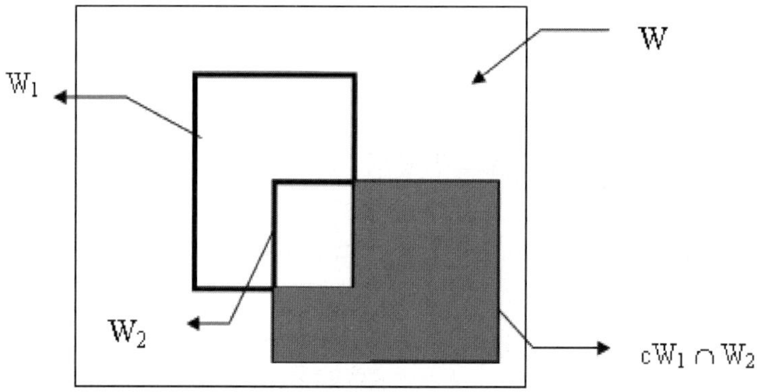


**Figure.13.3.** $cW_1 \cap W_2$ represents the part of $W_2$ that disagree with the set of responses of $W_1$.

The propositions $p_1$ is consonant respect to $p_2$ when $\| \neg p_2 \wedge p_1 \| ::= cW_2 \cap W_1 \neq \varnothing$ In this case $cW_2 \cap W_1$ is the part of $W_1$, which disagrees with $W_2$.(see Figure.13.4)

### *13.2.3. Consonant Propositions*

The propositions $p_2$ is consonant respect to $p_1$ when (see Figure.13.5)

$$\| \neg p_1 \wedge p_2 \| ::= cW_1 \cap W_2 = \varnothing \quad \text{or} \quad W_2 \subseteq W_1 \tag{13.4}$$

The propositions $p_1$ is consonant respect to $p_2$ when

$$\| \neg p_2 \wedge p_1 \| ::= cW_2 \cap W_1 = \varnothing \quad \text{or} \quad W_1 \subseteq W_2 \tag{13.5}$$

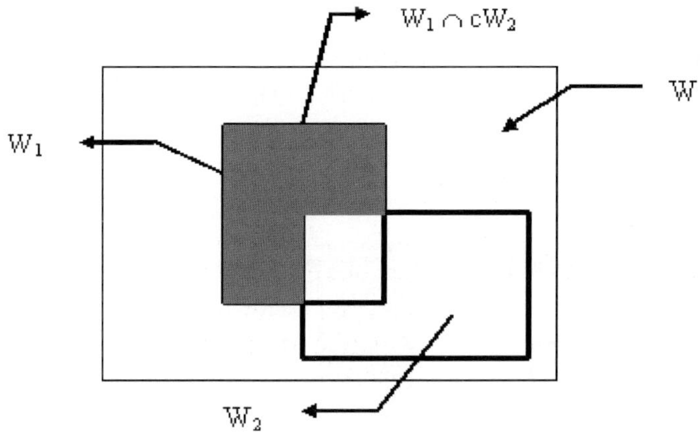


**Figure.13.4.** $cW_2 \cap W_1$ represents the part of $W_1$ that disagree with $W_2$

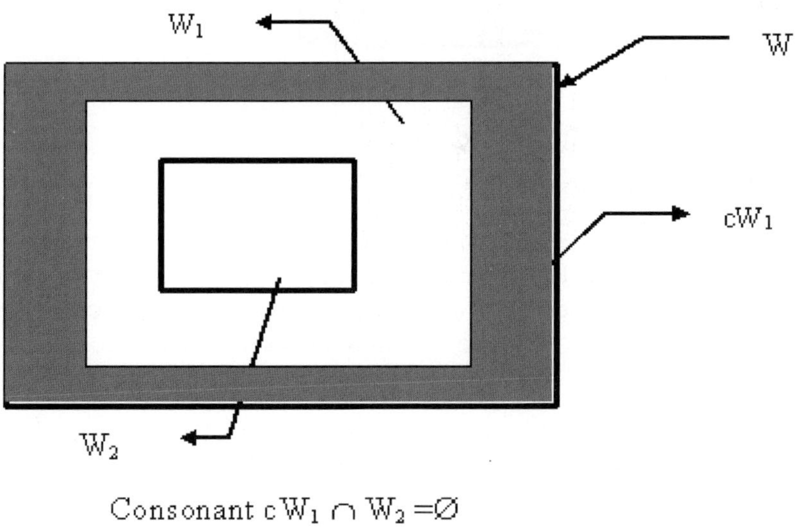


**Figure.13.5.** The part of the responses of $W_2$ that disagree with $W_1$ is empty.

**Example 0.** Given two persons George and Charles, let us assume that George has a point of view, i.e. perception, for a subject, i.e., a proposition p, but Charles has a proper subset of the point of view, i.e. a perception, of George for the same subject. Let us assume that $W_1$ is the set of world (samples) where the point of view of George for the given proposition p is true and $W_2$ is the set of worlds where p is true for Charles such that $W_2 \subseteq W_1$ . In this case Charles is not in conflict with George, i.e., the part of the responses of Charles that disagree with George is empty. Charles is in accord with George. But because the part of the response of George in relation to the response of Charles is not empty, George disagrees with Charles. This means that George is in conflict with Charles, i.e., George has dissonance with Charles. The conflict situation is not symmetric.

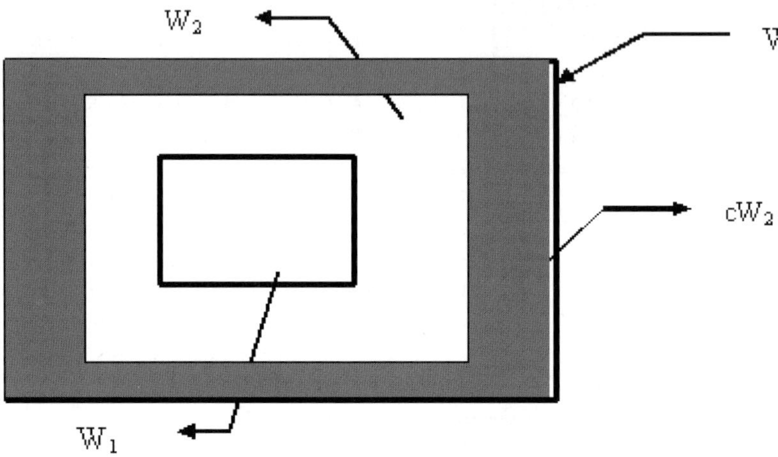


Consonant $cW_2 \cap W_1 = \varnothing$

**Figure.13.6.** The part of the responses of $W_1$ that disagree $W_2$ is empty.

***Remark.*** It should always be remembered that expressions in p's are (compound) propositions made up of atomic sentences. Whereas the corresponding expressions in W's are (compound) sets made up of crisp (traditional) sets that represent the compound propositions.

***Example 1***. Given the worlds, $W = \{w_1, w_2, w_3, w_4, w_5\}$ and $\|p_1\| ::= \{w_1, w_2, w_5\} = W_1$ , for all the worlds in $W_1$ , $p_1$ is true and $\|p_2\| ::= \{w_1, w_2\} = W_2$ , for all the worlds in $W_2$ the proposition $p_2$ is true. Then clearly, $W_2 \subset W_1$. It should be noted that $cW_1 = \{w_3, w_4\}$, thus $cW_1 \cap W_2 = \varnothing$ and $p_2$ is consonant respect to $p_1$. In turn, $cW_2 = \{w_3, w_4, w_5\}$ and thus $cW_2 \cap W_1 = \{w_5\}$ and $p_1$ is dissonant respect to $p_2$ .

It should also be noted that for this example, we have:

$$\mu_{p_1}(x) = \frac{|W_1|}{|W|} = \frac{3}{5}, \text{ and } \mu_{p_2}(x) = \frac{|W_2|}{|W|} = \frac{2}{5}.$$

## 13.3. "AND", "OR" and "COMPLEMENT"

Recall that linguistic operators (connectives), "AND", "OR", "NOT", etc., that are used in the combination of linguistic terms admit information granulation, i.e., they (connectives) are also fuzzy (imprecise) as well as the fuzzy sets that they combine. It should be recalled that Zimmermann and Zysno (1980) discovered in their experiments that human use of "AND", "OR" admit gradation. Thus they have introduced "Compensatory 'AND'". Later Türkşen (1994) showed that the degree of "ANDness" and "ORness" are represented within the interval of DNF(AND) $\subseteq$ CNF(AND) and DNF(OR) $\subseteq$ CNF(OR), respectively, as well as the mixtures of "ANDness" and "ORness" in the interval of CNF(AND) $\subseteq$ DNF(OR). In this section, preliminary concepts and formulas are developed with the terminology of modal logic for the computation of FDCF and FCCF expressions of fuzzy truthood sets which will be further discussed in Sections 13.4 and 13.5. These FDCF and FCCF expressions are the basic formulas that determine how two fuzzy truthood sets are to be combined in the formation of, and representation of, our knowledge within the grand paradigm shift caused by the fuzzy theory, which forms the foundation for "Computing With Words" and "Computing With Perceptions". [Zadeh, 1996, 2000]

### *13.3.1. "AND", "OR" Operations*

We are next interested in investigating the combinations of truthood propositions of any two truthood propositions, such as (1) '"John is tall" is true' AND '"John is heavy" is true', (2)'"John is tall" is true' OR '"John is heavy" is true' given that we know the membership functions of $\mu_{p_1}(x)$ for '"John is tall" is true', and $\mu_{p_2}(x)$ for '"John is heavy" is true'. It should be clear from Section 13.2 that if we know $W_1$ where $p_1$ is true and $W_2$ where $p_2$ is true then we can compute $\mu_{p_1}(x)$ and $\mu_{p_2}(x)$, i.e.,

$\mu_{p_1}(x) = |W_1| / |W|$ and $\mu_{p_2}(x) = |W_2|/|W|$.

**Proposition 1.** The membership value of the intersection of two fuzzy truthood sets associated with the propositions $p_1$ and $p_2$ is defined as:

$$\mu_{p_1 \wedge p_2}(x) = \frac{|W_1 \cap W_2|}{|W|} \tag{13.6}$$

in the meta-theory. Alternatively, when $\mu_{p_1}(x) \geq \mu_{p_2}(x)$, this expression can be written as

$$\mu_{p1\wedge p2}(x) = \min[\mu_{p1}(x), \mu_{p2}(x)] - \frac{|\,cW_1 \cap W_2\,|}{|\,W\,|}$$

(13.6')

where

$$\mu_{p_1}(x) = \frac{|W_1|}{|W|} \text{ and } \mu_{p_2}(x) = \frac{|W_2|}{|W|}.$$

**Proof.** Because the sets $T_1 = W_1 \cap W_2$, $T_2 = cW_1 \cap W_2$, $T_3 = W_1 \cap cW_2$, $T_4 = cW_1 \cap cW_2$ are disjoint sets and $W_2 = T_1 \cup T_2$ we have $T_1 = (T_1 \cup T_2) - T_2$.

Since it is assumed that $\mu_{p_1}(x) \geq \mu_{p_2}(x)$. We have

$$\min[\mu_{p_1}(x), \mu_{p_2}(x)] = \mu_{p_2}(x) = \frac{|W_2|}{|W|} = \frac{|T_1 \cup T_2|}{|W|}$$

and

$$\frac{|W_1 \cap W_2|}{|W|} = \frac{|T_1|}{|W|} =$$

$$\min[\mu_{p_1}(x), \mu_{p_2}(x)] - \frac{|T_2|}{|W|} = \min[\mu_{p_1}(x), \mu_{p_2}(x)] - \frac{|cW_1 \cap W_2|}{|W|}.$$

It is to be noted that because we assumed $\mu_{p1}(x) \geq \mu_{p2}(x)$, if $W_2 \subseteq W_1$ then we get

$$\mu_{p1\wedge p2}(x) = \min[\mu_{p1}(x), \mu_{p2}(x)] - \frac{0}{|W|} = \mu_{p2}(x)$$

In other words, we note that $\neg W_1 \cap W_2 = \varnothing$, i.e., The propositions $p_2$ is consonant respect to $p_1$, when $\mu_{p_1}(x) \geq \mu_{p_2}(x)$, i.e, the truthood of "John is tall" is larger than or equal to the truthood of "John is heavy"; that is the worlds in which "John is tall" includes the worlds in which "John is heavy".

**Proposition 2.** The membership value of the union of two fuzzy truthoods associated with the propositions $p_1$ and $p_2$ in the meta-theory is defined as:

$$\mu_{p_1 \vee p_2}(x) = \frac{|W_1 \cup W_2\,|}{|\,W\,|} \tag{13.7}$$

With a similar set of arguments, when $\mu_{p_1}(x) \geq \mu_{p_2}(x)$, this expression can alternately be written as:

$$\mu_{p_1 \vee p_2}(x) = \max\,[\mu_{p_1}(x)\,,\,\mu_{p_2}(x)\,] + \frac{|\,cW_1 \cap W_2\,|}{|\,W|} \qquad (13.7'^{\prime})$$

**Proof.** Because the sets $T_1 = W_1 \cap W_2$ , $T_2 = cW_1 \cap W_2$ , $T_3 = W_1 \cap cW_2$ , $T_4 = cW_1 \cap cW_2$ are disjoint set and $W_2 = T_1 \cup T_2$ and $W_1 \cup W_2 = T_1 \cup T_2 \cup T_3$ $W_1 \cup W_2 = (T_1 \cup T_3) + T_2$ Since it is assumed that $\mu_{p_1}(x) \geq \mu_{p_2}(x)$
We have

$$\max[\mu_{p_1}(x), \mu_{p_2}(x)] = \mu_{p_1}(x) = \frac{|W_1|}{|W|} = \frac{|T_1 \cup T_3|}{|W|}$$

and

$$\frac{|W_1 \cup W_2|}{|W|} = \max[\mu_{p_1}(x), \mu_{p_2}(x)] + \frac{|T_2|}{|W|} = \max[\mu_{p_1}(x), \mu_{p_2}(x)] + \frac{|cW_1 \cap W_2|}{|W|}.$$

When $\mu_{p_1}(x) < \mu_{p_2}(x)$

$$\mu_{p1 \wedge p2}(x) = \min[\mu_{p1}(x), \mu_{p2}(x)] - \frac{|\,cW_2 \cap W_1\,|}{|W|} \quad \text{and}$$

$$\mu_{p_1 \vee p_2}(x) = \max\,[\mu_{p_1}(x)\,,\,\mu_{p_2}(x)\,] + \frac{|\,cW_2 \cap W_1\,|}{|\,W|}$$

**Proof.** Since it is assumed that $\mu_{p_1}(x) < \mu_{p_2}(x)$

$$\min[\mu_{p_1}(x), \mu_{p_2}(x)] = \mu_{p_1}(x) = \frac{|W_1|}{|W|} = \frac{|(W_1 \cap W_2) \cup (W_1 \cap cW_2)|}{|W|} =$$

$$= \frac{|(W_1 \cap W_2)|}{|W|} + \frac{|(W_1 \cap cW_2)|}{|W|}$$

and since it is assumed that $\mu_{p_1}(x) < \mu_{p_2}(x)$ we have $\max(\mu_{p_1}(x), \mu_{p_2}(x)) = \mu_{p_2}(x)$ and

$$\mu_{p_2}(x) + \frac{|(W_1 \cap cW_2)|}{|W|} = \frac{|W_2|}{|W|} + \frac{|(W_1 \cap cW_2)|}{|W|}$$

$$= \frac{|W_2 \cup (W_1 \cap cW_2)|}{|W|} = \frac{|(W_2 \cup W_1)|}{|W|}$$

**Remarks.** Originally, in fuzzy theory, the intersection and union operation with two "descriptive" fuzzy sets were defined by Zadeh(1965) with the formulas:

$$\mu_{A \cap B}(x) = \min[\mu_A(x), \mu_B(x)] = \min[\mu_B(x), \mu_A(x)],$$

$$\mu_{A \cup B}(x) = \max[\mu_A(x), \mu_B(x)] = \max[\mu_B(x), \mu_A(x)],$$

without considering the order of the pair $(\mu_A(x), \mu_B(x))$. This is the result of an early, "myopic", generalisation from the classical two-valued theory where the order is not considered to be important within the confines of "reductionist" view of the world. Thus, the original generalization did not inquire about the order of the pair $(\mu_A(x), \mu_B(x))$. It should be recalled that Türkşen (1986) questioned the early generalization of these formulas arguing and showing that in two-valued theory we had DNF ≡ CNF but in fuzzy theory we have FDCF ≠ FCCF. Again recall that in fuzzy theory we call these expressions Fuzzy Disjunctive Canonical Form, FDCF, and Fuzzy Conjunctive Canonical Form, FCCF respectively, since in general they are not equal to each other. (Türkşen, 1999) Hence to call them "Normal" is not appropriate. It should be recalled that it is shown with the development and the construction of Fuzzy Truth Tables (Türkşen, 1994, 1999) that in fact FDCF and FCCF are equivalent in *form only* to DNF and CNF but in content FDCF ≠ FCCF due to the order of $\mu_A(x) \geq \mu_B(x)$ or $\mu_A(x) < \mu_B(x)$ in the combination of "descriptive" concepts. The order of the pair $(\mu_A(x), \mu_B(x))$ generates an eight entry Fuzzy Truth Table in analogy to the four entry Crisp Truth Table (Türkşen, 1999, 2000).

In this Chapter, it is shown that FDCF ≠ FCCF for fuzzy truthood statements, i.e., fuzzy "verification" statements in analogy to fuzzy "descriptive" statements (Türkşen, 1999, 2000). The approach here is from the perspective of meta-theory of modal logic, which is based on the verification, i.e., truthood of "descriptive" atomic sentences. In particular, in the next few sections, it is to be shown that synchronicity as well as the changes of the worlds (permutations) cause the containment relationships between FDCF and FCCF expressions. The main reason for this, as it will be shown, is the fact that synchronicity as well as the changes of the worlds combined with negation operation affect the computations when the order of the pairs ($\mu_{p_1}(x)$ and $\mu_{p_2}(x)$) are taken into account.

In preparation for the further developments, we stress here the fact that, in the meta-theory based on modal logic, *the order is very important.* This is the reason for the definitions given in **Propositions 1 and 2** where the intersection and union of fuzzy truth sets are stated as (13.6) and (13.7) when $\mu_{p_1}(x) \geq \mu_{p_2}(x)$. But when the order is changed, i.e., $\mu_{p_1}(x) < \mu_{p_2}(x)$ , then we have

$$\mu_{p_1 \wedge p_2}(x) \neq \min[\mu_{p_1}(x), \mu_{p_2}(x)] - \frac{|cW_2 \cap W_1|}{|W|}$$

$$\mu_{p_1 \vee p_2}(x) \neq \max[\mu_{p_1}(x)\mu_{p_2}(x)] + \frac{|cW_2 \cap W_1|}{|W|}$$

**Example 2.** Given $W = \{w_1, w_2, w_3, w_4, w_5\}$ and $||p_1|| ::= \{w_1, w_2, w_5\}$ and $||p_2||$ $::= \{w_1, w_2\}$, we have

$$\mu_{p_1}(x) = \frac{|W_1|}{|W|} = \frac{3}{5} \quad \text{and} \quad \mu_{p_2}(x) = \frac{|W_2|}{|W|} = \frac{2}{5}$$

as shown in Example 1 above. It is clear that $W_2 \subset W_1$, where $||p_1|| ::= W_1$ and $||p_2|| ::= W_2$. Since we have $\mu_{p_1}(x) \geq \mu_{p_2}(x)$ and $W_2 \subset W_1$, we get

$cW_1 = \{w_3, w_4\}$, and thus $cW_1 \cap W_2 = \varnothing$. Therefore, $\frac{|cW_1 \cap W_2|}{|W|} = 0$.

$$\mu_{p_1 \wedge p_2}(x) = \frac{|W_1 \cap W_2|}{|W|} = \frac{2}{5}.$$

Alternately, we find

$$\mu_{p_1 \wedge p_2}(x) = \min[\mu_{p_1}(x), \mu_{p_2}(x)] - \frac{|cW_1 \cap W_2|}{|W|}$$

$$\mu_{p_1 \wedge p_2}(x) = \frac{2}{5} - 0 = \frac{2}{5}$$

The proposition $p_2$ is consonant respect to the proposition $p_1$.

### *13.3.2. Worlds and Synchronisation*

In general, the conjunction, disjunction and negation operations are subject to synchronism. Given the time t , the set of worlds $S_t \subseteq W$ , where the proposition

$$s_t(w) = \text{“ the world w is active at the time t “}$$

is true is composed of worlds that are synchronised with one another, i.e., any operation in the synchronised worlds are execute at the same time. (see Figure 13.7)

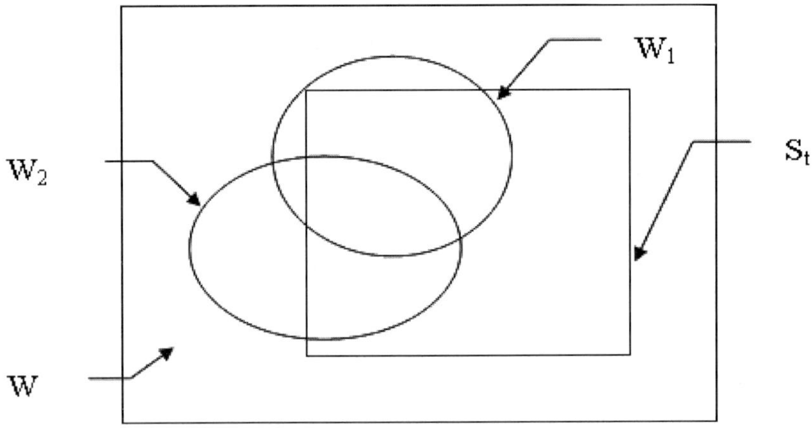


**Figure.13.7.** Any operation is execute at the same time t in the worlds inside the set $S_t$

#### *13.3.2.1. Conjunction and synchronisation*

Let the value of the conjunction of a proposition $p_1$ and $p_2$ be represented by the expression

$$( p_1 \wedge p_2 \wedge s_t(w) ) \vee ( p_1 \wedge \neg s_t(w) ) \tag{13.8}$$

that we show in the Table 13.2.

**Remark.** To simplify the notation we write the proposition $s_t(w)$ only with s and the set $S_t$ with S.

The proposition s in (13.8) generate the synchronous set S , $\| s \| ::= S$.

We remark that when s = 1 the expression (13.8) is the ordinary conjunction $p_1 \wedge p_2$, when s = 0 the expression (13.8) became the proposition $p_1$. This means that when s = 0, we cannot compose $p_1$ with $p_2$ so we have only the proposition $p_1$. Note that the following are the disjoint sets:

$W_1 \cap W_2 \cap S = B_1$, $W_1 \cap cW_2 \cap S = B_2$, $cW_1 \cap W_2 \cap S = B_3$, $cW_1 \cap cW_2 \cap S = B_4$

$W_1 \cap W_2 \cap cS = B_5$, $W_1 \cap cW_2 \cap cS = B_6$, $cW_1 \cap W_2 \cap cS = B_7$, $cW_1 \cap cW_2 \cap cS = B_8$.

**Table 13.2.** Conjunction with the proposition s which causes synchronism.

| $p_1$ | $p_2$ | s | $(p_1 \wedge p_2 \wedge s) \vee (p_1 \wedge \neg s)$ |
|---|---|---|---|
| 0 | 0 | 0 | 0 |
| 1 | 0 | 0 | 1 |
| 0 | 1 | 0 | 0 |
| 1 | 1 | 0 | 1 |
| 0 | 0 | 1 | 0 |
| 1 | 0 | 1 | 0 |
| 0 | 1 | 1 | 0 |
| 1 | 1 | 1 | 1 |

Therefore, we can write:

$$\| (p_1 \wedge p_2 \wedge s) \vee (p_1 \wedge \neg s) \| = B_1 \cup B_5 \cup B_6$$

When $\mu_{p_1 \wedge s} \geq \mu_{p_2 \wedge s}$, let the membership value of the conjunction of a proposition $p_1$ and $p_2$ be represented by the expression :

$$\min(\mu_{p_1 \wedge s}, \mu_{p_2 \wedge s}) = \mu_{p_2 \wedge s} = \frac{|B_1 \cup B_3|}{|W|} \quad \text{and}$$

$$B_1 \cup (B_5 \cup B_6) = (B_1 \cup B_3) - B_3 + (B_5 + B_6).$$

Therefore, we have:

$$\mu_{(p_1 \wedge p_2 \wedge s) \vee (p_1 \wedge \neg s)} = \min(\mu_{(p_1 \wedge s)}, \mu_{(p_2 \wedge s)}) - \frac{|B_3|}{|W|} + \frac{|B_5 + B_6|}{|W|} =$$

$$= \min(\mu_{(p_1 \wedge s)}, \mu_{(p_2 \wedge s)}) - \frac{|c(W_1 \cap S) \cap (W_2 \cap S)|}{|W|} + \frac{|W_1 \cap cS|}{|W|} \tag{13.9}$$

and when $\mu_{p_1 \wedge s}(x) < \mu_{p_2 \wedge s}(x)$ with the expression:

$$\mu_{(p_1 \wedge p_2 \wedge s) \vee (p_1 \wedge \neg s)} = \min(\mu_{(p_1 \wedge s)}, \mu_{(p_2 \wedge s)}) - \frac{|B_2|}{|W|} + \frac{|B_5 + B_6|}{|W|} =$$

$$= \min(\mu_{(p_1 \wedge s)}, \mu_{(p_2 \wedge s)}) - \frac{|W_1 \cap S \cap c(W_2 \cap S)|}{|W|} + \frac{|W_1 \cap cS|}{|W|} \qquad (13.9')$$

where

$$\mu_{p_1 \wedge s} = \frac{|\mathrm{W}_1 \cap \mathrm{S}|}{|\mathrm{W}|} = \frac{|B_1 \cup B_2|}{|\mathrm{W}|}, \quad \mu_{p_2 \wedge s} = \frac{|\mathrm{W}_2 \cap \mathrm{S})|}{|\mathrm{W}|} = \frac{|B_1 \cup B_3|}{|\mathrm{W}|}$$

### *13.3.2.2. Disjunction and synchronisation*

Let the membership value of the disjunction of a proposition $p_1$ and $p_2$ be represented with the Table 13.3.

**Table 13.3.** Disjunction with the proposition s which causes synchronism

| $p_1$ | $p_2$ | s | $((p_1 \vee p_2) \wedge s) \vee (p_1 \wedge \neg s)$ |
|---|---|---|---|
| 0 | 0 | 0 | 0 |
| 1 | 0 | 0 | 1 |
| 0 | 1 | 0 | 0 |
| 1 | 1 | 0 | 1 |
| 0 | 0 | 1 | 0 |
| 1 | 0 | 1 | 1 |
| 0 | 1 | 1 | 1 |
| 1 | 1 | 1 | 1 |

For the sets $B_k$ k =1,2,..8 We can write:

$$\| (p_1 \vee p_2 \wedge s) \vee (p_1 \wedge \neg s) \| = B_1 \cup B_2 \cup B_3 \cup B_5 \cup B_6$$

When $\mu_{p_1 \wedge s} \geq \mu_{p_2 \wedge s}$, let the membership value of the conjunction of a proposition $p_1$ and $p_2$ be represented by the expression :

$$\max(\mu_{p_1 \wedge s}, \mu_{p_2 \wedge s}) = \mu_{p_1 \wedge s} = \frac{|B_1 \cup B_2|}{|W|} \quad \text{and}$$

$$B_1 \cup B_2 \cup B_3 \cup (B_5 \cup B_6) = (B_1 \cup B_2) + B_3 + (B_5 + B_6)$$

So;

$$\mu_{(p_1 \vee p_2 \wedge s) \vee (p_1 \wedge \neg s)} = \max(\mu_{(p_1 \wedge s)}, \mu_{(p_2 \wedge s)}) + \frac{|B_3|}{|W|} + \frac{|B_5 + B_6|}{|W|} =$$

$$= \max(\mu_{(p_1 \wedge s)}, \mu_{(p_2 \wedge s)}) + \frac{|c(W_1 \cap S) \cap (W_2 \cap S)|}{|W|} + \frac{|W_1 \cap cS|}{|W|}$$

(13.10)

and when $\mu_{p_1 \wedge s}(x) < \mu_{p_2 \wedge s}(x)$ we have

$$\max(\mu_{p_1 \wedge s}, \mu_{p_2 \wedge s}) = \mu_{p_2 \wedge s} = \frac{|B_1 \cup B_3|}{|W|} \quad \text{and}$$

with the expression:

$$\mu_{(p_1 \wedge p_2 \wedge s) \vee (p_1 \wedge \neg s)} = \max(\mu_{(p_1 \wedge s)}, \mu_{(p_2 \wedge s)}) + \frac{|B_2|}{|W|} + \frac{|B_5 + B_6|}{|W|} =$$

$$= \max(\mu_{(p_1 \wedge s)}, \mu_{(p_2 \wedge s)}) + \frac{|W_1 \cap S \cap c(W_2 \cap S)|}{|W|} + \frac{|W_1 \cap cS|}{|W|} \quad (13.10')$$

We apply these results in the following example.

**Example 3.** Given W = $\{w_1, w_2, w_3, w_4, w_5\}$ , $\|p_1\|$ ::= $\{w_1, w_2, w_5\} = W_1$ , $\|p_2\|$ ::= $\{w_1, w_2\} = W_2$ , $\|s\|$ ::= $\{w_1, w_2, w_4\} = S$ , $\|p_1 \wedge s\| = \{w_1, w_2\}$ , $\|p_2 \wedge s\| = \{w_1, w_2\}$, and thus we have:

$$\mu_{p_1 \wedge p_2} = \min[\frac{2}{5}, \frac{2}{5}] - \frac{0}{5} + \frac{1}{5} = \frac{3}{5},$$

$$\mu_{p_1 \vee p_2} = \max[\frac{2}{5}, \frac{2}{5}] + \frac{0}{5} + \frac{1}{5} = \frac{3}{5}$$

*13.3.2.3. Complement and synchronisation*

Let W′ be the truth set of any proposition p which includes a descriptive fuzzy set A. Then the complement of p can be represented as shown in Table 13.4.

When s = 0 the negation is impossible, i.e., there is an asynchrony. Thus the result is equal to p. When s = 1 the negation is possible (synchrony) and the result is the ordinary negation of p. With the expression of the complement of p $\equiv (\neg p \wedge s) \vee (p \wedge \neg s)$, for $\|p\|$:: = $W' \subseteq W$ and $\|s\|$ ::= S, the membership value of the complement of a proposition p can be represented with the following expression:

$$\mu_{\neg p}(x) = \frac{|(S \cap cW') \cup (cS \cap W')|}{|W|},$$

where $\neg p$ is the complement of the proposition p, S is the truth set of a proposition s which causes the synchronism of a proposition p by acting on the truth set W' of the given proposition p. It should be noted that when s is true in all the worlds, i.e., S = W and therefore cS = ∅, then we have:
$(W \cap cW') \cup (\emptyset \cap W') = cW'$, since $cW' \cup W' = W$.

**Table 13.4.** Complement formation with the proposition s

| p | s | Complement of p ≡ ( ¬ p ∧ s ) ∨ ( p ∧¬ s ) |
|---|---|---|
| 1 | 0 | 1 |
| 0 | 0 | 0 |
| 1 | 1 | 0 |
| 0 | 1 | 1 |

On the other hand, when s is false in all the worlds, i.e., S = ∅, and therefore cS = W, then we have:
$(\emptyset \cap cW') \cup (W \cap W') = W'$.

Thus, in this theory, the proposition s and thus its truth set S causes to change the truth-value of the worlds as a complementation operator.
When in all the worlds, s is true, then all the worlds change, i.e., they are complemented, in a synchronic manner in agreement with the classical view. Therefore, s is the synchronic proposition. When

$$W'' \equiv (S \cap cW') \cup (cS \cap W')$$

it can be shown that:

$$(S \cap cW'') \cup (cS \cap W'') \equiv W'$$

in general in general. Therefore, the operator

$$F_s(W') = (S \cap cW') \cup (cS \cap W') = W'$$

has the property of the complementation operator.It should be noted that when S =W, i.e., s is true in all the worlds, we get $F_s(W') = cW'$. It is to be further noted that $F_s(W')$ is an involutive complementation, i.e.,

$$F_s^2(W') = W'$$

Therefore, $F_s(W')$ has the classical property of the involutive complementation of a set W′ corresponding to a proposition p in order to generate ¬p. More specifically, we have:

$$\mu_{\neg\neg p}(x) = \frac{|(S \cap cF_s(W')) \cup (cS \cap F_s(W'))|}{|W|} = \frac{|W'|}{|W|} = \mu_p(x).$$

**Example 4.** Given W = $\{w_1, w_2, w_3, w_4, w_5\}$ and ||p|| ::= W′ = $\{w_1, w_4, w_5\}$ and ||s||::= S = $\{w_2, w_3, w_5\}$, which means that in the worlds $w_2$, $w_3$, $w_5$, the proposition p changes its truth value, i.e., the negation of the truth values are caused by the synchronic proposition s.

$$\mu_{\neg p}(x) = \frac{|(S \cap cW') \cup (cS \cap W')|}{|W|} = \frac{4}{5} > 1 - \mu_p(x) = \frac{2}{5}.$$

In Table 13.5, below, we observe ||p|| ::= $\{w_1, w_4, w_5\}$ = W′ and thus || ¬p|| ::= $\{w_2, w_3\}$ = cW′, |||s||::= $\{w_2, w_3, w_5\}$ = S, and thus ||¬ s|| ::= $\{w_1, w_4\}$ = cS, we get:

|| s∧¬p||::= $\{w_2,w_3,w_5\} \cap \{w_2,w_3\} = \{w_2,w_3\}$, and
|| ¬s∧p||::= $\{w_1,w_4\} \cap \{w_1,w_4,w_5\} = \{w_1,w_4\}$.

**Table 13.5.** Negation by the proposition s versus classical regation

| | $V(p,w_k)$ | | |
|---|---|---|---|
| $W_1$ | p | s | Complement of p ≡ ( ¬ p ∧ s ) ∨ ( p ∧¬ s ) |
| $W_2$ | 1 | 0 | 1 |
| $W_3$ | 0 | 0 | 0 |
| $W_4$ | 1 | 1 | 0 |
| $W_5$ | 0 | 1 | 1 |

Hence we have:

$$(S \cap cW') \cup (cS \cap W') = \{w_2, w_3) \cup \{w_1, w_4) = \{w_1, w_2, w_3, w_4\}.$$

$$\mu_{\neg p}(x) = \frac{|(S \cap cW') \cup (cS \cap W')|}{|W|} = \frac{4}{5} > 1 - \mu_p(x) = \frac{2}{5},$$

since

$$\mu_p(x) = \frac{|W'|}{|W|} = \frac{3}{5}.$$

Thus as a result of synchronic proposition s, we have obtained the complementation of p to be different from the standard complementation , i.e., it is larger.

Next consider a different synchronic proposition s such that $||s||::= \{w_1.w_2, w_3, w_4, w_5\} = S$. That is we have the set corresponding synchronic position operator S = W, i.e., all the worlds are negated. Hence its negation c S = ∅. Thus for the same proposition p given above, we get:

$(S \cap cW') = \{w_1, w_2, w_3, w_4, w_5\} \cap \{w_2, w_3\} = \{w_2, w_3\}$, and
$(cS \cap W') = \emptyset \cap \{w_1, w_4, w_5\} = \emptyset$.

Therefore we have:

$$(S \cap cW') \cup (cS \cap W') = \{w_2, w_3) \cup \emptyset = \{w_2, w_3\} = cW'.$$

Therefore as a result, we get

$$\mu_{\neg p}(x) = \frac{|cW'|}{|W|} = \frac{2}{5} = 1 - \mu_p(x)$$

That is, as a result of this synchronic proposition s which is the synronic identity operator, i.e., S = W, we obtain the negation of $\mu_p(x)$ to be the standard negation.

Thus, the standard negation is obtained only when the synchronic proposition s causes all the worlds to be complemented by S = W, i.e., the whole set of worlds under consideration is negated.

$$(S \cap cW') = \{w_2, w_1, w_4, w_5\} = \{w_2\},$$

$$(cS \cap W') = \{w_3\} \cap \{w_1, w_4, w_5\} = \emptyset.$$

Next consider S such that we have $cS = \{w_3\}$ and

$(S \cap cW') \cup (cS \cap W') = \{w_2\} \cup \emptyset = \{w_2\}$ as a result we get

$$\mu_{\neg p}(\mathrm{x}) = \frac{|(\mathrm{S} \cap \mathrm{cW}') \cup (\mathrm{cS} \cap \mathrm{W}')|}{|\mathrm{W}|} = \frac{1}{5} < 1 - \mu_{\mathrm{p}}(\mathrm{x}).$$

Thus as a result of this synchronic proposition s which is not the synchronic identity operator, i.e., S ≠W, we obtain the complementation of p to be once again different from the standard negation, but in this case it is smaller. In conclusion, when S = W, the negation can be written as:

$$\mu_{\neg p}(\mathrm{x}) = \frac{|\mathrm{cW}'|}{|\mathrm{W}|} = 1 - \mu_{\mathrm{p}}(\mathrm{x})$$

which is the standard negation.

*13.3.3. The Effect of Transformation Γ of the worlds*

Let Γ be the transformation of the worlds, i.e.,
$\Gamma(\mathrm{w}_1, \mathrm{w}_2, \ldots\ldots, \mathrm{w}_\mathrm{n}) = (\mathrm{w}_{\alpha 1}, \mathrm{w}_{\alpha 2}, \ldots\ldots., \mathrm{w}_{\alpha \mathrm{n}})$

Given the proposition p we know that || p || = $W_p$ ,where $W_p$ is the set of worlds where p is true. For the transformation Γ the set of worlds where p is true changes in this way :

$$W_p' = \Gamma\ W_p$$

So after the transformation the set of worlds where p is true is Γ $W_p$

$$\|\ p\ \| = \Gamma\ W_p$$

In different times for the same proposition p , the number of worlds where p is true is different.So we can write

$\|\ p\ \|_t = (\Gamma)^t\ W_p$ for t =1 ,2,……n

where t is the time.

For the AND operation when $\mu_{p_1}(x,t) \geq \mu_{p_2}(x,t)$ we have

$$\mu_{p1 \wedge p2}(x,t) = \min[\mu_{p1}(x,t), \mu_{p2}(x,t)] - \frac{|\,\mathrm{c}\,\Gamma_1{}^t \mathrm{W}_1 \cap \Gamma_2{}^t \mathrm{W}_2\,|}{|\mathrm{W}|}$$

where

$$\mu_{p1}(x,t) = \frac{|\Gamma_1^{\,t} W_1|}{|W|}\ ,\ \mu_{p2}(x,t)] = \frac{|\Gamma_2^{\,t} W_2|}{|W|}$$

and $\Gamma_1^{\,t} = \Gamma_1\Gamma_1....\Gamma_1$ for t times , $\Gamma_2^{\,t} = \Gamma_2\Gamma_2....\Gamma_2$ for t times

For the OR operation when $\mu_{p_1}(x,t) \geq \mu_{p_2}(x,t)$ we have

$$\mu_{p1\vee p2}(x,t) = \max[\mu_{p1}(x,t), \mu_{p2}(x,t)] + \frac{\left|c\Gamma_1^t W_1 \cap \Gamma_2^t W_2\right|}{\left|W\right|}$$

where

$$\mu_{p1}(x,t) = \frac{|\Gamma_1^{\,t} W_1|}{|W|}\ ,\ \mu_{p2}(x,t)] = \frac{|\Gamma_2^{\,t} W_2|}{|W|}$$

and $\Gamma_1^{\,t} = \Gamma_1\Gamma_1....\Gamma_1$ for t times , $\Gamma_2^{\,t} = \Gamma_2\Gamma_2....\Gamma_2$ for t times

For NOT operation

$$\mu_{\neg p}(x,t) = \frac{|(S \cap c\Gamma^t W') \cup (cS \cap \Gamma^t W')|}{|W|},$$

Let us next consider the absurd condition, i.e., $p \wedge \neg p$ which is written as:

$$\mu_{p\wedge\neg p}(x,1) = \frac{|W_p \cap c\,\Gamma\ W_p\ ]\,|}{|W|}$$

where $\| p \| = W_p$

**Proposition 3.** We can prove that $c\,\Gamma\ W_p = \Gamma\ c\ W_p$ , i.e., the complement operator c commute with the operator $\Gamma$ .

**Proof.** Given

$\Gamma(w_1, w_2, \ldots\ldots, w_n) = (w_{\alpha 1}, w_{\alpha 2}, \ldots\ldots, w_{\alpha n})$ and $W_p = (w_{a_1}, w_{a_2}, \ldots\ldots, w_{a_s})$

we have

$\Gamma(w_{a1}, w_{a2}, .., w_{as}) = (w_{\alpha a1}, w_{\alpha a2}, \ldots, w_{\alpha as})$ and

$c\Gamma(w_{a1}, w_{a2}, .., w_{as}) = (w_{\alpha b1}, w_{\alpha b2}, \ldots, w_{\alpha bk})$ with k+s =n

but we have also that

$c(w_{a1}, w_{a2}, .., w_{as}) = (w_{b1}, w_{b2}, \ldots, w_{bk})$ and

$\Gamma c\,(w_{a1}, w_{a2}, .., w_{as}) = (w_{\alpha b1}, w_{\alpha b2}, \ldots, w_{\alpha bk})$ with k+s =n .

*13.3.4. Transformation and the absurd condition.*

For the absurd condition, we have two special cases:
(i) If $\Gamma$ is the identity transformation, i.e., $[c\Gamma W] = cW$, then recalling Proposition 1, whenever $\mu_{\neg p}(x) \geq 1 - \mu_p(x)$ i.e.,

$$\mu_{p \wedge \neg p}(x) = \min[\mu_p(x)\ ,\ 1 - \mu_p(x)] - \frac{|\,cW_p \cap cW_p|}{|\,W|} = (1 - \mu_p(x)) - (1 - \mu_p(x))$$

$= 0$

Note that we have:

$$\frac{|\,cW_p|}{|W|} = 1 - \mu_p(x)$$

(ii) If, on the other hand, we have $[c\Gamma W_p] \subseteq W_p$, then we have the consonant case, and hence, we get:

$$\frac{|\,cW_p \cap [c\Gamma W_p]|}{|\,W|} = 0$$

and therefore

$$\mu_{p \wedge \neg p}(x,1) = \min[\mu_p(x), 1 - \mu_p(x)]$$

For dissonant case, we have from the proposition 1

$$\mu_{p \wedge \neg p}(x,1) < \min[\mu_p(x), 1 - \mu_p(x)]$$

Therefore, the effects of two extreme transformations, $\Gamma$'s, result in an interval in which the fuzzy truthood membership values of the composite proposition $p \wedge \neg p$ is contained. Thus, we have:

$$0 \leq \mu_{p \wedge \neg p}(x) \leq \min[\mu_p(x),\ 1 - \mu_p(x)]$$

This result is an alternate verification of the result obtained by Türkşen (1999, 2001) with the development of Fuzzy Truth Tables. That is, we have "Fuzzy Law of Contradiction" in fuzzy theory. It should be noted that in Türkşen, *et al.* (1999) the Fuzzy Law of Contradiction is developed for fuzzy descriptor sets for the case of the two-valued truth. Whereas, the development here is for fuzzy truthoods of descriptor sets which are verified by the two-valued truth theory. It is to be noted that fuzzy truthoods could be over either two or infinite valued descriptor sets.

**Example 5.** Given the set of worlds, W = {$w_1$, $w_2$, $w_3$, $w_4$, $w_5$}, and ||p||::= $W_p$ = {$w_1$, $w_2$, $w_5$} are verfied to be true in two-valued theory for some descriptor set A, whether it be two-valued or infinite (fuzzy) valued. Clearly, then we have the proposition || ¬ p || ::= c $W_p$ = {$w_3$, $w_4$}. It can be observed that the two propositions p and ¬ p are completely dissonant. $cW_p \cap cW_p = cW_p = \{w_3, w_4\}$

Therefore, we get with Proposition 1:

$$\mu_{p\wedge\neg p}(x)=\min[\mu_p(x),\ \mu_{\neg p}(x)]-\frac{|cW_p \cap cW_p|}{|W|}=\min[\frac{3}{5},\frac{2}{5}]-\frac{2}{5}=0,$$

But if we introduce the permutation

$$\Gamma\{w_1,w_2,w_3,w_4,w_5\}=\{w_3,w_4,w_1,w_2,w_5\},$$

then we obtain, $\Gamma W_p=\{w_3,w_4,w_5\}$ , $c\Gamma W_p=\{w_1,w_2\}$ because c Γ $W_p \subset W_p$ (consonant case). Thus, we get:

$$\mu_{p\wedge\neg p}(x,1)=\min\{\mu_p(x),1\text{-}\mu(x)\}\ -\ \frac{|cW_p \cap c\Gamma W_p\,|}{|W|}=\min(\frac{3}{5},\frac{2}{5})=\frac{2}{5}$$

Therefore, we have obtain an interval that contains

$$\mu_{p\wedge\neg p}(x)\ ,\ \text{i.e.}\ ,0\le\mu_{p\wedge\neg p}(x)\le\frac{2}{5}$$

## 13.4. Canonical Forms for the Synchronous Case

In logic, for any expression, we have two different forms (i) one is the DCF, Disjunctive Canonical Form, and (ii) the other is the CCF or the Conjunctive Canonical Form. For "A AND B", we have the DCF expression as A∩B, and the CCF expression as:

$$(A\cup B)\cap(A\cup cB)\cap(cA\cup B)$$

In the classical set theory, we have the equivalence of these two expressions due to idempotency and the classical law of contradiction, i.e., it can be shown that:

$$(A\cup B)\cap(A\cup cB)\cap(cA\cup B)\equiv A\cap B$$

Because of this equivalence DCF and CCN are known as DNF and CNF, Disjunctive Normal Form and Conjunctive Normal Form, respectively.

However, in the fuzzy set theory, it is shown (Türkşen, 1986-2001) that in general:

FDCF(A AND B) $\neq$ FCCF(A AND B), i.e., we have

$$A \cap B \neq (A \cup B) \cap (A \cup cB) \cap (cA \cup B)$$

This is due to the fact that the classical Law of Contradiction and its dual the classical Law of Excluded Middle as well as the axioms of distributivity, absorption and idempotency no longer hold in general for the class of t-norms and t-conorms. With the meta-theory, based on modal logic, we can now explain why, we get this inequality for the case of truthood sets associated with propositions.

Let us next consider each term of the FDCF and FCCF for the case of $p_1$ and $p_2$:

$$\mu_{p1}(x) = \frac{|W_1|}{|W|} \text{ and } \mu_{\neg p2}(x) = \frac{|\Gamma_2[cW_2]|}{|W|} = 1 - \mu_{p2}(x),$$

(1) First consider

$$\mu_{p_1 \vee \neg p_2}(x) = \frac{|W_1 \cup \Gamma_2[cW_2]|}{|W|}$$

which can be written in two ways:

(i) when $\mu_{p1}(x) \geq 1 - \mu_{p2}(x)$ or $\mu_{p1}(x) + \mu_{p2}(x) \geq 1$, we obtain

$$\mu_{p \vee \neg p2}(x) = \max[\mu_{p1}(x),\ 1 - \mu_{p2}(x)] + \frac{|cW_1 \cap \Gamma_2[cW_2]|}{|W|}; \text{ and}$$

ii) when $1 - \mu_{p2}(x) > \mu_{p1}(x)$ or $\mu_{p1}(x) + \mu_{p2}(x) < 1$, we obtain

$$\mu_{p1 \vee \neg p2}(x) = \max[1 - \mu_{p2}(x),\ \mu_{p1}(x)] + \frac{|\Gamma_2[cW_2] \cap W_1|}{|W|}.$$

(2) Secondly, consider

$$\mu_{\neg p1 \vee p2}(x) \leq \frac{|\Gamma_1[cW_1] \cup W_2|}{|W|}$$

which can be written in two ways:

(i) when $\mu_{p2}(x) \geq 1 - \mu_{p1}(x)$ or $\mu_{p1}(x) + \mu_{p2}(x) \geq 1$, we obtain:

$$\mu_{\neg p_1 \vee p_2}(x) = \max[\mu_{p2}(x),\ 1 - \mu_{p1}(x)] + \frac{|cW_2 \cap \Gamma_1[cW_1]|}{|W|}, \text{ and}$$

ii) when $1 - \mu_{p1}(x) > \mu_{p2}(x)$ or $\mu_{p1}(x) + \mu_{p2}(x) < 1$, we obtain:

$$\mu_{\neg p1\vee p2}(\mathrm{x}) = \max[1-\mu_{p1}(\mathrm{x}), \mu_{p2}(\mathrm{x})] + \frac{|\Gamma_1[\mathrm{cW}_1] \cap \mathrm{W}_2|}{|\mathrm{W}|}.$$

(3) Finally consider

$$\mu_{\mathrm{p1\vee p2}}(\mathrm{x}) = \frac{|\mathrm{W}_1 \cup \mathrm{W}_2|}{|\mathrm{W}|}$$

which can also be written in two ways:

i) when $\mu_{\mathrm{p1}}(\mathrm{x}) \geq \mu_{\mathrm{p2}}(\mathrm{x})$ or $\mu_{\mathrm{p1}}(\mathrm{x}) - \mu_{\mathrm{p2}}(\mathrm{x}) \geq 0$, , we obtain:

$$\mu_{\mathrm{p1\vee p2}}(\mathrm{x}) = \max\,[\mu_{\mathrm{p1}}(\mathrm{x}), \mu_{\mathrm{p2}}(\mathrm{x})] + \frac{|\mathrm{cW}_1 \cap \mathrm{W}_2|}{|\mathrm{W}|},$$ and

(ii) when $\mu_{\mathrm{p2}}(\mathrm{x}) > \mu_{\mathrm{p1}}(\mathrm{x})$ or $\mu_{\mathrm{p1}}(\mathrm{x}) - \mu_{\mathrm{p2}}(\mathrm{x}) < 0$, we obtain:

$$\mu_{\mathrm{p1\vee p2}}(\mathrm{x}) = \max[\mu_{\mathrm{p2}}(\mathrm{x}), \mu_{p1}(\mathrm{x})] + \frac{|\mathrm{cW}_2 \cap \mathrm{W}_1|}{|\mathrm{W}|}.$$

If we first join the expressions $(\neg \mathrm{p}_1 \vee \mathrm{p}_2)$ and $(\mathrm{p}_1 \vee \neg \mathrm{p}_2)$ and obtain

$$\mu_{(\neg\mathrm{p1\vee p2})\wedge(\mathrm{p1}\vee\neg\mathrm{p2})}(\mathrm{x}) = \frac{|\{\Gamma_1[\mathrm{cW}_1]) \cup \mathrm{W}_2\} \cap \{\mathrm{W}_1 \cup \Gamma_2[\mathrm{cW}_2]\}|}{|\mathrm{W}|}$$

which can be written again in two ways:

$$\mu_{(\neg\mathrm{p1\vee p2})\wedge(\mathrm{p1}\vee\neg\mathrm{p2})}(\mathrm{x}) =$$

(i) $$\min[\mu_{\neg\mathrm{p1\vee p2}}(\mathrm{x}), \mu_{\mathrm{p1}\vee\neg\mathrm{p2}}(\mathrm{x})] - \frac{|\mathrm{c}\{\Gamma_1[\mathrm{cW}_1] \cup \mathrm{W}_2\} \cap \{\mathrm{W}_1 \cup \Gamma_2[\mathrm{cW}_2]\}|}{|\mathrm{W}|},$$

when $\mu_{\neg\mathrm{p1\vee p2}}(\mathrm{x}) \geq \mu_{\mathrm{p1}\vee\neg\mathrm{p2}}(\mathrm{x})$, and

$$\mu_{(\neg\mathrm{p1\vee p2})\wedge(\mathrm{p1}\vee\neg\mathrm{p2})}(\mathrm{x}) =$$

(ii) $$\min[\mu_{\neg\mathrm{p1\vee p2}}(\mathrm{x}), \mu_{\mathrm{p1}\vee\neg\mathrm{p2}}(\mathrm{x})] - \frac{|\mathrm{c}\{\mathrm{W}_1 \cup \Gamma_2[\mathrm{cW}_2]\} \cap \{\Gamma_1[\mathrm{cW}_1] \cup \mathrm{W}_2\}|}{|\mathrm{W}|}$$

when $\mu_{p1\vee\neg p2}(\mathrm{x}) > \mu_{p1\vee p2}(\mathrm{x})$.

Next, we join $(\neg \mathrm{p}_1 \vee \mathrm{p}_2) \wedge (\mathrm{p}_1 \vee \neg \mathrm{p}_2)$ and $(\mathrm{p}_1 \vee \mathrm{p}_2)$ and obtain

$$\mu_{(\neg p1\vee p2)\wedge(p1\vee\neg p2)\wedge(p1\vee p2)}(\mathrm{x}) = \frac{|\{\{\Gamma_1[\mathrm{cW}_1] \cup \mathrm{W}_2\} \cap \{\mathrm{W}_1 \cup \Gamma_2[\mathrm{cW}_2]\}\} \cap \{\mathrm{W}_1 \cup \mathrm{W}_2\}|}{|\mathrm{W}|}$$

which can be written again in two ways:

(i) $\mu_{(\neg\mathrm{p1\vee p2})\wedge(\mathrm{p1}\vee\neg\mathrm{p2})\wedge(\mathrm{p1\vee p2})}(\mathrm{x}) = \min[\mu_{(\neg\mathrm{p1\vee p2})\vee(\mathrm{p1}\vee\neg\mathrm{p2})}(\mathrm{x}), \mu_{(\mathrm{p1\vee p2})}(\mathrm{x})] +$

$$-\frac{|c\{\{\Gamma_1[cW_1]\cup W_2\}\cap\{W_1\cup\Gamma_2[cW_2]\}\}\cap\{W_1\cup W_2\}|}{|W|}$$

when $\mu_{\neg p1\vee p2)\wedge(p1\vee\neg p2)}(x)\geq\mu_{p1\vee p2}(x)$, and

(ii) $\mu_{(\neg p1\vee p2)\wedge(p1\vee\neg p2)\wedge(p1\vee p2)}(x)=\min[\mu_{(\neg p1\vee p2)}(x),\mu_{(\neg p1\vee p2)\wedge(p1\vee\neg p2)}(x)]+$

$$-\frac{|c\{W_1\cup W_2\}\cap\{\{\Gamma_1[cW_1]\cup W_2\}\cap[W_1\cup\Gamma_2[cW_2]\}\}|}{|W|}$$

when $\mu_{p1\vee p2}(x)>\mu_{(\neg p1\vee p2)\wedge(p1\vee\neg p2)}(x)$

*13.4.1. Equivalence or Containment*

Next we need to investigate:
i) if we get the equivalence between FDCF and FCCF when we have the identity transformation of the worlds;
ii) alternately, we want to show if we get containment of FDCF in FCCF when we have non-identity transformations of the worlds.

*13.4.1.1. Equivalence of FDNF and FCNF*

Recall from section 13.3.1 and Proposition 1 that

$$\mu_{p1\wedge p2}(x)=\frac{|W_1\cap W_2|}{|W|}=\min[\mu_{p1}(x),\mu_{p2}(x)]-\frac{|cW_1\cap W_2|}{|W|}=\mu_{p2}(x)$$

whenever $\mu_{p1}(x)\geq\mu_{p2}(x)$, since $|cW_1\cap W_2|=0$.

This gives us the FDCF $(p_1 \text{ AND } p_2)=p_1\wedge p_2$.

To evaluate FCCF $(p_1$ AND $p_2)$, $=(p_1\vee\neg p_2)\wedge(\neg p_1\vee p_2)\wedge(p_1\vee p_2)$, we need to evaluate

(a) $(p_1\vee\neg p_2)$

(b) $(\neg p_1\vee p_2)$

(c) $(p_1\vee p_2)$

This is done next.

(a) Thus, for the computation of $(p_1\vee p_2)$, we have two cases:

(i) when $\mu_{p1}(x)\geq 1-\mu_{p2}(x)$, we get

$$\mu_{p1\vee\neg p2}(x)=\max[\mu_{p1}(x),1-\mu_{p2}(x)]+\frac{|cW_1\cap\Gamma_2[cW_2]|}{|W|}=\mu_{p1}(x)$$

if $\Gamma_2$ is the identity transformation; since $|cW_1 \cap cW_2| = 0$;
(iii) on the other hand, when $1 - \mu_{p2}(x) > \mu_{p1}(x)$, we get

$$\mu_{p1\vee\neg p2}(x) = \max[1 - \mu_{p2}(x),_{p1}(x)] + \frac{|\Gamma_2[cW_2] \cap W_1|}{|W|}$$
$$= 1 - \mu_{p2}(x)$$

if $\Gamma_2$ is the identity transformation since $|cW_2 \cap W_1| = 0$.
(b) For the computation of $(\neg p_1 \vee p_2)$, we have again two cases:
(i) when $\mu_{p2}(x) \geq 1 - \mu_{p1}(x)$, we get

$$\mu_{\neg p1\vee p2}(x) = \max[\mu_{p2}(x), 1 - \mu_{p1}(x)] + \frac{|\neg W_2 \cap \Gamma_2[\neg W_2]|}{|W|} = \mu_{p2}(x)$$

if $\Gamma_1$ is the identity transformation; since $|cW_2 \cap cW_1| = 0$;
(ii) on the other hand when $1 - \mu_{p1}(x) > \mu_{p2}(x)$, we get:

$$\mu_{\neg p1\vee p2}(x) = \max[1 - \mu_{p1}(x), \mu_{p2}(x)] + \frac{|\Gamma_2[\neg W_1] \cap W_2|}{|W|} = 1 - \mu_{p1}(x)$$

if again $\Gamma_1$ is the identity transformation since $|cW_1 \cap W_2| = 0$.
(c) For the computation of $(p_1 \vee p_2)$, we have also two cases:
(i) when $\mu_{p1}(x) \geq \mu_{p2}(x)$, we get

$$\mu_{p1\vee p2}(x) = \max[\mu_{p1}(x), \mu_{p2}(x)] + \frac{|cW_1 \cap \Gamma_2[cW_2]|}{|W|} = \mu_{p1}(x)$$

if $\Gamma_2$ is the identity transformation;
(ii) on the other hand, when $\mu_{p2}(x) > \mu_{p1}(x)$, we get since $|cW_1 \cap W_2| = 0$;

$$\mu_{p1\vee p2}(x) = \max[\mu_{p2}(x), \mu_{p1}(x)] + \frac{|cW_2 \cap \Gamma_1[W_1]|}{|W|} = \mu_{p2}(x)$$

if $\Gamma_1$ is the identity transformation since $|cW_2 \cap W_1| = 0$.
In summary, therefore, we have the following results:

(a) (i) $\mu_{p1}(x) \geq 1 - \mu_{p2}(x)$ : $\mu_{p1\vee\neg p2}(x) = \mu_{p1}(x)$

(b) (i) $\mu_{p2}(x) \geq 1 - \mu_{p1}(x)$ : $\mu_{\neg p1\vee p2}(x) = \mu_{p2}(x)$

(c) (i) $\mu_{p1}(x) \geq \mu_{p2}(x)$ : $\mu_{p1\vee\neg p2}(x) = \mu_{p1}(x)$

Now to determine FCCF

$(p_1 \text{ AND } p_2) = (p_1 \vee \neg p_2) \wedge (\neg p_1 \vee p_2) \wedge (p_1 \vee p_2)$, we conjunct (a), (b), (c) for case (i), and obtain:

$$[\mu_{p1\vee\neg p2}(x)] \wedge [\mu_{\neg p1\vee p2}(x)] \wedge [\mu_{p1\vee p2}(x)] = \mu_{p1}(x) \wedge \mu_{p2}(x) \wedge \mu_{p1}(x) = \mu_{p2}(x)$$

since the combination of conditions are satisfied with $\mu_{p1}(x) \geq \mu_{p2}(x)$.

Hence we get the equivalence for the case of the identity transformation, i.e., FDCF $(p_1 \text{ AND } p_2) =$ FCCF $(p_1 \text{ AND } p_2) = \mu_{p2}(x)$.

Let us next try the other case.

Again as the summary of the computations from above, we have the following:

(a) (ii) $1 - \mu_{p2}(x) > \mu_{p1}(x) : \mu_{p1\vee\neg p2}(x) = 1 - \mu_{p2}(x)$:

(b) (ii) $1 - \mu_{p1}(x) > \mu_{p2}(x)$: $\mu_{\neg p1\vee p2}(x) = 1 - \mu_{p1}(x)$

(c) (ii) $\mu_{p2}(x) > \mu_{p1}(x)$ : $\mu_{p1\vee p2}(x) = \mu_{p2}(x)$

Let us again determine the conjunction of these, thus we get FCCF ($p_1$ AND $p_2$) by the conjunction of (a), (b), (c) for case (ii). We get:

$$[\mu_{p1\vee\neg p2}(x)] \wedge [\mu_{\neg p1\vee p2}(x)] \wedge [\mu_{p1\vee p2}(x)]$$
$$= (1 - \mu_{p2}(x)) \wedge (1 - \mu_{p1}(x)) \wedge (\mu_{p2}(x)) = \mu_{p2}(x)$$

Since the combination of the conditions are satisfied when $\mu_{p2}(x) < 1 - \mu_{p2}(x)$.

Hence again we obtain the equivalence for the case of the identity transformation.

FDCF $(p_1 \text{ AND } p_2) =$ FCCF $(p_1 \text{ AND } p_2) = \mu_{p2}(x)$.

Similar investigations of all the remaining feasible combinations show that FDNF(.) = FCNF (.) for the identity transformation. It should be recalled that under the identity transformation the worlds are consonant under the given conditions.

*13.4.1.2. Containment of FDNF and FCNF*

Now if $\Gamma_1$ and $\Gamma_2$ are generic permutations of the worlds that is different form the identity permutation, then we obtain:

$$|\{\Gamma_1[cW_1] \cup W_2\} \cap \{W_1 \cup \Gamma_2[cW_2]\} \cap [W_1 \cup W_2]| \supseteq |W_1 \cap W_2|$$

And hence FDCF(.) $\subseteq$ FCCF(.). This is now demonstrated in an example.

Example 6: Given

$W = \{w_1, w_3, w_4, w_5\}$, $\|p_1//::=$
$W_1 = \{w_1, w_2, w_4\}$, and $\|p_2//::= W_2 = \{w_1, w_2, w_5\}$, which are verified to be true

in two-valued theory, we have

$$\mu_{p1}(x) = \frac{|W_1|}{|W|} = \frac{3}{5}, \mu_{p2}(x) = \frac{|W_2|}{|W|} = \frac{3}{5}, \text{ and } \mu_{p1\wedge p2}(x)$$
$$= \frac{|W_1 \cap W_2|}{|W|} = \frac{2}{5}$$

Alternately, because $\mu_{p1}(x) = \mu_{p2}(x)$ and $cW_1 \cap W_2 = \{w_5\}$, which are verified to be true in two-valued theory, we have

$$\mu_{p1\wedge p2}(x) = \min[\mu_{p1}(x), \mu_{p2}(x)] - \frac{|cW_1 \cap W_2|}{|W|} = \frac{3}{5} - \frac{1}{5} = \frac{2}{5}$$

which is the result of FDCF ($p_1$ AND $p_2$) = $p_1 \wedge p_2$.

Note that the value of $\frac{1}{5}$ is the difference with respect to the Zadeh operations.

For the computation of FCCF, we have:

FCCF ($p_1$ AND $p_2$) = $(p_1 \vee p_2) \wedge (p_1 \vee \neg p_2) \wedge (\neg p_1 \vee \neg p_2)$
suppose we have non-identity permutations as:
$\Gamma_1\{w_1, w_2, w_3, w_4, w_5\} = \{w_5, w_1, w_2, w_3, w_4\}$,

$\Gamma_2\{w_1, w_2, w_3, w_4, w_5\} = \{w_4, w_5, w_1, w_2, w_3\}$,
then because $cW_2 = \{w_3, w_4\}$ we get $\Gamma_2[\neg p_2] = \{w_1, w_2\}$ under the transformation $\Gamma_2$.

Let us compute each parenthetical term of FCCF in two different ways.
(a) First let us compute ($p_1 \vee \neg p_2$) in two ways:

(i) $$\mu_{p1\vee\neg p2}(x) = \frac{|W_1 \cup \Gamma_2[cW_2]|}{|W|} = \frac{3}{5}$$

since $cW_2 = \{w_3, w_4\}$ and $\Gamma_2[cW_2] = \{w_1, w_2\}$ and thus for $W_1 \cup \Gamma_2[cW_2]$, we have

$\{w_1, w_2, w_4\} \cup \{w_1, w_2\} = \{w_1, w_2, w_4\}$

ii) Alternately, we compute:

$$\mu_{p1\vee\neg p2}(x) = \max[\mu_{p1}(x), 1-\mu_{p2}(x)] + \frac{|cW_1 \cap \Gamma_2[cW_2]|}{|W|} = \frac{3}{5} - 0 = \frac{3}{5}.$$

Because $\mu_{p1}(x) = \frac{3}{5}$, $1-\mu_{p2}(x) = \frac{2}{5}$, and for $cW_1 \cap \Gamma_2[cW_2]$, we have $\{w_3, w_5\} \cap \{w_1, w_2\} = \varnothing$.

(b) Next, let us compute $(\neg p_1 \vee p_2)$ again in two ways:

(i) $\mu_{\neg p1\vee p2}(x) = \frac{|\Gamma_1[cW_1] \cup W_2|}{|W|} = \frac{4}{5}$,

since for $\Gamma_1[cW_1] \cup W_2$, we have $\{w_2, w_4\} \cup \{w_1, w_2, w_5\} = \{w_1, w_2, w_4, w_5\}$.

(ii) $\mu_{\neg p1\vee p2}(x) = \max[\mu_{p2}(x), 1-\mu_{p1}(x)] + \frac{|cW_2 \cap \Gamma_1[cW_1]|}{|W|} = \frac{3}{5} + \frac{1}{5} = \frac{4}{5}$,

Because $\mu_{p1}(x) = \frac{3}{5}$ and $1-\mu_{p1}(x) = \frac{2}{5}$, and for $cW_2 \cap \Gamma_1[cW_1]$, we have $\{w_3, w_4\} \cap \{w_2, w_4\} = \{w_4\}$.

(c) Finally, let us compute $W_1 \cup W_2$ in two different ways:

(i) $\mu_{p1\vee p2}(x) = \frac{|W_1 \cup W_2|}{|W|} = \frac{4}{5}$,

because for $W_1 \cup W_2 =$ we get $\{w_1, w_2, w_4, w_5\}$.

(ii) Alternately, we compute

$$\mu_{p1\vee p2}(x) = \max[\mu_{p1}(x), \mu_{p2}(x)] + \frac{|cW_1 \cap W_2|}{|W|} = \frac{3}{5} + \frac{1}{5} = \frac{4}{5}.$$

because we have $\mu_{p1}(x) = \frac{3}{5}, \mu_{p2}(x) = \frac{3}{5}$ and $cW_1 \cap W_2$ = $\{w_3, w_5\} \wedge \{w_1, w_2, w_5\} = \{w_5\}$.

(d) Next let us first join conjunctively the terms $(\neg p_1 \vee p_2)$ and $(p_1 \vee \neg p_2)$.

(i) For this, we compute

$\{\Gamma_1[cW_1] \cup W_2\} \cap \{W_1 \cup \Gamma_2[cW_2]\}$

$= [\{w_2, w_4\} \cup \{w_1, w_2, w_5\}] \cap [\{w_1, w_2, w_4\} \cup \{w_1, w_2\}]$

$= \{w_1, w_2, w_4, w_5\} \cap \{w_1, w_2, w_4\} = \{w_1, w_2, w_4\}.$

Therefore, we get

$$\mu_{(\neg p1 \vee p2) \wedge (p1 \vee \neg p2)}(x) = \frac{|[\Gamma_1[cW_1] \cup W_2] \cap [W_1 \cup \Gamma_2[cW_2]]|}{|W|} = \frac{3}{5}$$

(i) Alternately, we compute the same as:

$$\mu_{(\neg p1 \vee p2) \wedge (p1 \vee \neg p2)}(x) = min\ [\mu_{\neg p1 \vee p2}(x), \mu_{p1 \vee \neg p2}(x)]$$

$$= \frac{|c[\Gamma_1[cW_1] \cup W_2] \cap [W_1 \cup \Gamma_2[cW_2]]|}{|W|}$$

$$= \frac{3}{5} - 0 = \frac{3}{5}$$

Because

$\mu_{p1 \vee \neg p2}(x) = \frac{3}{5}, \mu_{\neg p1 \vee p2}(x) = \frac{4}{5}$ and thus min[ $\mu_{p1 \vee \neg p2}(x) \mu_{\neg p1 \vee p2}(x)] = \frac{3}{5}$

and furthermore

$c\ \{\Gamma_1\ [cW_1] \cup W_2\} \cap \{W_1 \cup \Gamma_2\ [cW_2]\} = c\{w_1, w_2, w_3, w_4, w_5\} \cap \{w_1, w_2, w_4\}$
$= \{w_3\} \cap \{w_1, w_2, w_4\} = \varnothing$ .

Next let us conjunct the tree terms of the FCCF, i.e., $(\neg p_1 \vee p_2), (p_1 \vee \neg p_2)$, and $(p_1 \vee p_1)$ to compute the FCCF value.

(e) Again we compute this in two ways

(i) $\mu_{(\neg p1 \vee p2) \wedge (p1 \vee \neg p2) \wedge (p1 \vee p2)}(x)$

$$= \frac{|\{[\Gamma_1[cW_1] \cup W_2] \cap [W_1 \cup \Gamma_2[cW_2]]\} \cap [W_1 \cup W_2\}|}{|W|} = \frac{3}{5}$$

Since

$$\{[\Gamma_1[cW_1] \cup W_2] \cap [W_1 \cup \Gamma_2[cW_2]]\} \cap \{W_1 \cup W_2\}$$
$$= \{w_1, w_2, w_4\} \cap [w_1, w_2, w_4, w_5\} = \{w_1, w_2, w_4\}.$$

Note that:

$[\Gamma_1[cW_1] \cup W_2] \cap [W_1 \cup \Gamma_2[cW_2]] = \{w_1, w_2, w_4, w_5\} \cap \{w_1, w_2, w_4\}$
$= \{w_1, w_2, w_4\}$, and

$\{p_1 \vee p_2\} = \{w_1, w_2, w_4, w_5\}$.

(ii) Alternately, we compute:

$$\mu_{(\neg p1 \vee p2) \wedge (p1 \vee \neg p2) \wedge (p1 \vee p2)}(x) = \min[\mu_{p1 \vee p2}(x), \mu_{(\neg p1 \vee p2) \wedge (p1 \vee \neg p2)}(x)]$$

$$= \frac{|c\{W_1 \vee W_2\} \cap \{[\Gamma_1[cW_1] \cup W_2] \cap [W_1 \cup \Gamma_2[cW_2]]\}|}{|W|}$$

$$= \frac{3}{5} - 0 = \frac{3}{5},$$

Since $\mu_{(\neg p1 \vee p2) \wedge (p1 \vee \neg p2)}(x) = \frac{3}{5}$ and $\mu_{p1 \vee p2}(x) = \frac{4}{5}$ and

thus $\mu_{p1 \vee p2}(x) > \mu_{(\neg p1 \vee p2) \wedge (p1 \vee \neg p2)}(x)$ and furthermore;

$c\,\{W_1 \cup W_2\} \cap \{[\Gamma_1\,[cW_1] \cup W_2] \cap [W_1 \cup \Gamma_2\,[cW_2]]\,\}$
$= \{w_3\} \cap \{w_1, w_2, w_4\} = \varnothing.$

Thus, we have

$$\mu_{FDCF(p1ANDp2)}(x) = \mu_{p1 \wedge p2}(x) = \frac{2}{5}, and$$

$$\mu_{FCCF(p1ANDp2)}(x) = \mu_{(p1 \vee p2) \wedge (p1 \vee \neg p2) \wedge (\neg p1 \vee p2)}(x) = \frac{3}{5}$$

and therefore, it is shown that

FDCF ($p_1$ AND $p_2$) $\subseteq$ FCCF ($p_1$ AND $p_2$).

## 13.5. Canonical Forms for the Asynchronous Case

The expression of FCCF can be written also as

$\{\Gamma_1\ [S_1 \cap cW_1] \cup \{cS_1 \cap W_1\} \cap W_1\} \cup \{\Gamma_2[S_2 \cap cW_2] \cup \{cS_2 \cap W_2\} \cap W_2\} \cup \{W_1 \cap W_2\}$

where $\Gamma_1$, $\Gamma_2$, $S_1$, $S_2$ are the transformations and the synchronic propositions for the truth propositions $p_1$ and $p_2$.

**Example 7.** Given $W = \{w_1, w_2, w_3, w_4, w_5\}$, let the truth sets of the truth proposition sets be:

$p_1 = \{w_1, w_4, w_5\}$,

$S_1 = \{w_2, w_3, w_5\}$, $\Gamma_1\{w_1, w_2, w_3, w_4, w_5\} = \{w_5, w_1, w_2, w_3, w_4\}$, and

$p_2 = \{w_1, w_4\}$,
$S_2 = \{w_1, w_2, w_3, w_4\}$, and $\Gamma_2\{w_1, w_2, w_3, w_4, w_5\} = \{w_4, w_5, w_1, w_2, w_3\}$.
Under this set of conditions, we have:

$$\Gamma_1[S_1 \cap cW_1] \cup \{cS_1 \cap W_1\} \cap W_1 = \{w_1, w_4\}$$

$$\Gamma_2[S_2 \cap cW_2] \cup \{cS_2 \cap W_2\} \cap W_2 = \varnothing$$

and the FCCF can be completed as:

$$\{\Gamma_1[S_1 \cap cW_1] \cup \{cS_1 \cap W_1\} \cap W_1\} \cup \{\Gamma_2[S_2 \cap cW_2] \cup \{cS_2 \cap W_2\} \cap W_2\} \cup \{W_1 \cap W_2\}$$

and therefore, FDCF $\subset$ FCCF.

With the transformations and the synchronic propositions, it is observed that we can cause a change in the value of the FCCF that turns out to be greater then the value of the FDCF.

## 13.6. Soft computing example

Consider the prototype flow diagram as in Figure 13.8.

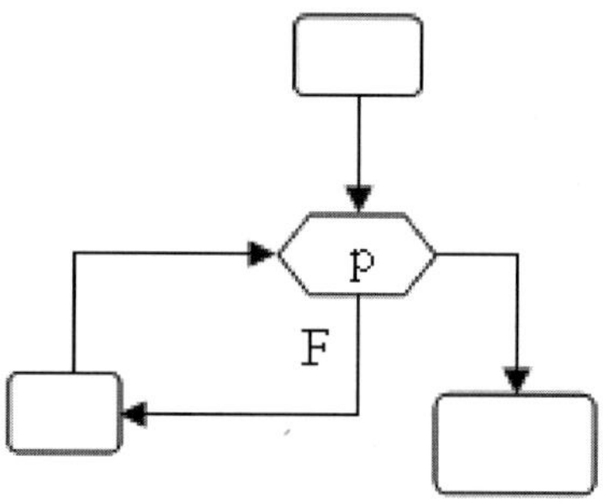


**Figure.13.8.** Prototype of flow diagram

It is composed by a decision cycle controlled by the proposition p. When the proposition p is false, we stay in the cycle, when p is true we go out to the cycle and the flow diagram has a stop condition. The flow diagram is locate in one world, thus there is no uncertainty nor fuzziness. The value of the proposition p is True or False. But when the sentence p is "John is tall" the linguistic term "tall" be represented by the fuzzy set A. For the meta – theory based upon modal logic, "John is tall" in a world wk $\in$ W. The sentence p in Fig.13.8 is true in a world wk $\in$ W. The fuzzy set A control a set of worlds and in any world we have the copy of the prototype cycle in Fig.13.8. For example in Fig.13.9, we have four worlds. In the universe of the worlds W = { w1 , w2 ,w3, w4 } we have V ( p , w1 ) = T , V ( p , w4 ) = T and || p || = { w1 , w4 } = Wp .In this situation the membership function value is

$$\mu_p = \frac{|\,W_p\,|}{|\,W\,|} = \frac{2}{4} = \frac{1}{2} \tag{13.11}$$

Two worlds are out of the cycle but other two worlds are in the cycle. This situation is shown in Fig.13.9.

With the same membership function value, we have always two flow diagrams in the cycle state. But we can have many different states as we shown next in Table 13.6.

Among these worlds, we can use the linguistic AND , OR and NOT operations. In this way, we can compose the four flow diagrams with the fuzzy rules obtained in this chapter. In Fig.13.10 we show the AND operation.

In Fig.13.10, we show the synchronic composition with the same worlds. But we know that we can have also no synchronic composition and we can change the worlds position.

### 13.7. Conclusion

It is shown that fuzzy canonical forms of fuzzy logic operations developed by the meta-theory based upon Modal Logic generates Fuzzy Disjunctive and Fuzzy Conjunctive Canonical Forms, FDCF and FCCF respectively. It is also shown that FDCF $\neq$ FCCF in general. The inequality is a result of synchronicity and transformation of worlds.

It is further shown that generally worlds do not change their truth-value at the same time, i.e., generally, they are asynchronous. As well, the worlds may be subject to transformation. Thus non-identity transformation of worlds should also be investigated. Therefore both the synchronicity and permutations of the worlds represent a better modal of the truth-value evaluations in meta-theory based on Modal Logic.

This alternate approach further verifies Türkşen's previous results(1986-2002) that show that FDCF$\subseteq$FCCF based on Fuzzy Truth Table and Canonical Form derivation algorithm

Therefore, it is important once more to suggest that fuzzy researchers should pay attention to FDCF and FCCF expressions. Otherwise, they are working with a theory that captures only part of the expressions and hence they are using a "myopic" approximation of what the rich fuzzy theory offers.

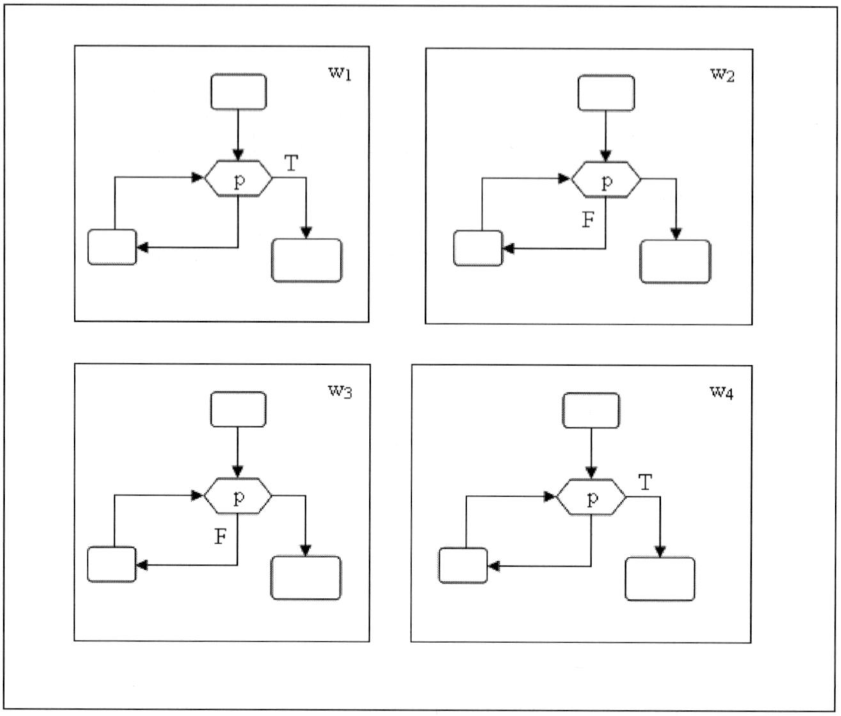


**Figure 13.9.** Inside the four worlds, we have four flow diagrams with the same logic proposition p with different values.

**Table 13.6.** States of the worlds with the membership value equal to 0.5

| $V(p, w_1)$ | $V(p, w_2)$ | $V(p, w_3)$ | $V(p, w_4)$ |
|---|---|---|---|
| T | T | F | F |
| No cycle | No cycle | cycle | cycle |
| T | F | F | T |
| No cycle | cycle | cycle | No cycle |
| F | T | T | F |
| cycle | No cycle | No cycle | cycle |
| F | T | F | T |
| cycle | No cycle | cycle | No cycle |
| F | F | T | T |
| cycle | cycle | No cycle | No cycle |
| T | F | T | F |
| No cycle | cycle | No cycle | cycle |

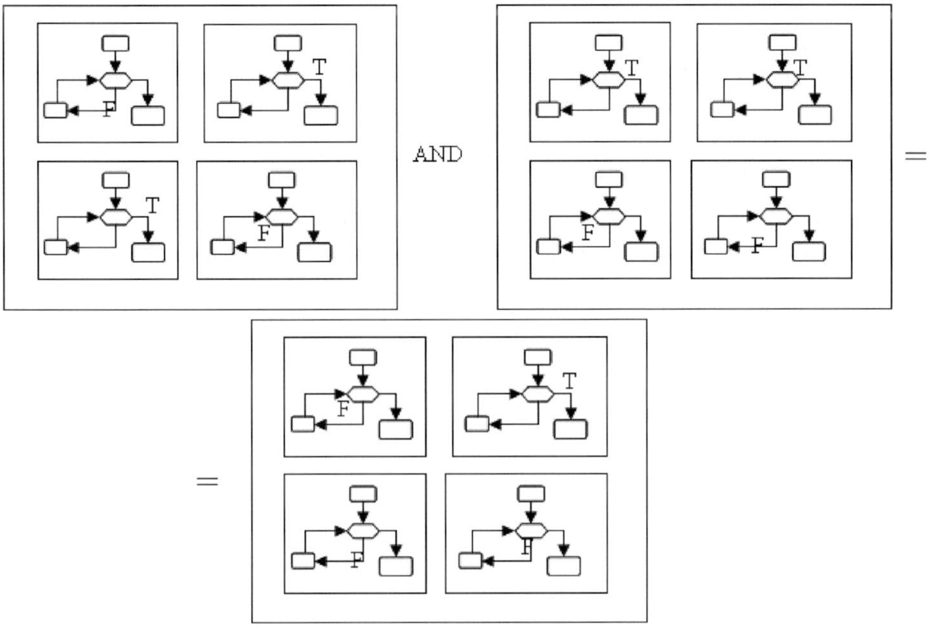


**Figure 13.10.** AND operation between the four worlds

The use of the modal logic is justified as semantic definition by worlds of the uncertainty process inside the fuzzy sets. Possible extensions to evidence theory generate accessible relations which are more complex than the simple reflexive connection in the fuzzy set theory. Hence we suggest that in the future we should use the non-trivial accessibility relations in order to form modal logic models of more complex world phenomena. In such models, we would be able to represent embedded uncertainty contained in fuzzy reasoning.

In this chapter we used the simplest Kripke model where any world can have access only to itself. In future, we can improve the Kripke model with the introduction of non-trivial accessibility relations and can formulate computational expression for the cases of the (fuzzy) linguistic operations AND, OR and NOT.

## References


1. Klir, G., Bo Yuan, Fuzzy Sets and Fuzzy Logic, *Prentice Hall*, 1995.
2. Klir, G., D. Harmanec (1994), "On Modal Logic Interpretation of Possibility Theory", Int. *J. of Uncertainty Fuzziness and Knowledge-Based Systems*, Vol. 2, 237-245.
3. Resconi, G., G.J. Klir and U. St. Clair (1992), "Hierarchical Uncertainty Metatheory Based Upon Modal Logic", Int. J. of General Systems, Vol. 21, 23-50.

4. Resconi, G., I.B. Türkşen (2001), "Canonical Forms of Fuzzy Truthoods by Meta-Theory Based Upon Modal Logic", *Information Sciences*, 131, 157-194.
5. Resconi, G., I.B. Türkşen (1999), "Normal Forms in Fuzzy Operation by Metatheory Based Upon Modal Logic", (in preparation).
6. Resconi, G., G.J. Klir, D. Harmanec and U. St. Clair (1996), "Interpretations of Various Uncertainty Theories Using Models of Modal Logics: A Summary", *Fuzzy Sets and Systems*, Vol. 80, 7-14.
7. Resconi, G., G.J. Klir, U. St. Clair and D. Harmanec (1993), "On the Integration of Uncertainty Theories", *Int. J. of Uncertainty, Fuzziness, and Knowledge-based Systems*, Vol. 1, 1-18.
8. Resconi, G., R. Rovetta (1994), "Fuzzy Sets and Evidence Theory in a Metatheory Based Upon Modal Logic", *Quaderni del Seminario Matematico di Brescia*, n.5.
9. Resconi, G., T.Murai (2000), "Field Theory and Modal Logic by Semantic Field to Make Uncertainty Emerge from Information", *Int.J.General System*.
10. Türkşen, I.B., A. Kandel, Y-Q. Zhang (1999), "Normal Forms of Fuzzy Middle and Fuzzy Contradiction", *IEEE-SMC, 29-2, Part B, Cybernetics*, 237-253.
11. [ 11 ] Türkşen I.B. (1999), "Type I and Type II Fuzzy System Modeling", *FSS*, 106, 11-34.
12. [ 12 ] Türkşen, I.B. (1991), "Measurement of Membership Functions and Their Acquisitions", *Fuzzy Sets and Systems,* 40, 5-38.

# Chapter 14

# APPROXIMATE REASONING*

In this chapter, we discuss approximate reasoning schemas for $\{D_{[0,1]}V_{\{0,1\}}\}$ model in detail. The schemas that are applicable to $\{D_{[0,1]}V_{\{0,1\}}\}$ are directly applicable to $\{\{D_{\{0,1\}}, D_{[0,1]}\}, V_{[0,1]}, V'_{\{0,1\}}\}$ for short, when system description $D_{[0,1]}$ is replaced by truthood description $V_{[0,1]}$ no matter how the base system description is stated, i.e., $D_{\{0,1\}}$ or $D_{[0,1]}$ before the fuzzy truthoods are assigned to them.

Approximate reasoning schemas are generalization of classical schemas. For example, one of the most often applied approximate reasoning schema is known as Generalized Modus Ponens, GMP which was proposed by Zadeh as Compositional Rule of Inference, CRI. GMP is infact a generalization of the classical Modus Ponens, MP. It should be clear from the developments, we have exposed in the previous chapters that there are two formulas known as FDCF and FCCF for implication expression.

Most fuzzy research work is based on the fact that the implication is expressed as FCCF of implication, which is the core of Type 1 fuzzy reasoning! In classical MP only CNF of implication was used because CNF≡DNF in classical theory. Thus most of the approximate reasoning GMP's were based just on FCCF.

The break down of the equivalences, i.e., FDCF(.)TFCCF(.), and the containment of FDCF⊆FCCF require that we should develop at least two versions of GMP. This is yet another example that demonstrates that there is a second alternate formula in fuzzy theory once again attesting to its extra richness. Therefore the approximate reasoning schemas will be review from the perspective of both FDCF and FCCF.

In particular, in this chapter, we investigate Type 1 GMP which is a myopic generalization of MP. In chapter 15, we investigate Interval-Valued Type 2 GMP which is a special restricted subset of Full Type 2 GMP. The Full Type 2

* In part extracted from Türkşen and Demirli(1993), Demirli and Türkşen(1994) and Demirli(1995).

GMP will be investigated in future works. Interval-valued Type 2 GMP have two version:

(a) A Rule antecedent and its consequent are represented as Type 1 fuzzy sets.
(b) A Rule antecedent and its consequent are represented as Interval-Valued Type 2 fuzzy sets.

These two versions of Internal-Valued Type 2 GMP will be investigated in the sequel. But first we will review classical reasoning and in particular classical MP and then Type 1 (fuzzy) GMP.

## 14.1. Classical Reasoning Methods

In classical reasoning, there are nine basic inference schemas which are known as "Rules of Inference" as shown in Table 14.1, where p, q, are crisp propositions, "→" is the crisp implication, and "¬" is the complementation symbol which is applied to propositions. (∨, ∧) are the classical disjunction and conjunction symbols; and "∴" stands for the "consequence". Furthermore, there are ten "Rules of Replacements" as shown in Table 14.2 which are the basic equivalences.

Amongst the rules of inference most of the fuzzy research is conducted on the Modus Ponens. It is the one particular reasoning methods that has been extensively investigated but only within the myopic framework of Type 1 representation and reasoning. Even in this myopic framework, very few works address other classical rules of inference, for example "Modus Tollens" ,etc., (Bandler and Kohout, 1980)

Next we will review the classical Modus Ponens and then discuss its extention to myopic Type 1 Generalized Modus Ponens.

## 14.2. Classical Modus Ponens

The classical Modus Ponens schema is known as:

A→B
A
-------
B

where A→B is a crisp rule, a first premise, made up of a crisp antecedent A and a crisp consequent B and a crisp second premise A which results in a crisp consequence B when the two premise are combined. In classical, two-valued set and logic theory, the truthood of both premises are implicitly assumed resulting in an implicitly assumed truthood of the consequence.

For clarity, we express these implicitly assumed truthoods explicitly and write:
If A→B is True (T)
and A is True (T)
then B is True (T)

In detail, the reasoning is executed with the following explicit steps:

$A \cap (A \rightarrow B) = B \; (T \cap T = T)$
$A \cap (c(A) \cup B) = B \; (T)$
$A \cap c(A) \cup A \cap B = B \; (T)$
$\varnothing \cup A \cap B = B \; (T)$

Since it is given that A is T, we get:

$T \cap B = B \; (T)$

In the treatment above, A's, B's are crisp sets. Let us look at the expressions in more detail going to the level of the elements of the sets A and B. Let us also leave the truth verifications implicit as it generally assumed and treat it at the set level:

A→B
A
-------
B

We need to introduce composition operator "o" to connect with Zadeh's proposal for the CRI. For this reason, we re-state MP as:
$A o (A \rightarrow B) = A \; o \; (\neg A \cup B)$

Let us consider a three element set such that
$a_i \in A, \; b_j \in B \qquad a_i, \; b_j \in \{0,1\}$
$A = (1/1, \; 1/2, \; 0/3) \qquad B = (0/1, \; 1/2, \; 1/3)$
where "1/2", for example indicates that the second element has the full membership in the set as per usual conversion.

Zadeh's CRI as applied to crisp sets turn out to be:
$b_j = \vee_i \; a_i \wedge (a_i \rightarrow b_j)$

For example $a_1 = 1$, $a_2 = 1$, $a_3 = 0$, and $b_1 = 0$.
Then we get:

$$b_1 = \vee_i a_i \wedge (a_i \to b_1) = \begin{bmatrix} (1 \wedge (0 \vee 0)) \\ \vee (1 \wedge (0 \vee 0)) \\ \vee (0 \wedge (1 \vee 0)) \end{bmatrix} = (0 \vee 0 \vee 0) = 0$$

In the relation matrix form, we have

$$c(A) \cup B = \begin{array}{c} c(A) \\ 0 \\ 0 \\ 1 \end{array} \begin{array}{c} \begin{array}{ccc} & B & \\ 0 & 1 & 1 \end{array} \\ \begin{bmatrix} 0 & 1 & 1 \\ 0 & 1 & 1 \\ 1 & 1 & 1 \end{bmatrix} \end{array}$$

$$\text{Ao}(A \to B) = (1, 1, 0) \begin{bmatrix} 0 & 1 & 1 \\ 0 & 1 & 1 \\ 1 & 1 & 1 \end{bmatrix} = (0, 1, 1)$$

This demonstrates that Zadeh's CRI has its origins in the classical MP.

$$b_j = \vee_i a_i \wedge ((1 - a_i) \vee b_j)$$

Zadeh proposed to generalize it by introducing fuzzy sets, i.e., setting $a_i$, $b_j \in [0,1]$.

In practice, when we apply the classical MP, we say, if the second premise(observation) A is an exact match of the antecedent A of the rule, A→B, then the rule is fired. This procedure is the one that is applied in two valued set and logic based expert systems. As well this the foundation on which classical functional and statistical analysis, etc., are build. That is classical models are build on the foundation of classical reasoning methods and usually some of the proofs are executed with the use of "contradiction" using the Law of Contradiction, etc.

The case of two-valued set and logic based MP is depicted in Figure 14.1.

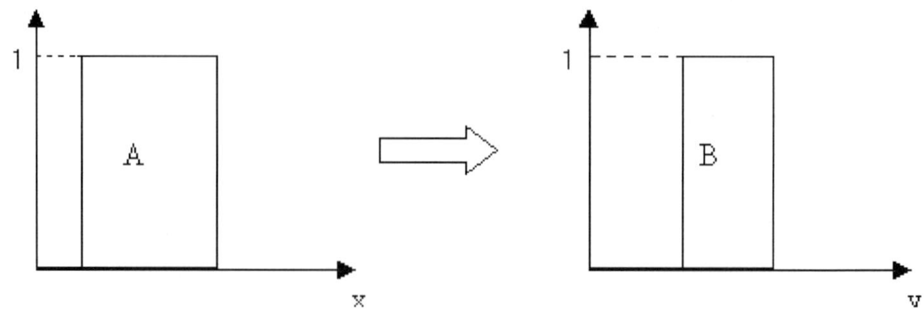


**Figure 14.1**. MP: Ao(A→B) = B in two-valued set and logic theory.

**Table 14.1.** Rules of Inference

1. Modus Ponens

$p \rightarrow q$

$p$

$\therefore q$

2. Modus Tollens

$p \rightarrow q$

$\neg p$

$\therefore \neg q$

3. Hypothetical Syllogism

$p \rightarrow q$

$q \rightarrow r$

$\therefore p \rightarrow r$

4. Disjunctive Syllogism

$p \vee q$

$\neg p$

$\therefore \neg q$

5. Constructive Dilemma

$(p \rightarrow q) \wedge (r \rightarrow s)$

$p \vee r$

$\therefore q \vee s$

6. Destructive Dilemma

$(p \rightarrow q) \wedge (r \rightarrow s)$

$\neg q \vee \neg s$

$\therefore \neg p \vee \neg r$

7. Simplification

$p \wedge q$

$\therefore p$

8. Conjunction

$p$

$q$

$\therefore p \wedge q$

9. Addition

$p$

$\therefore p \vee q$

**Table 14.2.** Rules of Replacement

| | |
|---|---|
| 10. De Morgan's Theorems<br>$\neg(p \wedge q) \equiv (\neg p \vee \neg q)$<br>$\neg(p \vee q) \equiv (\neg p \vee \neg q)$ | 15. Transposition<br>$(p \rightarrow q) \equiv (\neg q \rightarrow \neg p)$ |
| 11. Commutation<br>$(p \wedge q) \equiv (q \wedge p)$<br>$(p \vee q) \equiv (q \vee p)$ | 16. Material Implication<br>$(p \rightarrow q) \equiv (\neg p \vee q)$ |
| 12. Association<br>$[p \wedge (q \wedge r)] \equiv [(p \wedge q) \wedge r]$<br>$[p \vee (q \vee r)] \equiv [(p \vee q) \vee r]$ | 17. Material Equivalence<br>$(p \equiv q) \equiv [(p \rightarrow q) \wedge (q \rightarrow p)]$<br>$(p \equiv q) \equiv [(p \wedge q) \vee (\neg p \wedge \neg q)]$ |
| 13. Distribution<br>$[p \wedge (q \vee r)] \equiv [(p \wedge q) \vee (p \wedge r)]$<br>$[p \vee (q \wedge r)] \equiv [(p \vee q) \wedge (p \vee r)]$ | 18. Material Equivalence<br>$[(p \wedge q) \rightarrow r] \equiv [(p \rightarrow (q \rightarrow r)]$ |
| 14. Double Negation<br>$p \equiv \neg\neg p$ | 19. Tautology<br>$p \equiv (p \vee p)$<br>$p \equiv (p \wedge p)$ |

When we have two antecedents and a consequent then in two-valued set and logic theory, MP schema is as follows:

$A_1$ AND $A_2 \rightarrow B$
$A_1 \cap A_2 \rightarrow B = c(A_1 \cap A_2) \cup B$
$\mu_{A1}(x_1) \wedge \mu_{A2}(x_2) \rightarrow \mu_B(y)$
$[1-(\mu_{A1}(x_1) \wedge \mu_{A2}(x_2))] \vee \mu_B(y)$

This ease is depicted in Figure 14.2.

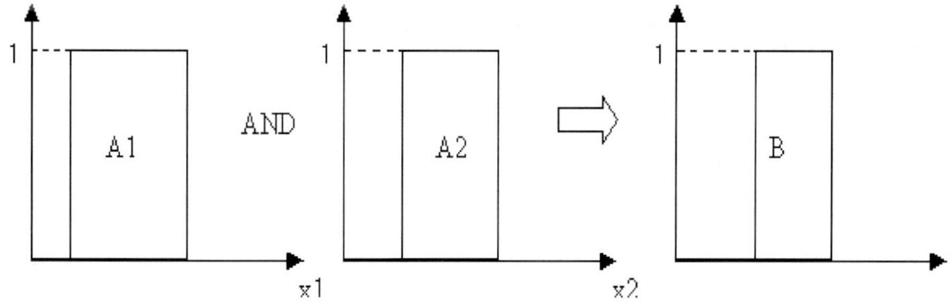


**Figure 14.2.** MP: $A_1$ AND $A_2 \rightarrow B$ in two-valued set and logic theory.

## 14.3. Generalized Modus Ponens

In fuzzy rule based expert systems, the rule, A→B, the first premise, is represented by a fuzzy relation made up of a fuzzy antecedent set and a fuzzy consequent set. The second premise A′ is a fuzzy set and in general A′≠A, i.e., it is not an exact match as in the classical case. A′ is similar, to A in some sense to be identified with a fuzzy similarity measure.

Alternately, A′ is near or around A in a fuzzy manner, with a degree of closeness. Thus, by explicitly specifying the truth verification of the rule and A→B the observation A′, we have the GMP schema as:

A → B (T)
A′ (T)
-------------
B* (T)
B* = A′o(A→B) (T∧T=T)

where "o" is interpreted as Max-t composition depending on the De Morgan Triple that is chosen.

**Remark:**

In some writings, we observe that B′ is used instead of our use of B*. However, we ought to use B* to indicate that the consequence is a result of the similarity between A and A′ but as well it depends on the t-norms and t-conorms chosen and applied in the GMP.

For example, if we have a∈A, b∈B, a′∈A′, b*∈B* a, a′, b, b*∈[0,1] as follows:

$a_i = (1 , .7, 0)$
$1\text{-}a_i = (0, .3, 1)$
$a_i' = (1, .6, 1)$
$b_j = (0, .6, 1)$

(1) If we choose, (∨, ∧, -) as our De Morgan Triple, then we get A→B as:

$$[(1-a_i)\vee b_j] = \begin{array}{c} {\scriptstyle 1-a_i} \\ 0 \\ .3 \\ 1 \end{array}\overset{\begin{array}{ccc} & b & \\ 0 & .6 & 1\end{array}}{\begin{bmatrix} 0 & .6 & 1 \\ .3 & .6 & 1 \\ 1 & 1 & 1 \end{bmatrix}}$$

(2) But if we choose, (L⊕, L□, -) Lukasiewicz De Morgan Triple, then we get A→B as:

$$[Min(1,(1-a_i)+b_j)] = \begin{array}{c} {\scriptstyle 1-a_i} \\ 0 \\ .3 \\ 1 \end{array}\overset{\begin{array}{ccc} & b & \\ 0 & .6 & 1\end{array}}{\begin{bmatrix} 0 & .6 & 1 \\ .3 & .9 & 1 \\ 1 & 1 & 1 \end{bmatrix}}$$

(1) Thus for (∨, ∧, -) De Morgan Triple we get:
A'o(A→B) = B*

$$b_j^* = \underset{i}{\vee} a_i' \wedge ((1 - a_i) \vee b_j)$$

$$(1,.6,.2) \wedge \begin{bmatrix} 0 & .6 & 1 \\ ..3 & .6 & 1 \\ 1 & 1 & 1 \end{bmatrix} = (.3,.6,1) = B^*$$

(2) For Lukasiewicz (L⊕, L□, -) De Morgan triple, we get: $b^* = \underset{i}{\vee} \text{Max}\{0, a_i' + \text{Min}\{1, 1 - a_i + b\} - 1\}$

$$(1, .6, .2)\ L_\square \begin{bmatrix} 0 & .6 & 1 \\ ..3 & .6 & 1 \\ 1 & 1 & 1 \end{bmatrix} = (1, .6, .2) = B^*$$

**Remark**:
We observe therefore that B* is dependent on both the similarity of A and A′ as well as the t-norm, t-conorm De Morgan Triple that is chosen for a particular application.

Therefore, the choice of t-norms and t-conorms and hence the implication is an important aspect of the GMP.

For this reason we next turn our attention to fuzzy relations of implications, i.e., fuzzy rules.

## 14.4. Type 1 Fuzzy Rules

Fuzzy rules are usually expressed in the following natural linguistic form:

(i) IF $x \in X$ isr A, THEN $z \in Z$ isr C,

(ii) IF $x \in X$ is A AND $y \in Y$ isr B, THEN $z \in Z$ is C,

where the antecedents "$x \in X$ isr A", "$y \in Y$ isr B" and the consequent "$z \in Z$ isr C" are fuzzy propositions; and A, B, C are fuzzy sets that are represented with Type 1 membership functions. The "isr" indicates that we are in fuzzy theory as opposed to crisp theory where "is" is used.

In short form notation we express these as:

(i) $A \rightarrow C$,

(ii) $A \cap B \rightarrow C$.

where in general ($\rightarrow$) is interpreted as a Type 1 implication and ($\cap$) corresponds to a t-norm.

### *14.4.1. Implication Functions*

There are many implication functions. But we will basically consider S and R-implication functions for representing fuzzy "if ...then ..." rules in Type 1 fuzzy set representation first. Later we will discuss certain Interval-Valued Type 2 representations.

"S" stands for strong implication functions which are derived from the classical definition of implication functions ($c(A) \cup B$) where $c(A)$ is a standard negation represented as $n(a)=(1-a)$ in the membership domain.
"R" stands for residuation implication.

The basic set operators $\cap$, $\cup$ are represented by a t-norm $\Delta$ and t-conorm $\nabla$. Three well known continuous t-(co)norms are:

| $\Delta(a,b)$ |
|---|
| Min{a,b} |
| ab |
| max{0,a+b-1} |

| $\nabla(a,b)$ |
|---|
| Max{a,b} |
| a+b-ab |
| min{1,a+b} |

*14.4.1.1. S-Implications*

**Definition 14.1.** S-implications are functions $f_\nabla$: $[0,1] \times [0,1] \rightarrow [0,1]$ defined by $f_\nabla(a.b) = \nabla(1 - a, b)$ where $\nabla$ is a continuous t-conorm.

**Examples:**

| $\nabla$(a,b) | $f_\nabla$(a,b) | Name |
|---|---|---|
| $a \vee b$ | $(1-a) \vee b$ | Kleene-Dienes |
| a+b-ab | 1-a+ab | Reichenbach |
| $1 \wedge a+b$ | $1 \wedge 1-a+b$ | Lukasiewicz |

These S-implications are demonstrated in Figures 14-3, 14-4, 14-5.

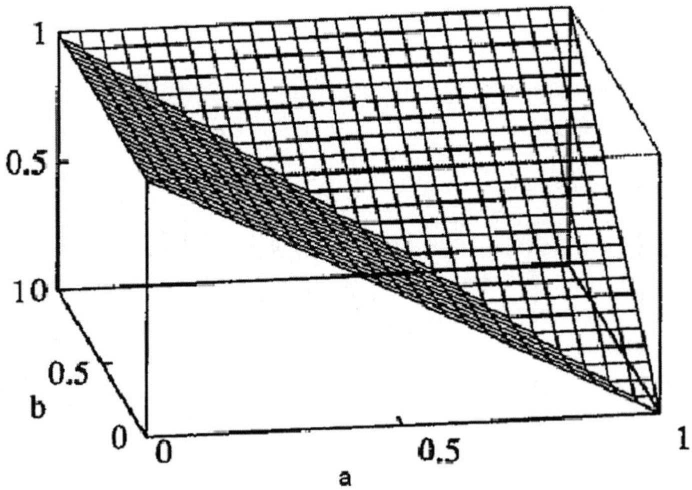


**Figure 14.3.** Kleene-Dienes Implication: $f_\nabla(a.b) = (1 - a) \vee b$

It is also possible to enlarge the class of t-conorms to non-commutative disjunction operators ($\nabla^*$) to generate the R-implications though the form $\nabla$(1-a,b).

The general form of the implication functions, then becomes:

$$f_{\nabla^*}(a.b) = \nabla^*(1 - a, b) \tag{14.1}$$

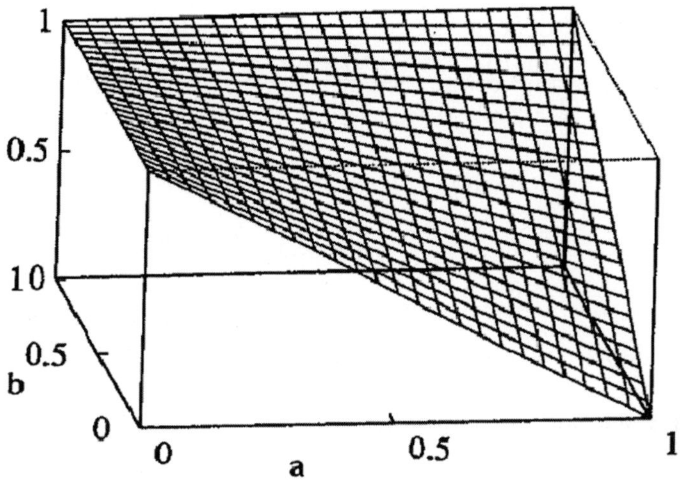


**Figure 14.4.** Reinchenbach Implication: $f_{\nabla}(a.b) = 1 - a + ab$

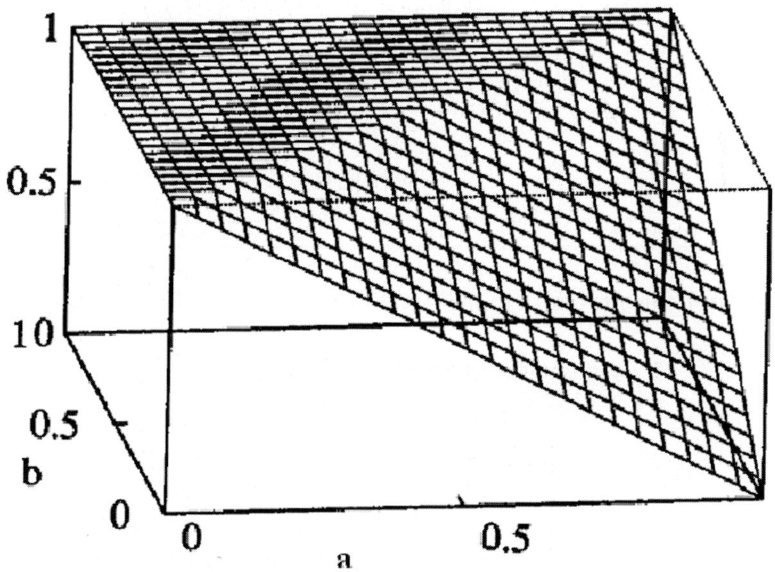


**Figure 14.5.** Lukasiewicz Implication: $f_{\nabla/\nabla^*}(a, b) = 1 \wedge 1 - a + b$

If $\nabla^*$ is chosen as a t-conorm $\nabla$ in (14-1) then the implication is an S-implication.
If $\nabla^*$ is chosen as a pseudo-disjunction $\nabla'$ in (14-1) then the implication is an R-implication.

| t-conorm $\nabla(a, b)$ | Pseudo-disjunction $\nabla(a, b)$ |
|---|---|
| $a \vee b$ | $\begin{cases} 1, & if \quad a+b \geq 1; \\ b, & otherwise. \end{cases}$ |
| $a + b - ab$ | $\begin{cases} 1, & if \quad a=1; \\ 1 \wedge \frac{b}{1-a}, & otherwise. \end{cases}$ |
| $1 \wedge a + b$ | $1 \wedge a + b$ |

*14.4.1.2. R-Implications*

Definition 14.2. R-implications are functions $f_{\nabla'}$: $[0,1] \times [0,1] \rightarrow [0,1]$ defined by $f_{\nabla'}(a.b) = \sup\{x \in [0,1] \mid \Delta(a,x) \leq b\}$ for which a continuous t-norm $\Delta$ exists.

**Examples:**

| $\Delta(a,b)$ | $f_{\nabla'}(a,b)$ | Name |
|---|---|---|
| $a \wedge b$ | $\begin{cases} 1, & if \quad a \leq b; \\ b, & otherwise. \end{cases}$ | Gödel |
| Ab | $\begin{cases} 1, & if \quad a=0; \\ 1 \wedge \frac{b}{a}, & otherwise. \end{cases}$ | Goguen |
| $0 \vee a+b-1$ | $1 \wedge 1-a+b$ | Lukasiewicz |

These R-implications are demonstrated in Figure 14.6, 14.7.

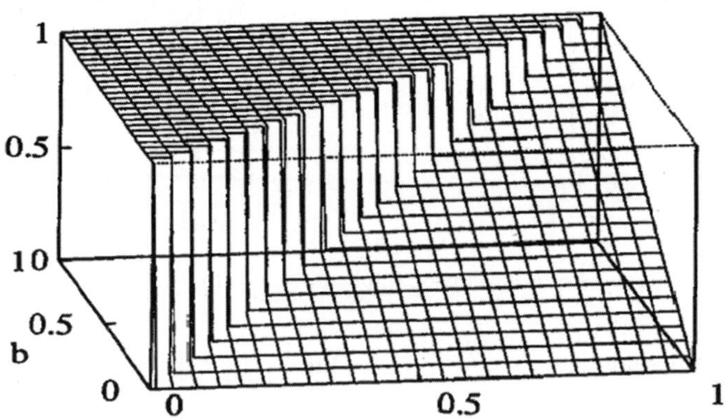


**Figure 14.6.** Gödel Implication: $f_{\nabla'}(a.b) = \begin{cases} 1, & if \quad a \leq b; \\ b, & otherwise. \end{cases}$

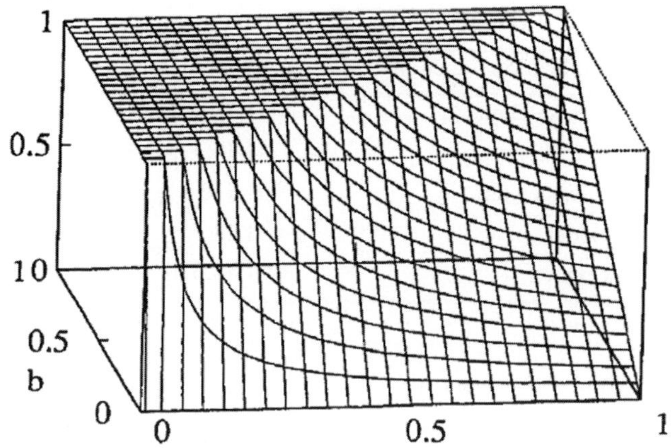


**Figure 14.7.** Gougen Implication: $f_{\nabla *}(a.b) = \begin{cases} 1, & \text{if } a = 0; \\ 1 \wedge \dfrac{b}{a}, & \text{otherwise.} \end{cases}$

From an axiomatic point of view, the following properties have been investigated for a 2-place operation I that represents an implication function, Dubois and Prade (1991):

I1: if $a \le a'$ then $I(a.b) \ge I(a',b)$,
I2: if $b \ge b'$ then $I(a.b) \ge I(a,b')$,
I3: if $I(0,b) = 1$ (falsity implies anything),
I4: if $I(1,b) = b$ (tautology cannot justify anything),
I5: $I(a,b) \ge b$,
I6: $I(a,a) = 1$ (identity principle),
I7: $I(a,I(b,c)) \le I(b,I(a,c))$ (exchange principle),
I8: $I(a,b) = 1$ iff $a \le b$ (implication defines an ordering),
I9: $I(a,b) = I(n(a), n(b))$ for some strong negation n (contraposition law),
I10: I is continuous.

Properties of some implication functions are shown in Table 14-3.

## 14.5. Type 1 Fuzzy Inference: Single Antecedent GMP

We first investigate the inference schema which consists of a fuzzy rule that has a single antecedent and a single consequent together with an observation that result in a consequence.

**Table 14.3.** Properties of some Implication

| Implication Function | Properties |
|---|---|
| Kleene-Dienes | I1-I5, I7, I9-I10 |
| Reinchenbach | I1-I5, I7, I9-I10 |
| Lukasiewicz | I1-I10 |
| Gödel | I1-I8 |
| Goguen | I1-I8, I10 |

Fuzzy Rule: (A→ C)
Observation: A′
----------------------------
Consequence:

Or in short form we express it as:

$$C^* = A' \circ_M (A \rightarrow C),$$

where $\circ_M$ is the Max-*M* composition.
In membership domain,

$C^* = A' \circ_M (A \rightarrow C)$, becomes
$\mu_{C^*}(k) = \vee_i\, M\, [\mu_{A'}(i), \mu_A(i) \rightarrow \mu_C(k)], \forall k.$

Dubois and Prade (1984) and Trillas and Valverde (1984) identified the functions for → and *M* that should be used together so hat C* will satisfy the following four desired properties P1-P4 where M refers to t-norms and to their associated pseudo conjunctions.
Desired Properties of Inference:

P1 - $C \subseteq C^*$
P2 - $C^* = C$ when $A' = A$,
P3 - If $A' \subseteq A''$ then $C^{*\prime} \subseteq C^{*\prime\prime}$,
P4 - For $A' = c(A)$, $C^* = I$.

where c(.) is the complement.

Based on this analysis, the suggested implication and *M* composition operation are listed in Table 14.4.

**Table 14.4.a.** Suggested operator selection table.

| **Implication: →** | **Operator: *M*** |
|---|---|
| Gödel | minimum |
| Goguen | product |
| Lukasiewicz | bold-intersection |
| Kleene-Dienes | *M* (minimum) |
| Reinchenbach | *M* (product) |

where the associated implication expressions are shown in Table 14.4b and the M-composition operators are shown in Table 14.4c.

**Table 14.4.b.** Implication operators

| **Implication: →** | **a→b** |
|---|---|
| Gödel | $\begin{cases} 1 & \text{if } a \le b; \\ b & \text{otherwise.} \end{cases}$ |
| Goguen | $\begin{cases} 1, & \text{if } a = 0; \\ 1 \wedge \frac{b}{a}, & \text{otherwise.} \end{cases}$ |
| Lukasiewicz | $1 \wedge 1\text{-a+b}$ |
| Kleene-Dienes | $(1 - a) \vee b$ |
| Reinchenbach | $1 - a + ab$ |

Remark: Max-min composition with Kleene-Dienes implication is also among the most popular operator combinations, in spite of the fact that it violates property P2.

## 14.6. Information Gap in Type 1 GMP

The Max-*M* composition was essentially created by Dubois and Prade (1984) and Trillas and Valverde (1984) to show that there is an information gap generated in GMP due to the structure of approximate reasoning that is caused by imprecisions embedded in fuzzy sets as opposed to crisp sets, i.e., two-valued.

Let us return to our examples again and compute GMP when A′=A. Recall we had $a_i = (1, .7, 0)$ $b_j = (0, .6, 1)$, i.e., A = (1/1, .7/2, 0/3) and B = (0/1, .6/2, 1/3). In case one with ($\vee$, $\wedge$, -) De Morgan Triple, we obtained

**Table 14.4.c.** M composition operators

| Operator | ***M* (a,b)** |
|---|---|
| Minimum | a∧b |
| Product | Ab |
| bold-intersection | 0 ∨ a + b - 1 |
| *M* (minimum) | $\begin{cases} 0 & if \;\; a+b \le 1; \\ b & otherwise. \end{cases}$ |
| *M* (product) | $\begin{cases} 0, & if \;\; a = 0; \\ 0 \vee \dfrac{a+b-1}{a}, & otherwise. \end{cases}$ |

$$A \to B: [(1 - a_i) \vee b_j] = \begin{pmatrix} 0 & .6 & 1 \\ .3 & .6 & 1 \\ 1 & 1 & 1 \end{pmatrix}$$

Let us now compute

$$Ao(\, A \to B): (1, .7, 0) \wedge \begin{pmatrix} 0 & .6 & 1 \\ .3 & .6 & 1 \\ 1 & 1 & 1 \end{pmatrix} = (.3, .6, 1) \neq (0, .6, 1) = b_j$$

But if we use *M*(minimum) as suggested, we compute and obtain

$$Ao(\, A \to B): (1, .7, 0)\, M_{\min} \begin{pmatrix} 0 & .6 & 1 \\ .3 & .6 & 1 \\ 1 & 1 & 1 \end{pmatrix} = (0, .6, 1)$$

$$\text{recall } M_{\min}(a,b) = \begin{cases} 0 & if \;\; a+b \le 1; \\ b & otherwise. \end{cases}.$$

Later, we will investigate how to remedy this information gap in Interval-Valued Type 2 reasoning.

## 14.7. Type 1 Fuzzy Inference: Two Antecedent GMP

We next investigate the inference schema which consists of a fuzzy rule that has two antecedent and a single consequent together with an observation that also has two fuzzy sets that correspond to two antecedents and where the inference results in a consequence

Fuzzy Rule: $A \cap B \to C$

Observation(s): $A' \cap B' \to C$

----------------

$C^*$

Or in short hand notation; it is stated as:

$C^* = (A' \cap B')\ o\ (A \cap B \to C)$
Desired properties of two-antecedent inference investigated by Dubois and Prade(1984) and Trillas and Valverde(1984) are now expressed as:
P1 - $C \subseteq C^*$
P2 - $C^* = C$ when $A' = A$ and $B' = B$,
P3 - If $A' \cap B' \subseteq A'' \cap B''$ then $C^{*\prime} \subseteq C^{*\prime\prime}$,
P4 - For $A' \cap B' = c(A \cap B)$, $C^* = I$.
where c(.) is the complement and conjunctions are represented by t-norms, $\Delta$.
In set notation, the inference schema that is written as
$C^* = (A' \cap B')\ o_M\ ((A \cap B) \to C)$,
is expressed in membership domain as:
$\mu_{C^*}(k) = \vee_i \vee_j\ M\ [\Delta(\mu_{A'}(i), \mu_{B'}(i)), \Delta(\mu_A(i), \mu_B(j)) \to \mu_C(k)], \forall k.$
where $M$ is the composition (conjunction, operator between the observations ($A' \cap B'$) and the rule ($A \cap B \to C$).
The question for this case is stated as:
"Given the results of the previous section, which t-norm, $\Delta$, should be used for the set conjunction, "$\cap$", so that P1-P4 are satisfied in two antecedent inference schema,
$C^* = (A' \cap B')\ o\ (A \cap B \to C)$.
In answer to the question, it is found that regardless of the t-norm selection, P1-P4 are satisfied. However, from the measurement theoretic point of view, the appropriate t-norm, $\Delta$, can be selected based on the selection of implication function and composition operator. Thus, a suggested operator selection table as shown in Table 14.5.

**Table 14.5.** Operator Selection for two antecedent inference schema.

| $\cap$ | **Implication: $\to$** | **M** |
|---|---|---|
| ? | Gödel | minimum |
| ? | Goguen | product |
| ? | Lukasiewicz | bold-intersection |
| ? | Kleene-Dienes | M (minimum) |
| ? | Reinchenbach | M (product) |

## 14.8. Decomposition

We next investigate the possibility of rule and operation decomposition techniques for fuzzy rules that consists of two or more antecedents.
For this purpose, we first state a couple of lemmas and corollaries and then investigate them in detail.

**Lemma 1.** Rules of the form $A \cap B \rightarrow C$ can be decomposed into $(A \rightarrow C) \cup (B \rightarrow C)$ for any S or R-implication functions when the conjunction of antecedents is to be by a minimum operator at the membership values level, i.e., corresponds to "$\cap$" "minimum".

**Corollary 1.** Lemma 1 is true for
$A = A_1 \cap A_2 \cap \ldots \cap A_{n-1}$ and $B = A_n$.
[K.Demirli. 1995]

**Lemma 2.** Rules of the form $A \cap B \rightarrow C$ can be represented as $A \rightarrow (B \rightarrow C)$ for any S-implications and for those R-implications that are generated from continuous t-norms. [Dubois and Prade, 1984]

**Corollary 2.** Lemma 2 is true for
$A = A_1 \cap A_2 \cap \ldots \cap A_{n-1}$ and $B = A_n$.
[K.Demirli. 1995]

### *14.8.1. Inference with rule decomposition*

We first investigate inference schema with the rule decomposition.
From lemma 1, we have
$(A \cap B \rightarrow C) = (A \rightarrow C) \cup (B \rightarrow C)$
Let us re-state the components of the rule decomposition as:
$C^* = (A' \cap B') \, o_M \, (A \cap B \rightarrow C)$
$C^{*A} = A' \, o \, (A \rightarrow C)$,
$C^{*B} = B' \, o \, (B \rightarrow C)$,
Our concern is to determine whether or not:
$C^* = C^{*A} \cup C^{*B}$?

**Theorem 1.** $C^* = C^{*A} \cup C^{*B}$ if "minimum" corresponds to "$\cap$" and "any S or R-implication", corresponds to $\rightarrow$ where "$\cup$" = "maximum".

**Table 14.6.** The correspondence between conjunction ($\cap$) of antecedent and the conjunction ($M$) between the rule and observations for various implication operators ($\rightarrow$) in rule decomposition.

| $\cap$ | Implication: $\rightarrow$ | $M$ |
|---|---|---|
| minimum | Gödel | minimum |
| minimum | Goguen | product |
| minimum | Lukasiewicz | bold-intersection |
| minimum | Kleene-Dienes | $M$ (minimum) |
| minimum | Reinchenbach | $M$ (product) |

It has been shown by Mizumoto(1985) that $C^* = C^{*A} \cup C^{*B}$ for "$\cap$" = "$o_M$" ="minimum" and for eight specific implication functions "$\rightarrow$" five of which are S and three are R-implications.

K.Demirli (1995) has shown that: rule decomposition is possible for the families of S and R-implication function generated from continuous t-norms.

*14.8.2. Inference with operation decomposition*

Inference schema with operation decomposition were investigated by K.Demirli (1995).

From Lemma 2, we have
$(A \cap B \rightarrow C) = A \rightarrow (B \rightarrow C)$
Let us re-state the components of the operation decomposition as:

$$
\begin{aligned}
C^* &= (A' \cap B') \text{ o } (A \cap B \rightarrow C) \\
C^{**} &= A' \text{ o } (A \rightarrow (B' \text{ o } B \rightarrow C) \\
&= A' \text{ o } (A \rightarrow C^{*B})
\end{aligned}
$$

We then ask whether or not: $C^* = C^{**}$?

Theorem 2. $C^* = C^{**}$ if t-norm that corresponds to "$\cap$" is "*M*"= "*M* (Product)" and "$\rightarrow$" corresponds to "Reichenbach".

**Table 14.7.** The correspondence between conjunction operator ($\cap$) of antecedents versus the composition conjunction ($o_M$) between the rule and observations for various implication operators in decomposition.

| $\cap$ | Implication: $\rightarrow$ | M |
|---|---|---|
| - | Gödel | minimum |
| - | Goguen | product |
| - | Lukasiewicz | bold-intersection |
| - | Kleene-Dienes | *M* (minimum) |
| Product | Reinchenbach | *M* (product) |

On the other hand, if we re-state the components of the inference with operation decomposition as follows:

$$
\begin{aligned}
C^* &= (A' \cap B') \text{ o } (A \cap B \rightarrow C) \\
C^{*A} &= A' \text{ o } (A \rightarrow C), \\
C^{*B} &= B' \text{ o } (B \rightarrow C),
\end{aligned}
$$

$C^{**} = A' \text{ o } (A \rightarrow C^{*B}),$

then we can state an alternate theorem.

**Theorem 3.** $C^{*A} \cup C^{*B} \subseteq C^* \subseteq C^{**}$ if t-norm that corresponds to "$\cap$" is "*M*" "$\rightarrow$"="product" and "$\rightarrow$" corresponds to "Gougen", where "$\cup$" = "maximum".
Theorem 4. $C^{*A} \cup C^{*B} \subseteq C^* \subseteq C^{**}$ if t-norm that corresponds to "$\cap$" is "*M*" ="bold-intersection" and "$\rightarrow$" corresponds to "Lukasiewicz", where "$\cup$" = "maximum".

**Table 14.8.**

| $\cap$ | Implication: $\rightarrow$ | o |
|---|---|---|
| - | Gödel | Minimum |
| product | Goguen | Product |
| bold-intersection | Lukasiewicz | bold-intersection |
| - | Kleene-Dienes | *M* (minimum) |
| - | Reinchenbach | *M* (product) |

## 14.9. Computational Complexity

Next rule and operation decomposition schemas are investigated in order to reduce the computational complexity. Here, we show very briefly that computational complexity is reduced quite a bit by the decomposition schemas discussed above.

Recall that

$$C^* = (A' \cap B') \text{ o } (A \cap B \rightarrow C)$$
$$C^{*A} = A' \text{ o } (A \rightarrow C),$$
$$C^{*B} = B' \text{ o } (B \rightarrow C),$$
$$C^{**} = A' \text{ o } (A \rightarrow C^{*B}),$$

It is found that (K.Demirli, 1995) computational complexity is related to:
(i) number of points used to represent the fuzzy sets, and
(ii) number of antecedent variables.

For simplicity, if one assumes that all fuzzy sets are represented by m discrete points then the computational complexity analysis show that the order of complexity is reduced to polynomial complexity as shown in Table 14.9.

## 14.10. Implementation with Type 1 Reasoning

Despite all the extensive mathematical investigations, we find that there are essentially two basic prototypes of approximate reasoning in applications. These are known as:

(i) Material Implication, MI, based reasoning schemas, and
(ii) Product (Mamdani) Implication, PI, based reasoning.

**Table 14.9.** Computational complexity analysis.

| | two antecedent case | N antecedent case |
|---|---|---|
| C* | $O(m^3)$ | $O(m^{N+1})$ |
| $C^{*A} \cup C^{*B}$ | $O(m^2)$ | $O(Nm^2)$ |
| C** | $O(m^2)$ | $O(Nm^2)$ |

In material implication based reasoning schemas, one starts with $A \to B = c(A) \cup B$ and in product implication based reasoning one starts with $A \to B = A \times B$. Afterwards one choses a t-norm and t-conorm that is suitable for an application and proceeds with GMP.

If there are $A_i \to B_i$, $i=1,\ldots,k$ rules, thus, one starts out with (i) $A_i(x) \to B_i(y) = n(A_i(x)) \nabla B_i(y)$ or (ii) $A_i(x) \to B_i(y) = A_i(x) \Delta B_i(y)$ where $A_i(x)$ and $B_i(y)$ are membership values obtained from membership functions and $\nabla$ and $\Delta$ are chosen t-conorm and t-norm. This fact is depicted in Figure 14.8.

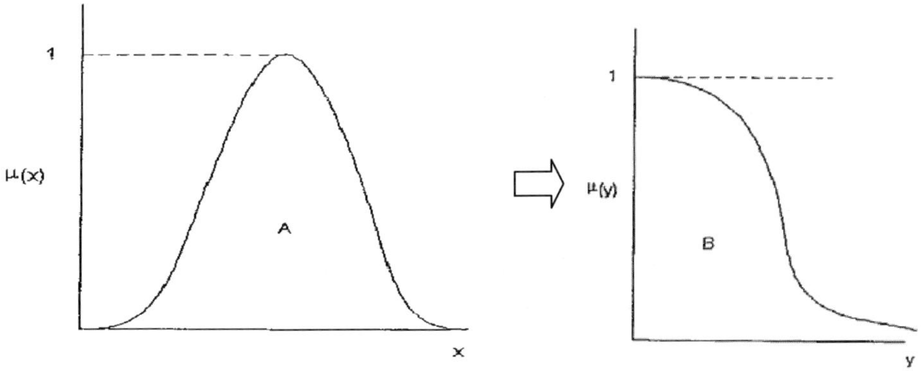


**Figure 14.8.** An illustration of Type 1 fuzzy set implication with a single antecedent, i.e., $A_i \to B_i$, $i=1,\ldots,k$.

Naturally, if there are for example two antecedents, when, the depiction is as shown in Figure 14.9.

Furthermore, in most applications, we usually have:

$$A'(x) = \begin{cases} 1 & x = x' \; ; \\ 0 & \text{otherwise} \end{cases}$$

that is the observation is taken to be a singleton. Thus, we obtain

(i) $B^{*}_{MI}(y) = \vee_{i} c(A_i(x')\nabla B_i(y)$, and

(ii) $B^{*}_{PI}(y) = \vee_{i} c(A_i(x')\Delta B_i(y)$, $i = 1,\ldots,k$.

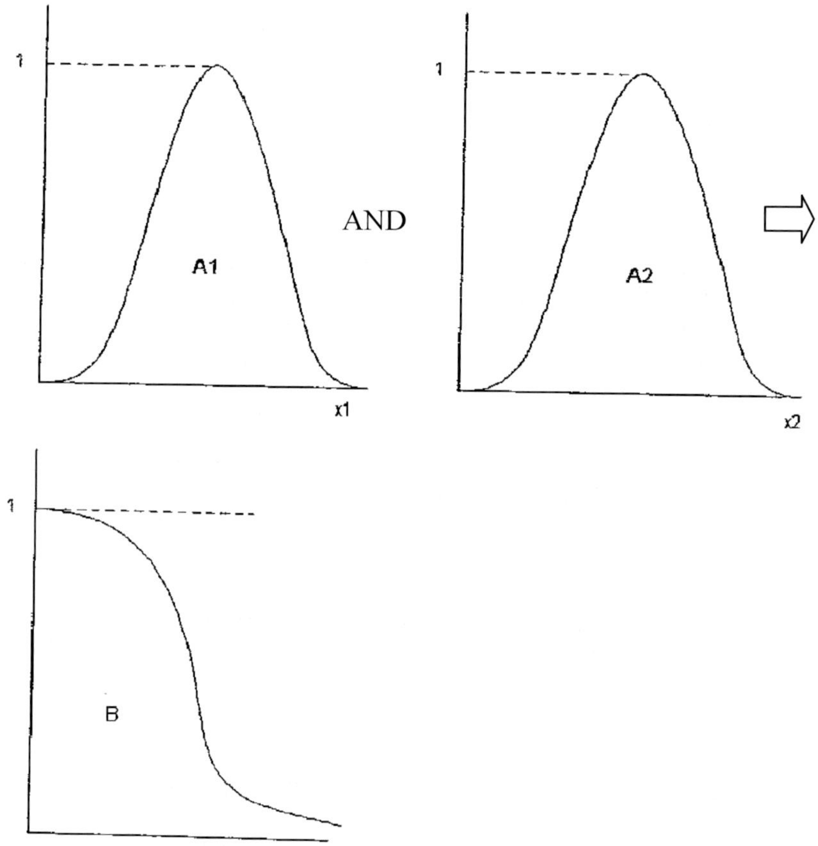


**Figure 14.9**. An illustration of Type 1 fuzzy set implication with two antecedents, i.e., $A_{1i}$ AND $A_{2i} \rightarrow B_i$, i=1,…,k..

There are also "compromise reasoning" schemas that obtain a weighted average of these two reasoning techniques by either utilizing a convex or exponential combination of these techniques:

That is, one computes:

$$B^*_{CLC}(y) = \gamma B^*_{PI}(y) + (1-\gamma)B^*_{MI}(y), \text{ or}$$

$$B^*_{EXC}(y) = \left[B^*_{PI}(y)\right]^{\gamma} + \left[B^*_{MI}(y)\right]^{1-\gamma}$$

where CLC and EXC stand for Convex Linear Combination and Exponential Combination, respectively.
Finally, one obtains a parameterized deffuzzification technique:

$$y^* = \frac{\sum y(B^*_{(.)}(y))^{\alpha}}{\sum (B^*_{(.)}(y))^{\alpha}}, \text{ where } 0 \le \alpha \le \infty.$$

Proposed by Yager and Filev(1994) which is known as BADD method.
In continues domain, it is expressed as:

$$y^* = \int_{y_0}^{y_1} y(B^*_{(.)}(y))^{\alpha} dy \Big/ \int_{y_0}^{y_1} (B^*_{(.)}(y))^{\alpha} dy, \text{ where } 0 \le \alpha \le \infty.$$

### 14.11. Type 1 Fuzzy System Modeling

Zadeh's proposal [1973] of linguistic approach as the model of human thinking introduced the "fuzziness" into systems theory. Fuzzy systems models are the outcome of such a linguistic approach. This idea was further developed by Tong [1979], Pedrycz [1984], Sugeno and Yasukawa [1993], and Yager and Filev [1994], among others.

Current literature on modeling and reasoning with fuzzy systems are essentially concentrated on multi-input single-output (MISO) systems. For this reason, we will consider in this section a fuzzy rule base that contains MISO rules as follows:

$R_1$: IF $X_1$ isr $A_{11}$ AND $X_2$ isr $A_{12}$ AND … AND $X_{ND}$ isr $A_{1ND}$ THEN Y isr $B_1$,
$R_2$: IF $X_1$ isr $A_{21}$ AND $X_2$ isr $A_{22}$ AND … AND $X_{ND}$ isr $A_{2ND}$ THEN Y isr $B_2$,

.

.

.

$R_n$: IF $X_1$ isr $A_{n1}$ AND $X_2$ isr $A_{n2}$ AND … AND $X_{ND}$ isr $A_{nND}$ THEN Y isr $B_n$,

where $X_1 \ldots X_{ND}$ are input variables, and Y is the output variable $A_{ij}$, i=1,…,n, j=1,...,ND and $B_i$, i=1,…,n, are fuzzy sets of the universes of discourse $X_1, X_2, \ldots, X_{ND}$ and Y, respectively.
Such rule bases are determined either via expert interview or fuzzy clustering methods as discussed in previous chapters.
Furthermore, it is either a matter of choice to represent the rule base with Type 1 or 2 membership functions which is dictated by system analyst or it is dictated

by the nature of the problem at hand. But in general, problems that contain a high degree of uncertainty should be represented with Type 2 membership functions.

Next to the generation of a fuzzy rule base, the second essential component of fuzzy modeling is the reasoning mechanism. In current methods of fuzzy modeling, the connectives of the inference mechanism are selected a priori before the identification procedure without any theoretical basis. In order to improve the objectivity of fuzzy modeling one needs to introduce an integrated parameterized formulation to the reasoning process in fuzzy systems. For this purpose, four reasoning parameters, p, q, $\alpha$ and $\beta$, are introduced whose values will cause a continuous range of variation in the reasoning mechanism. Therefore, we are no longer restricted to stay at the extremes in any step of the reasoning process. Consequently, rather than selecting the connectives a priori for fuzzy modeling, the fuzzy model be adjusted to depict effectively the "hidden" behavior of a system by adjusting these parameters and hence the selection of its connectives according to input-output data [Türkşen, et al 1998; Türkşen, 1999; Emami, et al, 1999, 2001].

### *14.11.1. FATI and FITA Representations*

In approximate reasoning we have essentially two methods:
1. First infer then aggregate, FITA, and
2. First aggregate then infer, FATI [Türkşen, et al 1993]
In FITA method, we need to:

(a) determine fuzzy aggregation of antecedents in each rule (AND connective);
(b) determine implication relation for each individual rule (IF-THEN connective);
(c) make inference with each rule of the set of rules, using the observed input;
(d) aggregate the fuzzy outputs obtained in (c) above to get the final fuzzy output (ALSO connective);
(e) defuzzify the final fuzzy output.
Whereas in FATI method, the steps (c) and (d) are modified as follows:
(c') aggregate all the rules of the rule base to determine the overall rule (ALSO connective);
(d') make inference with the overall rule using the observed input.
It can be shown that the two methods of inference with a rule set, i.e., FATI and FITA, always give the same fuzzy output for a singleton input [Emami, et al, 1996]. This will be reviewed in the sequel. Moreover, we propose a reasoning formulation that combines conjunctive and disjunctive reasoning methods for Type 1 and Type 2 knowledge representation. This then forms the foundation of a integrated parameterized fuzzy reasoning formulation.

### *14.11.2. Crisp connectives of fuzzy theory*

In order to cover various types of set operators used in fuzzy set theory, a parameterized families of t-norms(Δ) and t-conorms(∇) should be chosen from among the many parametrized family of t-norms and t-conorms. In our experiments, the following parametric Schweizer and Skalar's family[1983] is chosen for our modeling:

$$\Delta(a,b) = 1 - [(1-a)^p + (1-b)^p - (1-a)^p(1-b)^p]^{1/p}; \qquad (14.2)$$
$$\nabla(a,b) = [a^p + b^p - a^p b^p]^{1/p},\ p>0.$$

The main features of this parameterized family of t-norms and t-conorms are that they are analytically simple and symmetric and when p changes continuously as a positive real number, this family covers all t-norms and t-conorms.

Analogous to the classical set theory, De Morgan laws establish a link between union and intersection via complementation. The conjoint t-norms and t-conorms introduced in Eq. (14.2) are duals when considered with the required negation n(a) as discussed previously.

Although t-norm and t-conorm functions were defined as binary operators on [0,1], their associativity property allows them to be extended to n-ary operations as

$\Delta_n:[0,1]^n \rightarrow [0,1]$, $\qquad \nabla_n:[0,1]^n \rightarrow [0,1]$,
$\Delta(a_1,a_2,\ldots,a_n) = \Delta(\Delta_{n-1}(a_1,a_2,\ldots,a_{n-1}), a_n)$, $\qquad \nabla(a_1,a_2,\ldots,a_n) = \nabla(\nabla_{n-1}(a_1,a_2,\ldots,a_{n-1}), a_n)$,

where, Δ(...) and ∇(...) are the binary operators. It is proved that the n-ary operators $\Delta_n$ and $\nabla_n$ satisfy similar properties as the original binary Δ and ∇ [Emami, et al, 1996, Ruan, et al, 1993].

The extension of these operators to n arguments can be computed with a fast algorithm that has computational complexity of O(n).

That is, the original formulation, i.e.,

$$\nabla(a_1,a_2,\ldots,a_n) = \left[\sum_{i=1}^{n} a_i^p - \sum_{i=1}^{n}\sum_{\substack{j=1\\ j\neq i}}^{n} a_i^p a_j^p \sum_{i=1}^{n}\sum_{\substack{j=1\\ j\neq i}}^{n}\sum_{\substack{k=1\\ k\neq j\\ k\neq i}}^{n} a_i^p a_j^p a_k^p \pm \ldots \pm \prod_{i=1}^{n} a_i^p\right]^{1/p} \qquad (14.3)$$

is transformed into a fast computational algorithm as follows:

$$\nabla(a_1,a_2,\ldots,a_n) = [a_1^p + (1-a_1^p)[[a_2^p + (1-a_2^p)[\ldots[[a_{n-2}^p + (1-a_{n-2}^p)[a_{n-1}^p + (1-a_{n-1}^p a_n^p)]\ldots]]]^{1/p}, \qquad (14.4)$$

where for both of the formulations Δ(.) is computed as

$\Delta(a_1,a_2,\ldots,a_n) = 1-\nabla((1-a_1),(1-a_2),\ldots,(1-a_n)$, n>2.
It should be noted that Eq. (14.3) has an exponential complexity, i.e., $O(2^n)$, whereas the modified Eq.(14.4) has a complexity of O(n). Clearly, depending on the choice of the parametric family of t-norm and t-conorm a different fast algorithm may be developed.

### *14.11.3. Type 1 Implication and Aggregation*

Within the Type 1 theory of representation of memberships is Type 1 and reasoning is carried out in a version of "Sup-Δ" computation where each fuzzy rule is in the form

$$R_i - \text{IF } X_1 \text{ isr } A_{11} \text{ AND... AND } X_{ND} \text{ isr } A_{iND} \text{ THEN } Y \text{ isr } B_i, \qquad (14.5)$$

such that $R_i$ is a fuzzy relation defined on the Cartesian product universe of $X_1 \times X_2 \times \ldots X_{ND} \times Y$, is transformed into a computational formula with the use of myopic Boolean normal form expressions as indicated earlier.

Let us now discuss this single rule case in a bit more detail. In agreement with the discussion presented in Section 14.2, t-norm operators, Δ, are used to define conjunctions in the antecedent of the multi-input rule. Furthermore, modeling of implication relation is not unique in fuzzy logics. Two extreme paradigms for forming the implication relation are conjunctive(Mamdani)(product) and disjunctive(material implication) methods which are identified as cases a) and b) in Type 1 theory in section 14.8. In case (a), the fuzzy relation $R_i$ is simply the conjunction of the aggregated antecedent, i.e., the left-hand side with the consequent as shown in Eq. (14.5). With the Boolean disjunctive normal form of "AND", i.e., DNF, we have the fuzzified predicate version of $R_i$ expressed as $D_i(.)$:

$$D_i(x_1,x_2,\ldots,x_{ND}, y) = \Delta(A_i(x), B_i(y)), \qquad (14.6)$$

where the aggregated antecedent in predicate form is $A_i(x)=\Delta'(A_{i1}(x_1),A_{12}(x_2),\ldots,A_{iND}(x_{ND}))$ and $x=(x_1,x_2,\ldots,x_{ND})$.
This method of aggregation was introduced by Mamdani [1974] which originates with Zadeh' s interpolation reasoning [1973].

In Eq. (14.6) $A_{ij}(x_j)$, and $B_i(y)$ are the membership grades of the fuzzy sets $A_{ij}$, and $B_i$, Δ is t-norm operator (with parameter p) for the rule implication, and Δ′ is t-norm operator (with parameter q) for the antecedent aggregation for each of the rules, i = 1,2,…,n. On the other hand, in case (b) the material implication for the fuzzy relation $R_i$ is defined as a strong implication in the classical set theory, i.e., $A \rightarrow B \equiv c(A) \cup B$, where c(A) is the complement of A in the propositional domain. Thus this is nothing more than the fuzzified Boolean conjunctive

normal form, CNF, of "CLASSICAL IMPLICATION". Therefore, we have the fuzzified predicate form of $R_i$, expressed as $C_i(.)$:
$C_i(x_1,x_{2,...,}x_{ND},y) = \nabla(n(A_i(x)), B_i(y))$

where, $\nabla$ and $\nabla'$ are t-conorm operators with parameters p and q, respectively, $A_i(x) = \Delta'(A_{i1}(x)),\ldots,A_{iND}(x_{ND}))$ as indicated in 14.6 above and n is the negation in the predicate domain.
Next, selection of the rule aggregation function depends on the selection of implication function for individual rules.[Türkşen, et al, 1993]. For case (a) (Eq. (14.6), the aggregation of the rules should be with a maximum operator. In other words, the ALSO connective should be maximum operator. Recall from (14.6), $x=(x_1,x_{2,...,}x_{ND})$, then more compactly, we have the predicate expression:

$$D(x,y) = \underset{i}{\vee} D_i(x,y).$$

On the other hand, if each basic proposition is regarded as "[X isr $c(A_{ij})$] $\cup$ [Y isr $B_i$]", which is the Boolean conjunctive normal form of "CLASSICAL IMPLICATION", then the knowledge "(X $\rightarrow$Y) isr R" should be aggregated with the minimum operator [Türkşen, et al, 1993]. In other words, the ALSO connective is a minimum operator. In its compact form we have the predicate expression,

$$C(x,y) = \underset{i}{\wedge} C_i(x,y).$$

It is to be recalled that both of these methods are myopic since they are nothing more than the fuzzified versions of the Boolean normal forms, i.e., DNF and CNF.

### *14.11.4. Type 1 inference with a rule set*

Consider a single-input and a single-output (SISO) system, with a given implication relationship "(X $\rightarrow$Y) isr R" and the information that X belongs to a fuzzy set A$'$ is observed. Then the problem of inference is to find the fuzzy value for Y. According to Zadeh's compositional rule of inference (CRI)[1973], the membership function of the output in its predicate form is defined as

$$B^*(y)= \underset{x}{\vee} [\Delta'' (A'(x),R(x, y))]$$

in which $\underset{x}{\vee}$ means the maximum for all values of $x\in X$ and $\Delta''$ is a t-norm,
It is known that two different methods of reasoning, i.e., FITA or FATI, in a rule base R lead to two different reasoning results [Türkşen, et al, 1991], because t-norm and t-conorm operators are not distributive in general.
When it is executed with Mamdani's approximate reasoning which is based on Boolean disjunctive normal form, DNF, of "AND" then we have

$$B_D^*(y) = \underset{x}{\vee}[\Delta''[A'(x), D(x,y)]]$$

or with Boolean conjunctive normal form, CNF, of "CLASSICAL IMPLICATION" based approximate reasoning, i.e.,

$$B_C^*(y) = \underset{x}{\vee}[\Delta''[A'(x), C(x,y)]]$$

For this reason, it is necessary first to combine all the rules to get the relation R of the set of rules and then compose the combined aggregate rule with the fuzzy input A′(x). This method called first-aggregate-then-infer(FATI), is quite inefficient in terms of computational time and memory [Türkşen, et al, 1993]. However, in most applications of fuzzy modeling and control the input x* is crisp and then, the input fuzzy set A′(x*) is a fuzzy singleton. In such a case, it is possible to fire each single rule first and calculate individual fuzzy reasoning output $B_i$*(y) and finally aggregate fuzzy output of all the rules to obtain the inferred fuzzy output B*(y). This method, called first-infer-then-aggregate (FITA), is computationally more efficient than FATI but equivalent to it.
Next, let us review very briefly the equivalence of FITA and FATI for this special case when $A'(x^*) = 1$. For multi-input single-output systems, the antecedent of each rule i is a conjunction of ND fuzzy sets $A_{i1}, A_{i2,...,}A_{iND}$. For crisp input $x^* = (X_1^*, x_2, \ldots, x^*_{ND})$, with the given i-th rule:

$R_i$ – IF $X_1$ isr$A_{i1}$ AND ... AND $X_{ND}$ isr $A_{iND}$, THEN Y isr $B_i$,
we first compute:

$$A_i(x^*) = \Delta'(A_{i1}(x_1^*), A_{i2}(x_2^*), \ldots, A_{iND}(X_{ND}^*)),$$

where $A_i(x^*)$ is called the degree of firing (DOF) for the rule i. Since x=x* is a singleton, i.e., A′(x*)=1, it can be shown that the fuzzy output membership function may be computed equivalently by two different methods, i.e., FITA FATI, in both of the reasoning mechanisms as

$$B_D^*(y) = \underset{i}{\vee}\underset{x}{\vee} D_i(x^*, y) = \underset{i}{\vee} D_i(x^*, y) = \underset{i}{\vee} D(y)$$ since x* is a singleton,

$$B_C^*(y) = \underset{i}{\wedge}\underset{x}{\vee} C_i(x^*, y) = \underset{i}{\wedge} C_i(x^*, y) = \underset{i}{\wedge} C(y)$$ since **x* is** a singleton.

With these preliminaries, the integrated reasoning mechanism is proposed as a combination of the two extremes, i.e., the myopic Boolean normal form-based methods as

$$B^*(y) = \beta B_D(y) + (1-\beta) B_C^*(y), \ 0 \le \beta \le 1,$$

in order to get a convex linear combination of the two different methods which produce two boundary values that are used in Type 1 theory in the current practice.

### *14.11.5. Defuzzification*

Yager and Filev [1994] suggest a general defuzzification method, based on the probabilistic nature of the selection process among the values of a fuzzy set, called basic defuzzification distribution (BADD) method:

$$y^* = \int_{y_0}^{y_1} y(B^*(y))^x dy / \int_{y_0}^{y_1} (B^*(y))^x dy, \quad 0 \leq \alpha \leq \infty.$$

As it is well known, BADD method is essentially a family of defuzzification methods parameterized by α. By varying α continuously in the real interval, it is possible to have more appropriate mappings from the fuzzy set to the crisp value depending on the system behavior.

### *14.11.6. Parameter optimization*

Finally, a non-linear optimization program is executed in order to minimize performance index, E, i.e.,

$$\text{Min } E(p,q,\alpha,\beta) = \sum_{k=1}^{K} (y_k^* - y_k)^2 / K$$

subject to: $0<p$, $q<\infty$, $0\leq\alpha\leq\infty$ and $0\leq\beta\leq1$,

where $y_k$ is the actual observed output value and $y^*_k$ is the computed value from the parametrized model, whether it be computed with Type 1 formulas or Type 2 formulas and parameters, p, q, α, β, as defined in the previous sections, for the kth sample observed data vector $(x_k^*, y_k) = (x_{1k}^*, x_{2k}^*, ..., x_{NDk}^*, y_k)$, k=1,2,…,K. This could be computed with a "sequential quadratic programming" algorithm.
That is when we have k=1,2,…,K, training data sample vectors, we can conduct a supervised learning through the non-linear optimization programming as specified above. In this manner, the t-norm and t-conorm appropriate for a system behavior can be found with the use of sample data vectors through supervised learning. Hence, no need to determine t-norm and t-conorm "a priori" for the system model development.

## 14.12. Case studies

The case studies are conducted with Type 1 theory formulas are as follows.

### *14.12.1. Case 1 - An Industrial Process*

A real-life, industrial, case study is briefly discussed in this section to demonstrate the effectiveness of the integrated knowledge representation and inference model presented in the previous sections of this chapter.
Real life process data were provided by an industrial company. It consists of 93 input-output data vectors. Each data vector has 46 input variables and one output variable.
First, the a modified fuzzy clustering method discussed Chapter 5 is applied with the data. After proper analyses of the cluster validity index as a function of c within the reliable domain, it is found that m = 2.5 with c = 7 are suitable values for this industrial data. Next, the most important input variables that affect the output significantly are determined among all of the 46 input variables. It is found that variables $\mathbf{x}_4$, $x_9$, $x_{10}$, $x_{11}$, $x_{12}$, $x_{17}$, $x_{18}$, $x_{20}$, $x_{21}$, $x_{28}$, $x_{34}$, $x_{44}$, $x_{45}$ have a significant effect on the output variable which is the throughput efficiency measure of the process. It is found that there are seven effective rules, rule 0-6, that control the hidden behavior of this industrial process.
The integrated fuzzy inference model of this industrial system is developed next. The combined Type 1 myopic model with parameters, $p = 1.6190$, $q = 0.8438$, $\alpha = 8.2935$ and $\beta = 0.001$ produces the best fit for this industrial process with the mean square error equal to 0.0886 for the test data.

### *14.12.2. Case 2 - A nonlinear system*

The fuzzy modeling process of the non-linear system discussed in Sugeno and Yasukawa [1993] and Nakanishi et at. [1993] and reinvestigated by Emami et al [1996].

The fuzzy system identification is based on 50 input-output data. Two dummy input variables $x_3$ and $x_4$ have been added to check the input selection strategy.
As in Sugeno-Yasukawas's approach, in our approach, the significant input variables are identified as the first two input variables, i.e., $x_1$ and $x_2$. After approximating the input and output fuzzy clusters by suitable trapezoidal functions, in the first-step, we determine the fuzzy model of the system. In the next step, we conduct the inference schema and the sequential quadratic programming.

The set of parameters of the fuzzy model, p, q, $\alpha$, $\beta$, are identified as

$p=19.9959$, $q=0.4536$, $\alpha=5.3715$, $\beta=0.0255$

After 10 iterations the performance index becomes $E = 0.0040$.

The performance index of Sugeno-Yasukawa's fuzzy model of the same type (position type) starts from E=0.318, and after 20 iterations, it reaches to E=0.079.

*14.12.3. Case 3 - Gas furnace model*

Consider the system identification given by Box and Jenkins [1970]. The process is a gas furnace with single input u(t) (gas flow rate) and single output y(t) ($CO_2$ concentration). For this dynamic system, 10 input candidates y(t-1),…,y(t-4), u(t-1),...,u(t-6) are considered. We use the same 296 data vectors as in Sugeno and Yasukawa [1993] analysis. With the Emami, Türkşen, Goldenberge, ETG, method [1996], it is found that in = 2.5 and c = 6 are suitable for the system. The significant input variables are determined as y(t-1), u(t-2), and u(t-3). The optimum inference parameters for the gas furnace system are derived as

$p = 102.0147$, $q = 85918$, $\alpha = 27.9338$, $\beta = 0.3782$.

After 10 iterations, the performance index reduces to a value of E= 0.1584.

The Box-Jenkins linear model has E=0.193, and the Sugeno-Yasukawa position-gradient fuzzy model has E=0.190.

## Appendix 14.1. Brief explanation of the proposed stages of fuzzy system modeling

1. Structure identification

Structure identification consists of rule generation, input selection and input membership assignment as follows:

**Rule generation**

Rule generation steps are as follows:

(a) Determination of initial location of output cluster centers. Initially cluster centers should be determined, for example, as hard cluster centers with Agglomerative Hierarchical Clustering Algorithm.

(b) Determination of c and m values,

(i) First, curves of the trace of the fuzzy total scatter matrix, $S_T$ vs. different weighting exponent values (m) should be formed by using FCMEANS algorithm.

(ii) Secondly, the optimal (local optimal) number of clusters (c) should be determined with the analysis of cluster validity index.

(iii) Finally, the reliable region of the values of m (level of fuzziness of the model) should be determined with the analysis of the trace of the fuzzy total scatter matrix $S_T$ vs. different values of m as in step b(i) above.

(c) Formation of membership functions.

(i) First, the classification of the data points should be made, for example, with the application of K-Nearest Neighbor Method in order to extend the assigned fuzzy clusters to the entire output space.

(ii) Secondly, the projection method proposed by Sugeno and Yasukawa should be used to project the output membership values into the space of input variables.

**Input selection and input membership assignment**

It should be recalled that “One” is the neutral element of t-norm operators. Based on this property of t-norm operators, it is to be realized that an ineffective input candidates’ variable’s membership values has to be equal to “one” across its total domain. Therefore such input variable candidates should be removed from the model. The remaining input variables are selected as the significant input variables. A significance measure is defined in Emami et. al. [1996].

2.1. Selection of formulas

It is to be noted that both knowledge representation and approximate reasoning may be executed with either myopic Type 1 or fuzzy normal form based Type 2 expressions.

2.1. (a) Type 1 formula selection

Type 1 formulas are selected which are fuzzified Boolean DNF or CNF expressions depending on which is the sorted form.

2.1. (b) Type 2 formula selection

Type 2 formulas are selected which are FDCF and FCCF representations Modeling with these formulas will be further discussed in the next chapter.

2.2. Parameter identification

Parameter identification consists of fuzzy inference parameter optimization and membership parameter tuning stages.

2.2. (a) Fuzzy inference and fuzzy inference parameter optimization

The inference schema is generally based on t-norm and t-conorm operators in the parametrized reasoning formulation for both the Type 1 approach. There are four parameters in the reasoning schema, p, q, $\alpha$, $\beta$, where the parameters p and q of t-norms and t-conorms are for left-hand side aggregation and implication, respectively, the parameter $\beta$ is for the convex linear combination of two alternate reasoning methods, i.e., disjunctive and conjunctive Boolean formulas of Type 1 approach. The parameter $\alpha$ is the power of a general defuzzification method suggested by Yager and Filev. The inference schema constructed with these four parameters is tuned with an optimization algorithm for a given training data set such that it provides the most appropriate mechanism for the system under investigation. Sequential Quadratic Programming algorithm is suggested for the solution of this non-linear boundary optimization problem.

2.2. (b) Membership parameter tuning

Since, the input-output membership functions have already been identified in the structure identification, it would be better to tune their parameters if the training data is a stable and good sample of the system behavior. In order to adjust the trapezoidal membership function parameters, a tuning algorithm is applied where both input and output membership functions are adjusted. It is to be noted however that if the training data are not a good and stable sample of the system behavior and/or if it is not a large enough sample to represent the total system behavior, then it appears, based on our experiments, that it would be better not to execute this membership parameter tuning step. Fine tuning adversely affects the robustness of the system model.

3. Test and validation of fuzzy system model with new set of input-output data
System model should be validated with a totally new set of input—output data in order to increase our confidence on the "goodness" of the model developed in stages 1 and 2, In this manner we would be further assured that the fuzzy system model responds effectively with better performance.

## References


1. Bandler, W., L.J. Kohout (1980a), "Fuzzy Power Sets and Fuzzy Implication Operators?", *Fuzzy Sets and Systems*, 4(1), 13-30.
2. [ 2 ] Bandler, W., L.J. Kohout (1980b), "Semantics of Implication Operators and Fuzzy Relational Products", *Intern. J. of Man-Machine Studies*, 12(1), 89-116.
3. [ 3 ] Demirli, K., I.B. Türkşen, (19945), "A Review of Implications and Generalized Modus Ponens", FSS, Vol 2, 1440-1445.
4. [ 4 ] Demirli, K. (1995), A Unified and Extended Framework for Operator Selection in Generalized Modus Ponens Type Fuzzy Reasoning", Ph.D. Thesis, University of Toronto, (I.B. Türkşen, Supervisor).

5. [ 5 ] Dubois, D., H. Prade (1991), "Random Sets and Fuzzy Interval Analysis", *Fuzzy Sets and Systems,* 42, 87-101.
6. [ 6 ] Dubois, D., H. Prade (1984), "Fuzzy Logics and the Generalized Modus Ponens Revisited", *International Journal of Cybernetics and Systems*, 283-331.
7. [ 7 ] Sugeno, M., T. Yasukawa (1993), "A Fuzzy Logic Based Approach to Qualitative Modelling", *IEEE Transactions on Fuzzy Systems*, 1(1), 1-24
8. [ 8 ] Tong, R.M. (1979), The construction and evaluation of fuzzy models; Gupta M.M., Regade, R.K., Yager, R.R., editors.: *Advances in Fuzzy Set Theory and Application*, North-Holland, Amsterdam.
9. [ 9 ] Trillas, E., L. Valverde (1985), "On Mode and Implication in Approximate Reasoning", in W. Bandler, M.M. Gupta, A. Kandel and J.B. Kista (eds.), *Approximate Reasoning in Expert Systems*, Elsevier Sci B.V. (North Holland), 157-167.
10. [ 10 ] Türkşen, I.B. (1999), "Theories of Set and Logic with Crisp or Fuzzy Information Granules", *J.of Advanced Computational Intelligence*, 3,4, 264-273.
11. [ 11 ] Türkşen, I.B., A. Kandel, Y-Q. Zhang (1998), "Universal Truth Tables and Normal Forms", *IEEE-Fuzzy Systems*, 6-2, 295-303.
12. [ 12 ] Türkşen, I.B., Y. Tian (1993), "Combination of Rules and Their Consequences in Fuzzy Expert Systems", *Fuzzy Sets and Systems*, 58: 3-40.
13. [ 13 ] Türkşen, I.B., C. Lucas (1991), "A Pattern Matching Inference Method and its Comparison with Known Inference Methods", *Proceedings of IFSA'91*, July 7-12, Brussels, 231-234.

# Chapter 15

# INTERVAL-VALUED TYPE 2 GMP

In chapter 14, we have reviewed myopic Type 1 reasoning which is the type of reasoning schema that is investigated and applied by most of the fuzzy researchers. In previous chapters we have treated extensively the fact that the combination of concepts generate Interval-Valued Type 2 fuzzy sets even if the input to the combination of concepts are Type 1 fuzzy set representations. We now investigate the consequences of generating Interval-Valued Type 2 reasoning even if we start out with Type 1 representation of the fuzzy sets that make up a fuzzy rule. Towards the end of this chapter, we will discuss the Interval-Valued Type 2 system modeling. As will be observed in this chapter, the break down of classical equivalence in fuzzy theory causes the generation of two GMP in interval valued-Type 2 fuzzy theory. This helps us capture uncertainty in reasoning due to the break down of classical equivalences.

## 15.1. Interval-Valued Type 2 Fuzzy Rules

Let us start with the basic linguistic representation of the rules as before:

i- IF x∈X isr A, THEN z∈Z isr C
ii- IF x∈X isr A AND y∈Y isr B, THEN z∈Z isr C

where the antecedents "x∈X isr A", "y∈Y isr B" and the consequent "z∈Z isr C" are fuzzy propositions; and A, B, C are fuzzy sets that are represented with Type 1 membership functions.In short form notation, we express these now as

(i) - "A IMP C"
(ii) - "A AND B IMP C"

where now in general "IMP" is to be interpreted as a linguistic implication connective that generates an interval-valued Type 2 expression and "AND" is to be interpreted as a linguistic conjunction connective that generates an interval-valued Type 2 expression.

Recall that the linguistic interpretation of "IMP" and "AND" generate FDCF(.) and FCCF(.) as we have discussed in the previous chapters in detail with the property that FDCF(.)⊆FCCF(.) for certain classes of t-norm and co-norm De

Morgan Triples that include strict and nilpotent Archimedean t-norms and co-norms. That is, we have:

$$\text{(i) - "A IMP C"} = \begin{cases} \text{FDCF(A IMP C)} \\ \text{FCCF(A IMP C),} \end{cases}$$

$$\text{(ii) - "(A AND B) IMP C"} = \begin{cases} \text{FDCF((A AND B) IMP C)} \\ \text{FCCF((A AND B) IMP C),} \end{cases}$$

In more detail, we have

$$\text{(i) - "A IMP C"} = \begin{cases} (A \cap C) \cup (c(A) \cap C) \cup (c(A) \cap c(C)) \\ c(A) \cup C, \end{cases}$$

(ii) - “(A AND B) IMP C = “D IMP C” where “D=A AND B” such that

$$D = \begin{cases} \text{FDCF(A AND B)} \\ \text{FCCF(A AND B),} \end{cases}$$

$$= \begin{cases} A \cap B \\ (A \cup B) \cap (c(A) \cup B) \cap (A \cup (c(B)), \end{cases} \tag{15.1}$$

Let FDCF(A AND B) = $D_L$

FDCF(A AND B) = $D_U$ (15.2)

together with

$$\text{"D IMP C"} = \begin{cases} (D \cap C) \cup (c(D) \cap C) \cup (c(D) \cap c(C)) \\ c(D) \cup C, \end{cases} \tag{15.3}$$

we get four expressions as:

$$\text{"D IMP C"} = \begin{cases} \textbf{(1)}\ (D_L \cap C) \cup (c(D_L) \cap C) \cup (c(D_L) \cap c(C)) \\ \textbf{(2)}\ (D_U \cap C) \cup (c(D_U) \cap C) \cup (c(D_U) \cap c(C)) \\ \textbf{(3)}\ c(D_L) \cup C \\ \textbf{(4)}\ c(D_U) \cup C, \end{cases} \tag{15.4}$$

It should be noted that

$D_L \subseteq D_U$
$c(D_L) \supseteq c(D_U)$

$\therefore$ **(3)** dominates **(4)** meaning:

$c(D_L) \cup C \supseteq c(D_U) \cup C$

as well, we have $D_L \cap C \subseteq D_U \cap C$

$c(D_L) \cap C \supseteq c(D_U) \cap C$
$c(D_L) \cap c(C) \supseteq c(D_U) \cap c(C)$.

But this does not in general show a dominance of **(1)** or **(2)** or vise versa. Therefore, a simple procedure is to complete first FDCF(D) = $D_L$ and FCCF(D) = $D_U$ as in expression (15.1) and substitute these results in expression (15.4). But just compute expression (3) to determine the upper bound of the implication and then compute the minimum of expressions (1) and (2) to determine the lower bound of the implication.

## 15.2. Some Properties Interval-Valued Type 2 Implication

There are some properties of the interval-valued implication operators. A property of the interval-valued implication, i.e., FDCF(A IMP B) $\subseteq$ FCCF(A IMP B), is useful in demonstrating that certain well known implication expressions accepted in Type 1 representations are contained in [FDCF(A IMP B), FCCF(A IMP B)].

This means that, in the membership domain, we have:

$$\mu_I(a,b;\Delta,\nabla,n) = \{\mu_{FDCF(I)}(a,b;\Delta,\nabla,n) \leq \mu_I' \leq \mu_{FCCF(I)}(a,b;\Delta,\nabla,n)\}$$

where I is a well known linguistic implication that is contained in between the lower bound defined by as FDFC(I) and the upper bound defined by FCCF(I) for given $(\Delta,\nabla,n)$ De Morgan Triple and I is a well known Type 1 implication expression.
Consider for example Type 1 implications I that is investigated in part, in Bandler and Kohout [9] and elsewhere.

| | | |
|---|---|---|
| $I'_5$: | Lukasiewicz | $a \rightarrow_5 b = \min(1, 1 - a + b)$ |
| $I'_{5.5}$: | Kleene-Dienes-Lukasiewicz | $a \rightarrow_{5.5} b = 1 - a + ab$ |
| $I'_6$: | Kleene-Dienes | $a \rightarrow_6 b = (1 - a) \vee b$ |
| $I'_7$: | Early Zadeh | $a \rightarrow_7 b = (a \wedge b) \vee (1-a)$ |
| $I'_8$: | Willmott | $a \rightarrow_8 b = (a \rightarrow_7 b) \vee (b \wedge (1-b))$ |

It is to be observed that:

1. The Lukasiewicz implication $I'=a\to_5 b$ is the upper bound of the interval-valued fuzzy implication $I(A\to B)$ defined by bold operations, $(L_\odot, L_\oplus, -)$, i.e.,
$(aL_\odot b)\, L_\oplus\, [(1-a)L_\odot b]\, L_\oplus\, [(1-a)\, L_\odot\, (1-b)] \le \mu'_I \le (1-a)\, L_\oplus\, b$

where $(aLxb)\overset{\Delta}{=}\max(0,a+b-l)$ and $(aL\ b)\overset{\Delta}{=}\min(l, a+b)$ are the Lukasiewicz (bold) intersection and (bold) union, respectively.

2. Kleene-Dienes implication $I'=a\to_6 b$ is the upper bound of the interval-valued fuzzy implication $I(A\to B)$ defined by min-max standard negation operators $(\wedge,\vee,-)$, i.e.,
$(a\wedge b) \vee [(1-a) \wedge b] \vee [(1-a) \wedge (1-b)] \le \mu_I \le (1-a) \vee b$

Upon further investigation, it can be shown that:
$\mu_{FDNF(A\to B)}(a,b,\wedge,\vee,-) \le \{I'_6, I'_7, I'_8\} \le \mu_{FCCF(A\to B)}(a,b,\wedge,\vee,-)$

where, for example, $I'_6$ stands for $I'=a\to_6 b$ and so on.

3. It is to be observed that:
$\mu_{FDCF(A\to B)}(\odot,\oplus,-) \le \{I'_5\} \le \mu_{FCCF(A\to B)}(\odot,\oplus,-)$
where $a\oplus b = a+b-ab$ and $a\odot b = ab$ which are known as Algebraic sum and product operators.

4. It is to be observed that:
$\mu_{FDCF(A\to B)}(L_\odot, L_\oplus,-) \le \{I'_5, I'_{5.5}, I'_6\} \le \mu_{FCCF(A\to B)}(L_\odot, L_\oplus,-)$

5. Naturally it can be observed that
$\mu_{FDCF(A\to B)}(D_\odot, D_\oplus,-) \le \{I'_5, I'_{5.5}, I'_6\} \le \mu_{FCCF(A\to B)}(D_\odot, D_\oplus,-)$
where

$$D_\odot(a,b) = \begin{cases} a \wedge b & \text{if } a \vee b = 1 \\ 0 & \text{otherwise} \end{cases}$$

and

$$D_\oplus(a,b) = \begin{cases} a \vee b & \text{if } a \wedge b = 1 \\ 0 & \text{otherwise} \end{cases}$$

which are known as the Drastic intersection and unions operators.These results show that Interval-valued Type 2 set representation of implication is a cover for certain Type 1 fuzzy implication expressions. Thus, it may be more appropriate to use the Interval-valued Type 2 set representations of implication for certain domains of application where there exists a second-order imprecision as in the cases of human decision-making structures that are extracted by fuzzy data mining techniques from data bases that contain both imprecision and uncertainty.

## 15.3. Information Gap in Interval-Valued Type 2 GMP

Recall the discussion on the information gap in Section 14.6 related Type 1 GMP. A resolution was suggested by Dubois, and Prade (1984), as well as Trillas and Valverde (1984) with Max-*M* composition.

We now discuss an alternate resolution of the information gap (Türkşen, 1989) within the context of interval-valued Type 2-GMP reasoning schema. It is to be realize that this information gap resolution is infact a resolution of uncertainty generated by Type 2 – GMP. Recall that information gap was created in GMP as follows:

$$A o (A \rightarrow B) \neq B$$

Within the contexts of interval-valued Type 2 GMP schema, we need to consider both FDCF(A→B) and FCCF(A→B) .
Therefore, we get

$$B_L^* = A o \, \mathrm{FDCF}(A \rightarrow B)$$
$$B_U^* = A o \, \mathrm{FCCF}(A \rightarrow B)$$

That is, we have

$$B_L^* = A \cap [(A \cap B) \cup (c(A) \cap B) \cup (c(A) \cap c(B))]$$
$$B_U^* = A \cap [c(A) \cup B]$$

in the set notation and in the membership notation, we have

$$b_{jL}^* = \vee_i \, a_i \wedge [(a_i \wedge b_j) \vee ((1-a_i) \wedge b_j) \vee ((1-a_i) \wedge (1-b_j))]$$

$$b_{jU}^* = \vee_i \, a_i \wedge [(1-a_i) \vee b_j], \quad \forall \, i,j$$

when we choose (∨,∧,-) De Morgan Triple, where $a_i$ and $b_j$ are the membership values in the finite support sets of the fuzzy sets A and B, respectively. Furthermore, $b_{jL}^*$ and $b_{jU}^*$ are the lower and upper bounds of the membership values, respectively, for the consequence B*. Thus the information gap is created because $b_{jL}^* \neq b_{jU}^*$ in general.

A resolution can be obtained when we determine the values of A and B that general B*=B. That is we set $b^*_{jL} = b^*_{jU}$ .

It can be shown that the following two sufficient conditions are required for B*=B:

(1) A must be normalized, and
(2) We must take into account only those values of B* that are above the $\alpha$-cut which is defined as

$$\alpha = \max_i (a_i \wedge (1 - a_i))$$

The proof is trivial and is left for an exercise!
We next demonstrate this with 3 examples.

**Example 1.** Recall the results obtained from Ao(A→B) that was discussed in Section 14.6, i.e., $b^*_j$=(.3,.6,1)≠(0,.6,1)=$b_j$.
Recall also that we have $a_i$=(1,.7,0), b=(0,.6,1).
In the interval-valued Type 2 GMP schema, we first compute

$$A \rightarrow B = \begin{cases} \text{FDCF}(A \rightarrow B) \\ \text{FCCF}(A \rightarrow B), \end{cases}$$

$$= \begin{cases} [(a_i \wedge b_j) \vee ((1 - a_i) \wedge b_j) \vee ((1 - a_i) \wedge (1 - b_j))] \\ [(1 - a_i) \vee b_j] \end{cases}$$

Recall that: $[(1 - a_i) \vee b_j] = \begin{pmatrix} 0 & .6 & 1 \\ .3 & .6 & 1 \\ 1 & 1 & 1 \end{pmatrix}$ and $(1,.7,0) \wedge \begin{pmatrix} 0 & .6 & 1 \\ .3 & .6 & 1 \\ 1 & 1 & 1 \end{pmatrix}$ =(.3,.6,1)

which is the upper bound of the interval-valued Type 2 GMP. Next let us compute the lower bound of the interval-valued Type 2 GMP.
$\text{FDCF}(A \rightarrow B) = [(a_i \wedge b_j) \vee ((1 - a_i) \wedge b_j) \vee ((1 - a_i) \wedge (1 - b_j))]$

$$\begin{bmatrix} b_j & 0 & .6 & 1 \\ 1 - b_j & 1 & .4 & 0 \end{bmatrix}$$

$$= \begin{bmatrix} \overset{a_i}{1} & \overset{(1-a_i)}{0} \\ .7 & .3 \\ 0 & 1 \end{bmatrix} \begin{pmatrix} 0 & .6 & 1 \\ .3 & .6 & .7 \\ 1 & 1 & 1 \end{pmatrix}$$

and therefore, we get

$$\underset{i}{\vee} a_i \wedge [(a_i \wedge b_j) \vee ((1-a_i) \wedge b_j) \vee ((1-a_i) \wedge (1-b_j))]$$

$$=(1,.7,0) \wedge \begin{pmatrix} 0 & .6 & 1 \\ .3 & .6 & .7 \\ 1 & 1 & 1 \end{pmatrix} =(.3,.6,1)$$

Note that $\alpha = \underset{i}{\vee} (a_i \wedge (1-a_i)) = .3$.

Therefore, we get $b^*_{j\alpha} = (0,.6,1) = b_j$.

It is observed that in this example both the upper and the lower bounds of the implication turns out to be the same.

**Example 2.** Let us consider another case:
$a_i \in A = (.2,1,.4)$
$b_j \in B = (.3,1,.1)$

In the interval-valued Type 2 GMP schema, we first compute

$$A \to B = \begin{cases} \text{FDCF}(A \to B) = \to_L \\ \text{FCCF}(A \to B) = \to_U \end{cases}$$

For the $(\vee,\wedge,-)$ De Morgan Triple, we have

$$A \to B = \begin{cases} [(a_i \wedge b_j) \vee ((1-a_i) \wedge b_j) \vee ((1-a_i) \wedge (1-b_j))] \\ [(1-a_i) \vee b_j] \end{cases}$$

$$\begin{bmatrix} b_j & (.3) & 1 & (.1) \\ 1-b_j & (.7) & 0 & (.9) \end{bmatrix}$$

$$A \to B = \overset{\quad a_i \quad (1-a_i)}{\begin{bmatrix} .2 & .8 \\ 1 & 0 \\ .4 & .6 \end{bmatrix}} \begin{pmatrix} [.7,.8] & [.8,1] & [.8,.8] \\ [.3,.3] & [1,1] & [.1,.1] \\ [.6,.6] & [.6,1] & [.6,.6] \end{pmatrix}$$

Next let us compute

$$Ao(A \to B) = \begin{cases} \underset{i}{\vee} a_i \wedge \text{FDCF}(A \to B) \\ \underset{i}{\vee} a_i \wedge \text{FCCF}(A \to B) \end{cases}$$

$$(.2,1,.4) \wedge \begin{pmatrix} [.7,.8] & [.8,1] & [.8,.8] \\ [.3,.3] & [1,1] & [.1,.1] \\ [.6,.6] & [.6,1] & [.6,.6] \end{pmatrix} = ([.2,.2]\ [1,1]\ [.4,.4])$$

Clearly $B^* \neq B$, but $\alpha=.4$, we get $B^* \subseteq B$ for $\mu_B(2)=1$.

It is to be observed that in this case the upper and lower bounds of the implication turn out to be different.

**Example 3.** Consider $a_i \in A = (1,.7,.4)$

$b_j \in B = (.3,.8,1)$

In the interval-valued Type 2 GMP schema, we first compute

$$A \rightarrow B = \begin{cases} \text{FDCF}(A \rightarrow B) = \rightarrow_L \\ \text{FCCF}(A \rightarrow B) = \rightarrow_U \end{cases}$$

For the $(\vee,\wedge,-)$ De Morgan Triple, we have

$$A \rightarrow B = \begin{cases} [(a_i \wedge b_j) \vee ((1-a_i) \wedge b_j) \vee ((1-a_i) \wedge (1-b_j))] \\ [(1-a_i) \vee b_j] \end{cases}$$

$$\begin{matrix} & & & \begin{bmatrix} b_j & .3 & .8 & 1 \\ 1-b_j & .7 & .2 & 0 \end{bmatrix} \\ & a_i & (1-a_i) & \\ A \rightarrow B = & \begin{bmatrix} 1 & 0 \\ .7 & .3 \\ .4 & .6 \end{bmatrix} & & \begin{pmatrix} [.3,.3] & [.8,.8] & [1,1] \\ [.3,.3] & [.7,.8] & [.7,1] \\ [.6,.6] & [.6,.8] & [.6,1] \end{pmatrix} = [\rightarrow_L, \rightarrow_U]_{ij} \end{matrix}$$

Once again, we observe that the upper and lower values of implications are different.

In classical logic $Ao(A \rightarrow B) = B$, where as in fuzzy logic we have obtained $Ao(A \rightarrow B) = B^* \neq B$ since for $A \rightarrow B$ we get $[\rightarrow_L, \rightarrow_U]$ as shown in the above example.Let us compute the reasoning schema:

$$Ao(A \rightarrow B) = (1,.7,.4) \wedge \begin{pmatrix} [.3,.3] & [.1,.8] & [1,1] \\ [.3,.3] & [.7,.8] & [.7,1] \\ [.6,.6] & [.6,.8] & [.6,1] \end{pmatrix} = ([.4,.4]\ [.8,.8]\ [1,1])$$

Note $\alpha = \bigvee_i (a_i \wedge (1-a_i)) = .7$, therefore $B^*_{.8} \subseteq B_{.8}$, $b^*_{j\alpha} = (0,.8,1)$.

## 15.4. Implementations of Interval-Valued Type 2 Reasoning

As we discussed in Section 14.7, there are two main reasoning schemas that are used in most application studies. Recall that they are:

(i) Material Implication, MI, based reasoning schemas, and
(ii) Product (Mamdani) Implication, PI, based reasoning schemas

within interval-valued Type 2 framework, these two schemas have to consider FDCF(→) and FCCF(→) for each case.
That is, we have:

(i) First, $A \rightarrow B = \begin{cases} FDCF(A \rightarrow B) \\ FCCF(A \rightarrow B) \end{cases}$

for material implication it turns out to be:

$$A \rightarrow_{MI} B = \begin{cases} (A \cap B) \cup (c(A) \cap B) \cup (c(A) \cap c(B)) \\ c(A) \cup B. \end{cases}$$

For its realization we have for various (∇,Δ,n) De Morgan Triples, such as, for example:

(1) for $(\vee,\wedge,\text{-})$

$$A \rightarrow_{MI} B = \begin{cases} [(a \wedge b) \vee ((1-a) \wedge b) \vee ((1-a) \wedge (1-b))] \\ [(1-a) \vee b] \end{cases}$$

(2) for $(\odot,\oplus,\text{-})$

$$A \rightarrow_{MI} B = \begin{cases} [(a \odot b) \oplus ((1\text{-}a) \odot b) \oplus ((1\text{-}a) \odot (1\text{-}b))] \\ [(1-a) \oplus b] \end{cases}$$

(3) for $(L_{\odot},L_{\oplus},\text{-})$

$$A \rightarrow_{MI} B = \begin{cases} [(a L_{\odot} b)\ L_{\oplus}\ ((1\text{-}a)\ L_{\odot} b)\ L_{\oplus}\ ((1\text{-}a)\ L_{\odot}(1\text{-}b))] \\ [(1-a) L_{\oplus} b] \end{cases}$$

(ii) Next $A \rightarrow B = \begin{cases} FDCF(A \rightarrow B) \\ FCCF(A \rightarrow B) \end{cases}$

for the product implication to turns out to be:

$$A \rightarrow_{PI} B = \begin{cases} A \cap B \\ (A \cup B) \cap (c(A) \cup B) \cap (A \cap c(B)). \end{cases}$$

For its realization, we have for various $(\nabla,\Delta,n)$ De Morgan Triples, such as, for example:

(1) for $(\vee,\wedge,-)$

$$A \rightarrow_{PI} B = \begin{cases} a \wedge b \\ (a \vee b) \wedge ((1-a) \vee b) \wedge (a \vee (1-b)) \end{cases}$$

(2) for $(\odot,\oplus,-)$

$$A \rightarrow_{PI} B = \begin{cases} a \odot b \\ (a \oplus b) \odot ((1\text{-}a) \oplus b) \odot (a \oplus (1\text{-}b)) \end{cases}$$

(3) for $(L_{\odot},L_{\oplus},-)$

$$A \rightarrow_{PI} B = \begin{cases} aL_{\odot}b \\ (aL_{\oplus}b)L_{\odot}((1\text{-}a)\, L_{\oplus}b)L_{\odot}\,(aL_{\oplus}(1\text{-}b)) \end{cases}$$

## 15.5. Interval-Valued Type 2 System Modeling

As it was indicated at the start of the Approximate Reasoning Methods in Chapter 14, there are two schemas of interval-valued Type 2 Systems modeling.

(1) Input and output fuzzy sets are expressed as Type 1 representation but the reasoning is executed with Type 2 formulas.

(2) Input and output fuzzy sets are expressed as Type 2 representation and the reasoning is executed with Type 2 formulas.

We next review each of these cases.

### *15.5.1. Type 1 Representation and Type 2 Reasoning*

In this case, once again we start out with the meta-linguistic, canonical, form of a multi-input single output, MISO, system rule base as discussed in Section 14.9, i.e.,

$R_1$: IF $X_1$ isr $A_{11}$ AND $X_2$ isr $A_{12}$ AND … AND $X_{ND}$ isr $A_{1N}$ THEN Y isr $B_1$,
$R_2$: IF $X_1$ isr $A_{21}$ AND $X_2$ isr $A_{22}$ AND … AND $X_{ND}$ isr $A_{2N}$ THEN Y isr $B_2$,
.

.

.

$R_n$: IF $X_1$ isr $A_{n1}$ AND $X_2$ isr $A_{n2}$ AND … AND $X_N$ isr $A_{nN}$ THEN Y isr $B_n$,

where $X_1 \ldots X_{ND}$ are input variables, and Y is the output variable $A_{ij}$, i=1,…,n, j=1,...,N and $B_i$, i=1,…,n, are fuzzy set represented as Type 1 sets in the universes of discourse $X_1, X_2, \ldots, X_{ND}$ and Y, respectively.

Now, however, we have to first express the aggregate antecedent $A_i$ in both of its FDCF and FCCF formulas as

$$A_i = \begin{cases} \text{FDCF}(A_{i1} \text{ AND } A_{i2} \text{AND} \ldots \text{AND } A_{iN}) \\ \text{FCCF}(A_{i1} \text{ AND } A_{i2} \text{AND} \ldots \text{AND } A_{iN}) \end{cases}$$

Let $\text{FDCF}(A_i)$ and $\text{FCCF}(A_i)$ denote the lower and upper membership representations, respectively, of the left hand side, i.e., aggregate antecedent, such that $\text{FDCF}(A_i) \subseteq \text{FCCF}(A_i)$ for De Morgan Triples of strict and nilpotent Archimedean t-norms and t-conorms that are duals of each other for a suitable negation.

Recall from Chapter 8 and particular its generation in Section 8.2.4 that we can directly write $\text{FDCF}(A_i)$ and $\text{FCCF}(A_i)$ with the use of basic information granules. i.e., T-formalism. Therefore, we next assess the results of two Fuzzy canonical forms of $A_i$ for each rule, i.e.,
$\text{FDCF}(A_i) \rightarrow B_i$, and
$\text{FCCF}(A_i) \rightarrow B_i$

This means that next we have to form four expressions. i.e.,

(1) $\text{FDCF}[\text{FDCF}(A_i) \rightarrow B_i]$
(2) $\text{FCCF}[\text{FDCF}(A_i) \rightarrow B_i]$
(3) $\text{FDCF}[\text{FCCF}(A_i) \rightarrow B_i]$
(4) $\text{FCCF}[\text{FCCF}(A_i) \rightarrow B_i]$

Thus, one generates a Type 2 representation of the aggregate antecedent and then a Type 2 representation of each rule. That is we obtain four discrete point for the Type 2 representation of each rule. However, recall that we have the option of representing the implication as either product implication, PI, or material implication, MI.

Therefore we have two alternative choices to represent each of these four forms:

For PI, we get

(1′) FDCF[FDCF($A_i$) AND $B_i$]
(2′) FCCF[FDCF($A_i$) AND $B_i$]
(3′) FDCF[FCCF($A_i$) AND $B_i$]
(4′) FCCF[FCCF($A_i$) AND $B_i$]

Whereas for MI, we get

(1″) FDCF[c(FDCF($A_i$)) OR $B_i$]
(2″) FCCF[c(FDCF($A_i$)) OR $B_i$]
(3″) FDCF[c(FCCF($A_i$)) OR $B_i$]
(4″) FCCF[c(FCCF($A_i$)) OR $B_i$]

For PI, we know that

$(1') \subseteq (2')$,

and

$(3') \subseteq (4')$ in addition, it is clear that

$(1') \subseteq (3')$.

Therefore, we can write

$(1') \subseteq \text{Max}\{(2'), (4')\}$

which establishes, the upper and lower membership values for each rule i.

For MI, we know that

$(1'') \subseteq (2'')$, and

$(3'') \subseteq (4'')$ in addition, it is clear that

$(4'') \subseteq (2'')$. Therefore, we can write

$$\text{Min}\{(1''),(3'')\} \subseteq (2'')$$

which establishes, the upper and lower membership values for each rule i.

### *15.5.2. Type 2 Representation and Type 2 Reasoning*

As indicated in Chapters 4 and 5 under the headings of "Elicitation Methods" and "Fuzzy Clustering", respectively, the methods of the acquisition of membership functions in general produce Type 2 membership representations of the fuzzy sets $A_{ij}$ and $B_i$.

Let us denote the upper and lower membership functions as $A_{ij}^{L}$, $A_{ij}^{U}$, $B_i^{L}$ and $B_i^{U}$.

We next analyse FDCF and FCCF formulas for aggregate antecedents in terms of their upper and lower memberships. That is, we write:

FDCF($A_i^U$) and FDCF($A_i^L$)

FDCF($A_i^U$) and FCCF($A_i^L$) as:

(1) $FDCF(A_i^U) = FDCF(A_{i1}^U \; AND \; A_{i2}^U \; AND \; ... \; A_{iNV}^U)$

(2) $FCCF(A_i^U) = FCCF(A_{i1}^U \; AND \; A_{i2}^U \; AND \; ... \; A_{iNV}^U)$

(3) $FDCF(A_i^L) = FDCF(A_{i1}^L \; AND \; A_{i2}^L \; AND \; ... \; A_{iNV}^L)$

(4) $FCCF(A_i^L) = FCCF(A_{i1}^L \; AND \; A_{i2}^L \; AND \; ... \; A_{iNV}^L)$

First note that $FDCF(A_i^U) \subseteq FCCF(A_i^U)$. Secondly, we have $FDCF(A_i^L) \subseteq FCCF(A_i^L)$.

Thirdly, we have $FDCF(A_i^L) \subseteq FDCF(A_i^U) \subseteq FCCF(A_i^U)$ and therefore, we get $FDCF(A_i^L) \subseteq Max\{FCCF(A_i^L), FCCF(A_i^U)\}$.

Thus, we can compute a lower and an upper membership values for the aggregate left hand side (antecedent). Let these be $A_i^{LL}$ and $A_i^{UU}$, i=1,…,n. Next we need to compute:

(1′) $FDCF(A_i^{LL} \rightarrow B_i^L) \subseteq FCCF(A_i^{LL} \rightarrow B_i^L)$

(2′) $FDCF(A_i^{LL} \rightarrow B_i^U) \subseteq FCCF(A_i^{LL} \rightarrow B_i^U)$

(3′) $FDCF(A_i^{UU} \rightarrow B_i^L) \subseteq FCCF(A_i^{UU} \rightarrow B_i^L)$

(4′) $FDCF(A_i^{UU} \rightarrow B_i^U) \subseteq FCCF(A_i^{UU} \rightarrow B_i^U)$

Once again, we need to choose depending on the context of the system to be analysed whether we should apply PI or MI type of implication.

These two alternatives are expressed as follows.

a) For PI, we have:
(1″) FDCF( $A_i^{LL} \to B_i^L$ ) $\subseteq$ FCCF ( $A_i^{LL} \to B_i^L$ )
(2″) FDCF( $A_i^{LL} \to B_i^U$ ) $\subseteq$ FCCF ( $A_i^{LL} \to B_i^U$ )
(3″) FDCF( $A_i^{UU} \to B_i^L$ ) $\subseteq$ FCCF ( $A_i^{UU} \to B_i^L$ )
(4″) FDCF( $A_i^{UU} \to B_i^U$ ) $\subseteq$ FCCF ( $A_i^{UU} \to B_i^U$ )
Let LHS and RHS stand for "left-hand side" and "right-hand side", respectively.

It is clear that

LHS (1″) $\subseteq$ LHS {(2″),(3″),(4″)}

Next, it is clear that

LHS (1″) $\subseteq$ Max{RHS{(1″),(2″),(3″),(4″)}}

Hence, we can determine the upper and lower membership values of each rule I in Interval-Valued Type 2 representation. Naturally, if we wanted to, we can capture eight point discrete Type 2 representation of each rule i.
In a similar manner, for MI, we can write

(1‴) FDCF [ $c(A_i^{LL}) \to B_i^L$ ] $\subseteq$ FCCF [ $c(A_i^{LL}) \to B_i^L$ ]
(2‴) FDCF [ $c(A_i^{LL}) \to B_i^U$ ] $\subseteq$ FCCF [ $c(A_i^{LL}) \to B_i^U$ ]
(3‴) FDCF [ $c(A_i^{UU}) \to B_i^L$ ] $\subseteq$ FCCF [ $c(A_i^{UU}) \to B_i^L$ ]
(4‴) FDCF [ $c(A_i^{UU}) \to B_i^U$ ] $\subseteq$ FCCF [ $c(A_i^{UU}) \to B_i^U$ ]

In this case, in a similar analysis, it is clear that

RHS{(1‴),(3‴),(4‴)} $\subseteq$ RHS(2‴).

As well, we have

Min{LHS{(1‴),(2‴),(3‴),(4‴)} $\subseteq$ RHS(2‴)

Thus, once again, we can determine the upper and the lower membership values of each rule I in Interval-Valued Type 2 representation. Again naturally, if it need be, we can capture eight point discrete Type 2 representation of each rule i∈I, etc. Note that, we have a prototypical appearance as shown in Figure 15.1. Naturally, the discussion on Type 2 rule representation must be re asseded depending whether or not the observation comes from a singleton, or Type 1 or Type 2 representation.

In Figures 15.2, 15.3, 15.4 we show a skeleton of how these formulas represented so far should be further assessed for a two antecedent case to determine the aggregate antecedent.

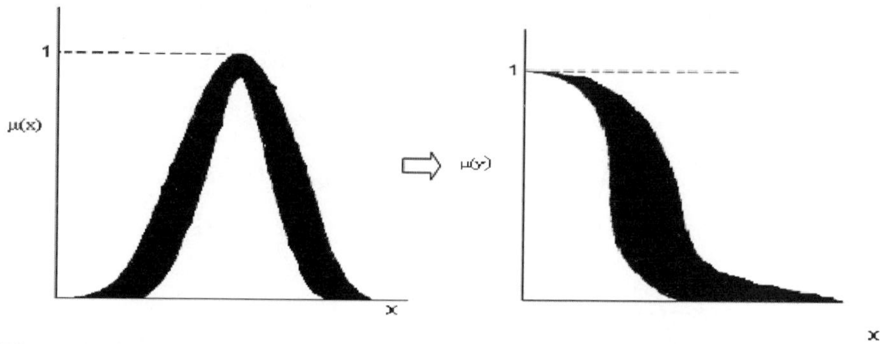


**Figure 15.1.** Type 2 representation that generates an eight layered Type 2 representation in X×Y space based on Type 2 reasoning.

Three Input-output data sets are considered in this investigation. They are:

(i) Example (a) - A non-linear system

(ii) Example (b) - Human operation of a chemical plant and

(iii) Example (c) — Daily price of a stock in a stock market.

The specifications of these data sets are shown in Table 15.1. Each data set is partitioned into three data groups, i.e., the first and the second sets are the data group A and the data group B which are used for model identification, i.e., training respectively, and the third set is the data group C used for the test, verification and comparison of the three different reasoning methods namely, two version of Sugeno's "Position-Gradient", Sugeno's Type 1 reasoning known as "Position and Türkşen's Interval-Valued GMP.

## 15.6. Application to Case Studies with Interval Valued Type 2 Reasoning

To be able to compare the performance characteristics of these three methods, we chose three sets of system data sets provided in Sugeno, Yasukawa (1993) investigations.

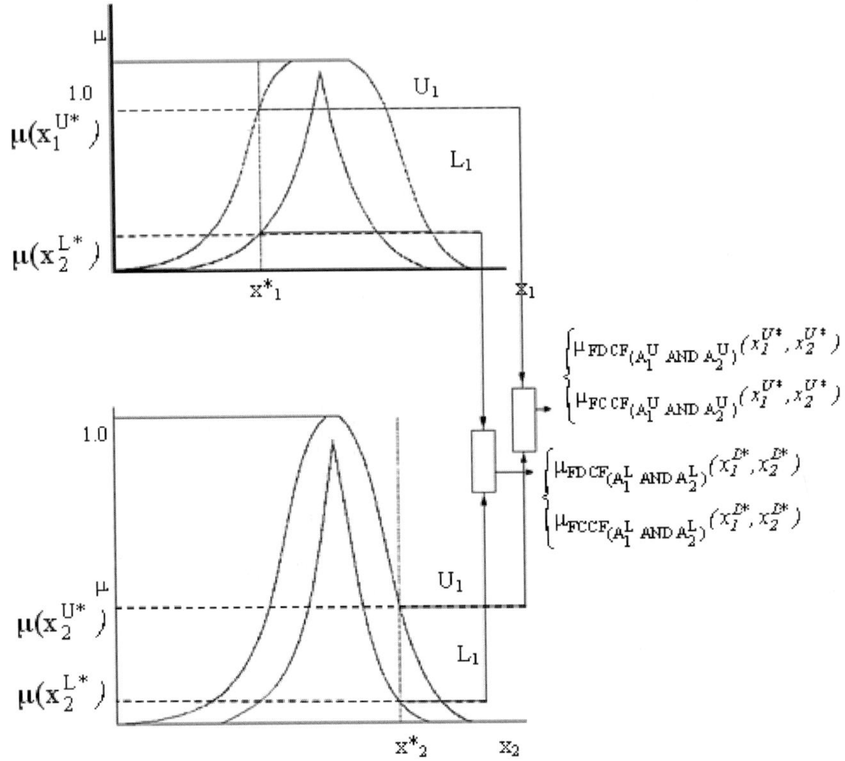


**Figure 15.2.** $x^*_1$, $x^*_2$: Singleton observation.

NOTE: The detail of these investigations were carried out and published in a review paper [Nakanishi, Türkşen, Sugeno, 1993], which included six methods of reasoning and their comparison. Here we extract certain results pertinent to this writing from that investigation.

The elements of the data sets and the specifications of model variables and rules are shown in Tables 15.1 and 15.2.

The rule base is formed with the application of Sugeno-Yasukawa method (1993).

The input and output variables and the actual data sets that are used in these examples are shown in Appendix 15.1.

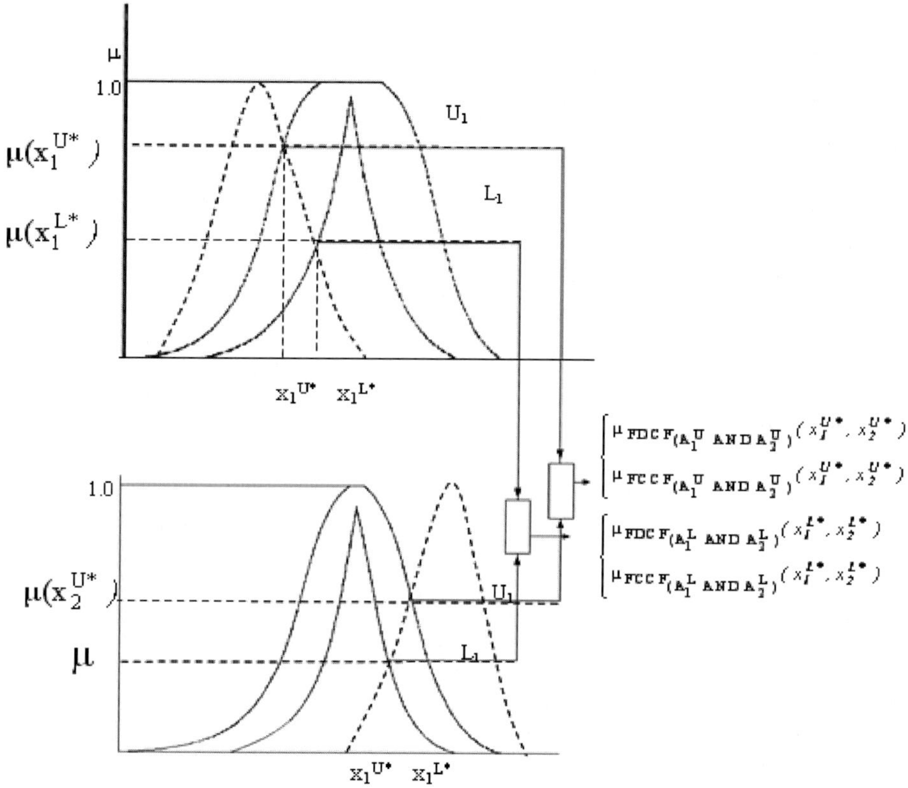

**Figure 15.3**. $x^*_1$, $x^*_2$: Type 1 observations.

**Table 15.1.** Elements of data sets

| Number of data points | Number of candidate input variable | Data sets | | |
|---|---|---|---|---|
| | | Group A | Group B | Group C |
| Example (a) | 4 | 12 | 13 | 25 |
| Example (b) | 5 | 17 | 18 | 35 |
| Example (c) | 10 | 25 | 25 | 50 |

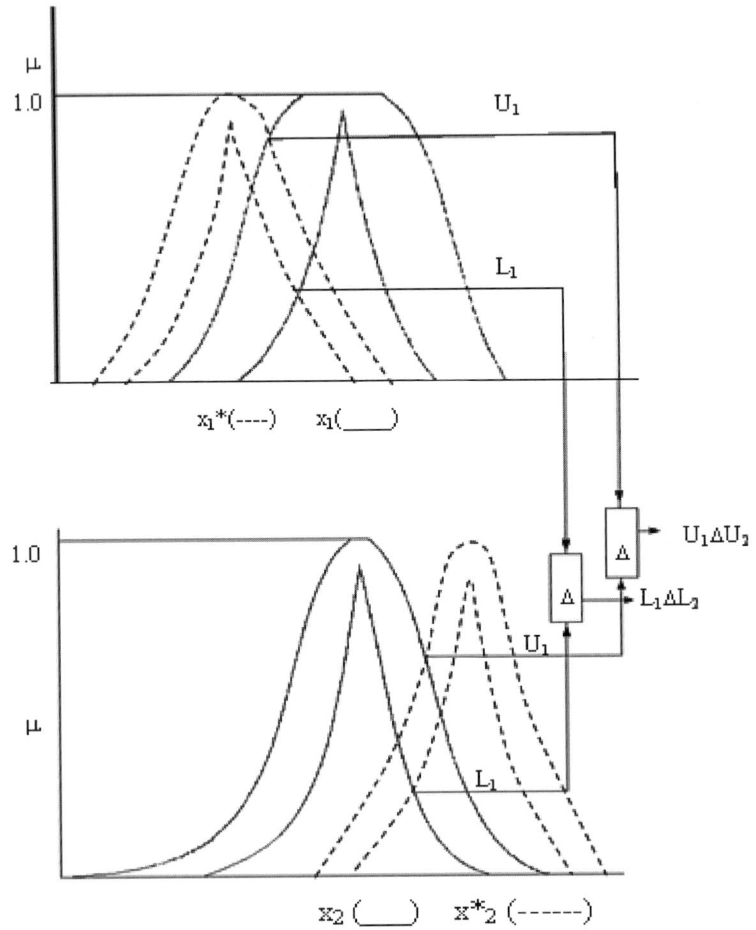


**Figure 15.4.** $x^*_1$, $x^*_2$: Type 2 observations.

**Table 15.2.** Specification of model variables and rules*.

| | Number of variables | Number of rules |
|---|---|---|
| Example (a) | 2 | 6 |
| Example (b) | 2 | 5 |
| Example (c) | 3 | 4 |

*The rules are shown in Appendix 15.2.

Test models of each example are formed through structure and parameter identification methods of FCM using the data groups A and B and the membership functions were determined with curve-fitting techniques.

### *15.6.1. Comparison of reasoning methods*

In Appendix 15.3 the actual output data and the results of there reasoning methods are shown. In the comparison, we discuss Type 1 reasoning and interval-valued Type 2 reasoning results separately.

The comparison of the reasoning methods is investigated by common data group C of each example data set. The result of root-mean-square error, RMSE, between a reasoning consequence and the actual output data is computed to determine the reasoning precision (pr). The reasoning precision together with the calculation time (t) and the number of data sets (n) for which the reasoning is carried out are shown in a Table 15.3. In Table 15.3, 'Sugeno (P&P-G)' is the reasoning method with Position type and Position-Gradient type which are used alternatively based on the maximum membership values of rules. This method is used for system modeling during the structure and parameter identification phase of our investigation. P&P-G results for pr, t, and n form the base against which the others are compared. 'Sugeno (P)' is the Position type reasoning method, 'Sugeno (P-G)' is the Position-Gradient type reasoning method. Because of the inherent bias of the model indicated above, each value of pr and t are computed as the relative value with respect to 'Sugeno (P-G)' method.

As we have argued throughout if the linguistic connectives in the reasoning process include imprecision, as in most expressions of human communication, it is then more appropriate to use the interval-valued fuzzy reasoning. Türkşen's IVG MP is able to get good results for all data points in all three examples. It needs to be pointed out that interval-valued reasoning maintains the uncertainty associated with human communications at the expense of additional computational costs. Such additional costs might be justifiable in certain cases where flexibility of interval-valued Type 2 fuzziness might be an advantage such as management decision-making models.

In Figures 15.5 and 15.6, examples of consequences of the reasoning methods are displayed graphically to show the Type 1 fuzziness associated with Sugeno reasoning methods and the Type 2 fuzziness associated with Türkşen's IVGMP reasoning method.

**Table 15.3.** Sugeno's Type 1 and Türkşen's IVGMP reasoning results

| | | Sugeno's (P&P-G) | Sugeno's (P) | Sugeno's (P-G) | Türkşen's (IVGMP) |
|---|---|---|---|---|---|
| Example (a) | pr | 1 | 1.58 | 1.17 | 1.45 |
| | t | 1 | 0.922 | 1.73 | 4.16 |
| | n | 25 | 24 | 25 | 25 |
| Example (b) | pr | 1 | 1.53 | 2.59 | 0.94 |
| | t | 1 | 0.990 | 1.88 | 3.49 |
| | n | 25 | 34 | 35 | 35 |
| Example (c) | pr | 1 | 0.837 | 1.34 | 0.99 |
| | t | 1 | 0.981 | 1.61 | 2.47 |
| | n | 30 | 50 | 50 | 50 |

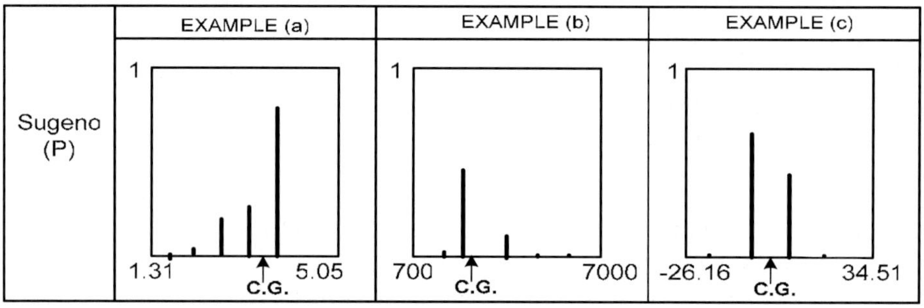


**Figure 15.5.** Examples of consequences by Type 1 and Type 2 reasoning methods.

### *15.6.2. Conclusion*

The three reasoning methods for fuzzy system models that are constructed based on three bench mark data sets have been reviewed. The reasoning precision, the calculation time and the number of valid input cases for reasoning differ according to each reasoning method, i.e.,

(a) Sugeno's methods(P), (P-G) which are Type 1 reasoning methods

(b) Türkşen's method (IVGMP) which is a Type 2 reasoning method.

Türkşen's method (IVGMP) gives a representation of the spread of vagueness that may be associated with linguistic connectives.

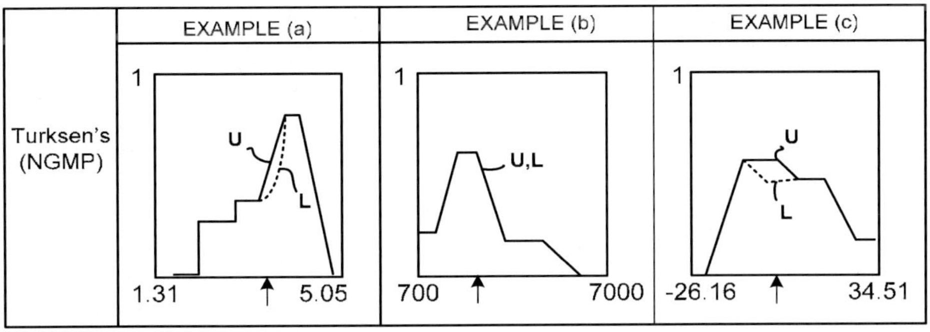


**Figure 15.6**. Examples of consequences by interval-valued reasoning methods.

Sugeno's methods which have an advantage in calculation speed. This method is quite useful in the case where rapid reasoning is necessary and no case data point should be missed in reasoning, for example, real-time control. However, in systems where human preferences play an important role such as managerial decision systems. In particular, this is clearly evident in the stock price model, i.e., Example (c) Table 15.5 and execute interval-valued Type 2 reasoning. For this real-life data set, where stock prices are greatly affected by the individual buyers preferences, or by institutional buying determined by management decision-making processes, the precision of interval valued IVGMP produced much better predictions.

The implementation of the reasoning methods for systems modeling should be oriented toward obtaining a good precision, a short calculation time and flexibility to cope with all case data points. However, different methods need to be used depending on whether the system is electro-mechanical or humanistic decision-making process.

## Appendix 15.1.

The candidate input variables and output variable of each example data set:

(a) Non-linear system
$y = (1 + x_1^{-2} + x_2^{1.5})^2 \, (1 \le x_1, x_2 \le 5)$.

(b) Human operation of a chemical plant
- x1 : monomer concentration,
- x2: charge of monomer concentration,
- x3: monomer flow rate,
- x4: local temperatures inside the plant,
- x5: local temperatures inside the plant,
- y: set point for monomer flow rate.

(c) Daily stock price data of a stock in a stock market

xl: past change of moving average (1) over a middle period,
x2: present change of moving average (1) over a middle period
x3: past separation ratio (1) with respect to moving average over a middle period,
x4: present separation ratio (1) with respect to moving average over a middle period,
x5: present change of moving (2) over a short period,
x6: past change of price (1), for instance, change on one day before,
x7: present change of price (1),
x8: past separation ratio (2) with respect to moving average, over a short period,
x9: present change of moving average (3) over a long period,
x10: present separation ratio (3) with respect to moving average over a short period,
y prediction of stock price.
The data sets of three examples

**Example a**. Non-linear system

Group C

| N | X1 | X2 | X3 | X4 | Y |
|---|---|---|---|---|---|
| 1 | 4.85 | 4.66 | 4.11 | 3.74 | 1.30 |
| 2 | 4.44 | 4.71 | 1.53 | 2.08 | 1.32 |
| 3 | 4.84 | 4.32 | 2.75 | 1.70 | 1.33 |
| 4 | 4.35 | 3.90 | 2.55 | 1.65 | 1.40 |
| 5 | 3.04 | 4.97 | 2.77 | 2.63 | 1.44 |
| 6 | 3.38 | 3.70 | 4.65 | 1.28 | 1.51 |
| 7 | 2.71 | 4.13 | 4.38 | 3.21 | 1.58 |
| 8 | 2.24 | 3.74 | 4.25 | 3.26 | 1.79 |
| 9 | 1.92 | 4.25 | 3.24 | 3.89 | 1.92 |
| 10 | 1.84 | 4.43 | 4.20 | 1.38 | 1.97 |
| 11 | 3.37 | 2.13 | 2.42 | 3.95 | 1.99 |
| 12 | 1.81 | 3.18 | 3.31 | 2.07 | 2.20 |
| 13 | 4.10 | 1.71 | 2.54 | 1.76 | 2.27 |
| 14 | 2.83 | 1.77 | 2.61 | 4.50 | 2.40 |
| 15 | 3.66 | 1.60 | 3.44 | 3.33 | 2.46 |

Group C

| N | X1 | X2 | X3 | X4 | Y |
|---|---|---|---|---|---|
| 16 | 4.51 | 1.52 | 4.53 | 2.54 | 2.51 |
| 17 | 2.03 | 1.88 | 1.41 | 1.10 | 2.66 |
| 18 | 1.96 | 1.90 | 1.71 | 1.59 | 2.70 |
| 19 | 4.51 | 1.37 | 3.97 | 1.70 | 2.80 |
| 20 | 1.18 | 4.29 | 1.60 | 3.80 | 3.35 |

| 21 | 2.84 | 1.24 | 4.42 | 1.21 | 3.42 |
|---|---|---|---|---|---|
| 22 | 1.40 | 1.80 | 3.00 | 3.80 | 3.70 |
| 23 | 1.65 | 1.38 | 4.57 | 4.03 | 3.94 |
| 24 | 1.05 | 2.55 | 3.03 | 2.02 | 4.63 |
| 25 | 1.19 | 1.53 | 2.54 | 3.22 | 4.99 |

**Example b**. Human operation of chemical plant

Group A

| N | X1 | X2 | X3 | X4 | X5 | Y |
|---|---|---|---|---|---|---|
| 1 | 6.59 | -0.21 | 464.00 | -0.10 | 0.10 | 700.00 |
| 2 | 6.59 | 0.00 | 703.00 | -0.10 | 0.10 | 900.00 |
| 3 | 5.51. | -0.29 | 1782.00 | -0.10 | 0.00 | 1900.00 |
| 4 | 5.62 | 0.11 | 3562.00 | -0.40 | 0.10 | 3700.00 |
| 5 | 5.79 | -0.03 | 3895.00 | 0.20 | -0.10 | 3900.00 |
| 6 | 5.65 | -0.14 | 3887.00 | -0.10 | 0.00 | 3900.00 |
| 7 | 4.81 | -0.23 | 4462.00 | 0.00 | -0.10 | 4900.00 |
| 8 | 4.72 | 0.01 | 5668.00 | 0.00 | -0.10 | 6000.00 |
| 9 | 4,70 | 0.05 | 6368.00 | -0.10 | 0.00 | 6400.00 |
| 10 | 4.59 | 0.04 | 6250.00 | -0.20 | -0.10 | 6400.00 |
| 11 | 4,77 | 0.00 | 6569.00 | 0.00 | -0.10 | 6600.00 |
| 12 | 4,77 | 0.03 | 6779.00 | 0.00 | -0.10 | 6800.00 |
| 13 | 4.57 | -0.04 | 6832.00 | 0.00 | 0.10 | 6800.00 |
| 14 | 4.56 | -0.01 | 6832.00 | -0.10 | 0.10 | 6900.00 |
| 15 | 4.56 | 0.00 | 7022.00 | 0.00 | 0.00 | 7000.00 |
| 16 | 4.51 | -0.03 | 6958.00 | 0.10 | -0.10 | 7000.00 |
| 17 | 4.50 | 0.02 | 7006.00 | 0.00 | 0.10 | 7000.00 |

Group B

| N | Xl | X2 | X3 | X4 | X5 | Y |
|---|---|---|---|---|---|---|
| 1 | 6.50 | -0.09 | 797.00 | 0.10 | 0.10 | 700.00 |
| 2 | 6.45 | -0.09 | 784.00 | 0.00 | 0.10 | 800.00 |
| 3 | 6.02 | -0.18 | 1211.00 | 0.00 | 0.10 | 1400.00 |
| 4 | 5.44 | 0.01 | 2404.00 | -0.10 | -0.10 | 2500.00 |

Group B

| N | Xl | X2 | X3 | X4 | X5 | Y |
|---|---|---|---|---|---|---|
| 5 | 594 | 0.17 | 3701.00 | -0.20 | 0.10 | 3800.00 |
| 6 | 5.99 | -0.03 | 3896.00 | 0.20 | -0.10 | 3900.00 |
| 7 | 124 | -0.24 | 4048.00 | 0.10 | 0.00 | 4400.00 |
| 8 | 4.61. | -0.01 | 5284.00 | -0.10 | 0.20 | 5400.00 |
| 9 | 4.58 | -0.14 | 5844.00 | -0.20 | 0.10 | 6100.00 |
| 10 | 4.84 | 0.03 | 6412.00 | -0.10 | -0.10 | 6400.00 |

| | | | | | | |
|---|---|---|---|---|---|---|
| 11 | 4.77 | 0.01 | 6587.00 | -0.10 | 0.10 | 6600.00 |
| 12 | 4.77 | 0.00 | 6559.00 | 0.00 | 0.00 | 6700.00 |
| 13 | 4.73 | 0.00 | 6844.00 | -0.10 | 0.00 | 6800.00 |
| 14 | 4.71 | -0.06 | 6783.00 | 0.00 | 0.00 | 6800.00 |
| 15 | 4.48 | -0.02 | 7032.00 | 0.00 | 0.00 | 7000.00 |
| 16 | 4.48 | 0.00 | 6973.00 | 0.00 | 0.00 | 7000.00 |
| 17 | 4.56 | -0.01 | 7009.00 | -0.10 | 0.10 | 7000.00 |
| 18 | 4.47 | 0.00 | 6986.00 | -0.10 | 0.10 | 7000.00 |

GroupC

| N | Xl | X2 | X3 | X4 | X5 | Y |
|---|---|---|---|---|---|---|
| 1 | 6.80 | -0.05 | 401.00 | -0.20 | -0.10 | 500.00 |
| 2 | 6.48 | -0.02 | 717.00 | -0.10 | 0.10 | 700.00 |
| 3 | 6.45 | 0.00 | 794.00 | -0.20 | 0.10 | 800.00 |
| 4 | 6.54 | 0.06 | 706.00 | -0.20 | 0.10 | 800.00 |
| 5 | 6.20 | -0.25 | 792.00 | 0.00 | 0.00 | 1000.00 |
| 6 | 5.80 | -0.22 | 1557.00 | -0.20 | 0.00 | 1600.00 |
| 7 | 5.43 | -0.08 | 2206.00 | -0.10 | 0.10 | 2300.00 |
| 8 | 5.51 | 0.07 | 2685.00 | 0.10 | 0.00 | 2800.00 |
| 9 | 5.97 | 0.03 | 3775,00 | -0.10 | 0.00 | 3800.00 |
| 10 | 5.77 | 0.15 | 3629.00 | -0.10 | 0.00 | 3800.00 |
| 11 | 6.02 | 0.05 | 3829.00 | -0.10 | -0.10 | 3900.00 |
| 12 | 5.82 | -0.17 | 3920.00 | 0.20 | -0.10 | 3900.00 |
| 13 | 5.48 | -1.70 | 3930.00 | 0.20 | 0.00 | 4000.00 |
| 14 | 5.04 | -0.20 | 4448.00 | 0.00 | 0.00 | 4700.00 |
| 15 | 4.62 | -0.19 | 5078.00 | -0.30 | 0.30 | 5200.00 |
| 16 | 4.54 | -0.07 | 5225.00 | -0.30 | 0.1.0 | 5600.00 |
| 17 | 4.71 | 0.17 | 5391.00 | -0.10 | 0.00 | 6000.00 |
| 18 | 4.55 | -0.03 | 6068.00 | -0.20 | 0.00 | 6400.00 |
| 19 | 4.65 | 0.06 | 6358.00 | -0.10 | -0.10 | 6400.00 |
| 20 | 4.81 | 0.1,1 | 6379.00 | -0.30 | 0.00 | 6400.00 |
| 21 | 4.83 | -0.01 | 6416.00 | 0.10 | -0.10 | 6500.00 |
| 22 | 4.76 | -0.07 | 6514.00 | 0.00 | 0.00 | 6600.00 |
| 23 | 4.73 | -0.04 | 6672.00 | 0.00 | 0.00 | 6700.00 |
| 24 | 4.74 | 0.01 | 6775,00 | -0.20 | 0.00 | 6800.00 |
| 25 | 4.63 | -0.07 | 6849.00 | 0.00 | 0.00 | 6800.00 |
| 26 | 4.66 | -0.05 | 6816.00 | 0.00 | 0.00 | 6800.00 |
| 27 | 4.61 | -0.02 | 6803.00 | 0.00 | 0.00 | 6800.00 |
| 28 | 4.70 | 0.04 | 6812.00 | 0.00 | 0.00 | 6800.00 |
| 29 | *4.57* | 0.01 | 6998,00 | -0.10 | 0.00 | 7000.00 |
| 30 | 4.50 | 0.00 | 7027.00 | 0.00 | 0.00 | 7000.00 |
| 31 | 4.54 | -0.02 | 6862.00 | -0.10 | -0.10 | 7000.00 |
| 32 | 4.54 | 0.06 | 6995.00 | 0.00 | 0.00 | 7000.00 |
| 33 | 4.57 | 0.03 | 6986.00 | 0.10 | -0.10 | 7000.00 |

| | | | | | | |
|---|---|---|---|---|---|---|
| 34 | 4.48 | 0.01 | 6975.00 | 0.00 | 0.00 | 7000.00 |
| 35 | 4.47 | -0.04 | 6998.00 | 0.00 | 0.10 | 7000.00 |

Example c. Daily data of a stock in stock market

| Group A | N | X1 | X2 | X3 | X4 | X5 | X6 | X7 | X8 | X9 | X10 | Y |
|---|---|---|---|---|---|---|---|---|---|---|---|---|
| | 1 | 0.31 | 0.38 | 10.67 | 22.19 | 0.97 | 1.37 | 4.17 | 7.25 | -4.52 | 1.16 | -24.34 |
| | 2 | 0.30 | 0.46 | 11.83 | 26.04 | 1.15 | 0.65 | -0.64 | 1.16 | 5.48 | 2.60 | -22.09 |
| | 3 | 0.39 | 0.29 | 19.28 | 12.77 | 0.95 | -1.49 | 5.79 | 6.32 | -2.22 | 3.29 | -20.47 |
| | 4 | 0.23 | 0.36 | 2.82 | 24.10 | 0.65 | 0.00 | -1.90 | 0.78 | 0.00 | 0.65 | -19.24 |
| | 5 | 0.43 | 0.29 | 22.51 | 16.70 | 1.41 | 3.79 | 3.13 | 5.85 | 3.01 | 5.71 | -16.80 |
| | 6 | 0.27 | 0.18 | 4.65 | 10.55 | -0.91 | -6.04 | -0.64 | -4.52 | -10.26 | -8.62 | -13.58 |
| | 7 | 0.35 | 0.35 | 13.59 | 14.32 | 1.38 | -1.48 | -1.63 | 8.74 | 3.91 | 6.32 | -10.53 |
| | 8 | 0.41 | 0.15 | 25.78 | -2.63 | 0.66 | 0.00 | 0.00 | 3.72 | 9.74 | 5.72 | -6.46 |
| | 9 | -0.01 | 0.09 | -3.39 | 2.07 | 0.83 | 0.00 | 1.77 | 10.84 | 7.44 | 6.91 | -3.85 |
| | 10 | 0.35 | 0.30 | 19.45 | 7.85 | 0.08 | 3.36 | -2.54 | 0.76 | 3.36 | 3.63 | -2.44 |
| | 11 | 0.30 | -0.03 | 7.85 | -5.23 | -2.48 | -14.30 | 0.65 | -3.62 | -18.93 | -19.69 | 0.00 |
| | 12 | -0.11 | -0.04 | -16.89 | -4.68 | -0.17 | 5.18 | -1.68 | -3.12 | 7.02 | 3.83 | 0.82 |
| | 13 | 0.06 | -0.04 | -5.53 | -5.55 | 0.00 | 0.00 | -1.70 | 4.21 | 0.00 | 1.01 | 1.67 |
| | 14 | 0.43 | 0.12 | 28.08 | -6.90 | 0.88 | -1.67 | 3.51 | 0.27 | 6.31 | 3.39 | 1.70 |
| | 15 | 0.35 | 0.03 | 19.69 | -6.58 | -0.17 | 3.51 | -6.09 | -11.33 | 7.28 | 2.08 | 2.54 |
| | 16 | 0.18 | -0.03 | 10.55 | -5.70 | 0.08 | -2.42 | -3.15 | 2.44 | -3.97 | -2.58 | 3.31 |
| | 17 | -0.04 | -0.08 | -11.07 | -1.08 | 0.40 | -0.79 | -1.60 | 4.30 | 0.00 | -1.02 | 4.77 |
| | 18 | 0.12 | -0.01 | -6.90 | -5.62 | 0.51 | 3.45 | 0.83 | 0.85 | 5.27 | 1.96 | 5.00 |
| | 19 | 0.29 | -0.03 | 16.70 | -9.73 | -2.20 | 8.58 | 1.77 | -10.43 | -4.21 | -1.56 | 5.27 |
| | 20 | 0.05 | 0.03 | -2.18 | -7.23 | -0.16 | 2.59 | 0.00 | -3.07 | -1.65 | -3.10 | 5.89 |
| | 21 | 0.00 | -0.04 | -5.97 | -10.25 | -0.92 | 0.88 | -3.34 | -4.96 | -1.71 | -3.12 | 6.96 |
| | 22 | 0.17 | -0.04 | -0.56 | -9.39 | -0.25 | 1.77 | -3.23 | -2.60 | -6.51 | -3.12 | 7.83 |
| | 23 | -0.10 | -0.04 | -14.67 | -11.07 | -0.99 | -0.87 | -0.83 | -3.94 | -4.21 | -4.84 | 10.54 |
| | 24 | 0.29 | -0.11 | 12.77 | -16.89 | -3.59 | -4.55 | -8.88 | -6.01 | -2.78 | -11.33 | 16.21 |
| | 25 | 0.24 | 0.23 | 8.31 | 2.82 | -0.50 | 0.87 | -3.28 | -2.56 | -2.52 | -3.25 | 34.51 |

| Group B N | X1 | X2 | X3 | X4 | X5 | X6 | X7 | X8 | X9 | X10 | Y |
|---|---|---|---|---|---|---|---|---|---|---|---|
| 1 | 0.29 | 0.41 | 13.07 | 25.80 | 1.44 | 3.38 | 4.00 | 0.76 | -0.65 | 3.10 | -26.16 |
| 2 | 0.36 | 0.43 | 10.50 | 28.08 | 1.34 | -0.65 | 2.92 | 11.92 | 2.67 | 7.25 | -23.39 |
| 3 | 0.47 | 0.34 | 29.02 | 21.82 | 1.75 | 2.13 | -1.48 | 5.71 | 9.10 | 7.71 | -20.85 |
| 4 | 0.26 | 0.41 | 5.86 | 25.78 | 0.72 | -1.90 | -5.20 | 2.60 | 1.31 | 1.37 | -20.01 |
| 5 | 0.25 | 0.40 | 6.21 | 25.56 | 0.00 | -1.89 | 3.38 | 1.37 | -1.27 | 1.43 | -18.60 |
| 6 | 0.47 | 0.35 | 28.46 | 19.69 | 1.39 | 2.92 | 2.27 | 3.29 | 5.23 | 7.31 | -16.32 |
| 7 | 0.26 | 0.34 | 12.79 | 11.26 | 0.76 | 5.79 | 0.84 | 1.68 | 4.07 | 6.67 | -11.73 |
| 8 | 0.41 | 0.11 | 28.49 | -5.09 | 0.85 | -2.42 | -5.09 | 6.63 | 0.83 | 2.28 | -6.62 |
| 9 | 0.27 | 0.05 | 5.53 | -2.16 | -2.38 | -4.62 | -5.13 | -19.69 | -11.44 | -13.54 | -4.04 |
| 10 | 0.35 | 0.00 | 14.32 | -5.97 | -2.84 | 10.20 | 8.34 | -14.89 | 5.31 | -6.01 | -3.36 |
| 11 | 0.06 | 0.02 | -11.77 | -2.48 | 0.68 | 3.34 | -0.87 | 1.96 | 6.90 | 4.21 | -0.81 |
| 12 | 0.40 | 0.06 | 26.04 | -5.53 | 0.52 | 6.20 | 6.20 | -2.01 | 1.70 | 3.72 | 0.00 |
| 13 | 0.10 | 0.01 | -2.48 | -3.19 | 0.17 | 2.50 | 3.45 | 1.01 | -0.81 | 3.36 | 1.63 |
| 14 | -0.03 | -0.10 | -9.73 | -6.16 | -0.09 | -1.64 | -1.71 | -1.45 | 4.35 | 2.22 | 1.67 |
| 15 | 0.09 | -0.02 | -6.98 | -5.61 | 0.42 | 0.00 | -4.96 | -0.94 | 1.70 | 1.52 | 2.50 |
| 16 | 0.32 | 0.05 | 17.36 | -1.84 | 1.15 | 0.00 | -2.42 | 4.31 | 2.44 | 2.44 | 3.18 |
| 17 | 0.46 | 0.02 | 30.82 | -10.64 | 0.00 | 1.80 | -4.55 | 0.00 | -4.24 | 0.27 | 4.43 |
| 18 | -0.04 | 0.00 | -6.45 | -3.45 | 0.00 | -1.60 | 5.22 | 6.91 | -5.39 | 0.74 | 4.88 |
| 19 | -0.07 | -0.03 | -13.02 | -9.41 | -0.84 | 0.87 | 2.39 | -4.84 | 0.87 | -1.45 | 5.18 |
| 20 | 0.36 | 0.13 | 24.10 | -1.79 | 0.66 | 0.00 | 0.00 | 2.90 | -0.79 | 3.62 | 5.56 |
| 21 | 0.41 | 0.02 | 25.80 | -10.99 | 0.70 | 0.89 | 1.80 | 3.42 | -4.24 | -1.83 | 6.20 |
| 22 | 0.04 | -0.13 | -12.21 | -10.44 | -0.43 | -0.87 | 0.87 | 2.98 | -5.00 | -2.48 | 7.02 |
| 23 | -0.04 | -0.03 | -9.22 | -10.32 | -0.66 | -1.71 | -2.42 | 3.10 | -0.86 | -4.98 | 10.44 |
| 24 | 0.38 | 0.06 | 22.19 | -11.77 | 0.17 | -5.09 | -2.63 | 3.30 | -6.67 | -2.01 | 10.72 |
| 25 | 0.32 | 0.27 | 21.25 | 5.82 | -0.42 | 0.84 | -0.83 | -3.25 | 4.35 | 0.75 | 24.19 |

| **Group C** N | X1 | X2 | X3 | X4 | X5 | X6 | X7 | X8 | X9 | X10 | Y |
|---|---|---|---|---|---|---|---|---|---|---|---|
| 1 | 0.35 | 0.48 | 13.12 | 30.82 | 2.05 | 4.00 | -1.49 | 7.71 | 10.65 | 11.92 | -27.58 |
| 2 | 0.39 | 0.40 | 19.06 | 26.39 | 2.17 | 4.17 | 0.75 | 7.31 | 9.50 | 9.82 | -26.02 |
| 3 | 0.26 | 0.41 | 7.37 | 28.49 | 0.59 | 2.58 | 1.37 | 4.09 | 3.25 | 3.38 | -23.91 |
| 4 | 0.35 | 0.40 | 11.81 | 29.47 | 1.65 | -0.64 | 3.79 | 9.82 | 7.64 | 9.39 | -22.59 |
| 5 | 0.27 | 0.46 | 9.73 | 28.73 | 1.13 | 2.60 | -0.65 | 3.10 | 6.76 | 4.09 | -21.53 |
| 6 | 0.24 | 0.39 | 5.07 | 24.26 | 0.00 | -0.64 | 0.65 | 3.38 | 0.00 | 0.78 | -20.66 |
| 7 | 0.33 | 0.40 | 12.09 | 16.44 | 1.65 | 2.27 | 3.38 | 6.67 | 11.58 | 9.41 | -20.02 |
| 8 | 0.22 | 0.41 | 2.17 | 24.55 | 0.13 | 0.65 | 2.60 | 1.43 | -1.89 | 1.30 | -19.24 |
| 9 | 0.36 | 0.36 | 11.02 | 20.99 | 0.91 | -5.20 | 2.13 | 9.39 | -6.41 | 0.76 | -19.19 |
| 10 | 0.40 | 0.32 | 15.97 | 14.81 | 1.20 | 0.75 | 0.00 | 9.41 | | | |

**Group C**

| N | X1 | X2 | X3 | X4 | X5 | X6 | X7 | X8 | X9 | X10 | Y |
|---|---|---|---|---|---|---|---|---|---|---|---|
| | | | | | | | | | 1.52 | 5.85 | -17.92 |
| 11 | 0.27 | 0.24 | 5.82 | 17.87 | -0.26 | 0.68 | -1.89 | 0.65 | -4.49 | -3.62 | -16.79 |
| 12 | 0.25 | 0.32 | 5.22 | 17.36 | 0.00 | -5.13 | 2.58 | 1.30 | -4.52 | 4.52 | -14.88 |
| 13 | 0.30 | 0.38 | 12.45 | 14.31 | 1.17 | 3.13 | -0.83 | 1.60 | 9.10 | 8.74 | -12.89 |
| 14 | 0.27 | 0.07 | 5.81 | 2.61 | -1.67 | 8.34 | 0.00 | -8.62 | -12.76 | -11.51 | -10.78 |
| 15 | 0.40 | 0.17 | 25.56 | -0.56 | 1.18 | 4.96 | 0.89 | 5.72 | 2.42 | 6.10 | -9.46 |
| 16 | 0.39 | 0.15 | 24.26 | -3.83 | 0.25 | -3.15 | 6.20 | 2.28 | -0.81 | 2.50 | -6.51 |
| 17 | 0.13 | 0.15 | -1.79 | 4.62 | 1.10 | 9.93 | -5.70 | -3.04 | 15.67 | 10.84 | -5.27 |
| 18 | 0.41 | 0.17 | 24.55 | -1.66 | 0.67 | 2.44 | 3.34 | 6.10 | 4.14 | 4.31 | -3.97 |
| 19 | 0.05 | 0.10 | -1.84 | 2.16 | 0.50 | -2.26 | -2.59 | 1.94 | 13.05 | 7.80 | -3.85 |
| 20 | 0.40 | 0.10 | 29.47 | -5.20 | 0.44 | 6.20 | 8.58 | -1.51 | 5.27 | 6.01 | -3.34 |
| 21 | -0.01. | 0.03 | -8.78 | 2.72 | 1.29 | 3.97 | 0.00 | -0.57 | 6.51 | 4.30 | -1.53 |
| 22 | 0.46 | 0.10 | 28.73 | -2.48 | 0.52 | 3.34 | -1.67 | -1.83 | 10.72 | 6.63 | -0.81 |
| 23 | 0.38 | -0.04 | 14.31 | -9.22 | -2.88 | 1.77 | -6.04 | -13.54 | -11.55 | -14.89 | 0.00 |
| 24 | 0.03 | 0.04 | -7.23 | -1.18 | 1.06 | 3.28 | -2.26 | 0.74 | 0.80 | 1.61 | 0.79 |
| 25 | 0.34 | 0.05 | 21.82 | -9.80 | -1.39 | -3.39 | 10.20 | -1.56 | 8.58 | 0.00 | 0.88 |
| 26 | 0.03 | -0.15 | -6.58 | -5.23 | 0.09 | 0.83 | -0.87 | 3.83 | 4.31 | 2.98 | 1.65 |
| 27 | 0.02 | -0.04 | -10.99 | -5.58 | -0.17 | -3.23 | 3.51 | 1.52 | 0.00 | 1.01 | 1.67 |
| 28 | 0.36 | 0.09 | 20.99 | -6.98 | -0.09 | 0.00 | -3.39 | 6.01 | 4.43 | 3.42 | 1.70 |
| 29 | 0.25 | 0.27 | 11.16 | 5.53 | 0.08 | 0.00 | 2.59 | 3.63 | 1.68 | 1.60 | 2.48 |
| 30 | -0.03 | -0.04 | -5.23 | -6.45 | -0.32 | -0.83 | 2.44 | -0.08 | -4.77 | -3.07 | 2.50 |
| 31 | 0.40 | 0.40 | 26.39 | -12.21 | -1.14 | -2.63 | -7.57 | 2.08 | -2.63 | -1.51 | 2.71 |

**Group C**

| N | X1 | X2 | X3 | X4 | X5 | X6 | X7 | X8 | X9 | X10 | Y |
|---|---|---|---|---|---|---|---|---|---|---|---|
| 32 | -0.03 | 0.05 | -5.70 | -1.91 | 0.41 | -3.85 | 0.00 | 7.80 | -6.02 | 2.38 | 3.20 |
| 33 | 0.34 | -0.04 | 11.26 | -10.84 | -3.00 | -8.88 | 0.68 | -11.51 | -5.84 | -18.78 | 3.54 |
| 34 | -0.03 | -0.04 | -10.32 | -0.38 | 0.96 | -3.06 | -3.85 | 1.61 | 4.10 | 0.16 | 4.73 |
| 35 | 0.24 | -0.01 | 1.7.87 | -3.39 | 0.90 | -1.59 | 4.96 | 3.62 | -1.59 | -0.08 | 4.84 |
| 36 | 0.17 | 0.05 | -1.66 | -4.69 | 0.08 | 5.22 | 2.50 | -3.12 | 7.09 | 1.94 | 4.96 |
| 37 | 0.02 | -0.04 | -10.64 | -7.25 | 0.09 | 3.51 | 5.18 | -2.05 | -2.48 | 0.85 | 5.09 |
| 38 | 0.07 | -0.03 | 2.61 | -9.55 | -0.65 | -3.34 | 0.00 | -2.58 | -6.46 | -5.70 | 5.18 |
| 39 | 0.32 | -0.07 | 14.81 | -13.02 | -3.00 | -7.57 | -4.62 | -17.13 | -4.35 | -10.43 | 5.46 |
| 40 | 0.40 | -0.10 | 16.44 | -14.67 | -3.56 | -6.09 | -14.30 | -18.78 | -12.91 | -17.13 | 5.56 |
| 41 | 0.15 | -0.02 | -2.63 | -8.69 | 0.08 | -5.70 | 0.00 | 1.01 | -3.34 | -2.60 | 6.04 |
| 42 | -0.03 | 0.01 | -9.55 | -4.28 | 0.49 | -0.81 | 9.93 | 2.38 | -6.16 | -0.57 | 6.56 |
| 43 | 0.05 | -0.18 | -9.80 | -9.77 | -0.09 | -4.96 | 0.88 | 2.22 | -5.74 | -2.05 | 6.96 |
| 44 | 0.27 | 0.27 | 14.32 | 5.81 | 0.25 | -1.63 | 0.87 | 0.34 | 0.83 | 1.68 | 7.44 |
| 45 | 0.10 | -0.05 | -5.20 | -8.78 | 0.09 | -1.70 | -1.64 | -2.48 | 0.87 | -0.94 | 8.63 |
| 46 | 0.15 | -0.03 | -3.83 | -9.37 | -0.08 | 0.00 | 0.00 | -5.05 | -0.86 | -3.04 | 10.44 |
| 47 | 0.11 | -0.05 | -5.09 | -11.00 | -0.08 | -2.59 | 3.34 | 3.36 | -5.84 | -5.05 | 10.63 |
| 48 | -0.04 | -0.01 | -10.84 | -8.78 | -0.81 | -1.68 | -1.59 | -5.70 | -2.50 | -3.94 | 11.98 |
| 49 | 0.35 | 0.27 | 19.87 | 4.65 | -0.42 | -0.83 | -0.84 | -0.50 | 2.59 | 0.34 | 17.66 |
| 50 | 0.28 | 0.25 | 16.77 | 5.22 | -0.25 | 2.59 | 1.70 | -4.57 | 0.85 | -0.50 | 24.39 |

## Appendix 15.2. Representation of rules

### Example (a)

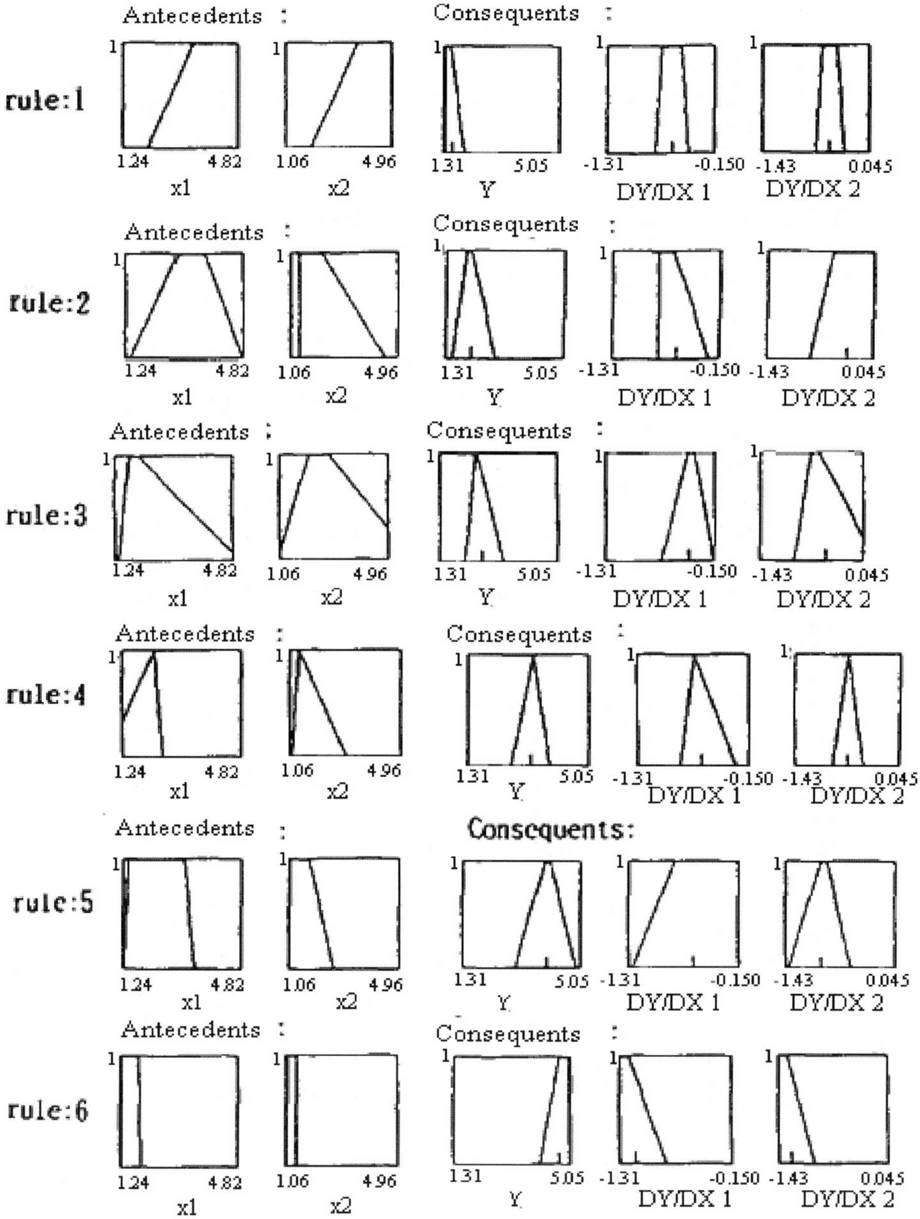

**Example (b)**

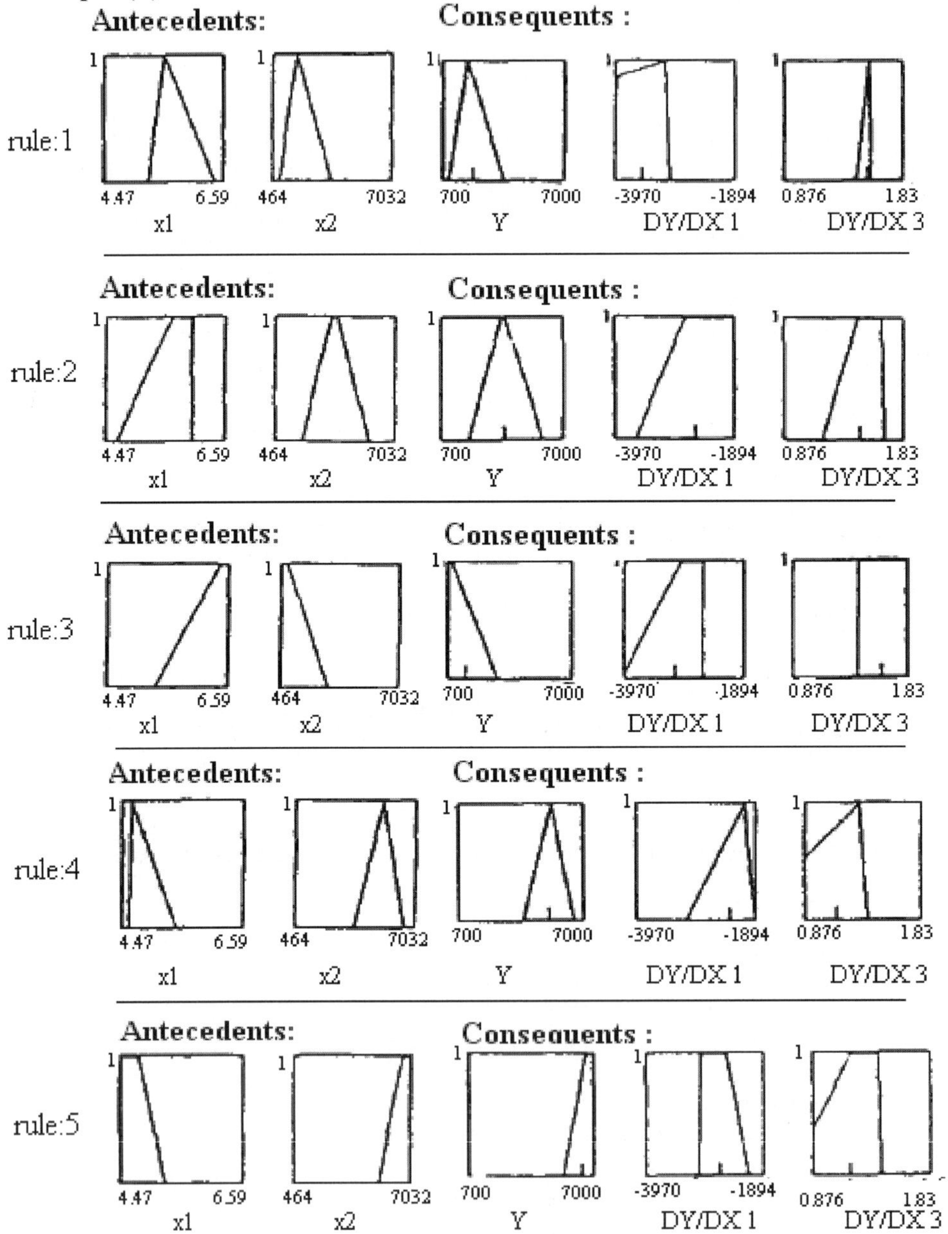
Antecedents:
Consequents :
rule:1
rule:2
rule:3
rule:4
rule:5
1
4.47
6.59
x1
464
7032
x2
700
7000
Y
-3970
-1894
DY/DX 1
0.876
1.83
DY/DX 3

Example (c)

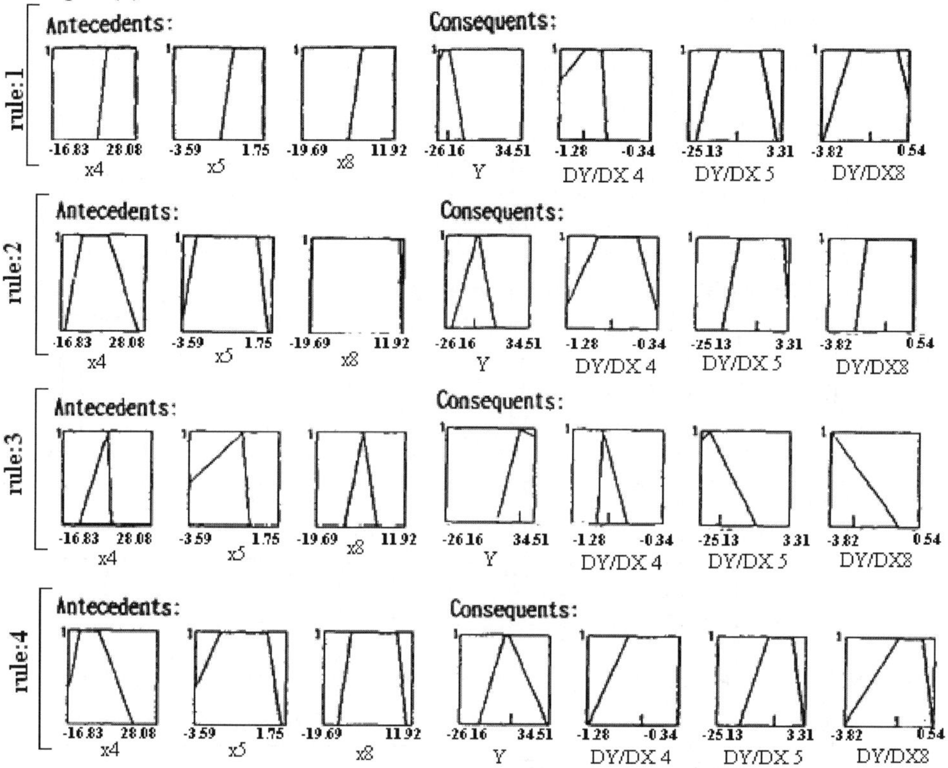


## Appendix 15.3. Actual output data vs the results of reasoning methods

Example (a)

| Output data | Result of reasoning methods | | | |
|---|---|---|---|---|
| Sugeno's (P&P-G) | | Sugeno's (P) | (P-G) | Sugeno's |
| 1 | 1.3 | 1.56 | 1.56 | 0.267 |
| 2 | 1.32 | 1.61 | 1.61 | 0.0729 |
| 3 | 1.33 | 1.57 | 1.57 | 0.419 |
| 4 | 1.4 | 1.72 | 1.72 | 0.645 |
| 5 | 1.44 | 1.77 | 1.77 | 0.836 |

| | | | | |
|---|---|---|---|---|
| 6 | 1.51 | 1.91 | 1.91 | 1.33 |
| 7 | 1.58 | 1.38 | 2.07 | 1.38 |
| 8 | 1.29 | 2.37 | 2.37 | 1.85 |
| 9 | 1.92 | 2.61 | 2.61 | 1.71 |
| 10 | 1.97 | 1.64 | 2.63 | 1.64 |
| 11 | 1.99 | 2.35 | 2.35 | 2.42 |
| 12 | 2.2 | 2.57 | 2.57 | 2.47 |
| 13 | 2.27 | 2.27 | 2.27 | 2.23 |
| 14 | 2.4 | 2.96 | 2.96 | 2.93 |
| 15 | 2.46 | 2.24 | 2.24 | 2.54 |
| 16 | 2.51 | 2.06 | 2.27 | 2.09 |
| 17 | 2.66 | 3.13 | 3.13 | 3.24 |
| 18 | 2.7 | 2.13 | 3.13 | 3.28 |
| 19 | 2.8 | 2.18 | 2.24 | 2.17 |
| 20 | 3.35 | 1.95 | - | 1.95 |
| 21 | 3.42 | 3.02 | 3.02 | 3.17 |
| 22 | 3.7 | 3.64 | 3.64 | 3.58 |
| 23 | 3.94 | 3.82 | 3.82 | 3.74 |
| 24 | 4.63 | 3.25 | 3.33 | 3.25 |
| 25 | 4.99 | 3.86 | 3.33 | 3.86 |

Example (b)

| | Output Data | Result of reasoning methods<br>Sugeno's (P&P-G) | Sugeno's (P) | Sugeno's (P-G) |
|---|---|---|---|---|
| 1 | 500 | 1553 | 1553 | -1710 |
| 2 | 700 | 1533 | 1533 | -1152 |
| 3 | 800 | 1553 | 1553 | -1029 |
| 4 | 800 | 1553 | 1553 | -1277 |
| 5 | 1000 | 1553 | 1553 | -92 |
| 6 | 1600 | 1982 | 1982 | 1445 |
| 7 | 2300 | 2460 | 2460 | 2806 |
| 8 | 2800 | 2906 | 2906 | 3008 |
| 9 | 3800 | 3935 | 3935 | 2936 |
| 10 | 3800 | 3898 | 3898 | 3220 |
| 11 | 3900 | 2962 | - | 2962 |

| | Output Data | Result of reasoning methods Sugeno's (P&P-G) | Sugeno's (P) | Sugeno's (P-G) |
|---|---|---|---|---|
| 12 | 3900 | 3935 | 3935 | 3323 |
| 13 | 4000 | 3935 | 3935 | 4044 |
| 14 | 4700 | 5471 | 4564 | 5471 |
| 15 | 5200 | 5305 | 5305 | 6805 |
| 16 | 5600 | 7060 | 6460 | 7060 |
| 17 | 6000 | 5464 | 5464 | 6909 |
| 18 | 6400 | 6460 | 6460 | 7778 |
| 19 | 6400 | 6345 | 6345 | 7890 |
| 20 | 6400 | 6377 | 6377 | 7563 |
| 21 | 6500 | 6408 | 6408 | 7550 |
| 22 | 6600 | 6460 | 6460 | 7862 |
| 23 | 6700 | 6460 | 6460 | 8075 |
| 24 | 6800 | 6460 | 6460 | 8137 |
| 25 | 6800 | 6460 | 6460 | 8383 |
| 26 | 6800 | 6460 | 6460 | 8305 |
| 27 | 6800 | 6460 | 6460 | 8672 |
| 28 | 6800 | 6460 | 6460 | 8234 |
| 29 | 7000 | 6460 | 6460 | 8602 |
| 30 | 7000 | 6460 | 6460 | 8747 |
| 31 | 7000 | 6460 | 6460 | 8549 |
| 32 | 7000 | 6460 | 6460 | 8653 |
| 33 | 7000 | 6460 | 6460 | 8594 |
| 34 | 7000 | 6460 | 6460 | 8741 |
| 35 | 7000 | 6460 | 6460 | 8777 |

Example (c)

| | Output data | Result of reasoning methods | | | |
|---|---|---|---|---|---|
| | | Sugeno's (P&P-G) | Sugeno's (P) | Sugeno's (P-G) | Türkşen's (IVCRI) |
| 1 | -27.6 | -18.9 | -18.9 | -34.4 | -11.4 |
| 2 | -26 | -18.9 | -18.9 | -32.4 | -14.3 |
| 3 | -23.9 | -18.9 | -18.9 | -21.9 | -1.54 |
| 4 | -22.6 | -18.9 | -18.9 | -31.1 | -13.7 |
| 5 | -21.5 | -18.9 | -18.9 | -23.6 | -3 |

| | Output data | Result of reasoning methods | | | |
|---|---|---|---|---|---|
| | | Sugeno's (P&P-G) | Sugeno's (P) (P-G) | Sugeno's (IVCRI) | Türkşen's |
| 6 | -20.7 | -18.2 | -18.2 | -16.4 | -1.4 |
| 7 | -20 | -18.9 | -18.9 | -23.2 | -18.9 |
| 8 | -19.2 | -18.5 | -18.5 | -15.6 | -1.4 |
| 9 | -19.2 | -16.4 | -16.4 | -22.5 | -5.35 |
| 10 | -17.9 | -15.3 | -15.3 | -20.3 | -4.58 |
| 11 | -16.8 | -13.4 | -13.4 | -8.67 | 1.75 |
| 12 | -14.9 | -14.6 | -14.6 | -9.66 | -1.17 |
| 13 | -12.9 | -14.3 | -14.3 | -14.8 | -3.2 |
| 14 | -10.8 | 2.68 | 2.68 | 22.7 | 5.71 |
| 15 | -9.46 | 1.05 | 1.05 | -6.88 | -3.41 |
| 16 | -6.51 | 1.94 | 1.94 | 2.53 | -1.54 |
| 17 | -5.27 | -0.0488 | -0.0488 | -3.33 | 3.56 |
| 18 | -3.97 | 1.94 | 1.94 | -4.26 | -1.54 |
| 19 | -3.85 | 1.25 | 1.25 | -2 | -1.54 |
| 20 | -3.34 | 1.94 | 1.94 | 4.63 | 2.24 |
| 21 | -1.53 | -5.92 | 0.164 | -5.92 | -1.06 |
| 22 | -0.81 | 1.94 | 1.94 | 3.58 | 2.6 |
| 23 | 0 | -1.25 | -1.25 | 34.9 | 3.72 |
| 24 | 0.79 | 1.26 | 1.26 | -3.03 | -1.93 |
| 25 | 0.88 | 3.79 | 3.79 | 17.5 | 8.4 |
| 26 | 1.65 | 1.94 | 1.94 | 2.79 | -1.54 |
| 27 | 1.67 | 1.99 | 1.99 | 4.84 | 0.779 |
| 28 | 1.7 | 2.35 | 2.35 | 1.76 | -0.558 |
| 29 | 2.48 | 0.363 | 0.363 | -3.81 | -1.54 |
| 30 | 2.5 | 2.47 | 2.47 | 7.51 | 3 |
| 31. | 2.71 | 5.24 | 5.24 | 14.8 | -4.74 |
| 32 | 3.2 | 0.909 | 0.909 | -4.18 | -1.87 |
| 33 | 3.54 | 2.97 | 2.97 | 34.8 | 5.77 |
| 34 | 4.73 | 1.19 | 1.19 | -2.8 | -1.91 |
| 35 | 4.84 | 1.35 | 1.35 | -2.54 | -1.78 |
| 36 | 4.96 | 1.94 | 1.94 | 8.03 | 4.02 |
| 37 | 5.09 | 2.47 | 2.47 | 7.52 | 2.87 |
| 38 | 5.18 | 3.64 | 3.64 | 13.9 | 9.96 |
| 39 | 5.46 | 41 | -4.87 | 41 | 2.75 |
| 40 | 5.56 | 49.1 | -4.87 | 49.1 | 9.82 |

| | Output data | Result of reasoning methods | | | |
|---|---|---|---|---|---|
| | | Sugeno's (P&P-G) | Sugeno's (P) | Sugeno's (P-G) | Türkşen's (IVCRI) |
| 41 | 6.04 | 3.16 | 3.16 | 5.84 | -1.22 |
| 42 | 6.56 | 1.94 | 1.94 | 1.34 | -1.54 |
| 43 | 6.96 | 3.78 | 3.78 | 6.46 | -0.558 |
| 44 | 7.44 | 0.279 | 0.279 | -2.62 | -0.567 |
| 45 | 8.63 | 3.21 | 3.21 | 8.9 | 3.37 |
| 46 | 10.4 | 3.54 | 3.54 | 12.4 | 5.83 |
| 47 | 10.6 | 4.6 | 4.6 | 6.68 | -0.668 |
| 48 | 12 | 3.21 | 3.21 | 17.2 | 9.45 |
| 49 | 17.7 | 5.46 | 5.46 | 3.46 | 4.78 |
| 50 | 24.4 | 5.99 | 5.99 | 7 | 8.52 |

## References


1. Dubois, D., H. Prade (1984), "Fuzzy Logics and the Generalized Modus Ponens Revisited", *International Journal of Cybernetics and Systems*, 283-331.
2. Nakanish, H., I.B. Türkşen and M. Sugeno (1993), "A Review and Comparison of Six Reasoning Methods, Fuzzy Sets and Systems 57, 257-294.
3. Sugeno, M., T. Yasukawa (1993), "A Fuzzy Logic Based Approach to Qualitative Modelling", *IEEE Transactions on Fuzzy Systems*, 1(1), 1-24.
4. Trillas, E., L. Valverde (1985), "On Mode and Implication in Approximate Reasoning", in W. Bandler, M.M. Gupta, A. Kandel and J.B. Kista (eds.), *Approximate Reasoning in Expert Systems*, Elsevier Sci B.V. (North Holland), 157-167.
5. Türkşen, I.B. (1989), "Four Methods of Approximate Reasoning with Interval-Valued Fuzzy Sets", *Internet. J. Approx. Reasoning,* 3, 121-142.

# Chapter 16

# A THEORETICAL APPLICATION OF INTERVAL-VALUED TYPE 2 REPRESENTATION*

In this chapter, we attempt to model a strict preference relation from a given subjective weak preference relation of the decision maker. Fuzzy relations and interval valued Type 2 fuzzy sets based on Fuzzy Disjunctive and Conjunctive Canonical Forms, FDCF and FCCF representations [Türkşen, 1986-2002] are used as basic tools in the modeling.

We start with a fuzzy weak preference relation which is assumed to be solicited from the decision maker(s). We assume that this fuzzy relation is an adequate representation of the decision maker's subjective weak preference.

There are at least three ways of getting this information from the decision maker(s) [Roberts, 1979]. One way is to ask the decision maker to compare pairs of alternatives over an over again in time under essentially the same experimental conditions. Here, there is the dubious assumption that the choice experiment can be performed under the same conditions without the interference of the previous choices [Türkşen, 1991]. Another way is to consider the preferences of a group of decision makers on the same set of alternatives and aggregate the opinions of the decision makers to obtain a fuzzy relation. Finally, if the alternatives have multi-attributes, preferences along the attributes can be aggregated to get a degree of preference for the alternative itself which gives rise to a fuzzy relation [Fodor and Rulens, 1994]. Here. we do not make any assumptions about how this information is solicited from the decision maker but assume that such a relation is available.

Throughout, it is assumed that the concept of weak preference is represented by a fuzzy relation. The concept of strict preference is defined in terms of the weak preference and this construction introduces the second order imprecision. We propose that the strict preference so constructed should be represented by an interval valued relation which represents the second order imprecision which is the uncertainty of imprecise information. Then, a crisp ordering relation is

---

* Extracted from I.B.Türkşen and T.Bilgic "Interval-valued Strict Preference with Zadeh triples", Fuzzy Sets and Systems 78 (1996) 183-195, and modified as needed.

defined in terms of the interval valued representation. The necessary and sufficient conditions on the initial representation are explored so that the crisp ordering relation is transitive and/or connected.It is also shown that if the initial weak preference relation satisfies a general linearity property, then the second order imprecision of strict preference disappears.

The concept of unfuzzy non-dominated alternatives introduced by Orlovski [1978] is translated to our setting and the conditions of its existence are analyzed.

Similar approaches to preference modeling with fuzzy relations appears in the literature [Barret. Et al., 1990; Basu, 1984; Fodor, 1991; Fodor and Rubans, 1994; Orlavski, 1978; Ovchinikov and Roubens, 1991, 1992; Perny, et.al., 1992; Roubens, et.al., 1985]. In all of these studies the higher constructions are point valued, i.e., Type 1, as opposed to our interval valued constructs which are Type 2.

It is argued (in [Lono, 1992] among others) that single values in the unit interval are a poor representation of degree of beliefs, because these fail to measure the degree of certainty or uncertainty that an individual or group has in their assessment. With a similar perspective, we attempt to represent strict preference as an interval that represents a second order *semantic uncertainty* by applying FDCF and FCCF.

The rationale of this approach can be explained as follows; one simply starts with an imprecise construct namely, the fuzzy weak preference of the decision maker. Further concepts defined from this construct induce a Type 2 imprecision, i.e., a second order semantic uncertainty (See [Goguen, 1967; Mizumoto, et.al., 1981] in this connection). This phenomena is well modeled by interval valued fuzzy sets [Türkşen, 1986-2002]. Applying interval valued fuzzy sets, Türkşen [1992] indicated and later proved that human use of "AND" and "OR" do not directly correspond to any triangular norm or conorm [Schiweizer and Sklar, 1960, 1961, 1963, 1983] that is used in the fuzzy set literature. This was also revealed experimentally in [Zimmermann and Zysno, 1980].

In Section 16.1 we give a brief account of fuzzy relations and interval valued fuzzy sets. A parametric definition of transitivity for fuzzy relations is given and its relationship to other definitions of transitivity is established.

In Section 16.2 the main results are discussed and a numerical example is shown. The properties of underlying nested family of crisp relations, associated with a fuzzy relation are analyzed in Section 16.2.3 and the conditions under which the higher order imprecision disappears are analyzed in Section 16.2.4. In

Section 16.3.5 the existence of unfuzzy non-dominated alternatives is discussed. Section 16.3 concludes this investigation and gives further research remarks.

## 16.1. Background

This section gives a brief introduction to fuzzy relations and interval valued fuzzy sets with references to related literature as a background.

### *16.1.1. Relations and Fuzzy Relations*

A (crisp) binary relation R on a set A is a subset of the Cartesian product A×A, i.e., R is a set of ordered pairs (a, b) for all a, b∈A. If the ordered pair (a, b) belongs to R, i.e., if a and b are in relation R, then the notation (a, b)∈R or aRb is used.

Let A be a classical set, called the universe of discourse. A fuzzy set X on A is defined as a function $\mu_X$: A → [0,1]. A fuzzy set is completely characterized by its membership function $\mu_X$.

Zadeh [1965] proposed to use the operators min(x, y) for intersection, max(z, y) for union and pseudo complement, n(x)=1-x, for negation of fuzzy sets for all x, y∈[0,11 where x=$\mu_X$ (a), y=$\mu_X$ (b) for some a,b∈A.

The functions min and max belong to the class of t-norms and t-conorms, respectively. A t-norm is a function, Δ: A×A→[0,1] that is associative, symmetric, monotonic and has boundaries on the unit interval. A t-conorm, ∇ can be constructed from a t-norm, Δ, by the following version of De Morgan identity: for all x, y∈[0, 1] ∇(x,y) = n(Δ(n(x),n(y)).

Such a triple, (Δ,∇,n), is called a De Morgan triple. Recall that apart from aforementioned properties of t-norms and t-conorms the min and max functions are distributive implying that they are in turn idempotent and absorptive [Weber, 1983] but do not fulfill the law of excluded middle and its dual, the law of contradiction in the classical sense but they fulfill these conditions to a degree [Türkşen, 1994; Türkşen et.al., 1999].

In this chapter, we do not consider triangular norms or conorms other than min and max and name the triple (Δ,∇,n) as Zadeh triple when Δ=min, ∇= max and n(x)=1-x for all x∈[0,1].

A fuzzy binary relation is a relation where every ordered pair has a value in the unit interval. Formally, if A is a (crisp) set, a fuzzy binary relation R on A is defined as a fuzzy subset of the product space A×A. This is a function of two variables on A×A with values in [0,1], i.e., R: A×A →[0, 1]. The value, R(a, b) of a fuzzy relation R denotes the degree to which a and b are in relation R, i.e., the function R(a, b) is the truth value of the proposition aRb.

Specifically, in the case of preferences, R(a, b) denotes the "truth degree with which an alternative a is not worse than an alternative b".

As a rule of notation, when we write aRb we mean a crisp binary relation, and when we write R(a, b) we are referring to a fuzzy relation and when we only write "relation" we mean "binary relation" unless otherwise specified.

Fuzzy relations are extensively studied by mathematical psychologists and fuzzy set researchers. Mathematical psychologists use the term valued relations [Dignon, et.al., 1986].

If a particular enumeration of the alternatives, A = {a1, a2,..., an), is given, then R is completely defined by a matrix [rij] where

$$rij = R(ai,aj) \qquad \text{for } 1 \leq i,j \leq n.$$

The basic properties of a crisp relation, R, are given as follows:

**Definition 16.1.1.** A crisp relation R on A is called (for all a, b$\in$A)

| | | |
|---|---|---|
| (i) | reflexive | if aRa |
| (ii) | irreflexive | if not(aRa) |
| (iii) | symmetric | if aRb $\Rightarrow$ bRa |
| (iv) | connected | aRb or bRa |

The corresponding properties of a fuzzy relation that are relevant for this study is given in the following definition (for more information see [Digman, et.al. 1986; Ovchinnikov, 1991]

**Definition 16.1.2.** The fuzzy relation R on A is called (for all a, b$\in$A)

| | | |
|---|---|---|
| (i) | reflexive | if R(a,a)=1 |
| (ii) | irreflexive | if R(a,a)=0 |
| (iii) | symmetric | if R(a,b)=R(b,a) |
| (iv) | connected | if max{R(a,b),R(b,a)} = 1 |

Notice that the properties of fuzzy relations are such that they yield the crisp properties as a limiting case.

Another important property of relations is the transitivity. For a crisp relation R, the transitivity is given by the linguistic expression:

if aRb and bRc then aRc (16.1)
for all a, b, c $\in$ A.

In the case of a fuzzy relation, R, there are many ways to generalize (1). The aim is to find conditions using the fuzzy relation R such that (16.1) is the limiting case when R attains only the values 0 and 1. One straightforward way is the following:

**Definition 16.1.3.** The fuzzy relation R on A is called min-transitive if for all a, b, c$\in$A, $\min\{R(a, b)\ R(b, c)\} \leq R(a, c)$.

This is one fuzzy representation of (16.1) and is the most common definition of transitivity in fuzzy set literature.

Another form of transitivity is given by (see [Kolodziejcyk, 1986; Billot, 1992] for example)

**Definition 16.1.4.** The fuzzy relation R on A is called weak-transitive if for all a, b, c$\in$A, the linguistic expression is written as:

$R(a, b) > R(b, a)$ and $R(b, c) > R(c, b)$ then $R(a, c) > R(c, a)$

We propose a parametric definition of transitivity as follows:

**Definition 16.1.5.** The fuzzy relation R on A is called parametric-transitive if there exists a $\vartheta \in (0,1)$ such that the linguistic expression reads as:

if $R(a, b) > \theta > R(b, a)$ and $R(b, c) > \theta > R(c, b)$ then $R(a, c) > \theta > R(c, a)$
for all a, b, c $\in$A.

Furthermore, if $\theta=1/2$, then R is called ½-transitive.

It should be noted that this is another fuzzy representation of (16.1). The following Theorem establishes the link between min-transitivity and parametric-transitivity. Specifically it shows that the class of parametric-transitive relations are larger than the class of min-transitive relations.

**Theorem 16.1.1.** If R is a min-transitive fuzzy relation, then it is a parametric transitive relation.

**Proof**: Assume R is a min-transitive fuzzy relation. If there does not exist a $\theta \in (0,1)$ such that $R(a, b) > \theta > R(b, a)$ and $R(b, c) > \theta > R(b, c)$ for some a, b$\in$A there is nothing to show. Assume that there exists a $\theta \in (0,1)$ such that R(a,

b) > θ > R(b, a) and R(b, c) > θ > R(c, b). We need to show that R(a, c) > θ > R(c, a).

$$\begin{aligned} R(a, c) &\geq \min\{R(a, b), R(b, c)\}, \text{ by min-transitivity} \\ &> \theta, \text{ since both } R(a, b) \text{ and } R(b, c) > \theta \end{aligned}$$

On the other hand,

$$\begin{aligned} \theta &> R(b,a), \text{ by hypothesis} \\ &\geq \min\{R(b, c), R(c, a)), \text{ by min-transitivity} \end{aligned}$$

Since R(b, c) > θ, in order for the above inequality to hold, it is necessary that R(c, a) < θ.

Clearly, the class of parametric transitive relations is a subset of the class of weak transitive relations. Hence, by Theorem 16.1.1. class of min-transitive relations is also a subset of the class of weak transitive relations. (See [Kolodziejczyk, 1986] in this connection.)

*16.1.2. Interval Valued Type 2 Fuzzy Sets*

The two (of the total sixteen) Fuzzy Type 2 Disjunctive and Conjunctive Canonical Forms that are of interest in this chapter are given in Table 16.1.

**Table 16.1.** Fuzzy Disjunctive and Conjunctive Canonical Forms of Linguistic Conjunction, "AND" and Non-Implication

| Concept | Combination | FDDF | FCCF |
|---|---|---|---|
| Conjunction | A AND B | (A∩B) | (A∪B)∩ (A∪ c(B))∩( c(A)∪B) |
| Non-Implication | A AND NOT B | (A∩c(B)) | (A∪B)∩ (A∪ c(B))∩( c(A)∪c(B)) |

Note that DNF(.)=CNF(.) in classical logic. Türkşen [1986-2002] shows that FDCF(.)⊆ FCCF(.) for certain families of De Morgan triples (and in particular for Zadeh triples) (Recall, that Bilgic (1995) shows that the containment hold for strict and nilpotent Archimedean t-norms and t-conorms and Kılıç (2002) show that it also hold for certain ordinal sums. It was proposed that the interval-valued fuzzy set (IVFS) to be defined as follows [Türkşen, 1986-2002]

$$IVFS(.) \stackrel{\Delta}{=} [FDCF(.), FCCF(.)]$$

In this chapter the same definition is applied to fuzzy relations in order to generalize the concept of a fuzzy relation.

**Definition 16.1.6.** If a new fuzzy relation is constructed from other fuzzy relations by means of linguistic connectives (such as “AND”, “OR”, “NOT” etc.), to model linguistic expression affectively it is defined as an interval valued fuzzy relation (IVFR) as

$$\text{IVFR}(.) \stackrel{\Delta}{=} [\text{FDCF}(.), \text{FCCF}(.)]$$

The basic motivation of Definition 16.1.6 that whatever higher concepts are constructed from a concept which is represented by a fuzzy relation via linguistic connectives, the resulting higher construct must exhibit the second order imprecision introduced by this construction. Hence, an interval valued fuzzy relation is a fuzzy relation which contains a second order imprecision and represents this imprecision as an interval of uncertainty on imprecise information.

Formally, since a crisp relation is a crisp subset of A×A and a fuzzy relation is a fuzzy subset of A×A, an interval valued relation is an interval valued fuzzy subset of A×A whenever it is constructed by linguistic connectives.

## 16.2. Strict Preference

A crisp binary relation R with respect to each pair of alternatives (a, b) of a given set of alternatives, A, is a weak preference relation and aRb models the linguistic expression.

“a is not worse than b“

Hence, the concept “is not worse than” is modeled by the crisp relation R. Usually, a strict preference P is constructed from R assuming that aPb depends only on aRb and bRa for all a, b∈A. This condition is known as Independence of Irrelevant Alternatives (IA). Hence, the strict preference is defined as the asymmetric part of R and aPb models and linguistically expressed as:

aRb and not bRa. (16.2)

in a PNL, Precisiated Natural Language.

When a fuzzy relation is used to represent the imprecise of preference, R(a,b) can be interpreted as the degree of credibility that “a is not worse than b”. There are many ways to construct a fuzzy strict preference relation P from R [Ovchinikov, 1991]. For example, the classical notions of strict preference and indifference can he extended as:

$P'(a, b) = \Delta(R(a,b), n(R(b, a)))$

$I'(a,b) = \Delta(R(a,b), n(b,a))$

where $\Delta$ is a t-norm, n is a strong negation, R is the fuzzy weak preference relation and $I'(a, b)$ captures the indifference between the elements a and b. Furthermore, the crisp construction has the following equivalence: $R' = P' \cup I'$. This yields the following functional equation (with $x = R(a, b)$, $y = R(b,a)$):

$$\nabla(\Delta (x, y),\Delta(x,n(y))) = x, \qquad \forall x, y \in [0,1] \tag{16.3}$$

where $\nabla$ is an n-dual t-conorm of $\Delta$ [Alsina, et.al., 1983].

However, it is proved that there is no De Morgan triple such that (16.3) is satisfied [Alsina, 1985]. In view of this negative result, there has been different ways of defining preference structures [Fodor and Roubens, 1994]. We propose a different approach [Türkşen, Bilgic, 1993].

First, it is assumed that a strict preference relation, P, for any two alternatives a and b is constructed such that the value of P(a, b) depends only on the values R(a, b) and R(b, a) and connectives of (16.2) are interpreted as linguistic connectives in the sense of interval valued fuzzy relations (cf. Section 16.1.2). Thus we model the concept of strict preferences linguistically as:

aRb AND NOT bRa (16.4)

That is the linguistic connectives such as "AND" do not isomorphically correspond in a one-to-one manner to a t-norm, $\Delta$.
It should be noted that, this is the "non-implication" displayed in Table 16.1. The interval valued strict preference P(a, b) can now be written as an interval valued fuzzy relation (see Definition 16.1.6).

$$P(a, b) = [\mu_{FDCF(aRb)\text{ AND NOT }bRa)}, \mu_{FCCF(aRb\text{ AND NOT }bRa)}]$$

To simplify the notation we use

$$\begin{aligned} P(a, b) &= [P_D(a, b), P_C(a, b)] \\ &= [\min\{R(a, b), n(R(b, a))\}, \min\{\max\{R(a,b, (b\ a)\}, \\ &\qquad \max\{R(a, b), n(R(b, a))\}, \max\{n(R(a, b)),(R(b,a))\}\}] \end{aligned} \tag{16.5}$$

for all $a, b \in A$ where
$P_D(a,b) = \min\{R(a,b), n(R(b,a))\}$
$P_C(a,b) = \min\{R(a,b), R(b, a)\}, \max\{R(a,b), n(R(b,a))\}, \max\{n(R(a,b)),n(R(b,a))\}\}$.

$P_D(a, b)$ and $P_C(a, b)$ are, respectively, the membership functions of the FDCF and FCCF parts of non-implication.

To simplify the notation further, P(a, b) can be rewritten using the notation x=R(a,b) and y = R(b, a) as;

$$P(a, b) = [P_D(a,b), P_C(a,b)]$$
$$= [\min\{x,n(y)\}, \min\{\max\{x,y\}, \max\{x,n(y)\}, \max\{n(x),n(y)\}] \quad (16.6)$$

for all $x,y\in[0,1]$.

All the results in the current relevant literature discuss only the Type 1 fuzzy strict preference. It should be noted that the Type 1 fuzzy strict preference is in the form of the lower bound, $P_D$, of the interval valued Type 2 strict preference $[P_D(a, b), P_C(a, b)]$ which we defined as an IVFR. However, there are many studies, which consider the construction of a strict preference in the form given by the lower bound but consider different t-norms and t-conorms in construction. The investigations of the proposed t-norms and t-conorms that are other than min and max can be done in a similar manner.

Figure 16.1 shows the FDCF and FCCF parts of the interval valued Type 2 strict preference, P(a, b), as defined by 16.5. In Figure 16.1, the y-axis shows the values of $P_D(a,b)$ and $P_C(a,b)$ in parts (a) and (b) respectively. It should be noted that for all $a, b\in A$, $P_D(a,b)\leq P_C(a,b)$.
From Figure 16.1 it should be observed that the interval valued strict preference, P(a,b), attains an interval value only when R(a, b) < R(b, a). This is expected since when the degree of preference of a over b is less than that of b over a, the imprecision inherent, in the interval valued Type 2 strict preference, P(a, b) should increase.
Next, we define a crisp ordering relation as follows:

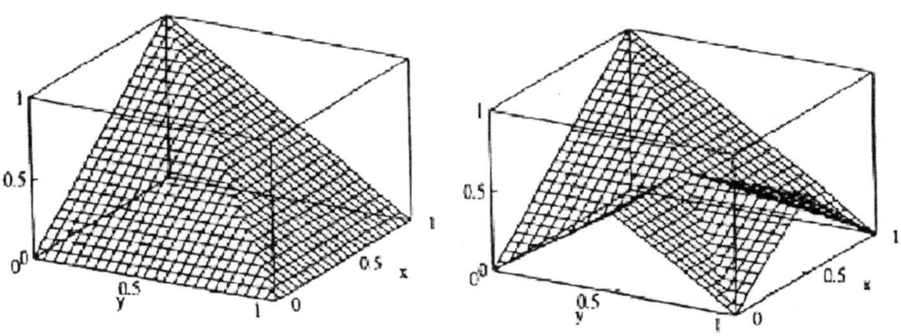


**Figure 16.1.** (a) $P_D(a, b)$ (b) $P_C(a, b)$

**Definition 16.2.1.** A (crisp) binary strict preference (order relation) "  " is defined as

$$a \succ b \text{ if and only if } P_D(a,b) > P_C(b,a) \qquad \text{for all } a,b \in A \qquad (16.7)$$

The order relation "$\succ$" means a is strictly preferred to b if and only if the interval P(a,b) is strictly to the right of the interval P(b,a). Furthermore, notice that "$\succ$" defined by Definition 16.1 is a crisp relation in the classical sense. The crisp order relation (or crisp strict preference) "$\succ$', is central to the development in this chapter. The aim is to find conditions on the fuzzy relation R so that "$\succ$" defines a crisp ordering of the alternatives.

It is easy to see that "$\succ$" is irreflexive (i.e., for all a, b$\in$A not(a $\succ$ a)) by Definition 16.1. Intuitively this makes sense, since a should not be strictly preferred to itself. In order for "$\succ$" to define an ordering on the space of alternatives, A, it needs to be transitive. This is investigated next.

*16.2.1. Transitivity of the Order Relation "$\succ$"*

In this section the transitivity "$\succ$" is investigated under different conditions imposed on R.

Specifically, it is shown that weak-transitivity is a necessary condition for "$\succ$" to be transitive but it is not sufficient. Then it is shown that ½-transitivity is a necessary and sufficient condition for a transitivity of the order relation "$\succ$". By Theorem 16.1.1 it follows that min-transitivity is a sufficient condition.

**Lemma 16.2.1.** If a   b, then R(a, b) > R(b, a) for all a, b $\in$ A.

**Proof:** By Definition 16.1 a$\succ$ b if and only if $P_D(a,b) > P_C(b,a)$ for all a,b $\in$ A. However $P_C(b,a) \geq P_D(b, a)$ (cf. Section 16.1.2). Hence $P_D(a, b) > P_D(b, a)$ which, as a result of (16.5) translates to

$$\min\{R(a,b), n(R(b, a))\} > \min\{R(b,a), n(R(a,b))\}$$

which can only hold when R(a,b) > R(b,a).

Lemma 16.2.1 ties the crisp order relation "$\succ$" to fuzzy order relation R. From Lemma 16.2.1, it is clear that weak-transitivity of R is a necessary condition for the transitivity of "$\succ$".

To show the insufficiency, consider that the weak-transitivity of R holds and a   b and b $\succ$ c. In order a $\succ$ c, by Definition 16.2.1 and (16.5) we must have:
$\min\{R(a, c), n(R(c,a))\} > \min\{\max\{R(c\ a), n(R(a, c))\},$

$$\max\{n(R(c, a)), n(R(a, c))\}, \max\{R(c, a), R(a, c)\}\}$$
$$= \min\{\max\{R(c, a), n(R(a, c))\}, \min\{R(a, c), n(R(c, a))\} \tag{16.8}$$

where the equality is obtained from weak-transitivity, i.e., R(a, c) > R(c, a) which also implies n(R(c, a)) > n(R(a, c)). Notice that, by De Morgan identity: min{R(a, c), n(R(c, a))} = n(max{R(c, a), n(R(a, c))}.
Hence, to have a $\succ$ c we must have:

$$\min\{R(a, c), n(R(c, a))\} > \max\{R(c, a), n(R(a, c))\} \tag{16.9}$$

which does not hold in general when R(a, c) > R(c, a) (the consequent of weak transitivity).

Notice that (16.9), by itself, is a sufficient condition for "$\succ$" to be transitive. This gives rise to the following

**Lemma 16.2.2.** For any a, c $\in$ A, the equation

min{R(a, a), n(R(c, a))} > max{R(c, a), n(R(a, c))}
holds if and only if R(a,c) > ½ > R(c,a).

**Proof.** Necessity is immediate. To show sufficiency consider two cases: (i) min{R(a, c), n(R(c,a))} = R(a, c). This holds if and only if max{n(R(a, c)), R(c, a)) = n(R(a, c)). Then in order for the hypothesis to hold R(a, c) > n(R(a, c)) which in turn holds if and only if R(a,c) > 1/2. Since n(R(c, a)) $\geq$ R(a, c) > 1/2, then R(c, a) < 1/2.

(ii) min{R(a,c), n(R(c,a)) = n(R(c,a))}. This holds if and only if max{n(R(a,c)),R(c,a))} = R(c,a). Then in order for the hypothesis to hold n(R(c, a)) > R(c, a) which in turn holds if and only if n(R(c, a)) > 1/2. Hence, R(c, a) < 1/2 and R(a, c) $\geq$ n(R(c, a)) > 1/2.

With Lemma 16.2.2, an equivalence is constructed between the initial fuzzy relation, R, and the order relation "$\succ$" as follows:

**Lemma 16.2.3.** For any a,b$\in$A a $\succ$ b if and only if
R(a, b) > ½ > R(b, a).

**Proof.** (sufficiency) By Lemma 16.1 a $\succ$ b implies R(a, b) > R(b, a). By Definition 16.1.6, Equation (16.5) and Lemma 16.2.1, we get:

a$\succ$ b$\Leftrightarrow$min{R(a,b), n(R(c,b))} >

$\min\{\max\{R(b,a),n(R(a,b))\},\min\{R(a,b),n(R(b,a))\}\}$
which holds if and only if $\min\{R(a,b),n(R(b,a))\} > \max\{R(b, a), n(R(a,b))\}$ which in turn holds if and only if $R(a, b) > ½ > R(b, a)$ by Lemma 16.2.2.

(necessity) $R(a, b) > 1/2 > R(b, a)$ implies $R(a, b) > R(b, a)$, and the rest follows exactly as in the proof of sufficiency shown above.

**Theorem 16.2.1.** The crisp ordering relation, “$\succ$” as defined by (16.7), is transitive if and only if R is a fuzzy relation satisfying ½ transitivity.

**Proof.** For any a, b, c $\in$ A, the two conditions;

(i) if $R(a,b) > ½ > R(b,a)$ and $R(b,c) > ½ > R(c,b)$ then $R(a,c) > ½ > R(c,a)$

(ii) if $a \succ b$ and $b \succ c$ then $a \succ c$

are equivalent by Lemma 16.2.3.

Using Theorem 16.1.1 the following Corollary can be stated.

**Corollary 16.2.1.** If R is a min-transitive fuzzy relation, then “$\succ$” as defined in (16.7) is a transitive relation.

**Proof.** Follows from Theorem 16.2.1 using Theorem 16.1.1.

Since “$\succ$” is irreflexive and transitive (but not necessarily complete) it is a partial order. In general, “$\succ$” is not a linear order when R satisfies the “1/2-transitivity” condition.

The following result establishes a necessary and sufficient condition for “ ” to be a linear order.

**Corollary 16.2.2.** The crisp ordering relation, “$\succ$” is a linear order if and only if the following condition is satisfied:

$$R(a,b) > ½ > R(b,a) \text{ or } R(b, a) > ½ > R(a,b) \qquad (16.10)$$

for all a, b $\in$ A.

**Proof.** By Lemma 16.2.3 $a \succ b \Leftrightarrow R(a, b) > 1/2 > R(b, a)$. In order for “$\succ$” to be a linear order
$a \succ b$ or $b \succ a$

must hold for all a, b $\in$ A, from which the result follows.

Equation 16.10 is called the linearity condition.

### *16.2.2. Numerical Example*

To illustrate the concepts developed this far, consider the following numerical example. Let the ½-transitive fuzzy weak preference relation, R, be represented as:

$$R = \left|\begin{array}{ccccccc} 1.00 & 0.03 & 0.00 & 0.03 & 0.00 & 0.90 & 0.00 \\ 0.60 & 1.00 & 0.00 & 0.75 & 0.00 & 0.60 & 0.00 \\ 0.35 & 0.30 & 1.00 & 0.30 & 0.30 & 0.96 & 0.30 \\ 0.10 & 0.10 & 0.00 & 1.00 & 0.00 & 0.10 & 0.00 \\ 0.10 & 0.10 & 0.06 & 0.53 & 1.00 & 0.10 & 0.06 \\ 0.35 & 0.03 & 0.00 & 0.03 & 0.00 & 1.00 & 0.00 \\ 0.35 & 0.35 & 0.35 & 0.53 & 0.75 & 0.35 & 1.00 \end{array}\right.$$

Then the strict preference P is represented as follows:

$$P = \left[\begin{array}{ccccccc} [0.00, 0.00] & [0.03, 0.40] & [0.00, 0.35] & [0.03, 0.10] & [0.00, 0.10] & [0.65, 0.65] & [0.00, 0.35] \\ [0.60, 0.60] & [0.00, 0.00] & [0.00, 0.30] & [0.75, 0.75] & [0.00, 0.10] & [0.60, 0.60] & [0.00, 0.35] \\ [0.35, 0.35] & [0.30, 0.30] & [0.00, 0.00] & [0.30, 0.30] & [0.30, 0.30] & [0.96, 0.96] & [0.30, 0.35] \\ [0.10, 0.10] & [0.10, 0.25] & [0.00, 0.30] & [0.00, 0.00] & [0.00, 0.47] & [0.10, 0.10] & [0.00, 0.47] \\ [0.10, 0.10] & [0.10, 0.10] & [0.06, 0.30] & [0.53, 0.53] & [0.00, 0.00] & [0.10, 0.10] & [0.06, 0.25] \\ [0.10, 0.35] & [0.03, 0.40] & [0.00, 0.04] & [0.03, 0.10] & [0.00, 0.10] & [0.00, 0.00] & [0.00, 0.35] \\ [0.35, 0.35] & [0.35, 0.35] & [0.35, 0.35] & [0.53, 0.53] & [0.75, 0.75] & [0.35, 0.35] & [0.00, 0.00] \end{array}\right]$$

and the corresponding “ ” is found to be (by Definition 16.2.1)

$$
= \begin{bmatrix} 0 & 0 & 0 & 0 & 0 & 1 & 0 \\ 1 & 0 & 0 & 1 & 0 & 1 & 0 \\ 0 & 0 & 0 & 0 & 0 & 1 & 0 \\ 0 & 0 & 0 & 0 & 0 & 0 & 0 \\ 0 & 0 & 0 & 1 & 0 & 0 & 0 \\ 0 & 0 & 0 & 0 & 0 & 0 & 0 \\ 0 & 0 & 0 & 1 & 1 & 0 & 0 \end{bmatrix}
$$

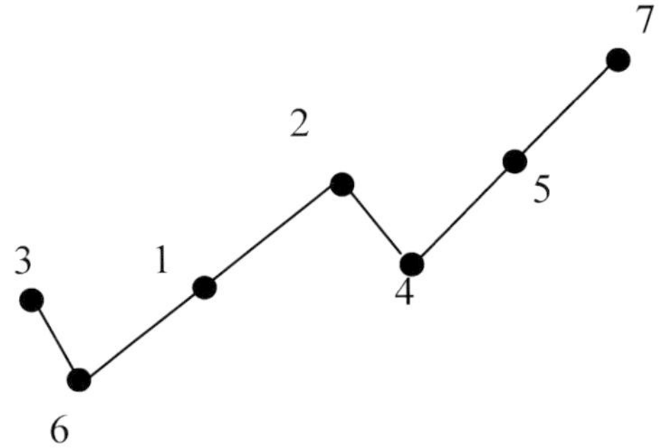


**Figure 16.2.** Hasse Diagram of " $\succ$ " 2-

The Hasse diagram corresponding to " $\succ$ " is given in Figure 16.2. In such a diagram, a is strictly preferred to b, i.e., a $\succ$ b if a is above b and there is a downward path from a to b.

*16.2.3. Threshold Relations*

A common way to study fuzzy relations is to consider 'threshold relations" associated with the fuzzy relation. Given $\alpha \in [0,1]$, the (crisp) relation $R_A$ is defined as

$$aR_\alpha b \Leftrightarrow R(a,b) \geq \alpha. \tag{16.11}$$

These $\alpha$-cut $R_\alpha$'s form a nested family of crisp relations (i.e., if $\alpha > \beta$ then $R_\alpha \subseteq R_\beta$).
The entire family of relations, $R_\alpha$, can be interpreted as a preference relation determined at a given level $\alpha$ of decisiveness or discriminability [Suppes, et.al., 1989, Chapter 17].

In Doignon et. al. (1986), it is shown that studying $R_\alpha$'s for $\alpha \in [0,1]$ and studying fuzzy relations are equivalent.

Various forms of transitivity of the fuzzy relation, R, impose a structure on the family $R_\alpha$. The min-transitivity of a fuzzy relation played an important role in the earlier sections. The following proposition shows the consequences of a min-transitive fuzzy relation en the underlying nested family of $\alpha$-cut relations.

**Proposition 16.2.1.** R is a reflexive and min-transitive fuzzy relation if and only if $R_\alpha$ is a reflexive partial ordering (reflexive and transitive) for all $\alpha \in (0,1)$.

**Proof.** (reflexivity,$\Rightarrow$) Reflexivity of fuzzy relation R is given by R(a, a) = 1 for all $a \in A$. By definition, (cf. equation (16.11)) $aR_\alpha a \Leftrightarrow R(a,a) \geq \alpha$. Hence $aR_\alpha a$ for all $\alpha \in [0,1]$.

(reflexivity, $\Leftarrow$): if $aR_\alpha a$ for all $\alpha \in [0,1)$, then $R(a,a) = \sup\{\alpha | aR_\alpha a, \forall \alpha \in [0,1]\} = 1$. (transitivity,$\Rightarrow$):

from min-transitivity of fuzzy relation R, $R(a, c) \geq \min\{R(a, b), R(b, c)\}$ for all a, b, $c \in A$. If $R(a, b) \geq \alpha$ and $R(b, c) \geq \alpha$, then $R(a, c) \geq \alpha$ by min-transitivity. Hence, for all a, b, $c \in A$ if $aR_\alpha b$ and $bR_\alpha c$, then $aR_\alpha c$. The result follows since $\alpha$ was arbitrary.

(transitivity, $\Leftarrow$): transitivity of $R_\alpha$, if $aR_\alpha b$, and $bR_\alpha c$, then $aR_\alpha c$ is equivalent to

$$\text{if } R(a,b) \geq \alpha \text{ and } R(b,c) \geq \alpha \text{ then } R(a,c) \geq \alpha$$

by equation (16.11). Assume to the contrary or min-transitivity that $R(a, c) < \min\{R(a, b), R(b, a)\}$ which implies $R(a, c) < \alpha$ contradiction.

The nested family of crisp relations, $R_\alpha$, is not any richer than a partial order when the fuzzy relation R is reflexive and min-transitive. In general, $R_\alpha$ are not connected for all $\alpha \in (0,1)$.

A reverse question can also be formulated. If $R_\alpha$ is connected for all $\alpha \in [0,1)$ what happens to the corresponding fuzzy relation R? $R_\alpha$ is connected if and only if $aR_\alpha b$ or $bR_\alpha a$ for all a, $b \in A$ which is equivalent to $(R(a, b) \geq \alpha$ or $(R(b, a) \geq \alpha)$ which is equivalent to: for all a,b $\in A$, $\max\{R(a, b), R(b, a)) = 1$ of Definition 16.1.2.

Consider the following generalization of the linearity condition (16.10); there exists a $\theta \in (0,1)$ such that:

$$\forall a, b \in A \qquad R(a, b) > \theta > R(b, a) \text{ or } R(b, a) > \theta > R(a, b) \qquad (16.12)$$

Clearly, (16.12) reduces to (16.10) when $\theta = 1/2$.

For a given fuzzy relation, R define:

$I_d = \{(a,b) \mid R(a,b) > R(b,a), \forall a, b \in A\}$
$I_n = \{R(b,a),R(a.b) \mid (a,b) \in I_d, \forall a, b \in A\}$
$I_\vartheta = \cap I_n.$

Here $I_d$ is the set of ordered pairs such that the first element dominates the second, $I_n$ is the set of well defined open intervals and $I_\vartheta$ is the intersection of all those open intervals.

It is easy to see that if $I_\theta=\theta$, then R does not satisfy (16.12). Furthermore, if $I_\theta$ is given as:

$$I_\theta = (\max_{(a,b)\in I_d} \{R(b, a)\}), \min_{(a,b)\in I_d}\{R(a, b)\}),$$

Then, an equivalent formulation of (16.12) is:

**Proposition 16.2.2.** If $I_d \neq \varnothing$, and $I_\theta \neq \varnothing$, and then a fuzzy relation, R, satisfies (16.12) for all $\theta \in I_\theta$.

**Proof.** Assume $I_\theta \neq \varnothing$. Then, by definition there exists $(a,b) \in I_\theta$ such that $\min_{(a,b)\in Id}\{R(a,b)\} > \max_{(a,b)\in Id}\{R(b, a)\}$. Hence, (16.12) is satisfied for all $\theta \in I_\theta$.

If $1/2 \in I_\theta$., then R also satisfies (16.10).

The following result shows that if $I_\theta$. is not empty, it contains a nested family of crisp relations that are weak orders.

**Proposition 16.2.3.** If R is a reflexive fuzzy relation which satisfies (16.12), then for all $\alpha \in I_\theta$. $R_\alpha$ is a weak ordering of A (reflexive, transitive and connected). Furthermore, the asymmetric part of $R_\alpha$ ($P_\alpha$)is a linear ordering of A for all $\alpha \in I_\theta$.

**Proof.** By the statement of the proposition, $I_d$ and $I_\theta$ are non-empty. Since the asymmetric part, $P_\alpha$ of the a reflexive crisp relation R is defined as:

$$aP_\alpha b \Leftrightarrow aR_\alpha b \text{ and not } bR_\alpha a$$

parametric transitivity implies the transitivity of both $R_\alpha$ and $P_\alpha$ whenever $\alpha \in I_\theta$. Reflexivity of $R_\alpha$ follows from reflexivity of fuzzy relation R. Irreflexivity of $P_\alpha$ follows from its definition and the reflexivity of $R_\alpha$. Assume to the contrary of connectedness; not $aR_\alpha b$ and not $bR_\alpha a$ which corresponds to ($R(a,b) < \alpha$ and $R(b, a) < \alpha$, but this contradicts parametric transitivity of R.

Hence, (16.12) guarantees that the family of nested relations, $R_\alpha$ is a weak order whenever $\alpha \in I_\theta$. Another interesting condition on the initial fuzzy relation R is the following: for all a, b$\in$A.

$$R(a,b)+R(b,a) = 1 \tag{16.13}$$

This is another form of conservation and hence invariance which we have discussed in Chapter 10.

Clearly, the class of fuzzy relations satisfying equ.(16.13) is a subset of fuzzy relations satisfying 1/2-transitivity, i.e., equ.(16.13) implies 1/2-transitivity, but the reverse implication does not necessarily hold.

As an example consider the following fuzzy relation, R:

$$R=\begin{bmatrix} 1 & 1/2+\varepsilon & 1/2+\varepsilon \\ 1/2-\varepsilon & 1 & 1/2+\varepsilon \\ 1/2-\varepsilon & 1/2-\varepsilon & 1 \end{bmatrix}$$

For $\varepsilon>0$, R is 1/2-transitive but equ.(16.3) does not hold since R(a, b) + R(b, a) $= 1/2+2\varepsilon+1/2-\varepsilon > 1$.

*16.2.4. The Case where the Interval Vanishes*

It is interesting to investigate the cases when the interval valued Type 2 strict preference (equation (16.5)) is point valued (Type 1), i.e., when the second order imprecision vanishes.

One sufficient condition is given by equ (16.12). The extra information one gains in this case is that whenever R(a, b) $>\theta$ then its inverse R(b, a) $<\theta$. This condition also implies that the underlying $\alpha$-cut relations are weak orders for a$\in$ $I_\theta$. (cf. Proposition 16.2.3). The sufficiency of equ (16.12) that causes the interval to vanish is established in the following.

**Proposition 16.2.4.** If R is a fuzzy relation satisfying equ (16.12), then $P_D = P_C$.

**Proof.** Given equ (16.12) let x = R(a, b) and y = R(b, a), then we new have $x > \theta > y$ and the strict preference can be rewritten as follows (using equ (16.6)):

$$\begin{aligned} P(a,b) &= [\min\{x, 1\text{-}y\}, \min\{\max\{x, 1\text{-} y\}, \min\{\max\{1\text{-}x, 1\text{-} y\}, \max\{x, y\}\}\}] \\ &= [\min\{x, 1\text{-}y\}, \min\{\max\{x, 1\text{-}y\}, \min\{x, 1\text{-}y\}\}] \\ &= [\min\{x, 1\text{-}y\}, \min\{x, 1\text{-}y\}\}] \end{aligned}$$

This completes the proof.

Since θ is arbitrary in equ (16.12), it is more general than equ (16.10). Hence equ (16.12) is implied by equ.(16.10) which in turn is implied by equ.(16.13) (see Section 16.2.3). Therefore, equ (16.10) and equ (16.13) are two other conditions for the interval to vanish.

Note that, when equ (16.13) holds, R(a, a) = 1/2 by convention. There are many studies in the mathematical psychology literature where equ (16.13) is employed and the R(a, b) is interpreted to be the observed probability of choosing a over b when confronted with a pair of alternatives {a, b} (see for example, [Fishburn, 1973; Roberts, 1979; Suppes, 1989] and the references therein).

The important information equ (16.13) brings is that from the knowledge of R(a,b) one can find R(b, a). (Note again, the notions of conservation and invariance)

In the specific case when equ (16.13) holds, the interval valued Type 2 strict preference reduces to the underlying fuzzy weak preference, R. To see this, given equ (16.13) let x = R(a,b) and y = R(b,a), we now have x + y =1 and strict preference can be rewritten as follows (with equ.(16.6):

$$\begin{aligned} P(a,b) &= [\min\{x, 1\text{-}y\}, \min\{\max\{x, 1\text{-}y\}, \max\{1\text{-}x, 1\text{-}y\}, \max\{x,y\}\}] \\ &= [\min\{x, x\}, \min\{\max\{x, x\}, \max\{1\text{-}x, x\}, \max\{x, 1\text{-}x\}\}] \\ &= [x, x] \end{aligned}$$

The strict preference, P, is no more interval valued and furthermore, P = R. The crisp ordering relation "   " as defined in equ.(16.7) becomes

$$a \succ b \Leftrightarrow P(a, b) > P(b, a) \Leftrightarrow P(a, b) > 1/2. \tag{16.4}$$

This gives rise to the following results.

**Corollary 16.2.3.** (i) If R satisfies equ (16.13), then "≻" as defined in equ (16.14) is a linear order if and only if R is also a 1/2-transitive relation.

(ii) If R satisfies equ (16.13) and is also a min-transitive fuzzy relation then "≻" as defined in equ (16.14) is a linear order.

**Proof.** It is trivial to show that "≻" is irreflexive. Connectedness follows from 1/2-transitivity of R in the first case and min-transitivity of R in the second case. Transitivity follows from Theorem 16.2.1 in the first case and from Theorem 16.1.1 in the second case.

*16.2.5. Non-unfuzzy Non-dominated Elements*

The concept of non-fuzzy non-dominated elements is introduced by Orlovski[1978]. The main idea can be translated to our case as follows: one can define the set of non-fuzzy non-dominated alternatives,

$A^{UND}$ as

$$A^{UND} = \{a \in A : P(a, b) = [1,1] \text{ for all } b \in A]. \quad (16.5)$$

The elements of $A^{UND}$ are such that they are strictly preferred to other elements of the set to the full extent.

**Proposition 16.2.5.** $A^{UND}$ as defined in equ (16.15) is non-empty if and only if there exists an alternative, $a \in A$ with $R(a,b) = 1$ and $R(b,a) = 0$ for all $b \in A$.

**Proof.** Sufficiency is immediate. To show necessity with equ (16.6) consider.
$P(a, b) = [\min\{x, n(y)\}, \min\{\max\{x,y\}, \max\{x, n(y)\}, \max\{n(x),n(y)\}\}]$
where $y = R(a, b)$ and $y = R(b, a)$. The lower bound of $P(a, b)$ becomes 1 if and only if $x=1$ and $n(y)=1-y=1$ which establishes the result.

This result means that a fuzzy non-dominated alternative, a, exists if and only if there is no initial imprecision about it in R. It is interesting to note that the linearity condition on R equ (16.10) and 1/2-transitivity, which ensures that " " is a linear order, are not sufficient for the existence of an fuzzy non-dominated alternative.

This is somewhat a negative result when compared to Orlovski's main result. In. his paper Orlovski shows that min-transitivity of R is sufficient for the existence of an non-fuzzy non-dominated alternative when the strict preference is defined as Type 1, i.e., point valued:

$$P(a, b) = \Delta\,(R(a, b), n(R(b, a)))$$

where $\Delta$ is the Lukasiewicz t-norm given by:
$\Delta(x,y) = \max\{x+y-1, 0\}$ for all $x,y \in [0, 1]$.

Here we employed an ordinal approach and defined the strict preference as interval valued Type 2 fuzzy set without resorting to addition or subtraction.

## 16.3. Conclusion

A modeling technique is presented that deals with imprecise of preferences. The basic assertion is that higher concepts constructed from imprecision preferences induce a second order imprecision. When the initial imprecision in the concept of weak preference is represented by a fuzzy relation, higher constructs of new concepts, such as strict preference, are represented by interval valued Type 2 relations to capture the second order imprecision introduced by this construction. A crisp ordering relation is defined by interpreting this second order.

Necessary and sufficient conditions on the initial imprecision are analyzed so that the crisp ordering relation is transitive and/or connected. When the crisp ordering relation is transitive the alternatives can be partially ranked; whereas if it is also connected the alternatives can be totally ranked. It is shown that even if the initial imprecision does not have a strong property like min-transitivity the alternatives can still be totally ranked.

Figure 16.3 shows a summary of the results. Specifically it shows that weak transitivity of R is necessary for the transitivity of the crisp ordering relation, "  ", whereas 1/2-transitivity is both necessary and sufficient which makes min-transitivity to be a sufficient condition as well. In order to have a complete ordering, the strict preference, "  ", must be connected. It is shown that a linearity condition (16.10) is necessary and sufficient for "  " to be connected, whereas a stronger condition (16.13) is only sufficient. Therefore, if the fuzzy weak preference relation, R, is both 1/2-transitive and the corresponding crisp ordering relation, "  ", is a linear order.

When the alternatives can only he partially ordered, this idea can be used to eliminate the dominated solutions and if there is no apparent choice between any two alternatives, more information regarding those alternatives can he demanded from the decision maker.

Fuzzy relations induce a family of nested crisp relations, The conditions imposed on the fuzzy relation have some consequences in terms of the family of nested crisp families. It is shown that a parametric linearity condition on the fuzzy relation endows the underlying nested family of crisp relations with connectivity in a certain region.

The case where the second order imprecision disappears is also analyzed. It is shown that a linearity condition (16.10) is sufficient for the second order imprecision to disappear.

The question of the existence of non-fuzzy non-dominated alternatives is formulated in this setting and a condition on the fuzzy relation R is derived to guarantee this existence. The condition requires that fuzzy relation should be crisp for the non-fuzzy non-dominated alternative.

This study considered the connectives min, max and pseudo complement. A generalization of this technique to strict and strong Archimedean De Morgan triples can be made in a similar manner. Recall that FDCF(.) $\subseteq$ FCCF(.) hold for those De Morgan triples as well.

Finally, the length of the interval for strict preference can be thought as a measure of confidence on the initial preference which can give rise to cardinal models of completeness of information in preference structures.

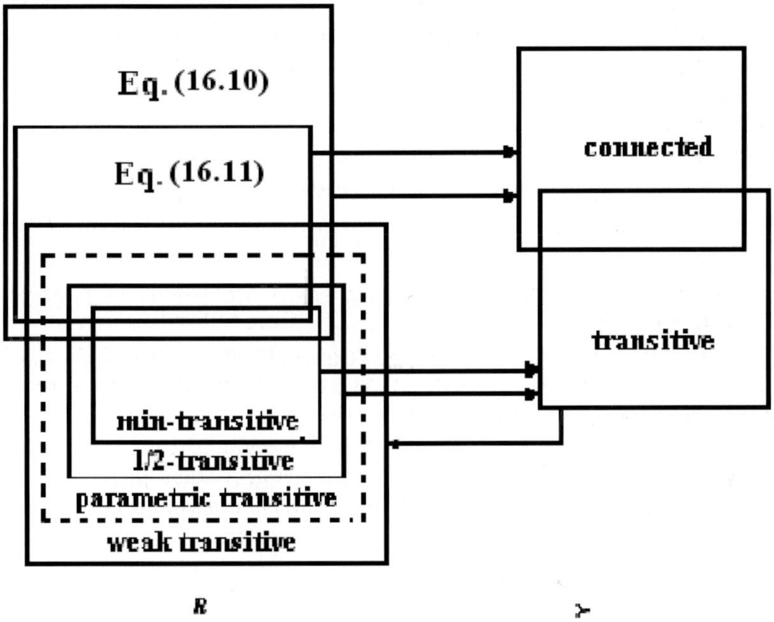


**Figure 16.3.** A Summary of Results: an arrow shows implication in the direction indicated, arrows on both sides indicate equivalence

## References

1. Alsina C., E. Trillas and I. Valverde (1983), "On Some Logical Connectives for Fuzzy Sets Theory", *Journal of Mathematical Analysis and Applications,* 93, 15-26.
2. Barrett, C.R., P.K. Pattanaik, and M. Salles (1999), "On Choosing Rationally When Preferences are Fuzzy", *Fuzzy Sets and* Systems 34, 197-212.
3. Basu, K. (1984), "Fuzzy Revealed Preference Theory", *Journal of Economic Theory*, 32, 212-217.
4. Billot, A. (1992), "Economic Theory of Fuzzy Equilibria: An Axiomatic Analysis", Vol. 373 of Lecture notes in Economics and Mathematical Systems. (Springer-Verlag, Berlin Heidelberg, Germany).

5. Fodor, J.C., M. Roubens (1994), "Fuzzy Preference Modelling and Multicriteria Decision Support", Vol. 14 of Theory and Decision Library. Series D, *System Theory, Knowledge Engineering, and Problem Solving*, Kluwer Academic, Dordecht, Boston.
6. Fodor, J.C. (1991), "Strict Preference Relations Based on Weak T-norms", *Fuzzy Sets and Systems,* **43** 321-366.
7. Goguen, J.A., L-fuzzy sets, Journal of mathematical analysis and applications 18(1967), 145-174.
8. Kolodziejczyk, W. (1986), "Orlovski's Concept of Decision Making with Fuzzy Preference Relation-further Results", *Fuzzy Sets and* Systems 19, 11-20.
9. Lano, K. (1992), "Formal Frameworks for Approximate Reasoning", *Fuzzy Sets and Systems,* 51 131-146.
10. Mizumoto, M., K. Tanaka (1981), "Fuzzy Sets of Type 2 Under Algebraic Product and Algebraic Sum", *Fuzzy* Sets *and Systems,* 5, 277-290.
11. Orlovski, S.A. (1978), "Decision Making with a Fuzzy Preference Relation", *Fuzzy Sets and* Systems 1, 155-167.
12. Ovchinnikov, S., M. Roubens (1992), "On fuzzy Strict Preference, Indifference and Incomparability Relations", *Fuzzy Sets and Systems* 47, 313-318. and also in Vol. 49, 15-20.
13. Ovchinnikov, S. (1991), "Similarity Relations, Fuzzy Partitions, and Fuzzy Orderings", *Fuzzy* Sets *and Systems,* 40, 107-126.
14. Ovchinnikov, S., M. Roubens (1991), "On Strict Preference Relations", *Fuzzy Sets and* Systems 43, 319-326.
15. Perny, P., B. Roy (1992), "The Use of Fuzzy Outranking Relations in Preference Modelling", *Fuzzy* Sets *and* Systems 49, 1, 33-53.
16. Roberts, F. (1979), Measurement Theory, Addison Wesley Pub. Co.
17. Roubens, M., P. Vincke (1985), Preference Modelling, Vol. 250 of *Lecture Notes in Economics and Mathematical* Systems. (Springer Verlag, Berlin, 1985).
18. Türkşen, I.B., A. Kandel, Y-Q. Zhang (1999), "Normal Forms of Fuzzy Middle and Fuzzy Contradiction", *IEEE-SMC, 29-2, Part B, Cybernetics*, 237-253.
19. Türkşen, I.B. (1994), "Interval Valued Fuzzy Sets and Fuzzy Measures", *Proceedings of the First International Conference of NAFIPS, IFIS, and NASA,* December 18-21, San Antonio, Texas, 317-321.
20. Türkşen I.B. (1994),"Interval-Valued Fuzzy Sets and Compensatory AND", *FSS,* 87-100.
21. Türkşen, I.B. (1994), "Fuzzy Normal Forms", *Fuzzy Sets and Systems,* 253-266.
22. Türkşen, I.B, (1994), "Interval-Valued Fuzzy Sets and 'Compensatory AND' ", *FSS*, 295-307.
23. Türkşen, I.B., T, Bilgiç (1993), Interval Valued Strict Preference, In *Proceedings of the First European Congress on Fuzzy and Intelligent Technologies* (September 1993), 593-599. EUFIT '93 Aachen, Germany.
24. Türkşen, I.B. (1992), "Interval Valued Fuzzy Sets and 'Compensatory AND' ", *Fuzzy Sets and Systems*, 51, 295-307.
25. Türkşen, I.B. (1991), "Measurement of Membership Functions and Their Acquisitions", *Fuzzy Sets and Systems,* 40, 5-38.
26. Weber, S. (1983), "A General Concept of Fuzzy Connectives, Negations and Implications Based on t-norms and t-conorms, *Fuzzy Sets and Systems* 11, 115-134.
27. Zimmermann, H.J., P. Zysno (1980), "Latent Connectives in Human Decision Making", *Fuzzy Sets and Systems*, 4, 37-51.

# Chapter 17

# A FOUNDATION FOR COMPUTING WITH WORDS: META-LINGUISTIC AXIOMS

## 17.1. Introduction

In this chapter, Meta-Linguistic axioms are proposed as a foundation for Computing With Words, CWW, (Zadeh; 1991, 2001) as an extension of fuzzy sets and logic theory. Over the last 40 plus years, we have discussed and made considerable progress on the foundations of fuzzy set and logic theory and their applications in domains of mainly fuzzy control and partially fuzzy decision support systems. But in all these works, we generally have started out with the classical axioms of classical set and logic theory which are expressed in set notation and then relaxed some of these axioms, such as distributivity absorption, idempotency, etc., in order to come up with the application of t-norms and t-conorms in various domains.

In all of this past work, we all have continued to use the classical axiomatic expressions of crisp set theory. That is, we said for example for certain t-norms and t-conorms, say distributivity, or idempotency, or, etc, does not hold. To say that a particular axiom "holds or does not hold" is contrary to the basic principle of fuzzy theory. If we are sincere in our basic principle which states that "all is a matter of degree" in fuzzy theory, then to say that a certain axiom "holds or does not hold" does not make sense. In fact, it speaks against our basic principle. In this regard therefore, we ought to say that all "meta-linguistic axioms hold as a matter of degree".

For this reason, in this paper, we propose that a unique foundation for CWW can be established by re-stating the original classical axioms in terms of meta-linguistic expressions where "AND", "OR" are expressed linguistically as opposed to their set theoretic symbols "$\cap$", "$\cup$", respectively. These meta-linguistic expressions can then be interpreted in terms of their Fuzzy Disjunctive and Conjunctive Canonical Forms, i.e., FDCF and FCCF. (Türkşen, 1986-2004)

We explore the consequences of this proposal when these meta-linguistic expressions are interpreted with their Fuzzy Disjunctive and Conjunctive Canonical Forms, FDCF, FCCF, respectively.

In our previous writings (Türkşen; 1986-2004), we have explored various aspects of FDCF and FCCF, including their generation, their non-equivalence, i.e., $FDCF_i(.)$ $FCCF_i(.)$, i=1,...,16, for the sixteen well known linguistic expressions that form the foundation of any set and logic theory. We have also explored that, for specific cases of t-norms and t-conorms, that are strict and nilpotent Archimedean, we get:

$FDCF_i(.) \subseteq FCCF_i(.)$ (Taner, 1995).

In this paper, in particular, we explore in detail, the consequences of re-stating the axioms of the classical theory as meta-linguistic expressions in the development of a foundation for CWW proposed by Zadeh (1999, 2001).

As a result, we show that new formulas are generated in fuzzy set and logic theory as a new foundation for CWW. This demonstrates the richness and expressive power of fuzzy set and logic theories and CWW that collapse into the classical theory under restricted assumptions of reductionism. Classical theory proposes that $\mu_A : X \to \{0,1\}$ in contrast to the basic principle of fuzzy set theory which proposes that $\mu_A : X \to [0,1]$. That is we obtain the equivalence of the Disjunctive and Conjunctive Normal Forms, $DNF_i(.) \equiv CNF_i(.)$, i=1,...,16, in the classical set and logic theory axioms because $\mu : X \to \{0,1\}$.

In our opinion, the break down of these classical equivalences, i.e., non-equivalences, are important in establishing the foundations of fuzzy set theories and the basic formulations of Computing With Words. This break-down and generation of additional formulas expose part of the uncertainty expressed in the combination of concepts that are generated by linguistic operators, "AND", "OR". As well, it allows us to state that the meta-linguistic axioms hold as "a matter of degree" staying true to the basic principle of fuzzy theory.
In the rest of this paper, we first state, in section 17.2, the meta-linguistic expression of the axioms for CWW in comparison to their set theoretic axioms of the crisp theory.

In Section 17.3, we explore the consequences of expressing the proposed meta-linguistic axioms in FDCF and FCCF, i.e., Fuzzy Disjunctive and Fuzzy Conjunctive Canonical Forms.

In Section 17.4, we present a basic Meta-Linguistic reasoning for GMP as ML-GMP.

In Section 17.5, we state our conclusions.

## 17.2. Meta-Linguistic Axioms

In order to form a sound foundation for the research to be conducted in Computing With Words, CWW, (Zadeh; 1999, 2001), we believe, it is rather necessary that we begin with a statement of the basic axioms stated in their meta-linguistic expressions.
In particular, we propose that we need to re-state the classical axioms shown in Table 17.1 in terms of their meta-linguistic expressions shown in Table 17.2.

**Table 17.1**. Axioms of Classical Set and Logic Theory, where A, B, C are crisp, two valued sets and c(.) is the standard complement, "$\cap$", "$\cup$" are set notations which stand for "AND", "OR" in a one-to-one correspond, respectively, X is the universal set and $\phi$ is the empty set.

| | |
|---|---|
| Involution | $c(c(A)) = A$ |
| Commutativity | $A\cup B = B\cup A$ |
| | $A\cap B = B\cap A$ |
| Associativity | $(A\cup B) \cup C = A\cup (B\cup C)$ |
| | $(A\cap B) \cap C = A\cap (B\cap C)$ |
| Distributivity | $A\cup(B\cap C) =(A\cup B) \cap (A\cup C)$ |
| | $A\cap(B\cup C) =(A\cap B) \cup (A\cap C)$ |
| Idempotence | $A\cup A = A$ |
| | $A\cap A = A$ |
| Absorption | $A\cup (A\cap B) = A$ |
| | $A\cap (A \cup B) = A$ |
| Absorption by X and $\phi$ | $A\cup X = X$ |
| | $A\cap\phi = \phi$ |
| Identity | $A\cup\phi = A$ |
| | $A\cap X = A$ |
| Law of contradiction | $A\cap c(A) = \phi$ |
| Law of excluded middle | $A\cup c(A) = X$ |
| De Morgan Laws | $c(A\cap B) = c(A) \cup c(B)$ |
| | $c(A\cup B) = c(A) \cap c(B)$ |

It is to be noted that in Table 17.1, in the Axioms of Classical Set and Logic Theory, A, B, C stand for classical sets such that, for example, $\mu_A(x)=a\in\{0,1\}$, where $\mu_A(x)=a$ is the crisp membership value of every $x\in X$, the universe of discourse X, c(.) is involutive complementation operator in the set domain which corresponds to the standard negation, n(a)=1-a, where n(.) is the involutive negation operator in the membership domain. Furthermore, "$\cap$", "$\cup$", are set theoretic "intersection", "union" operators and are taken in one-to-one correspondence to the linguistic operators "AND", "OR", respectively, in the classical reductionist perspective.

**Table 17.2.** Meta-Linguistic Expression of the Axioms for CWW where A, B, C are fuzzy sets and stand for linguistic terms of linguistic variables, NOT(.) is the complementation operator, "AND", "OR" are linguistic connectives that are not in a one-to-one correspondence with "$\cap$", "$\cup$", respectively.

| | |
|---|---|
| Involution: | NOT(NOT(A)) vs A |
| Commutativity: | A AND B vs B AND A |
| | A OR B vs B OR A |
| Associativity: | (A AND B) AND C vs A AND (B AND C) |
| | (A OR B) OR C vs A OR (B OR C) |
| Distributivity: | A AND (B OR C)vs(A AND B) OR (A AND C) |
| | A OR (B AND C)vs(A OR B) AND (A OR C) |
| Idempotency: | A AND A vs A |
| | A OR A vs A |
| Absorption : | A OR (A AND B) VS A |
| | A AND (A OR B) VS A |
| Absorption by X and $\varnothing$: | A OR X vs X |
| | A AND $\varnothing$ vs$\varnothing$ |
| Identity: | A OR $\varnothing$ vs A |
| | A AND X vs A |
| Law of contradiction: | A AND NOT(A) $\subseteq \varnothing$ |
| Law of excluded middle: | A OR NOT(A) $\subseteq$ X |
| De Morgan's Laws: | NOT(A AND B) vs NOT(A) AND NOT(B) |
| | NOT(A OR B) vs NOT(A) OR NOT(B) |

Whereas in Table 17.2, in the Meta-Linguistic expressions of the proposed Axioms for CWW, A, B, C stand for fuzzy sets which are linguistic terms of linguistic variables, such that, for example, $\mu_A(x)=a\in[0,1]$ where $\mu_A(x)=a$ is the fuzzy membership value of every $x\in X$, NOT(.) is a linguistic negation operator, which will be taken to be equivalent to the involutive negation of the classical theory, i.e., for the purposes of this paper, NOT(.)=c(.) and hence n(a)=1-a. However, the linguistic "AND", "OR" operators will be taken as linguistic connectives which do not map in a one-to-one correspondence to "∩", "∪", respectively, which are symbols of the classical set theory that map to t-norms and t-conorms, respectively, within the perspective of fuzzy theory.

This notion that linguistic "AND", "OR" do not correspond in a one-to-one mapping to "∩", "∪", respectively, is supported by Zimmermann, Zysno(1980) experiments and our investigations on "Compensatory 'AND' " (Türkşen,1992). It should be recalled that when meta-linguistic expressions are represented in terms of FDCF and FCCF, Fuzzy Disjunctive Canonical Forms and Fuzzy Conjunctive Canonical Forms, respectively in Table 17.3, they are no longer equivalent, i.e., $FDCF_i(.)$  $FCCF_i(.)$, i=1,…,16, for the sixteen basic natural language expressions of any two sets and logic theory (Türkşen, 1986, 2002) which are shown in Table 17.4. This is true in particular, for example, for 3rd and 6th expressions shown in Table 17.4, i.e., "A OR B", and "A AND B", respectively. These two expressions and their FDCF and FCCF expressions are essential in re-interpreting the Meta-Linguistic axioms stated in Table 17.2. They can be observed in Table 17.3 but it should be recalled that $FDCF_i(.)=DNF_i(.)$ and $FCCF_i(.)=CNF_i$, i=1,…,16 in form only but not in content.

**Table 17.3.** Classical Disjunctive Normal and Fuzzy Disjunctive Canonical Forms, DNF and FDCF and Classic Conjunctive Normal and Fuzzy Conjunctive Canonical Forms, CNF and FCCF, where ∩ is a conjunction, ∪ is a disjunction and c is a complementation operator in the set domain.

**Table 17.3.a**. Fuzzy Disjunctive Canonical Forms / Disjunctive Normal Forms

| No: | Fuzzy Disjunctive Canonical Forms/Disjunctive Normal Forms |
|---|---|
| 1 | (A∩B) ∪ (A∩c(B)) ∪ (c(A) ∩B) ∪ (c(A) ∩c(B)) |
| 2 | ∅ |
| 3 | (A∩B) ∪ (A∩c(B)) ∪ (c(A) ∩B) |
| 4 | (c(A) ∩c(B)) |
| 5 | (A∩c(B)) ∪ (c(A) ∩B) ∪ (c(A) ∩c(B)) |
| 6 | (A∩B) |
| 7 | (A∩B) ∪ (c(A) ∩B) ∪ (c(A) ∩c(B)) |
| 8 | (A∩c(B)) |
| 9 | (A∩B) ∪ (A∩c(B)) ∪ (c(A) ∩c(B)) |
| 10 | (c(A) ∩B) |
| 11 | (A∩B) ∪ (c(A) ∩c(B)) |
| 12 | (A∩c(B)) ∪ (c(A) ∩B) |
| 13 | (A∩B) ∪ (A∩c(B)) |
| 14 | (c(A) ∩B) ∪ (c(A) ∩c(B)) |
| 15 | (A∩B) ∪ (c(A) ∩B) |
| 16 | (A∩c(B)) ∪ (c(A) ∩c(B)) |

**Table 17.3.b.** Fuzzy Conjunctive Canonical Forms / Conjunctive Normal Forms

| No. | Fuzzy Conjunctive Canonical Forms/Conjunctive Normal Forms |
|---|---|
| 1 | I |
| 2 | (A∪B) ∩ (A∪c(B)) ∩ (c(A) ∪B) ∩ c(A) ∪c(B)) |
| 3 | (A∪B) |
| 4 | (A∪c(B)) ∩ (c(A) ∪B) ∩ c(A) ∪c(B)) |
| 5 | (c(A) ∪c(B)) |
| 6 | (A∪B) ∩ (A∪c(B)) ∩ (c(A) ∪B) |
| 7 | (c(A) ∪B) |
| 8 | (A∪B) ∩ (A∪c(B)) ∩ (c(A) ∪c(B)) |
| 9 | (A∪c(B)) |
| 10 | (A∪B) ∩ (c(A) ∪B) ∩ (c(A) ∪c(B)) |
| 11 | (A∪c(B)) ∩ (c(A) ∪B) |
| 12 | (A∪B) ∩ (c(A) ∪c(B)) |
| 13 | (A∪B) ∩ (A∪c(B)) |
| 14 | (c(A) ∪B) ∩ (c(A) ∪c(B)) |
| 15 | (A∪B) ∩ (c(A) ∪B) |
| 16 | (A∪c(B)) ∩ (c(A) ∪c(B)) |

## 17.3. Consequences of the Proposed Meta-Linguistic Axioms

In order to appreciate the consequences of the proposed Meta-Linguistic Axioms for CWW, we first very briefly review the classical axioms and the usual use of them in our investigation. After this, we state the consequences of the proposed Meta-Linguistic Axioms for CWW.

**Table 17.4.** Sixteen Possible Combinations of any two sets, A and B expressed in natural language.

| Number | Meta-Linguistic Expressions |
|---|---|
| 1 | UNIVERSE |
| 2 | EMPTY SET |
| 3 | A OR B |
| 4 | NOT A AND NOT B |
| 5 | NOT A OR NOT B |
| 6 | A AND B |
| 7 | A IMPLIES B |
| 8 | A AND NOT B |
| 9 | A OR NOT B |
| 10 | NOT A AND B |
| 11 | A IF AND ONLY IF B |
| 12 | A EXCLUSIVE OR B |
| 13 | A |
| 14 | NOT A |
| 15 | B |
| 16 | NOT B |

### *17.3.1. Classical Axioms*

It is well known that $DNF_i(.) \equiv CNF_i(.)$, i=1,…,16, in classical theory. Thus, in classical applications, one always use the shortest of these two forms.
For example:

(1) For "A AND B", we get:

$$\begin{cases} \mathrm{DNF(A\ AND\ B)} = \mathrm{A} \cap \mathrm{B}, \\ \mathrm{CNF(A\ AND\ B)} = (\mathrm{A} \cup \mathrm{B}) \cap (\mathrm{c(A)} \cup \mathrm{B}) \cap (\mathrm{A} \cup \mathrm{c(B)}), \end{cases}$$

together with the equivalence of DNF and CNF, i.e., DNF(A AND B) ≡ CNF(A AND B).

But in all our calculations, we use only "A∩B" for a representation of "A AND B" in the classical set domain. This is a conventional habit.

(2) For "A OR B", we get: $\begin{cases} \mathrm{DNF(A\ OR\ B)} = (\mathrm{A} \cap \mathrm{B}) \cup (\mathrm{c(A)} \cap \mathrm{B}) \cup (\mathrm{A} \cap \mathrm{c(B)}), \\ \mathrm{CNF(A\ OR\ B)} = \mathrm{A} \cup \mathrm{B}, \end{cases}$

together with equivalence of DNF(A OR B) ≡ CNF(A OR B).
But again in all our calculations, we use only "A∪B" for a representation of "A OR B" in the classical set domain. Again this is a conventional habit.

This habitual use of the short hand form is applied to all the remaining linguistic combination.

Furthermore, this habit of using the short hand form of these combinations was carried out by most fuzzy researchers in their applied as well as theoretical investigations without any inquiry of what happens to the longer versions, i.e., DNF(A OR B) and CNF(A AND B).

However, the investigations carried out by Türkşen (1986-2002) and Türkşen, et.al.(1998, 1999) Resconi and Türkşen (2001) indicate that we ought to use both the $FDCF_i(.)$ and $FCCF_i(.)$, i=1,…,16, because the equivalence no longer holds in fuzzy theory, i.e., $FDCF_i(.) \not\equiv FCCF_i(.)$.

For each of the proposed meta-linguistic axioms for CWW stated in Table 2, we next investigate the consequences of this non-equivalence which is an important issue known as the break-down of classical equivalences in fuzzy theory.

*17.3.2. Investigation of Meta-Linguistic Axioms*

In fuzzy theory and its applications in CWW, most researchers continue the usual habit of using the shortest form of classical axioms by directly fuzzifying all the classical axioms. That is as in the classical theory, "A AND B" is directly taken to be "A∩B" but fuzzified, "A OR B" is directly taken to be "A∪B" but fuzzified. But the other longer forms are ignored or not considered either because of habit or because most of us are generally short sighted.

In addition most of us used to state that certain axioms hold and others do not hold. But such a stance is not in the spirit of fuzzy theory. As all of us believe, or ought to believe in fuzzy theory, all are or ought to be a matter of degree. This usual habit of use continues to persist in most of the current research and applications despite the fact that the equivalences break down in fuzzy theory, e.g., $FDCF_i(.)$ $FCCF_i(.)$, i=1,...,16, that have been published in various paper over about the last eighteen years or so (Türkşen, 1986-2002). If we take into the account the fact that these non-equivalences of the Fuzzy Disjunctive and Conjunctive Canonical forms, then we have to realize that the interpretation of the proposed Meta-Linguistic Axioms must be expressed in two distinct forms in set symbolic notation and must give two distinct results in computational, numeric, domain with the application of t-norms and t-conorms. Furthermore these two distinct forms must be interpreted in the spirit of fuzz theory to state that they hold as a matter of degree.

We next express this realization for each of the proposed Meta-Linguistic Axioms.

*17.3.3. Fuzzy Set Theoretic Axioms for CWW*

Next we investigate, both the FDCF and FCCF versions of the whole set of Meta-Linguistic Axioms for CWW. It should be noted that there are four alternative forms we can investigate since we can form FDCF vs. FDCF, FDCF vs. FCCF, FCCF vs. FCCF, FCCF vs. FDCF. For some axioms we investigate all four forms while for some other we investigate only two of these alternatives. The remaining alternatives are left for exercises of the readers.

*17.3.3.1. Fuzzy Involution:*

For the purposes of this chapter, we take NOT(.) = c(.). Hence the involution axiom holds as specified, i.e., n(n(a)=a) and hence as a matter of degree. That is there is no new interpretation of this axiom at this writing. In the future, when we investigate other linguistic negation operators, this will probably produce some new results as it should.

For our purpose in this paper, an example is this: if a =0.4 then n(a)=0.6, n(n(a))=0.4.

*17.3.3.2. Fuzzy Commutativity:*

In Table 17.2, there are two Meta-Linguistic Commutativity axioms:
"A AND B = B AND A", and "A OR B = B OR A"

Now, we know that
FDCF(A AND B) ≢FCCF(A AND B), and FDCF(A OR B) ≢FCCF(A OR B). Therefore, we obtain two set theoretic axioms of the Commutativity in fuzzy theory for CWW for these two Meta-Linguistic Axioms, (MLA).

**Fuzzy Commutativity with "AND":**

(a) FDCF(A AND B) vs. FDCF(B AND A), by a substitution of their fuzzy set symbols, we get:
i.e., A∩B vs. B∩A.

(b) FCCF(A AND B) vs. FCCF(B AND A), again by a substitution, we get:
i.e., (A∪B)∩(c(A)∪B)∩(c(B)∪A) vs. (B∪A)∩(c(B)∪A)∩(B∪c(A))

**Fuzzy Commutativity with "OR":**

(a) FDCF(A OR B) vs. FDCF(B OR A),
i.e., (A∩B)∪(c(A)∩B)∪(A∩c(B)) vs. (B∩A)∪(c(B)∩A)∪(B∩c(A))

(b) FCCF(A OR B) vs. FCCF(B OR A),
i.e., A∪B vs. B∪A

Therefore, Fuzzy Commutativity holds fuzzily as a matter of degree in two separate forms of the MLA Commutativity axiom. This in turn exposes an uncertainty region for the Fuzzy Commutativity axioms.

**Numerical Illustrations**

We will illustrate the effect of interpreting meta-linguistic axioms with fuzzy theory for the following values of a=0.3, b=0.8, c=0.4 in all the numerical examples in the rest of the chapter.

**Illustration - Fuzzy "AND" Commutativity:**

For the well-known t-norm and t-conorm De Morgan Triples, we obtain:

{Max, Min, StN}:

$$\left.\begin{aligned}(a)\ \ &\mu[\text{FDCF(A AND B)}] = 0.3\\ (b)\ \ &\mu[\text{FCCF(A AND B)}] = 0.3\end{aligned}\right\}[0.3]$$

This is a singleton point valued result. In *{Max, Min, StN),* we sometimes get such a singleton result. But in general it is interval-valued Type2 fuzzy set.

{Sum,Product, StN}:

$$\left.\begin{aligned}(a)\ \ &\mu[\text{FDCF(A AND B)}] = 0.24\\ (b)\ \ &\mu[\text{FCCF(A AND B)}] = 0.355\end{aligned}\right\}[0.24, 0.355]$$

This is an interval-valued Type-2 result.

$\{L_{\oplus},, L\ \ , \text{StN}\}$:

$$\left.\begin{aligned}(a)\ \ &\mu[\text{FDCF(A AND B)}] = 0.1\\ (b)\ \ &\mu[\text{FCCF(A AND B)}] = 0.5\end{aligned}\right\}[0.1, 0.5]$$

This is also an interval-valued Type-2 result.

**Illustration - Fuzzy "OR" Commutativity**

For the well-known t-norm and t-conorm De Morgan Triples we obtain:

{Max, Min, StN}:

$$\left.\begin{aligned}(a)\ \ &\mu[\text{FDCF(A OR B)}] = 0.7\\ (b)\ \ &\mu[\text{FCCF(A OR B)}] = 0.8\end{aligned}\right\}[0.7, 0.8]$$

{Sum,Product, StN}:

$$\left.\begin{aligned}(a)\ \ &\mu[\text{FDCF(A OR B)}] \cong 0.7\\ (b)\ \ &\mu[\text{FCCF(A OR B)}] \cong 0.9\end{aligned}\right\}[0.7, 0.9]$$

$\{L_{\oplus},, L_{\odot}, \text{StN}\}$:

$$\left.\begin{array}{ll}(a) & \mu[\text{FDCF(A OR B)}] = 0.6 \\ (b) & \mu[\text{FCCF(A OR B)}] = 1.0\end{array}\right\}[0.6,1.0]$$

Again, we observe that they are interval-valued Type2 results.

**Interpretation:**

These interval-valued Type2 numerical results demonstrates that "Commutativity Axiom" holds in a narrow range for *{Max, Min, StN}* De Morgan Triple and it gets larger as we move toward Lukasiewicz De Morgan Triple of *{$L_{\oplus}$,L , StN}*. Thus depending on which De Morgan Triple is applicable in a real case problem domain, the range in which "the Commutativity Axiom" holds as a matter of degree increases from a narrow range to a wider range.

*17.3.3.3. Fuzzy Associativity*

There are two ML Associativity axioms.

(1) (A AND B) AND C vs. A AND (B AND C) and (2) (A OR B) OR C vs. A OR (B OR C). Recall that, FDCF ≢FCCF(A AND B), and FDCF(A OR B) FCCF(A OR B)

Therefore, we obtain two set theoretic axioms of associativity in fuzzy theory for CWW for these two MLA.

**Fuzzy Associativity with "AND"**

(a) Let us first investigate the fuzzy associativity with FDCF's:
FDCF[FDCF(A AND B) AND C] vs. FDCF[A AND FDCF (B AND C)]

We find that:

FDCF[FDCF(A AND B) AND C)] vs. FDCF[A AND FDCF(B AND C)]
i.e., we get $(A\cap B)\cap C = A\cap(B\cap C)$

This version holds to a fuzzy degree with the associativity property of t-norms and t-conorms.

(b) Let us next investigate the fuzzy associativity with FCCF's, i.e.,

FCCF[FCCF(A AND B) AND C)] vs. FCCF[A AND FCCF(B AND C)]
Recall that, we have:

FCCF(A AND B)
$= (A\cup B)\cap(c(A)\cup B)\cap (A\cup c(B))$

FCCF(B AND C) $= (B\cup C)\cap(c(B)\cup C)\cap(B\cup c(C))$

Therefore, with left hand side we get:

FCCF[FCCF(A AND B) AND C]
= [FCCF(A AND B)$\cup$C] $\cap$[c[FCCF(A AND B)]$\cup$C] $\cap$[FCCF(A AND B)$\cup$c(C)]

$=\{[(A\cup B)\cap(c(A)\cup B)\cap(A\cup c(B))]\cup C\}\cap\{c[(A\cup B)\cap(c(A)\cup B)\cap(A\cup c(B)]\cup C\}$
$\cap\{[(A\cup B)\cap(c(A)\cup B)\cap(A\cup c(B))]\cup c(C)\}$

With right hand side, we get:

FCCF[A AND FCCF(B AND C)]
=[A$\cup$FCCF(B AND C)]$\cap$[c(A)$\cup$FCCF(B AND C)] $\cap$\{A$\cup$c[FCCF(B AND C)]\}

$=\{A\cup[(B\cup C)\cap(c(B)\cup C)\cap(B\cup c(C))]\}\cap\{c(A)\cup[(B\cup C)\cap(c(B\cup C)\cap(B\cup c(C)]\}$
$\cap\{A\cup c[(B\cup C)\cap(c(B\cup C)\cap(B\cup c(C)]\}$

It should be noted and it is clear and straight forward to drive and observe that in the first case, i.e., (a) FDCF version, the interpretation of the associativity equality holds for all t-norms and t-co-norms to a fuzzy degree. However, it is also clear that in the second case, i.e., case (b)FCCF version, the interpretation of associativity does hold to an interval of a fuzzy degree.

**Illustrations - Fuzzy Associativity with "AND" in case (a):**

For the well-known t-norm De Morgan Triples we obtain:

{Max, Min, StN}:

$$\left.\begin{array}{l}\mu\left[\text{FDCF}\left[\text{FDCF(A AND B) AND C}\right]\right]=0.3\\ \mu\left[\text{FDCF}\left[\text{A AND FDCF(B AND C)}\right]\right]=0.3\end{array}\right\}\quad[0.3]$$

{Sum,Product, StN}:

$$\left.\begin{array}{l}\mu\left[\text{FDCF}\left[\text{FDCF(A AND B) AND C}\right]\right]=0.134\\ \mu\left[\text{FDCF}\left[\text{A AND FDCF(B AND C)}\right]\right]=0.134\end{array}\right\}\quad[0.134]$$

{L⊕,,L⊙, StN}:

$$\left.\begin{array}{l}\mu\left[\text{FDCF}\left[\text{FDCF(A AND B) AND C}\right]\right]=0.0\\ \mu\left[\text{FDCF}\left[\text{A AND FDCF(B AND C)}\right]\right]=0.0\end{array}\right\}\quad[0.0]$$

We note that in case (a) two alternate forms hold to the same fuzzy degree because the results and set expressions, $(A \cap B) \cap C$ vs. $A \cap (B \cap C)$ hold with the t-norm , t-conorm property of associativity.

**Illustration - Fuzzy Associativity with "AND" in case (b):**

For the well-known t-norm De Morgan Triples we get:

{Max, Min, StN}:

$$\left.\begin{array}{l}\mu\left[\text{FCCF}\left[\text{FCCF(A AND B) AND C}\right]\right]=0.4\\ \mu\left[\text{FCCF}\left[\text{A AND FCCF(B AND C)}\right]\right]=0.4\end{array}\right\}\quad[0.4]$$

This is again a singleton Type 1 result.

{Sum,Product, StN}:

$$\left.\begin{array}{l}\mu\left[\text{FCCF}\left[\text{FCCF(A AND B) AND C}\right]\right]=0.39..\\ \mu\left[\text{FCCF}\left[\text{A AND FCCF(B AND C)}\right]\right]=0.33\end{array}\right\}\quad[0.33, 0.39]$$

This is interval-valued Type2 result.

{L⊕,,L⊙, StN}:

$$\left.\begin{array}{l}\mu\left[\text{FCCF}\left[\text{FCCF(A AND B) AND C}\right]\right]=0.8\\ \mu\left[\text{FCCF}\left[\text{A AND FCCF(B AND C)}\right]\right]=0.6\end{array}\right\}\quad[0.6, 0.8]$$

This is also interval-valued Type2 result.

A Foundation For Computing With Words:Meta-Linguistic Axioms

Note that the resultant expressions in case (b) is not a property of the t-norms and t-conorms. Instead of the result holding to a specific fuzzy degree, they hold to fuzzy degrees in an interval that can be computed as shown above, i.e., we get Type 2 interval valued result which indicates that there is an interval of uncertainty in which the Fuzzy Associativity with "AND" holds to a fuzzy degree.

**Fuzzy Associativity with "OR"**

(a) First let us investigate the fuzzy associativity for ((A OR B) OR C)=(A OR (B OR C)) with FDCF's:

FDCF[FDCF(A OR B) OR C)] vs. FDCF[A OR FDCF(B OR C)]

Recall that, we have:

$$\text{FDCF(A OR B)} = (A\cap B)\cup(c(A)\cap B)\cup(A\cap c(B))$$
$$\text{FDCF(B OR C)} = (B\cap C)\cup(c(B)\cap C)\cup(B\cap c(C))$$

Therefore, with left hand side we get:

$$\begin{aligned}&\text{FDCF[FDCF(A OR B) OR C]}\\ &= [\text{FDCF(A OR B)}\cap C] \cup[c[\text{FDCF(A OR B)}]\cap C] \cup[\text{FDCF(A OR B)}\cap c(C)]\\ &=\{[(A\cap B)\cup(c(A)\cap B)\cup(A\cap c(B))]\cap C\}\\ &\cup\{c[(A\cap B)\cup(c(A)\cap B)\cup(A\cap c(B)]\cap C\}\\ &\quad\cup\{[(A\cap B)\cup(c(A)\cap B)\cup(A\cap c(B))]\cap c(C)\}\end{aligned}$$

As well with right hand side, we get:

$$\begin{aligned}&\text{FDCF[A OR FDCF(B OR C)]}\\ &=A\cap[(B\cap C)\cup(c(B)\cap C)\cup(B\cap c(C))]\}\cup\{c(A)\cap[(B\cap C)\cup(c(B)\cap C)\cup(B\cap c(C))]\\ &\} \cup\{A\cap c[(B\cup C)\cup(c(B)\cap C)\cup(B\cap c(C))]\}.\end{aligned}$$

It is clear that in general, in this case, i.e., case (a), interpretation of associativity hold to fuzzy degrees in an interval.

(b) Next let us investigate the fuzzy associativity for ((A OR B) OR C) and (A OR (B OR C)) with FCCF's, i.e.,:

FCCF[FCCF(A OR B) OR C)] and FCCF[A OR FCCF(B OR C)]

Since FCCF(A OR B) = $A\cup B$, and FCCF(B OR C) = $B\cup C$,

Thus, we get:

$(A \cup B) \cup C$ vs. $A \cup (B \cup C)$,

which holds in a straight forward manner but naturally to a fuzzy degree!

**Illustration - Fuzzy Associativity with "OR":**

In case (a), it holds to an interval of fuzzy degrees..

{Max, Min, StN}:

$$\left.\begin{aligned}\mu\big[\mathrm{FDCF}\big[\mathrm{FDCF(A\ OR\ B)\ OR\ C}\big]\big]=0.6\\ \mu\big[\mathrm{FDCF}\big[\mathrm{A\ OR\ FDCF(B\ OR\ C)}\big]\big]=0.6\end{aligned}\right\}\quad[0.6]$$

{Sum,Product, StN}:

$$\left.\begin{aligned}\mu\big[\mathrm{FDCF}\big[\mathrm{FDCF(A\ OR\ B)\ OR\ C}\big]\big]=0.651\\ \mu\big[\mathrm{FDCF}\big[\mathrm{A\ OR\ FDCF(B\ OR\ C)}\big]\big]=0.600\end{aligned}\right\}\quad[0.6,0.651]$$

{$L_{\oplus}$,,$L_{\odot}$, StN}:

$$\left.\begin{aligned}\mu\big[\mathrm{FDCF}\big[\mathrm{FDCF(A\ OR\ B)\ OR\ C}\big]\big]=0.2\\ \mu\big[\mathrm{FDCF}\big[\mathrm{A\ OR\ FDCF(B\ OR\ C)}\big]\big]=0.6\end{aligned}\right\}\quad[0.2,0.6]$$

Therefore for the case (a) of the Fuzzy Associativity with "OR", we again get point valued fuzzy degree for *{Max,Min,StN}* De Morgan Triple, whereas we get an interval valued Type2 fuzzy degrees which demonstrates an increasing size in intervals of fuzzy degrees as we move away from *{Max,Min,StN}* toward *{Sum,Product,StN)* and *{ $L_{\oplus}$,$L_{\odot}$, StN)* De Morgan Triples. In case (b), it holds to a fuzzy degree as follows:

{Max, Min, StN}:

$$\left.\begin{aligned}\mu\big[\mathrm{FCCF}\big[\mathrm{FCCF(A\ OR\ B)\ OR\ C}\big]\big]=0.8\\ \mu\big[\mathrm{FCCF}\big[\mathrm{A\ OR\ FCCF(B\ OR\ C)}\big]\big]=0.8\end{aligned}\right\}\quad[0.8]$$

{Sum,Product, StN}:

$$\left.\begin{array}{l}\mu\left[\text{FCCF}\left[\text{FCCF(A OR B) OR C}\right]\right]=0.916\\ \mu\left[\text{FCCF}\left[\text{A OR FCCF(B OR C)}\right]\right]=0.916\end{array}\right\}\quad[0.916]$$

{ $L_{\oplus}$,, $L_{\odot}$, StN }:

$$\left.\begin{array}{l}\mu\left[\text{FCCF}\left[\text{FCCF(A OR B) OR C}\right]\right]=1.0\\ \mu\left[\text{FCCF}\left[\text{A OR FCCF(B OR C)}\right]\right]=1.0\end{array}\right\}\quad[1.0]$$

This demonstrates that, the Fuzzy Associativity for case (b) holds to a fuzzy degree but the degree with which it holds gets larger as we move from *{Min,Max,StN}* toward the *{ $L_{\oplus}$, $L_{\odot}$, StN }* De Morgan Triple.

*17.3.3.4. Fuzzy Distributivity*

Fuzzy Distributivity has two meta-linguistic forms:

Form (1) : “A AND (B OR C)” vs. “(A AND B) OR (A AND C)”

Form (2) : “A OR (B AND C)” vs. “(A OR B) AND (A OR C)”

We will next investigate each in turn with their FDCF and FCCF expressions and numerically illustrate each with the three well known t-norms and t-co-norms.

**Fuzzy Distributivity in Form (1) with FDCF:**

First let us explore the FDCF expressions of both the left and right hand sides of the Fuzzy Distributivity in Form (1). The left hand-side of FDCF expression of the Fuzzy Distributivity in the meta-linguistic Form (1) is:

$$\begin{array}{l}\text{FDCF[A AND FDCF(B OR C)]}=\{A\cap[(B\cap C)\cup(c(B)\cap C)\cup(B\cap c(C)]\}\\ \cup\,\{c(A)\cap[(B\cap C)\cup(c(B)\cap C)\cup(B\cap c(C)]\}\\ \cup\,\{A\cap c[(B\cap C)\cup(c(B)\cap C)\cup(B\cap c(C)]\}\end{array}$$

The right hand-side FDCF expression of the Fuzzy Distributivity in the meta-linguistic form (1) is:

FDCF[FDCF(A AND B) OR FDCF(A AND C)]
$=(A\cap B)\cap(A\cap C)\cup c(A\cap B)\cap(A\cap C)\cup(A\cap B)\cap c(A\cap C)$

**Illustrations - Fuzzy Distributivity in Form (1) with FDCF:**

{Max, Min, StN}:

$$\left.\begin{aligned}&\mu\left[\text{FDCF}\left[\text{A AND FDCF(B OR C)}\right]\right]=0.6\\&\mu\left[\text{FDCF}\left[\text{FDCF(A AND B) OR FDCF(A AND C)}\right]\right]=0.3\end{aligned}\right\}[0.3,0.6]$$

Thus we obtain an interval of uncertainty: [0.3, 0.6].

{Sum,Product StN}:

$$\left.\begin{aligned}&\mu\left[\text{FDCF}\left[\text{A AND FDCF(B OR C)}\right]\right]=0.843\\&\mu\left[\text{FDCF}\left[\text{FDCF(A AND B) OR FDCF(A AND C)}\right]\right]=0.303\end{aligned}\right\}[0.3,0.8]$$

Thus, we obtain an interval of uncertainty: ~[0.3, 0.8]

{ $L_{\oplus}$,,$L_{\odot}$, StN }:

$$\left.\begin{aligned}&\mu\left[\text{FDCF}\left[\text{A AND FDCF(B OR C)}\right]\right]=1.0\\&\mu\left[\text{FDCF}\left[\text{FDCF(A AND B) OR FDCF(A AND C)}\right]\right]=0.1\end{aligned}\right\}[0.1,1.0]$$

Thus we obtain an uncertainty: [0.1, 1.0]. Hence, we observe that Fuzzy Distributivity in Form (1) assessed with FDCF, we get an interval of uncertainty where the axiom holds as a matter of degree but the interval gets larger as we move from {Min,Max,StN} toward *{L⊕,,L ,StN}*. It is also observed that the left-hand side degree of fuzziness is larger than the right hand side degree of fuzziness. This is due to the FDCF's realization in "AND"/ "OR" combinations.

**Fuzzy Distributivity in Form (1) with FCCF:**

Let us next explore the FCCF expressions of both the left and right hand sides of Fuzzy Distributivity in From (1).

The left hand side of FCCF expression of the Fuzzy Distributivity in the meta-linguistic Form (1) is:
FCCF[A AND FCCF(B OR C)] =[A∪ (B∪C)]∩[c(A)∪(B∪C]∩[A ∪ c(B∪C)]

The right hand side FCCF expression of the Fuzzy Distributivity in the meta-linguistic form (1) is

FCCF[FCCF(A AND B) OR FCCF(A AND C)]
=[(A∪B)∩(c(A)∪B)∩(A ∪ c(B)]
∪[(A∪C)∩(c(A)∪C)∩(A ∪ c(C)]

**Illustrations – Fuzzy Distributivity in Form (1) with FCCF:**

{Max, Min, StN}:

$$\left.\begin{array}{l}\mu\left[\text{FCCF}\left[\text{A AND FCCF(B OR C)}\right]\right]=0.3\\ \mu\left[\text{FCCF}\left[\text{FCCF(A AND B) OR FCCF(A AND C)}\right]\right]=0.4\end{array}\right\}\ [0.3, 0.4]$$

Thus Fuzzy Distributivity in Form (1) with FCCF formulas hold in an uncertain interval of [0.3, 0.4] when *{Max,Min,StN}* De Morgan Triple is applied.

{Sum,Product, StN}:

$$\left.\begin{array}{l}\mu\left[\text{FCCF}\left[\text{A AND FCCF(B OR C)}\right]\right]=0.339..\\ \mu\left[\text{FCCF}\left[\text{FCCF(A AND B) OR FCCF(A AND C)}\right]\right]=0.533..\end{array}\right\}\ \sim[0.3, 0.5]$$

Thus Fuzzy Distributivity in form (1) with FCCF formulas hold in an uncertain interval of ~[0.3, 0.5] when *{Sum,Prod,StN}* De Morgan Triple is applied.

{ $L_{\oplus}$,,$L_{\odot}$, StN }:

$$\left.\begin{array}{l}\mu\left[\text{FCCF}\left[\text{A AND FCCF(B OR C)}\right]\right]=0.0\\ \mu\left[\text{FCCF}\left[\text{FCCF(A AND B) OR FCCF(A AND C)}\right]\right]=0.6\end{array}\right\}\ [0.0, 0.6]$$

Thus Fuzzy Distributivity in Form (1) with FCCF formulas hold in an uncertain interval of ~[0.0, 0.6] when Lukasiewicz De Morgan Triple is applied.

**Fuzzy Distributivity in Form (1) with FDCF and FCCF Combinations:**

First let us investigate a particular FDCF and FCCF combination as follows:

FDCF[A AND FCCF(B OR C)] vs.
FCCF[FDCF(A AND B) OR FDCF(A AND C)]

By substituting known values of FDCF(.) and FCCF(.), we get
$A\cap(B\cup C)$ vs. $(A\cap B)\cup(A\cap C)$

**Illustration:**

{Max, Min, StN}:

$$\left.\begin{array}{l}\mu\left[\text{FDCF}\left[\text{A AND FCCF(B OR C)}\right]\right]=0.3\\ \mu\left[\text{FCCF}\left[\text{FDCF(A AND B) OR FDCF(A AND C)}\right]\right]=0.3\end{array}\right\}\ [0.3]$$

{Sum,Product, StN}:

$$\left.\begin{array}{l}\mu\left[\text{FDCF}\left[\text{A AND FCCF(B OR C)}\right]\right]=0.264\\ \mu\left[\text{FCCF}\left[\text{FDCF(A AND B) OR FDCF(A AND C)}\right]\right]=0.331\end{array}\right\}\sim[0.264,0.331]$$

{ $L_\oplus$,,$L_\odot$, StN }:

$$\left.\begin{array}{l}\mu\left[\text{FDCF}\left[\text{A AND FCCF(B OR C)}\right]\right]=0.3\\ \mu\left[\text{FCCF}\left[\text{FDCF(A AND B) OR FDCF(A AND C)}\right]\right]=0.1\end{array}\right\}\ [0.1,0.3]$$

Hence,

FDCF[A AND FCCF(B OR C)] vs.
FCCF[FDCF(A AND B) OR FDCF(A AND C)]

generates in general an interval of Type 2 fuzziness which holds to an uncertain fuzzy degree.

While it holds to a singleton fuzzy degree for *{Min,Max,StN}* it holds to an uncertain interval for the other two De Morgan Triple.

**Fuzzy Distributivity in Form (2) with FCCF and FDCF Combinations**

Again let us first investigate a particular FCCF and FDCF combination:

FCCF[(A OR FDCF(B AND C)] = A∪(B∩C)
FDCF[FCCF(A OR B) AND FCCF(A OR C)] = (A∪B)∩(A∪C)

Therefore, we get:

FCCF[A OR FDCF(B AND C)] vs. FDCF[FCCF(A OR B) AND FCCF(A OR C)]

By substituting known values FCCF(.) and FDCF(.), we get:

A∪(B∩C) vs. (A∪B) ∩ (A∪C)

**Illustration:**

{Max, Min, StN}:

$$\left.\begin{array}{l}\mu\left[\text{FCCF}\left[\text{A OR FDCF(B AND C)}\right]\right]=0.4\\ \mu\left[\text{FDCF}\left[\text{FCCF(A OR B) AND FCCF(A OR C)}\right]\right]=0.4\end{array}\right\}\ [0.4]$$

{Sum,Product, StN}:

$$\left.\begin{array}{l}\mu\left[\text{FCCF}\left[\text{A OR FDCF(B AND C)}\right]\right]=0.468\\ \mu\left[\text{FDCF}\left[\text{FCCF(A OR B) AND FCCF(A OR C)}\right]\right]=0.4988\end{array}\right\}\ \sim[0.46,0.49]$$

{ $L_{\oplus}$,,$L_{\odot}$, StN }:

$$\left.\begin{array}{l}\mu\left[\text{FCCF}\left[\text{A OR FDCF(B AND C)}\right]\right]=0.5\\ \mu\left[\text{FDCF}\left[\text{FCCF(A OR B) AND FCCF(A OR C)}\right]\right]=0.7\end{array}\right\}\ [0.5,0.7]$$

Hence Fuzzy Distributivity in Form (2) with FCCF and FDCF combinations generates in general an interval of Type 2 fuzziness which holds to an uncertain fuzzy degree while it holds a certain fuzzy degree for *{Max,Min,StN}*.

**Fuzzy Distributivity in Form (2) with FDCF:**

In FDCF expressions, the left hand side is:

FDCF[A OR FDCF(B AND C)]=$(A\cap(B\cap C)) \cup (c(A)\cap(B\cup C)) \cup A\cap c(B\cap C))$

In FDCF Expression the right hand side is:

FDCF[FDCF(A OR B) AND FDCF(A OR C)]
$=[(A\cap B)\cup(c(A)\cap B)\cup(A\cap c(B)]\cap[(A\cap C)\cup(c(A)\cap C)\cup(A\cap c(C)]$

**Illustrations – Fuzzy Distributivity in Form (2) with FDCF:**

{Max, Min, StN}:

$$\left.\begin{array}{l}\text{FDCF}\left[\text{A OR FDCF(B AND C)}\right]=0.4\\ \text{FDCF}\left[\text{FDCF(A OR B) AND FDCF(A OR C)}\right]=0.4\end{array}\right\}\ [0.4]$$

Thus Fuzzy Distributivity in Form (2) with FDCF hold with fuzzy degree of (0.4) when *{Max,Min,StN}* De Morgan Triple is applied.

{Sum,Product, StN}:

$$\left.\begin{array}{l}\text{FDCF}\left[\text{A OR FDCF(B AND C)}\right]=0.442..\\ \text{FDCF}\left[\text{FDCF(A OR B) AND FDCF(A OR C)}\right]=0.480..\end{array}\right\}\ \sim[0.44,0.48]$$

Thus Fuzzy Distributivity in Form (2) with FDCF formulas hold within an uncertainty interval of ~[0.44,0.48] as a matter of degree when *{Sum,Prod,StN}* De Morgan Triple is applied.

{ $L_{\oplus}$,,$L_{\odot}$, StN }:

$$\left.\begin{array}{l}\text{FDCF}\left[\text{A OR FDCF(B AND C)}\right]=0.1\\ \text{FDCF}\left[\text{FDCF(A OR B) AND FDCF(A OR C)}\right]=0.6\end{array}\right\}\ [0.1,0.6]$$

Thus Fuzzy Distributivity in Form (2) with FDCF hold within an uncertainty interval of [0.1,0.6] as a matter of degree when Lukasiewicz De Morgan Triple is applied.

Again we observe that interval of uncertainty is increased as we move from *{Max,Min,StN}* to $\{L_{\oplus}, L\ \ , StN\}$.

**Fuzzy Distributivity in Form (2) with FCCF:**

In FCCF expressions, the left hand side is:

FCCF[A OR FCCF(B AND C)] =A∪[(B∪C)∩(c(B)∪C)∩(C∪c(C))]

In FCCF expression, the right hand side is:

FCCF[FCCF(A OR B) AND FCCF(A OR C)]

=[(A∪B)∪(A∪C)]∩[c(A∪B)∪(A∪C)]∩[(A∪B)∪c(A∪C)]

Let us next numerically illustrate the degree with which they hold with the three well known De Morgan Triples.

**Illustrations – Fuzzy Distributivity in Form (2) with FDCF:**

{Max, Min, StN}:

$$\left.\begin{array}{l} FCCF[A\ OR\ FCCF(B\ AND\ C)] = 0.4 \\ FCCF[FCCF(A\ OR\ B)\ AND\ FCCF(A\ OR\ C)] = 0.4 \end{array}\right\} [0.4]$$

{Sum,Product, StN}:

$$\left.\begin{array}{l} FCCF[A\ OR\ FCCF(B\ AND\ C)] = 0.594.. \\ FCCF[FCCF(A\ OR\ B)\ AND\ FCCF(A\ OR\ C)] = 0.552.. \end{array}\right\} \sim[0.55, 0.59]$$

$\{ L_{\oplus}, L_{\odot}, StN \}$:

$$\left.\begin{array}{l} FCCF[A\ OR\ FCCF(B\ AND\ C)] = 0.9 \\ FCCF[FCCF(A\ OR\ B)\ AND\ FCCF(A\ OR\ C)] = 0.7 \end{array}\right\} [0.7, 0.9]$$

One again, we observe that the interval of uncertainty increases as we move from *{Max,Min,StN}* toward *{ $L_{\oplus}, L_{\odot}$ StN)*. We also observe that the left hand side fuzzy degree is larger than right hand side degree of fuzziness. This is also due in this case FCCF's realization in "OR/AND" combinations.

*17.3.3.5. Fuzzy Idempotency:*

There are two Fuzzy Idempotency expressions. We investigate each separately.

**Fuzzy Idempotency "A AND A vs. A"**

FDCF(A AND A) = $A \cap A$
FCCF(A AND A) = $(A \cup A) \cap (c(A) \cup A) \cap (A \cup c(A))$

These were investigated in Türkşen, et.al.(1999) in detail and in general expose a region of uncertainty between these two expressions of fuzzy idempotency. However, in that paper there is a duplication error which should be noted, and it was corrected in later papers Türkşen (2001, 2002).

**Illustrations - Fuzzy Idempotency "A AND A":**

(i) {Max,Min,StN}:

$$\left.\begin{array}{l}\mu[\text{FDCF(A AND A)}] = 0.3\\ \mu[\text{FCCF(A AND A)}] = 0.3\end{array}\right\}[0.3]$$

(ii) {Sum,Product,StN}:

$$\left.\begin{array}{l}\mu[\text{FDCF(A AND A)}] = 0.09\\ \mu[\text{FCCF(A AND A)}] = 0.31\end{array}\right\}[0.09, 0.31]$$

(iii) { $L_{\oplus}$,, $L_{\odot}$, StN }:

$$\left.\begin{array}{l}\mu[\text{FDCF(A AND A)}] = 0\\ \mu[\text{FCCF(A AND A)}] = 1\end{array}\right\}[0,1]$$

Once again, we observe that Fuzzy Idempotency interval of uncertainty increases as we move from {Max,Min,StN} towards { $L_{\oplus} L$ , $StN$ } De Morgan Triple.

**Fuzzy Idempotency "A OR A vs. A"**

FDCF(A OR A) = $(A \cap A) \cup (c(A) \cap A) \cup (A \cap c(A))$
FCCF(A OR A) = $A \cup A$

These Fuzzy idempotency laws were discussed under the heading of "Reaffirmation ..." in Türkşen, et.al.(1999) again there is a duplication error which is to be noted. These duplication errors were corrected in Türkşen (2001, 2002).

**Illustrations - Fuzzy Idempotency "A OR A vs.A"**

(i) {Max,Min,StN}:

$$\left.\begin{array}{l}\mu[\text{FDCF(A OR A)}] = 0.3\\ \mu[\text{FCCF(A OR A)}] = 0.3\end{array}\right\}[0.3]$$

(ii) {Sum,Product,StN}:

$$\left.\begin{array}{l}\mu[\text{FDCF(A OR A)}] \cong 0.4\\ \mu[\text{FCCF(A OR A)}] \cong 0.3\end{array}\right\}[0.4, 0.5]$$

(iii) $\{ L_{\oplus},, L_{\odot}, \text{StN} \}$:

$$\left.\begin{array}{l}\mu[\text{FDCF(A OR A)}] \cong 0.0\\ \mu[\text{FCCF(A OR A)}] \cong 0.6\end{array}\right\}[0.0, 0.6]$$

Once again we observe that the interval of uncertainty increases as we move from *{Max,Min,StN}* towards $\{ L_{\oplus}, L_{\odot}, StN \}$ De Morgan Triple.

*17.3.3.6. General Fuzzy Absorption*

General Fuzzy Absorption has two forms

Form (1): "A OR (A AND B) vs A" and

Form (2): "A AND (A OR B) vs A"

We next investigate each of these in turn.

**General Fuzzy Absorption in Form (1) with FDCF and FCCF :**

FDCF[A OR FDCF(A AND B)] = $[A\cap(A\cap B)] \cup [c(A) \cap (A\cap B)] \cup [A\cap c(A\cap B)]$, and

FCCF [A OR FCCF(A AND B)] = $[A \cup [(A \cup B) \cap (c(A) \cup B) \cap (A \cup c(B))]]$

**Illustration-General Fuzzy Absorption in Form (1) and FDCF and FCCF:**

{Max,Min,StN}:

$$\left.\begin{array}{l}\mu\big[\text{FDCF}\big[\text{A OR FDCF (A AND B)}\big]\big]=0.3\\ \mu\big[\text{FCCF}\big[\text{A OR FCCF (A AND B)}\big]\big]=0.3\end{array}\right\}\quad[0.3]$$

{Sum,Product,StN}:

$$\left.\begin{array}{l}\mu\big[\text{FDCF}\big[\text{A OR FDCF (A AND B)}\big]\big]=0.4\\ \mu\big[\text{FCCF}\big[\text{A OR FCCF (A AND B)}\big]\big]=0.54\end{array}\right\}\quad[0.4,0.54]$$

{ $L_{\oplus}$,,$L_{\odot}$, StN }:

$$\left.\begin{array}{l}\mu\big[\text{FDCF}\big[\text{A OR FDCF (A AND B)}\big]\big]=0.2\\ \mu\big[\text{FCCF}\big[\text{A OR FCCF (A AND B)}\big]\big]=0.8\end{array}\right\}\quad[0.2,0.8]$$

It can be observed that General Fuzzy absorption in Form (1) is preserved with *{Max,Min,StN},* De Morgan Triple since "A OR (A AND B) = A" with the fuzzy degree of A. However, it is not preserved but still is applicable for an interval of fuzzy uncertainty which gets larger and larger as we move toward *{$L_{\oplus}$,$L_{\odot}$, StN}* De Morgan Triple.

**General Fuzzy Absorption in Form (2) with FDCF and FCCF:**

FDCF[A AND FDCF(A OR B)] = $[A \cap [(A \cap B) \cup (c(A) \cap B) \cup (A \cap c(B))]]$,

and

FCCF[A AND FCCF(A OR B)] = $[A \cup (A \cup B)] \cap [c(A) \cup (A \cup B)] \cap [A \cup c(A \cup B)]$

**Illustration-General Fuzzy Absorption in Form (2) and FDCF and FCCF:**

{Max,Min,StN}

$$\left.\begin{array}{l}\mu\left[FDCF\left[A\ AND\ FDCF\ (A\ OR\ B)\right]\right]=0.3\\ \mu\left[FCCF\left[A\ AND\ FCCF\ (A\ OR\ B)\right]\right]=0.3\end{array}\right\}\quad[0.3]$$

{Sum,Product,StN}

$$\left.\begin{array}{l}\mu\left[FDCF\left[A\ AND\ FDCF\ (A\ OR\ B)\right]\right]\cong 0.2\\ \mu\left[FCCF\left[A\ AND\ FCCF\ (A\ OR\ B)\right]\right]\cong 0.3\end{array}\right\}$$

{ $L_{\oplus}$,,$L_{\odot}$, StN }:

$$\left.\begin{array}{l}\mu\left[FDCF\left[A\ AND\ FDCF\ (A\ OR\ B)\right]\right]=0.0\\ \mu\left[FCCF\left[A\ AND\ FCCF\ (A\ OR\ B)\right]\right]=0.0\end{array}\right\}$$

One again we observe that the General Fuzzy Absorption is preserved for *{Max,Min,StN}* De Morgan Triple to the fuzzy degree of "A". It is preserved to a week degree with *{Sum,Prod,StN}* De Morgan Triple in an uncertain fuzzy interval. While it is not preserved for $\{L_{\oplus}L\ \ ,\ StN\}$ De Morgan Triple in Form (2).

*17.3.3.7. Fuzzy Absorption*

There are two fuzzy absorption expressions, we investigate each separately.

**Fuzzy Absorption in Form "A OR X by X"**

FDCF(A OR X) = (A∩X)∪(c(A)∩X)∪(A∩c(X))

FCCF(A OR X) = A∪X = X

**Illustrations - Fuzzy Absorption in Form "A OR X"**

(i) {Max,Min,StN}:

$$\left.\begin{aligned}\mu[\text{FDCF(A OR X)}] &= 0.7\\ \mu[\text{FCCF(A OR X)}] &= 1.0\end{aligned}\right\}[0.7,1.0]$$

(ii) {Sum,Product,StN}:

$$\left.\begin{aligned}\mu[\text{FDCF(A OR X)}] &= 0.79\\ \mu[\text{FCCF(A OR X)}] &= 1.0\end{aligned}\right\}[0.79,1.0]$$

(ii) $\{ L_{\oplus},, L_{\odot}, \text{StN} \}$:

$$\left.\begin{aligned}\mu[\text{FDCF(A OR X)}] &= 1.0\\ \mu[\text{FCCF(A OR X)}] &= 1.0\end{aligned}\right\}[1.0,1.0]$$

It is to be observed that the interval of fuzzy uncertainty degree for Fuzzy Absorption with "OR" decreases as we move from *{Max,Min,StN}* toward *$\{ L_{\oplus}, L_{\odot}$, StN}* De Morgan Triple. But the degree to which it holds is increased which is natural.

**Fuzzy Absorption in Form "A AND ∅ by ∅"**

FDCF(A AND ∅) = A∩∅ = ∅

FCCF(A AND ∅) = (A∪∅)∩(c(A)∪∅)∩(A∪c(∅))

**Illustrations - Fuzzy Absorption in Form "A AND ∅ by ∅"**

(i) {Max,Min,StN}:

$$\left.\begin{aligned}\mu[\text{FDCF(A AND }\varnothing)] &= 0.0\\ \mu[\text{FCCF(A AND }\varnothing)] &= 0.3\end{aligned}\right\}[0.0,0.3]$$

(ii) {Sum,Product,StN}:

$$\left.\begin{array}{l}\mu[\text{FDCF(A AND }\varnothing)] = 0.0\\ \mu[\text{FCCF(A AND }\varnothing)] = 0.21\end{array}\right\}[0.0, 0.21]$$

(iii) {L⊕,,L⊙, StN}:

$$\left.\begin{array}{l}\mu[\text{FDCF(A AND }\varnothing)] = 0.0\\ \mu[\text{FCCF(A AND }\varnothing)] = 0.0\end{array}\right\}[0.0, 0.0]$$

Once again we observe that the interval of uncertainty for Fuzzy Absorption "A AND ∅" decreases. But in this case the degree to which it holds also decreases in contrast to Fuzzy Absorption "A OR X" which is as expected.

*17.3.3.8. Fuzzy Identity*

Fuzzy Identity has two form, we investigate each separately.

Fuzzy Identity by "A OR ∅ by ∅"

FDCF(A OR ∅) = (A∩∅)∪(c(A)∩ ∅)∪(A∩c(∅))
FCCF(A OR ∅) = A∪∅ = A

**Illustrations**:

Fuzzy Identity by "A OR ∅"

(i) {Max,Min,StN}:

$$\left.\begin{array}{l}\mu[\text{FDCF(A OR }\varnothing)] = 0.3\\ \mu[\text{FCCF(A OR }\varnothing)] = 0.3\end{array}\right\}[0.3]$$

(ii) {Sum,,Product,StN}:

$$\left.\begin{array}{l}\mu[\text{FDCF(A OR }\varnothing)] = 0.3\\ \mu[\text{FCCF(A OR }\varnothing)] = 0.3\end{array}\right\}[0.3]$$

(ii) $\{ L_{\oplus},, L_{\odot}, \text{StN} \}$:

$$\left.\begin{array}{l}\mu[\text{FDCF(A OR }\varnothing)] = 0.3\\ \mu[\text{FCCF(A OR }\varnothing)] = 0.3\end{array}\right\}[0.3]$$

As it can be observed Fuzzy Identity holds to a fuzzy degree.

**Fuzzy Identity "A AND X by X"**

FDCF(A AND X) = A∩X=A
FCCF(A AND X) = (A∪X)∩(c(A)∪X)∩(A∪c(X))

**Illustrations - Fuzzy Identity "A AND X"**

(i) {Max,Min,StN}:

$$\left.\begin{array}{l}\mu[\text{FDCF(A AND }X)] = 0.3\\ \mu[\text{FCCF(A AND }X)] = 0.3\end{array}\right\}[0.3]$$

(ii) {Sum,Product,StN}:

$$\left.\begin{array}{l}\mu[\text{FDCF(A AND }X)] = 0.3\\ \mu[\text{FCCF(A AND }X)] = 0.3\end{array}\right\}[0.3]$$

(iii) $\{ L_{\oplus},, L_{\odot}, \text{StN}\}$:

$$\left.\begin{array}{l}\mu[\text{FDCF(A AND }X)] = 0.3\\ \mu[\text{FCCF(A AND }X)] = 0.3\end{array}\right\}[0.3]$$

Once again we observe that fuzzy identity holds to a fuzzy degree.

*17.3.3.9. Law of Fuzzy Contradiction "A AND NOT(A) $\supseteq \varnothing$"*

The Law of Fuzzy Contradiction was discussed in detail in Tűrkşen, et.al.(1999). The duplication error in it was corrected later in Tűrkşen (2001, 2002).

FDCF(A AND NOT(A)) = A∩c(A) $\supseteq \varnothing$
FCCF(A AND NOT(A)) = (A∪c(A))∩(c(A)∪c(A))∩(A∪A) $\subseteq$ X

"The Law of Fuzzy Contradiction" was discussed in detail in Türkşen et.al.(1999).

**Illustrations - Law of Fuzzy Contradiction**

(i) {Max,Min,StN}:

$$\left.\begin{aligned}\mu[\text{FDCF(A AND } NOT(A))] &= 0.3\\ \mu[\text{FCCF(A AND } NOT(A))] &= 0.3\end{aligned}\right\}[0.3]$$

(ii) {Product,Sum,StN}:

$$\left.\begin{aligned}\mu[\text{FDCF(A AND } NOT(A))] &= 0.21\\ \mu[\text{FCCF(A AND } NOT(A))] &\cong 0.36\end{aligned}\right\}[0.21, 0.36]$$

(iii) $\{ L_{\oplus},, L_{\odot}, \text{StN} \}$:

$$\left.\begin{aligned}\mu[\text{FDCF(A AND } NOT(A))] &= 0.0\\ \mu[\text{FCCF(A AND } NOT(A))] &\cong 0.6\end{aligned}\right\}[0.0, 0.6]$$

It is observed that the Law of Fuzzy Contradiction holds as a matter of degree in an interval-valued Type2 Fuzzy Set. Furthermore, the fuzzy interval of uncertainty increases as we move from *{Max,Min,StN}* toward $\{L_{\oplus}, L\ \ , StN\}$

De Morgan Triple.

*17.3.3.10. Law of Fuzzy Middle "A OR NOT(A)⊆X"*

FDCF(A OR NOT(A)) $\subseteq$ $(A \cap c(A)) \cup (c(A) \cap c(A)) \cup (A \cap A) \subseteq X$

FCCF(A OR NOT(A)) $= A \cup c(A) \subseteq X$

"The Law of Fuzzy Middle" was discussed in detail in Türkşen et.al.(1999). Once again, it is noted that in all these topics that were discussed in Türkşen, et.al.(1999) there are duplication errors which were corrected in later papers in Türkşen (2001, 2002).

**Illustrations - Fuzzy Law of Fuzzy Middle**

(i) {Max,Min,StN}:

$$\left.\begin{array}{l}\mu[\text{FDCF(A OR X)}] = 0.7 \\ \mu[\text{FCCF(A OR X)}] = 1.0\end{array}\right\}[0.7,1.0]$$

(ii) {Sum,Product,StN} :

$$\left.\begin{array}{l}\mu[\text{FDCF(A OR X)}] = 0.79 \\ \mu[\text{FCCF(A OR X)}] = \ 1.0\end{array}\right\}[0.79,1.0]$$

(iii) { $L_\oplus$,,$L_\odot$, StN }

$$\left.\begin{array}{l}\mu[\text{FDCF(A OR X)}] = 1.0 \\ \mu[\text{FCCF(A OR X)}] = 1.0\end{array}\right\}[1.0,1.0]$$

It is observed that the interval of uncertainty for the Law of Fuzzy Middle get smaller as we move from *{Max,Min,StN}* toward *{$L_\oplus$,L , StN}* and it holds with full certainty for *{ $L_\oplus$,$L_\odot$, StN }*.

*17.3.3.11. Fuzzy De Morgan Law in Form (1) "NOT(A AND B) vs. NOT(A) OR NOT(B)"*

Here we have four possible sub-forms:

Form (1.1) "NOT(FDCF(A AND B)) vs. FDCF(NOT(A) OR NOT(B))"

Form (1.2) "NOT(FCCF(A AND B)) vs. FCCF(NOT(A) OR NOT(B))"

Form (1.3) "NOT(FDCF(A AND B)) vs. FCCF(NOT(A) OR NOT(B))"

Form (1.4) "NOT(FCCF(A AND B)) vs. FDCF(NOT(A) OR NOT(B))"

Let us investigate each in turn separately:

**Form (1.1) "NOT(FDCF(A AND B)) vs. FDCF(NOT(A) OR NOT(B))":**

NOT(FDCF(A AND B)) = $c(A\cap B)$ vs.

FDCF(NOT(A) OR NOT(B)) = $(c(A)\cap c(B))\cup(A\cap c(B))\cup(c(A)\cap B)$

**Illustrations:**

(i) {Max,Min,StN}:

$$\left.\begin{aligned}\mu[\text{NOT(FDCF(A AND B)]} &= 0.7\\ \mu[\text{FDCF(NOT(A) OR NOT}(B))] &= 0.7\end{aligned}\right\}[0.7]$$

(ii) {Sum,Product,StN}:

$$\left.\begin{aligned}\mu[\text{NOT(FDCF(A AND B)]} &= 0.76\\ \mu[\text{FDCF(NOT(A) OR NOT}(B))] &= 0.64\end{aligned}\right\}[0.64, 0.76]$$

(iii) {$L_{\oplus}$,,L , StN}:

$$\left.\begin{aligned}\mu[\text{NOT(FDCF(A AND B)]} &= 0.9\\ \mu[\text{FDCF(NOT(A) OR NOT}(B))] &= 0.5\end{aligned}\right\}[0.5, 0.9]$$

Thus in Form (1.1), the Fuzzy De Morgan Law holds a matter of degree in *{Max,Min,StN}*, but it holds in an uncertainty interval of Type 2 Fuzziness for *{Sum,Product,StN}* and for *{$L_{\oplus}$,L , StN}* and the interval of uncertainty

increases toward { $L_{\oplus}$,,$L_{\odot}$, StN}. It is also noted that the left hand side is larger than right hand side due to properties of FDCF.

**Form (1.2) "NOT(FCCF(A AND B)) vs. FCCF(NOT(A) OR NOT(B))":**

NOT(FCCF(A AND B)) = c[(A∪B) ∩(c(A) ∪B) ∩(A∪c(B))] vs.
FCCF(NOT(A) OR NOT(B)) = c(A) ∪ c(B)

**Illustrations:**

(i) {Max,Min,StN}:

$$\left.\begin{aligned}\mu[\text{NOT(FCCF(A AND B)]} &= 0.7\\ \mu[\text{FCCF(NOT(A) OR NOT}(B))] &= 0.7\end{aligned}\right\}[0.7]$$

(ii) {Sum,Product,StN}:

$$\left.\begin{aligned}\mu[\text{NOT(FCCF(A AND B)]} &\cong 0.35\\ \mu[\text{FCCF(NOT(A) OR NOT}(B))] &= 0.76\end{aligned}\right\}\sim[0.35, 0.76]$$

(iii) { $L_{\oplus}$,,$L_{\odot}$, StN }:

$$\left.\begin{array}{ll}\mu[\text{NOT(FCCF(A AND B)}] & = 0.5\\ \mu[\text{FCCF(NOT(A) OR NOT(}B))] & = 0.9\end{array}\right\}[0.5,0.9]$$

Thus in Form (1.2) the Fuzzy De Morgan Law holds as a matter of degree in *{Max,Min,StN},* while it holds to an uncertain degree in interval-valued Type 2 fuzziness. For *{Sum,Product,StN}* and for *{$L_{\oplus}$,,$L_{\odot}$ StN}*. It is noted that the left hand side is larger than the right hand side due to FCCF property in combinations of "NOT", "AND", "OR" linguistic operations. Furthermore, the interval of uncertainty increases as we move toward *{Sum,Product,StN}* and *{$L_{\oplus}$,,$L_{\odot}$ StN}* De Morgan Triples.

**Form (1.3) "NOT(FDCF(A AND B)) vs. FCCF(NOT(A) OR NOT(B))":**

NOT(FDCF(A AND B)) = c(A∩B) vs.
FCCF(NOT(A) OR NOT(B)) = c(A) ∪ c(B)

Thus it is to be observed that in From (1.3) the Fuzzy De Morgan Law holds in the classical manner, i.e., c(A∩B) = c(A) ∪ c(B). However it is to be observed that it holds as a matter of degree, i.e., not absolutely.

**Illustrations:**

(i) {Max,Min,StN}:

$$\left.\begin{array}{ll}\mu[\text{NOT(FDCF(A AND B)}] & = 0.7\\ \mu[\text{FCCF(NOT(A) OR NOT(}B))] & = 0.7\end{array}\right\}[0.7]$$

(ii) {Sum,Product,StN}:

$$\left.\begin{array}{ll}\mu[\text{NOT(FDCF(A AND B)}] & = 0.76\\ \mu[\text{FCCF(NOT(A) OR NOT(}B))] & = 0.76\end{array}\right\}[0.76]$$

(iii) { $L_{\oplus}$,,$L_{\odot}$, StN}:

$$\left.\begin{array}{ll}\mu[\text{NOT(FDCF(A AND B)}] & = 0.9\\ \mu[\text{FCCF(NOT(A) OR NOT(}B))] & = 0.9\end{array}\right\}[0.9]$$

Thus it is observed that on Form (1.3) the Fuzzy De Morgan Law holds as a matter of degree. Furthermore, the degree with which it holds increases as we move from *{Max,Min,StN}* toward *{$L_\oplus$,L , StN}* De Morgan Triple

Form (1.4) "NOT(FCCF(A AND B)) vs. FDCF(NOT(A) OR NOT(B))":
NOT(FCCF(A AND B)) = c[(A∪B) ∩ (c(A)∪B) ∩ (A∪c(B))] vs.
FDCF(NOT(A) OR NOT(B)) = (A∩B) ∪ (c(A)∩B) ∪ (A∩c(B))

**Illustrations:**

(i) {Max,Min,StN}:

$$\left.\begin{array}{ll}\mu[\text{NOT(FCCF(A AND B)]} & = 0.7\\ \mu[\text{FDCF(NOT(A) OR NOT}(B))] & = 0.7\end{array}\right\}[0.7]$$

(ii) {Sum,Product,StN}:

$$\left.\begin{array}{ll}\mu[\text{NOT(FCCF(A AND B)]} & \cong 0.35\\ \mu[\text{FDCF(NOT(A) OR NOT}(B))] & \cong 0.68\end{array}\right\}\sim[0.35, 0.68]$$

(iii) {$L_\oplus$,,$L_\odot$, StN}:

$$\left.\begin{array}{ll}\mu[\text{NOT(FCCF(A AND B)]} & = 0.5\\ \mu[\text{FDCF(NOT(A) OR NOT}(B))] & = 0.6\end{array}\right\}[0.5, 0.6]$$

It is observed that in Form (1.4) the Fuzzy De Morgan Law holds to a degree for *{Max,Min,StN}*,but holds to an uncertain fuzzy degree in an interval-valued Type 2 fuzzy set. It is further observed that, while the interval is a bit large for *{Sum,Product,StN}* it becomes smaller for *{ $L_\oplus$,$L_\odot$ StN}*.

*17.3.3.12. Fuzzy De Morgan Law in Form (2) "NOT(A OR B) vs. NOT(A) AND NOT(B)"*

Here again, we have four possible sub forms.

Form (2.1) "NOT(FDCF(A OR B)) vs. FDCF(NOT(A) AND NOT(B))"
Form (2.2) "NOT(FCCF(A OR B)) vs. FCCF(NOT(A) AND NOT(B))"
Form (2.3) "NOT(FDCF(A OR B)) vs. FCCF(NOT(A) AND NOT(B))"
Form (2.4) "NOT(FCCF(A OR B)) vs. FDCF(NOT(A) AND NOT(B))"

Let us next investigate each in turn separately.

**Form (2.1) "NOT(FDCF(A OR B)) vs. FDCF(NOT(A) AND NOT(B))":**

NOT(FDCF(A OR B)) = c[(A∩B) ∪ (c(A)∩B) ∪ (A∩c(B))] vs.

FDCF(NOT(A) AND NOT(B)) = c(A) ∩ c(B)

**Illustrations:**

(i) {Max,Min,StN}:

$$\left.\begin{array}{ll}\mu[\text{NOT(FDCF(A OR B)}] & = 0.3\\ \mu[\text{FDCF(NOT(A) AND NOT}(B))] & = 0.2\end{array}\right\}[0.3, 0.2]$$

(ii) {Sum,Product,StN}:

$$\left.\begin{array}{ll}\mu[\text{NOT(FDCF(A OR B)}] & \cong 0.68\\ \mu[\text{FDCF(NOT(A) AND NOT}(B))] & = 0.14\end{array}\right\}[0.14, 0.68]$$

(iii) { $L_\oplus$,, $L_\odot$, StN }:

$$\left.\begin{array}{ll}\mu[\text{NOT(FDCF(A OR B)}] & = 0.4\\ \mu[\text{FDCF(NOT(A) AND NOT}(B))] & = 0.0\end{array}\right\}[0.0, 0.4]$$

Thus we observe that in From (2.1) Fuzzy De Morgan Law holds a matter of an uncertain degree in the interval-valued Type 2 fuzzy set such that the interval of uncertainty picks at *{Sum,Product,StN}*.

**Form (2.2) "NOT(FCCF(A OR B)) vs. FCCF(NOT(A) AND NOT(B))":**

NOT(FCCF(A OR B)) = c(A∪B) vs.

FCCF(NOT(A) AND NOT(B)) = (c(A)∪c(B)) ∩ (A ∪ c(B)) ∩ (c(A)∪B).

**Illustrations:**

(i) {Max,Min,StN}:

$$\left.\begin{array}{ll}\mu[\text{NOT(FCCF(A OR B)}] & = 0.2\\ \mu[\text{FCCF(NOT(A) AND NOT}(B))] & = 0.8\end{array}\right\}[0.2, 0.8]$$

(ii) {Sum,Product,StN}:

$$\left.\begin{array}{ll}\mu[\text{NOT(FCCF(A OR B)}] & = 0.14\\ \mu[\text{FCCF(NOT(A) AND NOT}(B))] & \cong 0.99\end{array}\right\}\sim[0.14, 0.99]$$

(iii) { L⊕,,L⊙, StN }:

$$\left.\begin{array}{ll}\mu[\text{NOT(FCCF(A OR B)}] & = 0.0\\ \mu[\text{FCCF(NOT(A) AND NOT}(B))] & \cong 0.4\end{array}\right\}[0.0, 0.4]$$

Thus in Form (2.2) Fuzzy De Morgan Law holds in an interval of uncertainty and the range of uncertainty is increased for *{Sum,Product,StN}* while the range is a bit less for *{Max,Min,StN}* and *{L⊕,L , StN}*. But they all demonstrate an interval-valued Type2 Fuzzy uncertainty.

**Form (2.3) "NOT(FDCF(A OR B)) vs. FCCF(NOT(A) AND NOT(B))":**

NOT(FDCF(A OR B)) = c[(A∩B) ∪ (c(A) ∩B) ∪(A∩c(B))] vs.
FCCF(NOT(A) AND NOT(B)) = [(A∪B) ∩ (c(A)∪B) ∩ (A∪c(B))]

**Illustrations:**

(i) {Max,Min,StN}:

$$\left.\begin{array}{ll}\mu[\text{NOT(FDCF(A OR B)}] & = 0.3\\ \mu[\text{FCCF(NOT(A) AND NOT}(B))] & = 0.7\end{array}\right\}[0.3, 0.7]$$

(ii) {Sum,Product,StN}:

$$\left.\begin{array}{ll}\mu[\text{NOT(FDCF(A OR B)}] & = 0.31\\ \mu[\text{FCCF(NOT(A) AND NOT}(B))] & \cong 0.61\end{array}\right\}\sim[0.31, 0.61]$$

(iii) { L⊕,,L⊙, StN }:

$$\left.\begin{array}{ll}\mu[\text{NOT(FDCF(A OR B)}] & = 0.4\\ \mu[\text{FCCF(NOT(A) AND NOT}(B))] & \cong 0.5\end{array}\right\}[0.4, 0.5]$$

Thus in Form (2.3) the Fuzzy De Morgan Law holds to an uncertain degree in an interval value of Type 2 fuzzy sets. It is of interest to note that in this case the uncertainty interval is largest when *{Max,Min,StN}* is applied, while the interval of uncertainty gets smaller as we move toward *{ $L_\oplus$, $L_\odot$, StN}*

.

**Form (2.4) "NOT(FCCF(A OR B)) vs. FDCF(NOT(A) AND NOT(B))":**

NOT(FCCF(A OR B)) = c(A $\cup$ B) vs.
FDCF(NOT(A) AND NOT(B)) = c(A) $\cap$ c(B)

This is the classical De Morgan Law but now expressed in fuzzy domain and it holds as a matter of a fuzzy degree, i.e., it has a singleton fuzzy degree.

**Illustrations:**

(i) {Max,Min,StN}:

$$\left.\begin{array}{ll}\mu[\text{NOT(FCCF(A OR B)}] & = 0.2\\ \mu[\text{FDCF(NOT(A) AND NOT(}B))] & = 0.2\end{array}\right\}[0.2]$$

(ii) {Sum,Product,StN}:

$$\left.\begin{array}{ll}\mu[\text{NOT(FCCF(A OR B)}] & = 0.14\\ \mu[\text{FDCF(NOT(A) AND NOT(}B))] & = 0.14\end{array}\right\}[0.14]$$

(iii) { $L_\oplus$,, $L_\odot$, StN }:

$$\left.\begin{array}{ll}\mu[\text{NOT(FCCF(A OR B)}] & = 0.0\\ \mu[\text{FDCF(NOT(A) AND NOT(}B))] & = 0.0\end{array}\right\}[0.0]$$

It is observed that while in Form (2.4) the Fuzzy De Morgan Law holds to a fuzzy degree, it is largest when *{Max,Min,StN}* is applied, and it does not hold when *{ $L_\oplus$, $L_\odot$, StN}* is applied.

## 17.4. Meta-Linguistic Reasoning

Suppose our usual approximate reasoning expressions is expressed Meta-Linguistically. That is instead of implementing the GMP(CRI) as usual, e.g.,

(1) A o (A → B) = B,

where recall that "o" indicates (Sup, Min) Composition.
We now consider the Meta-Linguistic MP, ML-GMP, as foundation for CWW.

That is, we now state

(1′) A COMP (A IMP B) vs. B

Where COMP stands for linguistic composition in place of "o" in GMP. Therefore "COMP" will be interpreted as "OR", "AND" composition linguistically.

Now implementing MLGM, we will get

(1) OR A AND (A IMP B) vs. B

in its four forms as:

Form (1.1) FDCF[A AND FDCF(A IMP B)]
Form (1.2) FCCF[A AND FCCF(A IMP B)]
Form (1.3) FDCF[A AND FCCF(A IMP B)]
Form (1.4) FCCF[A AND FDCF(A IMP B)]

Recall that we omit "OR" composition when we evaluate singletons. In general, we have to include "OR" composition with matrix operations. Thus we have;

FDCF(A IMP B) = $(A\cap B) \cup (c(A)\cap B) \cup (c(A)\cap c(B))$ and
FCCF(A IMP B) = $c(A) \cup B$

Therefore, we have:

Form (1.1) [FDCF[A AND FDCF(A IMP B)]

$$= [A\cap [(A\cap B) \cup (c(A)\cap B) \cup (c(A)\cap c(B))]]$$

Form (1.2) [FCCF[A AND FCCF(A IMP B)]]

$$= [[A\cup (c(A)\cup B)] \cap [c(A) \cup (c(A) \cup B)] \cap [A \cup c(c(A) \cup B)]]$$

Form (1.3) [FDCF[A AND FCCF(A IMP B)]]

$$= [A \cap (c(A) \cup B)]$$

Form (1.4) OR[FCCF[A AND FDCF(A IMP B)]]

$$\begin{aligned} = \vee \{&[A\cup[(A\cap B)\cup(c(A)\cap B)\cup(c(A)\cap c(B))]] \\ &\cap [c(A)\cup[(A\cap B)\cup(c(A)\cap B)\cup(c(A)\cap c(B))]] \\ &\cap [A\cup c[(A\cap B)\cup(c(A)\cap B)\cup(c(A)\cap c(B))]]\} \end{aligned}$$

Note that, amongst the four sub-forms only Form (1.3) corresponds to the usual GMP applied in the current approximate reasoning techniques.

**Illustration:**

We will now demonstrate the results of these four sub-forms for the singleton values of a=0.3, b=0.8 as we used before and hence forget about "OR", "$\vee$" operation that is needed for a complete fuzzy set realization which comes up in matrix operations. Thus with *{Max,Min,StN}* we get for all sub-forms:

Form (1.1) $\mu$ [FDCF[A AND FDCF(A IMP B)]] = 0.3
Form (1.2) $\mu$ [FCCF[A AND FCCF(A IMP B)]] = 0.3
Form (1.3) $\mu$ [FDCF[A AND FCCF(A IMP B)] = 0.3
Form (1.4) $\mu$ [FCCF[A AND FDCF(A IMP B)] = 0.3

Thus, with *{Max,Min,StN},* we get the same result in all four sub-forms which correspond to the well-known result, i.e., there is no uncertainty interval, just a realization as a matter of degree.

Next, let us investigate all four sub-forms with *{Sum,Product,StN}:*

Form (1.1) $\mu$ [FDCF[A AND FDCF(A IMP B)]] = 0.61
Form (1.2) $\mu$ [FCCF[A AND FCCF(A IMP B)]] $\cong$ 0.32
Form (1.3) $\mu$ [FDCF[A AND FCCF(A IMP B)] $\cong$ 0.28
Form (1.4) $\mu$ [FCCF[A AND FDCF(A IMP B)] $\cong$ 0.28

Thus it is observed that the ML-GMP results are in the uncertainty interval of ~[0.28, 0.61].

Finally, let us investigate all the sub-forms with $\{ L_{\oplus}, L_{\odot}, StN\}$:

Form (1.1) $\mu$ [FDCF[A AND FDCF(A IMP B)]] = 0.0
Form (1.2) $\mu$ [FCCF[A AND FCCF(A IMP B)]] = 1.0
Form (1.3) $\mu$ [FDCF[A AND FCCF(A IMP B)] = 0.0
Form (1.4) $\mu$ [FCCF[A AND FDCF(A IMP B)] = 0.6

It is to be observed that there is a maximum uncertainty interval of [0.0, 1.0] where the approximate reasoning result of ML-GMP would fall.

In summary, we observe that Meta-Linguistic interpretation of the axioms and their application in approximate reasoning result in an uncertainty interval of interval-valued Type 2 fuzziness as a matter of degree.

## 17.5. Conclusion

We have proposed a Meta-Linguistic Set of Axioms as a foundation for Computing With Words. In these meta-linguistic axioms, linguistic terms of linguistic variables are symbolized by A, B, C,...,etc, representing fuzzy sets that capture the meaning of the linguistic terms in a precisiated natural language expression where the meaning representation is precisiated via fuzzy set membership functions. In addition linguistic connectives "AND", "OR" are not mapped in a one-to-one correspondence to the set symbols "$\cap$", "$\cup$", respectively.

Rather, linguistic connectives are represented by interval-valued Type 2 fuzzy sets generated by FDCF and FCCF, Fuzzy Disjunctive and Fuzzy Conjunctive Canonical Forms, respectively.

As a result all the meta-linguistic axioms generate two or four distinct expressions to represent meaning specification of these axioms. This generally reveals an uncertainty interval in the interpretation of the meta-linguistic axioms except for most case of *{Max,Min,StN}* De Morgan Triple which generates a singleton degree of fuzziness. But provides a more expressive power that gets closer to the rich meaning representation associated with natural language expressions. For some cases of certain axioms we have investigated all FDCF and FCCF combinations which gives us four alternatives. This type of mixed FDCF and FCCF investigation could be studied for the others as well but is not treated for all axioms in this chapter it is left for readers to do.

It is expected that the proposed set of meta-linguistic axioms and their applications will provide a sound grounding for Computing With Words at this early stages of development in furthering fuzzy set theory to provide a good bases for future intellectual developments.

Finally, in the light of the analysis developed in this paper, it should be noted that it is not enough to start with the classical axioms and make statements that suggest certain axioms do not hold.

In turn, its myopic to state that "there are only four axioms for t-norms and t-conorms class of fuzzy theory, i.e., boundary, commutativity and associativity and monotoincity".

As we have shown all classical axiom hold as a matter of degree generally in an uncertain interval when the axiom are expressed in Meta-Linguistic From. Therefore, in practice, we should apply all axiom as they are needed but indicate the degree to which they hold.

As a result, we ought to say that all axioms hold "approximately" as a matter of degree in Fuzzy Theory. Finally, as we exemplified by the simple case of ML-GMP, we get results in uncertainty intervals of Type 2 fuzzy set. The interpretation of these results demonstrates that when we are confronted with

Computing with Words that are dependent on imprecise information or subjective expert opinions we would generally be able to infer results in an uncertain Interval-Valued Type 2 fuzziness. This means that human decision-making based on imprecise information can never be able to indicate any absolute result in general.

## References

1. Bilgic, T. (1995), Measurement-Theoretic Frameworks for Fuzzy Set Theory with Applications to Preference Modelling, PhD thesis, University of Toronto, Dept. of Industrial Engineering Toronto, Ontario M5S 1A4 Canada.(supervisor, I.B. Türkşen)
2. Dempster A.P (1967), "Upper and Lower Probabilities Induced by a Multivalued Mapping", In: Annals of Mathematical Statistics, 38, 325-339.
3. Resconi, G., I.B. Türkşen (2001), "Canonical Forms of Fuzzy Truthoods by Meta-Theory Based Upon Modal Logic", Information Sciences, 131, 157-194.
4. Türkşen I.B. (2002), "Interval-valued Type 2 Fuzzy Sets, Multi-valued Maps and Rough Sets", (in) A Grmela and N.E.Mastorakis (eds.), Advances in Intelligent Systems, Fuzzy Sets, Evolutionary Computation, WSEAS, 142-146.
5. Türkşen I.B. (2002), "Upper and Lower Set Formulas: Restriction and Modification of Dempster-Pawlak Formalism", Special Issue of the International Journal of Applied Mathematics and Computer Science, V.12, No.3, 101-111.
6. Türkşen I.B. (2001), "Computing with Descriptive and Veristic Words: Knowledge Representation and Reasoning", in: Computing With Words, P.P. Wang(ed.), Chapter 10, Wiley, New York, 297-328.
7. Türkşen, I.B. (1999), "Theories of Set and Logic with Crisp or Fuzzy Information Granules", J.of Advanced Computational Intelligence, 3,4, 264-273.
8. Türkşen, I.B., A. Kandel, Y-Q. Zhang (1999), "Normal Forms of Fuzzy Middle and Fuzzy Contradiction", IEEE-SMC, 29-2, Part B, Cybernetics, 237-253.
9. Türkşen, I.B., A. Kandel, Y-Q. Zhang (1998), "Universal Truth Tables and Normal Forms", IEEE-Fuzzy Systems, 6-2, 295-303.
10. Türkşen, I.B. (1992), "Interval Valued Fuzzy Sets and 'Compensatory AND' ", Fuzzy Sets and Systems, 51, 295-307.
11. Türkşen I.B. (1986), "Interval-Valued Fuzzy Sets based on Normal Forms", Fuzzy Sets and Systems, 191-210.
12. Zadeh, L.A. (2001), "From Computing with Numbers to Computing with Words - From Manipulation of Measurements to Manipulation of Perceptions", in: P.P. Wang(ed.) Computing With Words, Wiley Series on Intelligent Systems, Wiley and Sons, New York, 35-68.
13. Zadeh, L.A. (1999), "From Computing with Numbers to Computing with Words—From Manipulation of Measurements to Manipulation of Perceptions", IEEE-Trans on Curciuts and Systems, 45, 105-119.
14. Zadeh, L.A. (1979), "A Theory of Approximate Reasoning", in J. Hayes, D. Michie, and L.I. Mikulich (eds) Machine Intelligence, Halstead Press, New York, Vol. 9, 149-194.
15. Zadeh, L.A. (1978), "Fuzzy Sets as a Basis for a Theory of Possibility", Fuzzy Sets and Systems, 3-28.
16. Zadeh, L.A. (1973), "Outline of a New Approach to the Analysis of Complex Systems and Decision Processes", IEEE Trans. Systems Man Cybernet, 3, 28-44.

17. Zadeh, L.A. (1968), "Probability Measures of Fuzzy Events", J.Math. Analysis and Appl., 10, 421-427
18. Zadeh, L.A. (1965), "Fuzzy Sets", Information and Control Systems, Vol.8, 338-353.
19. Zimmermann, H.J., P. Zysno (1980), "Latent Connectives in Human Decision Making", Fuzzy Sets and Systems, 4, 37-51.

# EPILOGUE

In this book, we have discussed briefly certain issues of ontological and epistemological foundations of fuzzy theory. It was shown that certain equivalences that exist in classical theory break down in fuzzy theory.

In particular, it was shown that normal forms of the combination of concepts in classical theory, such as "A AND B", "A OR B", A IMP B", etc., which are known as DNF and CNF, Disjunctive Normal Form and Conjunctive Normal Form, respectively, break down in fuzzy theory. That is FDCF and FCCF, Fuzzy Disjunctive Canonical Form and Fuzzy Conjunctive Canonical Form, respectively, are not equivalent, i.e., $FDCF_i(.) \not\equiv FCCF_i(.)$, i=1,…,16, for t-norms and t-conorms.

The result of this non-equivalence in fuzzy theory, generate two canonical forms for the combination of concepts, and hence introduces a region of uncertainty in the representation of combined concepts between these two canonical forms.

Furthermore, additional formulas are generated for the laws of conservation, for t-norm and co-norms, for Belief, Plausibility and Probability as indicated in the body of this book.

We conjecture that there are potentially other equivalences that exist in classical theory, but they break down in fuzzy theory. Alternatively, they require a new formulation and hence enrichment with degrees of fuzziness.
These issues and related investigations are left for future works.

In this regard, in Chapter 17, we questioned all of the axioms of classical theory, and investigated the consequences of stating all these axioms in meta-linguistics form where the connectives are linguistic and the words are symbolized meta-linguistically as fuzzy sets to capture the meaning representation of words for a potential realization of Computing With Words, CWW, in agreement with Prof. Zadeh suggestions.

Our brief investigation reveals that there are two set theoretic representation for each meta-linguistic axiom which is a re-statement of the classical axioms in linguistic structure which could form a starting point of a foundation for computing with words, CWW.

Naturally, we need to conduct a more extensive investigation in this domain.

# REFERENCES


1. Aczél, J. (1966), Lecture on Functional Equation and Applications, *Academic Press*, New York.
2. Adams, E.W., R.F. Fagot and R.E. Robinson (1965), "A theory of Approximate Statistics", *Psychometrika*, 30, 99-127.
3. Alefeld, G., J., Herzberger (1983), Introduction of Interval Computations, New York, *Academic Press.*
4. Alsina C. (1985), "On a Family of Connectives for Fuzzy Sets, *Fuzzy Sets and Systems* 16, 231-235.
5. Alsina C., E. Trillas and I. Valverde (1983), "On Some Logical Connectives for Fuzzy Sets Theory", *Journal of Mathematical Analysis and Applications,* 93, 15-26.
6. Atanassov, K.T. (1994), "Operators Over Interval Valued Intuitionistic Fuzzy Sets", *Fuzzy Sets and Systems*, 64, 159-174.
7. Atanassov, K.T., G. Gargov (1989), "Interval Valued Intuitionistic Fuzzy Sets", *Fuzzy Sets and Systems*, 31(3), 343-349.
8. Backer, E., A.K. Jain (1981), "A Clustering Performance Measure Based on Fuzzy Set Decomposition, *IEEE Trans. Pattern Anal. Machine Intell.*, PAMI-3(1), 66-74.
9. Bandler, W., L.J. Kohout (1988), "Special Properties, Closures and Interiors of Crisp and Relations?", *Fuzzy Sets and Systems*, 26(3), 317-331.
10. Bandler, W., L.J. Kohout (1986a), "On new Types of Homomorphism and Congruences for Partial Algebraic Structures and n-ary Relations", *Intern. J. of General Systems*, 12(2), 149-157.
11. Bandler, W., L.J. Kohout (1986b), "On the General Theory of Relational Morphisms", *Intern. J. of General Systems,* 13(1), 47-68.
12. Bandler, W., L.J. Kohout (1985), "Probabilistic Versus Fuzzy Production Rules in Expert Systems", *Intern. J. of Man-Machine Studies*, 22(3), 347-353.
13. Bandler, W., L.J. Kohout (1980a), "Fuzzy Power Sets and Fuzzy Implication Operators?", *Fuzzy Sets and Systems*, 4(1), 13-30.
14. Bandler, W., L.J. Kohout (1980b), "Semantics of Implication Operators and Fuzzy Relational Products", *Intern. J. of Man-Machine Studies*, 12(1), 89-116.
15. Barrett, C.R., P.K. Pattanaik, and M. Salles (1999), "On Choosing Rationally When Preferences are Fuzzy", *Fuzzy Sets and* Systems 34, 197-212.
16. Basu, K. (1984), "Fuzzy Revealed Preference Theory", *Journal of Economic Theory*, 32, 212-217.
17. Bellman, R., M. Giertz (1973), "On Analytic Formalism of the Theory of Fuzzy Sets", *Information Sciences*, 5, 149-156.
18. Bellman, R., L.A. Zadeh (1970), "Decision-Making in a Fuzzy Environment", *Management Science*, 17:4, 141-164.
19. Bergmann, M., J. Moor and J.A. Nelson (1988), The Logic Book, Mcgraw Hill, New York.

20. Bezdek, J.C. (1994), "What is computational intelligence", in J.M. Zurada, R.J. Marks II, C.J. Robinson (eds), *Computational Intelligence* - Imitating Life, IEEE Press, New York, 1-12.
21. Bezdek, J.C. (1991), Self-organization and Clustering Algorithms, Technical Report, Defense Technical Information Center (DTIC), N91-21783.
22. Bezdek, J.C. (1981), Pattern Recognition with Fuzzy Objective Function Algorithms, NY, *Plenum Press*.
23. Bezdek, J.C., M.P. Windham and R. Ehrlich (1980), "Statistical Parameters of Cluster Validity Functionals", *International Journal of Comput. Inf. Sci.*, 9(4), 324-336.
24. Bezdek, J.C., J.D. Harris (1979), "Convex Decompositions of Fuzzy Partitions", *Journal of Mathematical Analysis and Applications*, 6, 490-512.
25. Bezdek, J.C. (1973), Fuzzy Mathematics in Pattern Classification, PhD. Dissertation, Cornell University, Ithaca, NY.
26. Bilgic, T., I.B. Türkşen (2000), "Measurement of Membership Functions: Theoretical and Empericial Work", in: *Handbook of Fuzzy Theory*, D. Dubois and H. Prade (eds.).
27. Bilgic, T., I.B. Türkşen (1997), "Measurement - Theoretical Frameworks in Fuzzy Theory", in: *Fuzzy Logic in Artificial Intelligence*, A. Ralescue, T. Martin (eds.), Springer-Verlag, Berlin, 552-565.
28. Bilgic, T., I.B. Türkşen (1997), "Elicitation of Membership Functions: How Far Can Theory Take us?", *Proceedings of Fuzzy-IEEE '97*, July 1-5, 1997, Barcelona, Spain, Vol. III, 1321-1325.
29. Bilgic, T. (1995), Measurement-Theoretic Frameworks for Fuzzy Set Theory with Applications to Preference Modelling, PhD thesis, University of Toronto, Dept. of Industrial Engineering Toronto, Ontario M5S 1A4 Canada.(supervisor, I.B. Türkşen)
30. Bilgic, T., I.B., Türkşen (1995), "Measurement-theoretic Frameworks for Fuzzy Set Theory". In T.P. Martin and A.L. Ralescu, editors, Fuzzy Logic in Artificial Intelligence: Towards Intelligent Systems, Volume 1188 of Lecture Notes in Artificial Intelligence, 252-265, Springer, 1997. Selected papers from IJCAI'95 Workshop Montreal, Canada, August.
31. Bilgic T., I.B. Türkşen (1995), "Measurement-theoretic Justification of Fuzzy Set Connectives", *Fuzzy Sets and Systems*, 289-308.
32. Berztiss, A.T. (1971), Data Structure, Theory and Practice, Academic Press, New York.
33. Black, M. (1937), "Vagueness: An exercise in logical analysis", *Philosophy of Science*, 4, 427-455.
34. Billot, A. (1992), "Economic Theory of Fuzzy Equilibria: An Axiomatic Analysis", Vol. 373 of Lecture notes in Economics and Mathematical Systems. (Springer-Verlag, Berlin Heidelberg, Germany).
35. Bolinger, D.L.M. (1972), "Degree Words", Vol. 53 of *Janua linguarum. Series maio*, The Hague Moutonr.
36. Bollmann-Sdorra, P., S.K.M. Wong and Y.Y. Yao (1993), "A Measurement-Theoretic Axiomatization of Fuzzy Sets", *Fuzzy Sets and Systems,* 60(3), 295-307.
37. Box, G.E.P., G.M. Jenkins (1970), Time Series Analysis, Forecasting and Control, Holden Day, San Francisco.
38. Brock, J.E. (1979), Principal Themes in Peirce's Logic of Vagueness, in K.L. Ketner and J.M. Ransdell (eds) Studies in Peirce's Semiotic, Institute for Studies in Pragmaticism.
39. Burillo, P., H. Bustince (1996), "Entropy on Intuitionistic Fuzzy Sets and on Fuzzy Sets and on Interval Valued Fuzzy Sets", *Fuzzy Sets Syst.* 78, 305-316.
40. Burillo, P., H. Bustince (1995), "Intuitionistic Fuzzy Relations Part I", *Mathware Soft Comput.* 2, 5-38.

41. Bustince, H., P. Burillo (1996), "Interval Valued Fuzzy Relations in a Set Structures", *J. Fuzzy Math.* 4, 765-785.
42. Carrol, L. (1865), Alice's Adventures in Wonderland, Penguin Books Ltd. London, England.
43. Chameau, J.L., J.C. Santamarina (1987a), "Membership Functions Part I: Comparing Method of Measurement", *International Journal of Approximate Reasoning,* 1, 287-301.
44. Chameau, J.L., J.C. Santamarina (1987b), "Membership Functions Part II: Trends in Fuzziness and Implications", *International Journal of Approximate Reasoning*, 1, 303-317.
45. Chellas, B.F. (1980), Modal logic: An Introduction, *Cambridge University Press*, Cambridge, UK.
46. Chen, J.E., K.N. Otta (1995), "Constructing Membership Functions Using Interpolation and Measurement Theory", *Fuzzy Sets and Systems*, 73, 313-327.
47. Christensen, R. (1982), Belief and Behaviour, Entropy Ltd., Lincoln, Mass,.
48. Codd, E. (1970), "A Relational Model for Large Shared Databanks", *CACM*, 13: 6.
49. Copy, I.M. (1982), Introduction to logic, *Macmillan Pub. Co. Inc.*, New York.
50. Curry, H.B. (1963), Foundations of Mathematical Logic, Mc Graw Hill, New York.
51. Davidson, D. (1990), "The Structure and Contents of Truth", Journal of Philosophy 87, 279-328.
52. Davidson, D. (1984), Inquiries into Truth and Interpretation, Clarendon Press, Oxford.
53. Davidson, D. (1969), "True to the Facts", Journal of Philosophy 66, reprinted in Davidson (1984), 37-54.
54. Davidson, D. (1967), "Truth and Meaning", Synthese 17, reprinted in Davidson (1984), 17-36.
55. Demirli, K., I.B. Türkşen, (1995), "A Review of Implications and Generalized Modus Ponens", FSS, Vol 2, 1440-1445.
56. Demirli, K. (1995), A Unified and Extended Framework for Operator Selection in Generalized Modus Ponens Type Fuzzy Reasoning", Ph.D. Thesis, University of Toronto, (I.B. Türkşen, Supervisor).
57. Dempster A.P (1967), "Upper and Lower Probabilities Induced by a Multivalued Mapping", In: *Annals of Mathematical Statistics*, 38, 325-339.
58. Doignon, J-P., B. Monjardet, M. Roubens, and Ph. Vincke (1986), "Biorder Families, Valued Relations, and Preference Modelling", *Journal of Mathematical Psychology,* 30, 435-480.
59. Dombi, J. (1990), "Membership function as an evaluation", *Fuzzy Sets and Systems,* 35, 1-22.
60. Dubois, D., H. Prade (2001), (eds), Special Issue on Optimization and Decision, Fuzzy Sets and Systems, 119, 1.
61. Dubois, D., H. Prade and R. Yager (1997), Fuzzy Information Engineering, *Wiley*, New York,.
62. Dubois, D., H. Prade (1993), "Fuzzy Sets and Probability: Misunderstandings, Bridges and Gaps", *Second IEEE International Conference on Fuzzy Systems*, IEEE, San Francisco, California March 28-April 1, 1059-1068.
63. Dubois, D., H. Prade (1991), "Random Sets and Fuzzy Interval Analysis", *Fuzzy Sets and Systems,* 42, 87-101.
64. Dubois, D., H. Prade (1989), "Fuzzy Sets, Probability and Measurement", *European Journal of Operational Research*, 40, 135-154.
65. Dubois, D. (1988), "Possibility Theory: Searching for Mormative Foundations", in B.R. Munier (ed.), *Risk, Decision and Rationality*, D. Reidel Publishing Company, 601-614.

66. Dubois, D., H. Prade (1988), Possibility Theory, *Plemun Press*.
67. Dubois, D. (1986), "Belief Structures, Possibility Theory and Decomposable Confidence Measures on Finite Sets", *Computers and Artificial Intelligence,* 5(5), 403-416.
68. Dubois, D., H. Prade (1984), "Fuzzy Logics and the Generalized Modus Ponens Revisited", *International Journal of Cybernetics and Systems*, 283-331.
69. Dubois, D., H. Prade (1982), "A class of Fuzzy Measures Based on Triangular Norms: A General Framework for the Combination of Information", *International Journal of General Systems*, 8, 43-61.
70. Dubois, D., H. Prade (1980), Fuzzy Sets and Systems: Theory and Applications, *Academic Press*, New York.
71. Duda, R.O., P.E. Hart (1973), Pattern Classification and Scene Analysis, *John Wiley & Sons, Inc.*
72. Dumitrescu D., D. Tatar (1994), "Normal Forms of Fuzzy Formulas and Their Minimization", *Fuzzy Sets and Systems*, 64, 113-117.
73. Dummett, M. (1978), Truth and other Enigmas, Duckworth, London.
74. Dummett, D. (1959), "Truth", Proceedings of the Aristotelian Society (New Series) 59, reprinted in Dummett (1978), 1-24.
75. Dunn, J.C. (1974) "A Fuzzy Relative of the ISODATA Process", *Cybernetics*, 32-57.
76. Emami, M.R., I.B. Türkşen, A.A. Goldenberg (1997), "Fuzzy Clustering and Fuzzy Modeling", IEEE-Trans on Fuzzy Systems, (to appear).
77. Emami, M.R., I.B. Türkşen and A.A. Goldenberg (1996) "An Improved Fuzzy Modeling Algorithm Part I: Inference Mechanism", *Proceedings of NAFIPS, 96*, Berkeley, CA, 289-293.
78. Emami, M.R., I.B. Türkşen and A.A. Goldenberg (1996) "An improved fuzzy modeling, Part II: Systems Identification, Proceedings of NAFIPS 96, Berkeley, CA. 294-298.
79. Emami, M.R., I.B. Türkşen and A.A. Goldenberg (1991) "A Unified Parameterized Formation on Reasoning Process in Fuzzy Modeling and Control", *Fuzzy Sets and Systems*.
80. Erickson, R.P., P.M.D. Lorenzo and M.A. Woodbury (1994), "Classification of Taste Responses in Brain Stem: Membership in Fuzzy Sets", *Journal of Neurophysiology,* 71(6): 2139-2150.
81. Fishburn, P.C. (1973), "Binary Choice Probabilities: On the Varieties of Stochastic Transitivity", *Journal of Mathematical Psychology* 10, 327-352.
82. Fishburn, P.C. (1970), Utility theory for decision making, Operations Research Society of America, *Publications in Operations Research*, No. 18, Wiley, New York.
83. Fodor, J.C., T. Keresztfalvi (1995), "Nonstandard Conjunctions and Implications in Fuzzy Logic", *International Journal of Approximate Reasoning*, 12: 69-84.
84. Fodor, J.C., M. Roubens, (1994), "Valued Preference Structures", *European Journal of Operational Research* (to appear).
85. Fodor, J.C., M. Roubens (1994), "Fuzzy Preference Modelling and Multicriteria Decision Support", Vol. 14 of Theory and Decision Library. Series D, *System Theory, Knowledge Engineering, and Problem Solving*, Kluwer Academic, Dordecht, Boston.
86. Fodor, J.C. (1993), "A New Look at Fuzzy Connectives", *Fuzzy Sets and Systems*, 57: 141-148.
87. Fodor, J.C. (1991), "Strict Preference Relations Based on Weak T-norms", *Fuzzy Sets and Systems,* 43, 321-366.
88. French, S. (1984), "Fuzzy Decision Analysis: Some Criticisms", in H. Zimmermann, L. Zadeh & B. Gaines (eds), *Fuzzy Sets and Decision Analysis*, North Holland, 29-44.
89. Fuchs, L. (1963), Partially Ordered Algebraic Systems, *Pergamon Press*, London.

90. Fukuyama, Y., M. Sugeno (1989), "A New Method of Choosing the Number of Clusters for Fuzzy C-means", *Proceedings of the 5th Fuzzy Systems Symposium* (in Japanese), 247-250.
91. Fung, L.W. & Fu, S.K. (1975), An axiomatic approach to rational decision making in a fuzzy environment, in H.J. Zimmermann, L.A. Zadeh & B.R. Gaines (eds), *Fuzzy Sets and Their Application to Cognitive and Decision Processes*, Academic Press, New York, pp. 227-256.
92. Gaines, B.R. (1978), "Fuzzy and Probabilistic Uncertainty Logics", *Information and Control*, 38(2), 154-169.
93. Gath, I., A.B. Geva (1989), "Unsupervised Optimal Fuzzy Clustering", *IEEE Trans. On Pattern Analysis and Machine Intelligence*, PAMI-11(7), 773-781.
94. Gerhrk, M., C. Walker and E. Walker (2000), "Fuzzy Normal Forms and Truth Tables", *Proceedings of JCIS- 2000*, Feb. 27-March 3, Atlantic City, New Jersey, 211-214.
95. Giles, R. (1988), The concept of grade of membership, Fuzzy Sets and Systems 25: 297-323.
96. Giles, R. (1976), "Lukasiewicz Logic and Fuzzy Set Theory", *International Journal of Man-Machine Studies*, 8(3), 312-327.
97. Goguen, J. (1969), The logic of inexact concepts, Synthese 19: 325-373.
98. Goguen, J.A., L-fuzzy sets, Journal of mathematical analysis and applications 18(1967), 145-174.
99. Gorzalczany, M. (1989), Interval valued fuzzy inference method - some basic properties, Fuzzy Sets and Systems, 31(2): 235-251.
100. Gupta, M.M. and Sanchez, E. (eds), "Approximate Reasoning in Decision Analysis", North-Holland, New York, (1982).
101. Harmanec, D., G.J. Klir and G. Resconi (1994), "On Modal Logic Interpretation of Dempster-Shefer Theory of Evidence", *International Journal of Intelligent Systems*, 9 [10] 941-951.
102. Henry E. Kyburg, J. (1992), Measuring errors of measurement, in C.W. Savage & P. Ehrlich (eds), Philosophical and Foundational Issues in Measurement Theory, Lawrence Erlbaum Associates, Inc. Publishers, New Jersey, USA, Chapter 5, pp. 75-92.
103. Hersh, H., A. Carmazza and H.H. Brownell (1979), "Effects of Context on Fuzzy Membership Functions", in M.M. Gupta, R.M. Ragade & R.R. Yager (eds), *Advances in Fuzzy Set Theory*, North-Holland, Amsterdam, 389-408.
104. Hersh, H. & Carmazza, A. (1976), A fuzzy set approach tomodifiers and vagueness in natural language, Journal of Experimental Psychology: General 105(3): 254-276.
105. Hintikka, J. (1975), "A Countereexample to Tarski-Type Truth-Definition as Applied to Natural Languages", Philosophia 5, 207-212.
106. Hisdal, E. (1988), "Are Grades of Membership Probabilities?", Fuzzy Sets and Systems, 25, 325-348.
107. Hisdal, E. (1985), "Reconciliation of the Yes-No Versus Grade of Membership Dualism in Human Thinking", in M.M. Gupta, A. Kandel, W. Bandler & J.B. Kiszka (eds), Approximate Reasoning in Expert Systems, North-Holland, Amsterdam, pp. 33-46. Also in Dubois, D., H. Prade, and R. Yager eds., Readings in Fuzzy Sets for Intelligent Systems, Morgan Kaufmann, pp. 854-860, 1993.
108. Hodge, R. (2001), "Key Terms in Fuzzy Logic Deep Roots and New Understanding", *University of Western Sydney*, Australia (Submitted).
109. Hoopes, J. (ed). (1991), Peirce on Signs: Writings on Semiotics, Chapel Hill and London, *University of North Carolina Press.*

110. Hughes, G.E., M.J. Cresswell (1968), An Introduction to Modal Logic. Methuen.
111. Ishibuchi, H., T. Murata and I.B. Türkşen (1997), "Single-objective and Two-objective Genetic Algorithms for Selecting Linguistic Rules for Pattern Classification Problems", *IIII-Trans SMC*, 17, 55-67.
112. Ishibuchi, H., K. Morioka and I.B. Türkşen (1995), "Learning of Fuzzified Neural Networks", *IJAR*, 13, 387-358.
113. Ichihashi, H., I.B. Türkşen (1993), "A Neruo-fuzzy Approach to Data Analysis of Pairwise Comparisons", *IJAR*, 9: 227-248.
114. Kamp, J.A.W. (1975), "Two Theories About Adjectives", in E.L. Keenan (ed.), *Formal Semantics of Natural Language*, Cambridge University Press, London, 123-155.
115. Kandel, A., Zhang, Y-Q & Miller, T. (1995), "Knowledge Representation by Conjunctive Normal Forms and Disjunctive Normal Forms Based on n-variable-m-dimensional Fundamental Clauses and Phrases", *Fuzzy Sets and Systems*, 76, 73-89.
116. Kandel, A. (1982), Fuzzy Techniques in Pattern Recognition, *Wiley*, New York.
117. Karnik, N.N., J.M. Mendel and Q. Liang (1999), "Type 2 Fuzzy Logic System", *IEEE Trans. Fuzzy Systems.* Vol. 7, No. 5, Oct.
118. Karnik, N.N., J.M. Mendel (1998), Introduction to Type 2 Fuzzy Logic Systems, *Presented at1998 IEEE FUZZ Conference,* May, Anchorage, AK.
119. Karnik, N.N., J.M. Mendel (1998), Type 2 Fuzzy Logic Systems: Type-Reduction, *Presented at 1998 IEEE SMC Conference,* October, San Diego, CA.
120. Karnik, N.N. (1998), Type 2 Fuzzy Logic Systems, *PhD Dissertation*, University of Southern California, Los Angeles, CA.
121. Katz, M. (1982), Real-valued Models with Metric Equality in Uniformly Continuous Predicates", *Journal of Symbolic Logic*, 47, 772-792.
122. Katz, M., Lukasiewicz (1981), Logic and the Foundations of Measurement, *Studia Logica*, 15 (3), 209-225.
123. Kaufman, L., P.J. Rousseeuw (1990), Finding Groups in Data, *Wiley*, New York.
124. Kaufmann, A., M.M. Gupta (1985), "Introduction to Fuzzy Arithmetic: Theory and Applications", *Van Nostrand Reinhold*, New York.
125. Kaufmann, A. (1975), Introduction to the Theory of Fuzzy Subsets, *Academic Press*, New York, Vol.1.
126. Keller, J.M., M.R. Gray, J.A. Givens (1985), "A fuzzy k-nearest Algorithm", *IEEE Trans. Systems, Man, and Cybernetics*, Vol. SMC-15, No. 4, 580-585.
127. Keyser, C.J. (1992), Mathematical Philosophy, *E.P. Dutton*, New York.
128. Klir, G., Bo Yuan, Fuzzy Sets and Fuzzy Logic, *Prentice Hall*, 1995.
129. Klir, G., D. Harmanec (1994), "On Modal Logic Interpretation of Possibility Theory", Int. *J. of Uncertainty Fuzziness and Knowledge-Based Systems*, Vol. 2, 237-245.
130. Klir, G.J. (1994), "Multivalued Logics Versus Modal Logics: Alternative Frameworks for Uncertainty Modeling", in P.P. Wang (ed.), *Advances in Fuzzy Theory and Technology*, Duke University, Durham, North Carolina USA, Vol. 2, 3-48.
131. Kneale, W.C. (1962), The development of logic, Oxford, *Clarendon Press*, England.
132. Kohonen, T.T. (1989), Self-organization and associative memory, *Springer-Verlag*, Berlin.
133. Kolodziejczyk, W. (1986), "Orlovski's Concept of Decision Making with Fuzzy Preference Relation-further Results", *Fuzzy Sets and* Systems 19, 11-20.
134. Korzybski, A. (1995), Science and Sanity, Fifth Edition, *Institute of General Semantics*, Eglewood, New Jersey.
135. Kosko, B. (1997), Fuzzy Engineering, *Prentice Hall*, New Jersey, USA.

136. Kosko, B. (1991), Neural Networks and Fuzzy Systems, *Prentice-Hall*, Eagle-wood Cliffs, NJ.
137. Krantz, D.H., R.D. Luce, P. Suppes and A. Tversky (1971), Foundations of Measurement, Vol. 1, *Academic Press*, San Diego.
138. Labov, W., (1973), The Boundaries of Words and Their Meanings", in C.J. Bailey & R.W. Shuy (eds), New Ways of Analyzing Variation in English, *Georgetown University Press*, Washington.
139. Lakoff, G. (1987), Women, Fire, and Dangerous Things: What Categories Reveal About the Mind, *The University of Chicago Press*, Chicago.
140. Lano, K. (1992), "Formal Frameworks for Approximate Reasoning", *Fuzzy Sets and Systems,* 51 131-146.
141. Laviolette, M., J.W. Seaman (1994), "The Efficacy of Fuzzy Representations of Uncertainty", *IEEE Transactions on Fuzzy Systems*, 2(1), 4-15. With comments from D. Dubois and H. Prade, E. Hisdal, G.J. Klir, B. Kosko, N. Wilson, D.V. Lindley.
142. Liang, Q., J.M. Mendel (2000), "Decision Feedback Equalizers for Non-linear Time Varying Channels using Type 2 Fuzzy Adaptive Filters", *Proceedings of the 9th IEEE International Conference on Fuzzy Systems*, 7-10 May, San Antonio, Texas, 883-888.
143. Liang, Q., J.M. Mendel (2000), "Interval Type 2 Fuzzy Logic System", *Proceedings of the 9th IEEE International Conference on Fuzzy Systems,* 7-10 May, San Antonio, Texas, 328-333.
144. Linde, Y., et. al. (1980), "An Algorithm for Vector Quantizer Design", *IEEE Trans. Commun.*, COM-28, 84-95.
145. Ling, C.H. (1965), "Representation of Associative Functions", Publicationes Mathematicae Debrecen, 12, 189-212.
146. Luce, R.D., Krantz, H.D., Suppes, P. & Tversky, A. (1990), Foundations of measurement, Vol. 3, Academic Press, San Diego.
147. Luce, R.D., H. Raiffa (1957), Games and Decisions, *Wiley*, New York.
148. Ludig, G. (1981), Impressions in physics, in A. Hartkamper and H.J. Schmidt (eds.), Structure and Approximation in Physical Theories, Plenum Press, New York.
149. Lukasiewicz, J. (1920), O Logice Trojwartosciowej (On three valued logic) in, N. Rescher, Many-Valued Logic (1969), *McGraw Hill*, New York.
150. Mamdani, E.H., S. Assilian (1981), "An Experiment in Linguistic Synthesis with a Fuzzy Logic Controller, (in) *Fuzzy Reasoning and Its Applications* (eds) E.H. Mamdani and B.R. Gaines, Academic Press, New York, 311-323.
151. Mamdani, E.H. (1977), "Applications of Fuzzy Logic to Approximate Reasoning Using Linguistic Systems", *IEEE Trans. Comput.*, 26, 1182-1191.
152. Mamdani, E.H. (1976), "Advances in the Linguistic Synthesis of Fuzzy Controllers", *J. Man-Mach. Studies*, 8, 669-678.
153. Mamdani, E.H., (1976) "Applications of Fuzzy Algorithms for Control of Simple Dynamical Plant", *Proceedings of IEEE,* 121, 1585-1588.
154. Mamdani, E.H., S. Assilian (1975), "An Experiment in Linguistic Synthesis with a Fuzzy Logic Controller", *Int. J. of Machine Studies*, 7,1, 1-13.
155. Marinos, P.N. (1969), "Fuzzy Logic and its Applications to Switching Systems", *IEEE Trans.Comp.*, Vol 4, 343-348.
156. McCall, S., K. Ajdukiewicz (eds) (1967), Polish logic, 1920-1939, Oxford, Clarendon P. papers by Ajdukiewicz [and others]; with, an introduction by Tadeusz Kotarbinski, edited by Storrs McCall, translated by B. Gruchman [and others].

157. McCloskey, M.E., S. Glucksberg (1978), "Natural Categories: Well Defined or Fuzzy Sets?", Memory and Cognition, 6, 462-472.
158. Mc Luhan, M. (1995), Essential Mc Luhan, E. Mc Luhan and F. Zingrone (eds.) Basic Books, New York.
159. Menger, K. (1951), "Probabilistic Theory of Relations", *Proceedings of National Academic Science*, USA, 37, 178-180.
160. Meystel, A. (1995), Semiotic Modeling and Situation Analysis: An Introduction, *Adrem Inc*.
161. Ming, M., I.B. Türkşen, A. Kandel (1998), "Fuzzy Partition of Fuzzy Rule Base", *IJ of Information Sciences*, 108, 109-121.
162. Mizumoto, M. (1989), "Method of Fuzzy Inference Suitable for Fuzzy Control", *J. Soc. Instrument and Control Engrs.,* 28, 959-963.
163. Mizumoto, M., K. Tanaka (1981), "Fuzzy Sets of Type 2 Under Algebraic Product and Algebraic Sum", *Fuzzy* Sets *and Systems,* 5, 277-290.
164. Murai, T., M. Nakata and M. Shimbo (1998), "Fuzzy Sets and Binary-Proximity-based Rough Sets", *Information Science*, Vol. 104, 49-80.
165. Murai, T., M. Nakata and M. Shimbo (1996), "Ambiguity, Inconsistency, and Possible-Worlds: A New Logical Approach", *Proceedings of the Ninth Conference on Intelligence Technologies in Human-Related Sciences*, Leòn, Spain, 177-184,.
166. Narazaki, H., I.B. Türkşen (1994), "An Integrated Approach for Syllogistic Reasoning and Knowledge Consistency Level Maintennance", *IEEE Trans on SMC*, 24, 4, 548-563.
167. Nakanish, H., I.B. Türkşen and M. Sugeno (1993), "A Review and Comparison of Six Reasoning Methods, Fuzzy Sets and Systems 57, 257-294.
168. Narens, L. (1986), Abstract Measurement Theory, *MIT Press*, Cambridge, Mass.
169. Negoita, C.V. (1985), "Expert Systems and Fuzzy Systems", *Benjamin/ Cummings*, Menlo Park, Cal.
170. Negoita, C.V., A.C. Stefanescu (1982), "On Fuzzy Optimization", in Gupta and Sanchez (eds), *Approximate Reasoning in Decision Analysis*, North Holland, New York, 247-250.
171. Negoita, C.V. (1981), "Fuzzy Systems", *Abacus*, Tunbridge Wells, UK.
172. Negoita, C.V., D.A. Ralescu (1975), "Applications of Fuzzy Sets to Systems Analysis", *Birkhauser*, Basel and Stuttgart,
173. Norwich, A.M., I.B. Türkşen (1984), "A Model for the Measurement of Membership and the Consequences of its Empirical Implementation", *Fuzzy Sets and Systems*, 12, 1-25.
174. Norwich, A.M. I.B. Türkşen (1982), "Stochastic Fuzziness", in: M.M. Gupta and E.E. Sanches (eds.), *Fuzzy Information and Decision Processes*. North Holland, Amsterdam, 13-22.
175. Norwich, A.M., I.B.Türkşen (1982a), "The Construction of Membership Functions", in: R.R. Yager (ed.), *Fuzzy Sets and Possibility Theory*. Pergamon Press, New York, 61-67.
176. Norwich, A.M., I.B. Türkşen (1982b), "The Fundamental Measurement of Fuzziness", in: R.R. Yager (ed.), *Fuzzy Sets and Possibility Theory*. Pergamon Press, New York, 49-60.
177. Norwich, A.M., I.B. Türkşen (1981), "Measurement and Scaling of Membership Functions", in: Lasker, G.E. (ed.), *Proceedings of International Conference on Applied Systems and Cybernetics*, Vol. VI, Acapulco, Mexico, Dec.1980, Pergamon Press, New York, 2851-2858.
178. Oden, G.C. (1979), Fuzzy propositional approach to psyholinguistic problems: an application of fuzzy set theory in cognitive science, in M.M. Gupta, R.K. Ragade & R.R. Yager (eds), Advances in Fuzzy Set Theory and Applications, North-Holland, Amsterdam, 409-420.

179. Orlovski, S.A. (1978), "Decision Making with a Fuzzy Preference Relation", *Fuzzy Sets and* Systems 1, 155-167.
180. Ovchinnikov, S., M. Roubens (1992), "On fuzzy Strict Preference, Indifference and Incomparability Relations", *Fuzzy Sets and Systems* 47, 313-318. and also in Vol. 49, 15-20.
181. Ovchinnikov, S. (1991), "Similarity Relations, Fuzzy Partitions, and Fuzzy Orderings", *Fuzzy* Sets *and Systems,* 40, 107-126.
182. Ovchinnikov, S., M. Roubens (1991), "On Strict Preference Relations", *Fuzzy Sets and* Systems 43, 319-326.
183. Ozawa, J., M. Sugeno (1990), "Linguistic Modeling of Numerical Data, *Proc. 6-th Fuzzy System Symposium*, 159-164.
184. Pal, N.R., J.C. Bezdek(1995), "On Cluster Validity for the Fuzzy c-means Model", IEEE Transactions on Fuzzy Systems, Vol.3, No.3, 370-379.
185. Palmer, F. (1981), Semantics, 2 edn, *Cambridge University Press*, New York.
186. Pawlak Z. (1991), Rough Sets, *Kluwer*, Netherlands.
187. Pawlak Z. (1982), "Rough Sets", *Int. J. of Computer and Information Sciences*, 11, 341-356.
188. Pedrycz, W. (1994), "Why Triangular Membership Functions?", *Fuzzy Sets and Systems*, 64(1), 21-30.
189. Pedrycz, W. (1984), "Identification in Fuzzy Systems", *IEEE Trans. on Systems, Man, and Cybernetics*, No. 14, 361-366.
190. Perny, P., B. Roy (1992), "The Use of Fuzzy Outranking Relations in Preference Modelling", *Fuzzy* Sets *and* Systems 49, 1, 33-53.
191. Pultr, A. (1984), "Fuzziness and Fuzzy Equality", in H.J. Skala, S. termini and E. Trillas (eds), Aspects of Vagueness, D. Reidel Publ. Co., 119-136.
192. Resconi, G., G.J. Klir and U. St. Clair (1992), "Hierarchical Uncertainty Metatheory Based Upon Modal Logic", Int. J. of General Systems, Vol. 21, 23-50.
193. Resconi, G., I.B. Türkşen (2001), "Canonical Forms of Fuzzy Truthoods by Meta-Theory Based Upon Modal Logic", *Information Sciences*, 131, 157-194.
194. Resconi, G., I.B. Türkşen (1999), "Normal Forms in Fuzzy Operation by Metatheory Based Upon Modal Logic", (in preparation).
195. Resconi, G., G.J. Klir, D. Harmanec and U. St. Clair (1996), "Interpretations of Various Uncertainty Theories Using Models of Modal Logics: A Summary", *Fuzzy Sets and Systems*, Vol. 80, 7-14.
196. Resconi, G., G.J. Klir, U. St. Clair and D. Harmanec (1993), "On the Integration of Uncertainty Theories", *Int. J. of Uncertainty, Fuzziness, and Knowledge-based Systems*, Vol. 1, 1-18.
197. Resconi, G., R. Rovetta (1994), "Fuzzy Sets and Evidence Theory in a Metatheory Based Upon Modal Logic", *Quaderni del Seminario Matematico di Brescia*, n.5.
198. Resconi, G., T.Murai (2000), "Field Theory and Modal Logic by Semantic Field to Make Uncertainty Emerge from Information", *Int.J.General System*.
199. Roberts, F. (1979), Measurement Theory, Addison Wesley Pub. Co.
200. Robinson, A. (1966), Non-standard Analysis, North-Holland Pub. Co., Amsterdam.
201. Rocha, L.M. (1994), "Cognitive Categorization Revisited: Extending Interval Valued Fuzzy Sets as Simulation Tools for Concept Combination", *Proceedings of the First International Conference of NAFIPS, IFS and NASA*, December 18-21, San Antonio, Texas, 400-404.

202. Rocha, A.F. (1982), "Basic Properties of Neural Circuits", *Fuzzy Sets and Systems,* 7, 109-121.
203. Rosser, J.B., A.R. Turquette (1977), Many-Valued Logics, *Creenwood Press*, Westport Connecticut.
204. Roubens, M., P. Vincke (1985), Preference Modelling, Vol. 250 of *Lecture Notes in Economics and Mathematical* Systems. (Springer Verlag, Berlin, 1985).
205. Ruan, D., E.E. Kerre (1993), "On the Extension of the Compositional Rule of Inference", *International Journal of Intelligent Systems*, Vol. 8: 807-817.
206. Rumelhart, D.E., J.L. McClelland (1986), The PDP Research Group, Parallel Distributed Processing (Vol.1), Cambridge, MA, *MIT Press*.
207. Ruspini, E.H. (1991), "On the Semantics of Fuzzy Logic", *International Journal of Approximate Reasoning,* 5(1), 45-88.
208. Ruspini, E.H. (1982), "Recent Developments in Fuzzy Clustering", in R.R. Yager (ed) *Fuzzy Sets and Possibility Theory: Recent Developments*, Pergamon Press, 133-147.
209. Ruspini, E.H. (1969), "A New Approach to Clustering", *Information and Control*, 15, 22-32.
210. Russel, B. (1923), "Vagueness", *Australian Journal of Philosophy*, 1, 83-92.
211. Saaty, T.L. (1986), "Scaling the Membership Function", *European Journal of Operational Research*, 25, 320-329.
212. Saaty, T.L. (1974), "Measuring the Fuzziness of Sets", *Journal of Cybernetics*, 4, 53-61.
213. Sapir, E. (1944), "Grading: a Study in Semantics", *Philosophy of Science*, 11, 93-116.
214. Savage, C.W., P. Ehrlich (eds) (1992), Philosophical and Foundational Issues in Measurement Theory, *Lawrence Erlbaum Associates Inc. Publishers*, New Jersey, USA.
215. Schwede, G.W., A. Kandel (1977), "Fuzzy Maps", *IEEE Trans. SMC* Vol 9, 669-674.
216. Schweizer, B., A. Sklar (1983), *Probabilistic Metric Spaces,* North Holland, Amsterdam.
217. Schweizer, B. A. Sklar (1963), "Associative Functions and Abstract Semigroups", *Publ. Math. Debrecen,* 10, 69-81.
218. Schweizer, B., A. Sklar (1961), "Associative Functions and Statistical Triangle Inequalities", *Publ. Math. Debrecen,* 8, 169-186.
219. ] Schweizer, B., A. Sklar (1960), "Statistical Metric Spaces", *Pacific Journal of Mathematics* 10, 313-334.
220. Scott, D. (1976), Does many-valued logic have any use?, in S. Korner (ed.), Philosophy of logic, *Camelot Press*, Southampton, Great Britain, Chapter 2, pp. 64-95, with comments by T.J. Smiley, J.P. Cleave and R. Giles, Bristol Conference on Critical Philosophy, 3d, 1974.
221. Shafer, G. (1976), A Mathematical Theory of Evidence, *Princeton University Press*.
222. Shanahan, J.G. (2000), Soft Computing for Knowledge Discovery: Introducing Cartesian Granule Features, *Kluwer*, Boston,
223. Sheriff, J.K. (1989), The fate of meaning: Charles Peirce, Structuralizm and Literature, *Princeton University Press,* Princeton.
224. Slowinski, R., M. Hapke (2000), Scheduling Under Fuzziness, *Physica-Verlag*, Hiedelbeg.
225. Sowa, J.F. (1999), Knowledge Representation: Logical, Philosophical, and Computational Foundation, *PWS Publishing Co*., Pacific Grove, CA.
226. Sugeno, M., T. Yasukawa (1993), "A Fuzzy Logic Based Approach to Qualitative Modelling", *IEEE Transactions on Fuzzy Systems*, 1(1), 1-24.
227. Sugeno, M., G.T. Rang (1988), "Structure Identification of Fuzzy Model", *Fuzzy Sets and Systems,* 28, 15-33.
228. Sugeno, M. (1988), Fuzzy Control, Nikkan Kogyo Shinbunsya, Tokyo, Japan.
229. Sugeno, M. (1985), "An Introductory Survey of Fuzzy Control", *Inf. Sci.* 36, 59-83.

230. Sugiyama, K. (1985), Analysis and Synthesis of Rule Based Self-Organizing Controller, Ph.D. thesis.
231. Suppes, P., DH. Krantz, R.D. Luce and A. Tversky (1989), Foundations of Measurement, Vol. 2, *Academic Press*, San Diego.
232. Suppes, P. (1974), "The Measurement of Belief", *Journal of Royal Statistical Society Series B,* 36, 2, 160-175.
233. Takagi, H. (1990), "Fusion Technology of Fuzzy Theory and Neural Network: Survey and Future Directions", *Proc. of International Conference on Fuzzy Logic and Neural Networks*, Iizuka, Japan, 13-26.
234. Takagi, T., M. Sugeno (1985), "Fuzzy Identification of Systems and its Applications to Modeling and Control", *IEEE Trans. Systems Man Cybernet*, 15, 116-132.
235. Tarski, A. (1956), Logic, Semantics, Metamathematics, Clarendon Press, Oxford.
236. Tarski, A. (1944), "The Semantic Conception of Truth", Philosophy and Phenomenological Research 4, 341-375.
237. Tarski, A. (1932), "Der Wahrheitsbegriff in den Sprachen der deduktiven Disziplinen", Akademischer Anzeiger der Akademie der Wissenschaften in Wien 69, translated as "The Concept of Truth in Formalized Languages" in Tarski (1956), 152-278.
238. Thole, U., H.J. Zimmermann, P. Zysno (1979), "On the Suitability of Minimum and Product Operators for the Interpretation of Fuzzy Sets", *Fuzzy Sets and Systems*, 2, 167-180.
239. Tong, R.M. (1979), The construction and evaluation of fuzzy models; Gupta M.M., Regade, R.K., Yager, R.R., editors.: *Advances in Fuzzy Set Theory and Application*, North-Holland, Amsterdam.
240. Tou, J.T., R.C. Gonzales (1974), Pattern Recognition Principles, Reading, MA: *Addison-Wesley Publishing Company*.
241. Triantaphyllou, E., S.H. Mann (1990), "An Evaluation of the Eigenvalue Approach for Determining the Membership Values in Fuzzy Sets", *Fuzzy Sets and Systems,* 35, 295-301.
242. Trillas, E., L. Valverde (1985), "On Mode and Implication in Approximate Reasoning", in W. Bandler, M.M. Gupta, A. Kandel and J.B. Kista (eds.), *Approximate Reasoning in Expert Systems*, Elsevier Sci B.V. (North Holland), 157-167.
243. Trillas, E. (1979), "Sobre Functiones de Negation en la Teoria de Conjunctos Diffuses", *Stochastica*, 3, 47-59.
244. Trojan, G.J., J.B. Kiszka, M.M. Gupta, P.N. Nikiforuk (1987), "Solution of Multivariable Fuzzy Equations", *Fuzzy Sets and Systems*, 22, 271-279.
245. Türkşen I.B. (2002), "Belief, Plausibility and Probability Measures over Type 2 Fuzzy Sets", *I.J. of Applied Mathematics and Computer Science*, (to appear).
246. Türkşen I.B. (2002), "Type 2 Representation and Reasoning for CWW", *Fuzzy Sets and Systems*, 127, 17-36.
247. Türkşen I.B. (2002), "Interval-valued Type 2 Fuzzy Sets, Multi-valued Maps and Rough Sets", (in) A Grmela and N.E.Mastorakis (eds.), Advances in Intelligent Systems, *Fuzzy Sets*, Evolutionary Computation, WSEAS, 142-146.
248. Türkşen I.B. (2002), "Upper and Lower Set Formulas: Restriction and Modification of Dempster-Pawlak Formalism", *Special Issue of the International Journal of Applied Mathematics and Computer Science,* V.12, No.3, 101-111.
249. Türkşen I.B. (2001), "Computing with Descriptive and Veristic Words: Knowledge Representation and Reasoning", *in: Computing With Words*, P.P. Wang(ed.), Chapter 10, Wiley, New York, 297-328.

250. Türkşen, I.B. (2001), "Sources, Measurements and Models of Type 2 Fuzziness in the New Millennium", *in: Fuzziness in the New Millennium*, V. Dimitrov (ed), Springer Verlag (in press).
251. Türkşen, I.B. (2000), "Theories of Set and Logic with Crisp and Fuzzy Information Granules", *J. of Advanced Computational Intelligence* (to appear in the Special Issue).
252. Türkşen, I.B. (1999), "Computing with Descriptive and Veristic Words, *Proceedings of NAFIPS'99* (Special Invited Presentation), New York, June 10-12, 13-17.
253. Türkşen, I.B. (1999), "Theories of Set and Logic with Crisp or Fuzzy Information Granules", *J.of Advanced Computational Intelligence*, 3,4, 264-273.
254. Türkşen, I.B., A. Kandel, Y-Q. Zhang (1999), "Normal Forms of Fuzzy Middle and Fuzzy Contradiction", *IEEE-SMC, 29-2, Part B, Cybernetics*, 237-253.
255. Türkşen I.B. (1999), "Type I and Type II Fuzzy System Modeling", *FSS*, 106, 11-34.
256. Türkşen, I.B., A. Kandel, Y-Q. Zhang (1998), "Universal Truth Tables and Normal Forms", *IEEE-Fuzzy Systems*, 6-2, 295-303.
257. Türkşen, I.B., M.H.F. Zarandi (1998), "Fuzzy System Models for Aggregate Scheduling Analysis", *IJAR* (to appear).
258. Türkşen, I.B. (1996), "Non-specificity and Interval Valued Fuzzy Sets", *Fuzzy Sets and Systems* (Invited Special Issue), 87-100.
259. Türkşen, I.B. (1996), "Fuzzy Truth Tables and Normal Forms", *Proceedings of BOFL'96*, December 15-18, TIT, Nagatsuta, Yokohama, Japan, 7-12,1996.
260. Türkşen, I.B. (1995), "Fuzzy Normal Forms", *Fuzzy Sets and Systems*, 69, 319-346.
261. Türkşen, I.B. (1995), "Type I and 'Interval-Valued' Type II Fuzzy Sets and Logics", in P.P. Wang (ed), *Fuzzy Theory and Technology*, III, Bookwrights Press, Durham, D.C, 31-82.
262. Türkşen, I.B. (1994), "Interval Valued Fuzzy Sets and Fuzzy Measures", *Proceedings of the First International Conference of NAFIPS, IFIS, and NASA,* December 18-21, San Antonio, Texas, 317-321.
263. Türkşen I.B. (1994),"Interval-Valued Fuzzy Sets and Compensatory AND", *FSS,* 87-100.
264. Türkşen, I.B. (1994), "Fuzzy Normal Forms", *Fuzzy Sets and Systems,* 253-266.
265. Türkşen, I.B, (1994), "Interval-Valued Fuzzy Sets and 'Compensatory AND' ", *FSS*, 295-307.
266. Türkşen, I.B., T, Bilgiç (1993), Interval Valued Strict Preference, In *Proceedings of the First European Congress on Fuzzy and Intelligent Technologies* (September 1993), 593-599. EUFIT '93 Aachen, Germany.
267. Türkşen, I.B., Y. Tian (1993), "Combination of Rules and Their Consequences in Fuzzy Expert Systems", *Fuzzy Sets and Systems*, 58: 3-40.
268. Türkşen, I.B., Y. Tian (1992), "A Fuzzy Expert System for a Service Centre of Spare Parts", Expert Systems with Applications, Vol. 5, 447-464.
269. Türkşen, I.B., Y. Tian (1992), "Interval Valued Fuzzy Set Representation on Multiple Antecedents Fuzzy s-implications and Reasoning", *Fuzzy Sets and Systems*, 52, 1-25.
270. Türkşen, I.B. (1992), "Interval Valued Fuzzy Sets and 'Compensatory AND' ", *Fuzzy Sets and Systems*, 51, 295-307.
271. Türkşen, I.B., C. Lucas (1991), "A Pattern Matching Inference Method and its Comparison with Known Inference Methods", *Proceedings of IFSA'91*, July 7-12, Brussels, 231-234.
272. Türkşen, I.B. (1991), "Measurement of Membership Functions and Their Acquisitions", *Fuzzy Sets and Systems,* 40, 5-38.

273. Türkşen, I.B., Z. Zhong (1990), "An Approximate Analogical Reasoning Scheme Based on Similarity Measures and Interval Valued Fuzzy Sets", *Fuzzy Sets and Systems,* 34, 323-346.
274. Türkşen, I.B. (1989), "Four Methods of Approximate Reasoning with Interval-Valued Fuzzy Sets", *Internet. J. Approx. Reasoning,* 3, 121-142.
275. Türkşen, I.B. (1988), "Stochastic Fuzzy Sets: A Survey, Combining Fuzzy Imprecision with Probabilistic Uncertainty in Decision Making", *Springer-Verlag*, New York, 168-183.
276. Türkşen I.B. (1986), "Interval-Valued Fuzzy Sets based on Normal Forms", *Fuzzy Sets and Systems,* 191-210.
277. Uncu, O., I.B. Türkşen (2001), "A Novel Fuzzy System Modeling Approach: Multidimensional Structure Identification and Inference", 10th IEEE - Int. Conference on Fuzzy Systems, Dec. 2-5'01, Melbourne, Australia, 557-561.
278. Von Bertalanffy, L. (1981), A system View of Man, P.A. LaViolette (ed), *Westview Press*, Boulder, Colorado.
279. Wang, P.P. (ed.) (2001), Computing With Words, Wiley, New York (to appear).
280. Ward, J.H.(1963), "Hierarchical Grouping to Optimize an Objective Function", *J. American Statistics Association*, 58, 236-244.
281. Weber, S. (1983), "A General Concept of Fuzzy Connectives, Negations and Implications Based on t-norms and t-conorms, *Fuzzy Sets and Systems* 11, 115-134.
282. Whitehead, A.N. (1925), Science and Modern World, *McMillan*, New York.
283. Windham, M.P. (1982), "Cluster Validity of the Fuzzy c-means Clustering Algorithm", *IEEE Trans. Pattern Analysis and Machine Intelligence*, PAMI-4(4), 357-363.
284. Yager, R.R., D.P. Filev (1994), Essentials of Fuzzy Modeling and Control, John Wiley and Sons.
285. Yager, R.R. (1980), "On a General Class of Fuzzy Connectives", Fuzzy Sets and Systems, 4, 235-242.
286. Yager, R.R. (1979), "A Measurement-informational Discussion of Fuzzy Union and Intersection", *International Journal of Man-Machine Studies,* 11, 189-200.
287. Yamakawa, T., M. Furukawa (1992), "A Design Algorithm of Membership Functions for a Fuzzy Neuron Using Example-based Learning", *Proceedings of the First IEEE Conference on Fuzzy Systems*, San Diego, 75-82.
288. Yasukawa, T. (1991), Linguistic Modeling Based on Numerical Data Using Fuzzy Theory, *TIT*, Master thesis.
289. Yasunobu, S. (1991), Fuzzy Engineering, Syokodo, Tokyo, Japan.
290. Younis, K.S., et. al. (1996), "Vector Quantization Based on Dynamic Adjustment of Mahalanobis Distance", *IEEE National Aerospace and Electronics Conference (NAE-CON)*, 2, 858-862.
291. Zadeh, L.A. (2002), "Toward a Perception-Based Theory of Probabilistic Reasoning with Imprecise Probabilities", J.of Statistical Planning and Inference, Elsevier Science, V.105, pp.233-264.
292. Zadeh, L.A. (2001), "From Computing with Numbers to Computing with Words - From Manipulation of Measurements to Manipulation of Perceptions", in: P.P. Wang(ed.) *Computing With Words*, Wiley Series on Intelligent Systems, Wiley and Sons, New York, 35-68.
293. Zadeh, L.A. (2000), Computing With Perceptions, Keynote Address, *IEEE- Fuzzy Theory Conference*, San Antonio, May 7-10.

294. Zadeh, L.A. (2000), "Toward a Perception-Based Theory of Probabilistic Reasoning", Key note address; *Fourth International Conference on Applications of Fuzzy Systems and Soft Computing*, June 27-29, Siegen, Germany.
295. Zadeh, L.A. (1999), "From Computing with Numbers to Computing with Words—From Manipulation of Measurements to Manipulation of Perceptions", *IEEE-Trans on Curciuts and Systems*, 45, 105-119.
296. Zadeh, L.A. (1997), "Toward a Theory of Fuzzy Information Granulation and its Centrality in Human Reasoning and Fuzzy Logic", *Fuzzy Sets and Systems,* 90, 111-127.
297. Zadeh, L.A. (1996), "Fuzzy Logic = Computing With Words", *IEEE-Transactions on Fuzzy Systems*, 4, 2, 103-111.
298. Zadeh, L.A. (1995), Keynote address, *IFSA Congress*, Sao Paulo, Brazil.
299. Zadeh, L.A., J. Kacprzyk (eds), (1992), "Fuzzy Logic for the Management of Uncertainty", *John Wiley and Sons*, New York.
300. Zadeh, L.A. (1986), “A Simple View on the Dempster-Shafer Theory of Evidence and Its Implication for the Rule of Combination”, *Al Magazine*, Vol. 7, 85-90.
301. Zadeh, L.A. (1985), "Syllogistic reasoning in Fuzzy Logic and its Application to Usuality and Reasoning with Dispositions", *IEEE-Trans.SMC*, 15, 754-765.
302. Zadeh, L.A. (1984), "A Theory of Common Sense Knowledge", *in H.J. Skala, S. Termini, and E. Trillas (eds), Aspects of Vagueness*, Dodrecht: D.Riedel, 257-296.
303. Zadeh, L.A. (1983), "A Computational Approach to Fuzzy Quantifiers in Natural Language", *Comp. and Maths. with Appls*., 9, 149-184.
304. Zadeh, L.A. (1983), "The Role of Fuzzy Logic in the Management of Uncertainty in Expert Systems", *Fuzzy Sets and Systems*, 11, 199-227.
305. Zadeh, L.A. (1979), "A Theory of Approximate Reasoning", *in J. Hayes, D. Michie, and L.I. Mikulich (eds) Machine Intelligence*, Halstead Press, New York, Vol. 9, 149-194.
306. Zadeh, L.A. (1978), "Fuzzy Sets as a Basis for a Theory of Possibility", *Fuzzy Sets and Systems*, 3-28.
307. Zadeh, L.A. (1976), "A Fuzzy-Algorithmic Approach to the Definition of Complex or Imprecise Concepts", *Int. J. Man-Machine Studies*, 8, 249-291.
308. Zadeh, L.A. (1976), "The Concept of a Linguistic Variable and its Application to Approximate Reasoning", Part 3, *Information Sciences*, 9, 43-80.
309. Zadeh, L.A., K-S. Fu, K. Tanaka, and M. Shimura (1975), "Fuzzy Sets and Their Applications to Cognitive and Decision Progresses", *Academic Press*, New York,
310. Zadeh, L.A. (1975), “Fuzzy Logic and Approximate Reasoning”, *Synthese*, 30, 407-428.
311. Zadeh, L.A. (1975), "The Concept of a linguistic variable and its application to approximate reasoning", Parts I, II, *Information Sciences*, 8, 199-249, 301-357, 43-80.
312. Zadeh, L.A. (1975), "Calculus of Fuzzy Restrictions", *Fuzzy Sets and Their Applications to Cognitive and Decision Progresses*, L.A. Zadeh, K.S Fu, K. Tanaka and M. Shimura (eds), Academic Press, New York, 1-39.
313. Zadeh, L.A. (1973), “Outline of a New Approach to the Analysis of Complex Systems and Decision Processes”, *IEEE Trans. Systems Man Cybernet*, 3, 28-44.
314. Zadeh, L.A. (1971), "Similarity Relations and Fuzzy Ordering", *Information Sciences*, 3, 177-200.
315. Zadeh, L.A. (1968), "Probability Measures of Fuzzy Events", *J.Math. Analysis and Appl.*, 10, 421-427
316. Zadeh, L.A. (1965), "Fuzzy Sets", *Information and Control Systems*, Vol.8, 338-353.
317. Zadeh, L.A., C.A. Desoer (1963), "Linear System Theory-The State Space Approach", *Mc Graw-Hill*, New York,

318. Zadeh, L.A., J.R. Ragazzini (1952), "The Analysis of Sampled-Data Systems", *Applications and Industry (AIEE)*, 1, 224-234.
319. Zadeh, L.A., J.R. Ragazzini (1950), "An Extension of Wieners Theory of Prediction", *J.Appl.Phys.*, 21, 645-655.
320. Zadeh, L.A. (1950), "Frequency Analysis of Variable Networks", *Proc.of IRE*, 291-299.
321. Zimmermann, H.J. (ed.) (1999), Practical Applications of Fuzzy Technologies, *Handbook of Fuzzy Set Series*, Kluwer, Boston.
322. Zimmermann, H.J. (1987), Fuzzy Sets, *Decision Making, and Expert Systems*, Kluwer, Boston.
323. Zimmermann, H.J., P. Zysno (1985), "Quantifying Vagueness in Decision Models", *European Journal of Operational Research*, 22, 148-154.
324. Zimmermann, H.J., P. Zysno (1980), "Latent Connectives in Human Decision Making", *Fuzzy Sets and Systems*, 4, 37-51.
325. Zimmermann, H.J. (1978), "Fuzzy Programming and Linear Programming with Several Objective Functions", *FSS*, 1, 45-55.
326. Zwick, R. (1987), "A Note on Random Sets and the Thurstonian Scaling Methods", *Fuzzy Sets and Systems*, 21, 351-356.
327. Zwick, R., E. Carlstein, D.V. Budescu (1987), "Measures of Similarity Among Fuzzy Concepts: A Comparative Analysis", *International Journal of Approximate Reasoning*, 1(2), 221-242.

# INDEX

# AUTHOR INDEX